Algorithmic Lie Theory for Solving Ordinary Differential Equations

PURE AND APPLIED MATHEMATICS

A Program of Monographs, Textbooks, and Lecture Notes

EXECUTIVE EDITORS

Earl J. Taft
Rutgers University
Piscataway, New Jersey

Zuhair Nashed
University of Central Florida
Orlando, Florida

EDITORIAL BOARD

M. S. Baouendi
University of California,
San Diego

Jane Cronin
Rutgers University

Jack K. Hale
Georgia Institute of Technology

S. Kobayashi
University of California,
Berkeley

Marvin Marcus
University of California,
Santa Barbara

W. S. Massey
Yale University

Anil Nerode
Cornell University

Freddy van Oystaeyen
University of Antwerp,
Belgium

Donald Passman
University of Wisconsin,
Madison

Fred S. Roberts
Rutgers University

David L. Russell
Virginia Polytechnic Institute
and State University

Walter Schempp
Universität Siegen

MONOGRAPHS AND TEXTBOOKS IN PURE AND APPLIED MATHEMATICS

Recent Titles

J. Galambos and I. Simonelli, Products of Random Variables: Applications to Problems of Physics and to Arithmetical Functions (2004)

Walter Ferrer and Alvaro Rittatore, Actions and Invariants of Algebraic Groups (2005)

Christof Eck, Jiri Jarusek, and Miroslav Krbec, Unilateral Contact Problems: Variational Methods and Existence Theorems (2005)

M. M. Rao, Conditional Measures and Applications, Second Edition (2005)

A. B. Kharazishvili, Strange Functions in Real Analysis, Second Edition (2006)

Vincenzo Ancona and Bernard Gaveau, Differential Forms on Singular Varieties: De Rham and Hodge Theory Simplified (2005)

Santiago Alves Tavares, Generation of Multivariate Hermite Interpolating Polynomials (2005)

Sergio Macías, Topics on Continua (2005)

Mircea Sofonea, Weimin Han, and Meir Shillor, Analysis and Approximation of Contact Problems with Adhesion or Damage (2006)

Marwan Moubachir and Jean-Paul Zolésio, Moving Shape Analysis and Control: Applications to Fluid Structure Interactions (2006)

Alfred Geroldinger and Franz Halter-Koch, Non-Unique Factorizations: Algebraic, Combinatorial and Analytic Theory (2006)

Kevin J. Hastings, Introduction to the Mathematics of Operations Research with *Mathematica*®, Second Edition (2006)

Robert Carlson, A Concrete Introduction to Real Analysis (2006)

John Dauns and Yiqiang Zhou, Classes of Modules (2006)

N. K. Govil, H. N. Mhaskar, Ram N. Mohapatra, Zuhair Nashed, and J. Szabados, Frontiers in Interpolation and Approximation (2006)

Luca Lorenzi and Marcello Bertoldi, Analytical Methods for Markov Semigroups (2006)

M. A. Al-Gwaiz and S. A. Elsanousi, Elements of Real Analysis (2006)

Theodore G. Faticoni, Direct Sum Decompositions of Torsion-Free Finite Rank Groups (2007)

R. Sivaramakrishnan, Certain Number-Theoretic Episodes in Algebra (2006)

Aderemi Kuku, Representation Theory and Higher Algebraic K-Theory (2006)

Robert Piziak and P. L. Odell, Matrix Theory: From Generalized Inverses to Jordan Form (2007)

Norman L. Johnson, Vikram Jha, and Mauro Biliotti, Handbook of Finite Translation Planes (2007)

Lieven Le Bruyn, Noncommutative Geometry and Cayley-smooth Orders (2008)

Fritz Schwarz, Algorithmic Lie Theory for Solving Ordinary Differential Equations (2008)

Algorithmic Lie Theory for Solving Ordinary Differential Equations

Fritz Schwarz

Fraunhofer Gesellschaft
Sankt Augustin, Germany

Boca Raton London New York

Chapman & Hall/CRC is an imprint of the
Taylor & Francis Group, an informa business

Chapman & Hall/CRC
Taylor & Francis Group
6000 Broken Sound Parkway NW, Suite 300
Boca Raton, FL 33487-2742

© 2008 by Taylor & Francis Group, LLC
Chapman & Hall/CRC is an imprint of Taylor & Francis Group, an Informa business

No claim to original U.S. Government works
Printed in the United States of America on acid-free paper
10 9 8 7 6 5 4 3 2 1

International Standard Book Number-13: 978-1-58488-889-5 (Hardcover)

This book contains information obtained from authentic and highly regarded sources. Reprinted material is quoted with permission, and sources are indicated. A wide variety of references are listed. Reasonable efforts have been made to publish reliable data and information, but the author and the publisher cannot assume responsibility for the validity of all materials or for the consequences of their use.

No part of this book may be reprinted, reproduced, transmitted, or utilized in any form by any electronic, mechanical, or other means, now known or hereafter invented, including photocopying, microfilming, and recording, or in any information storage or retrieval system, without written permission from the publishers.

For permission to photocopy or use material electronically from this work, please access www.copyright. com (http://www.copyright.com/) or contact the Copyright Clearance Center, Inc. (CCC) 222 Rosewood Drive, Danvers, MA 01923, 978-750-8400. CCC is a not-for-profit organization that provides licenses and registration for a variety of users. For organizations that have been granted a photocopy license by the CCC, a separate system of payment has been arranged.

Trademark Notice: Product or corporate names may be trademarks or registered trademarks, and are used only for identification and explanation without intent to infringe.

Library of Congress Cataloging-in-Publication Data

Schwarz, Fritz, 1941-
    Alogrithmic lie theory for solving ordinary differential equations / Fritz Schwarz.
        p. cm.
    Includes bibliographical references and index.
    ISBN 978-1-58488-889-5 (alk. paper)
    1. Differential equations--Numerical solutions. 2. Lie algebras. I. Title.

QA371.S387 2007
518'.63--dc22   2007020633

Visit the Taylor & Francis Web site at
http://www.taylorandfrancis.com

and the CRC Press Web site at
http://www.crcpress.com

To my Mother

Contents

1 Introduction — 1

2 Linear Differential Equations — 11
   2.1 Linear Ordinary Differential Equations — 11
   2.2 Janet's Algorithm — 40
   2.3 Properties of Janet Bases — 55
   2.4 Solving Partial Differential Equations — 77

3 Lie Transformation Groups — 105
   3.1 Lie Groups and Transformation Groups — 105
   3.2 Algebraic Properties of Vector Fields — 115
   3.3 Group Actions in the Plane — 121
   3.4 Classification of Lie Algebras and Lie Groups — 135
   3.5 Lie Systems — 143

4 Equivalence and Invariants of Differential Equations — 159
   4.1 Linear Equations — 162
   4.2 Nonlinear First Order Equations — 176
   4.3 Nonlinear Equations of Second and Higher Order — 186

5 Symmetries of Differential Equations — 193
   5.1 Transformation of Differential Equations — 194
   5.2 Symmetries of First Order Equations — 206
   5.3 Symmetries of Second Order Equations — 210
   5.4 Symmetries of Nonlinear Third Order Equations — 224
   5.5 Symmetries of Linearizable Equations — 236

6 Transformation to Canonical Form — 247
   6.1 First Order Equations — 247
   6.2 Second Order Equations — 249
   6.3 Nonlinear Third Order Equations — 279
   6.4 Linearizable Third Order Equations — 298

7 Solution Algorithms — 311
   7.1 First Order Equations — 312
   7.2 Second Order Equations — 319
   7.3 Nonlinear Equations of Third Order — 328
   7.4 Linearizable Third Order Equations — 345

x

8	Concluding Remarks	351
A	Solutions to Selected Problems	355
B	Collection of Useful Formulas	377
C	Algebra of Monomials	383
D	Loewy Decompositions of Kamke's Collection	387
E	Symmetries of Kamke's Collection	403
F	ALLTYPES Userinterface	417
	References	419
	Index	431

Chapter 1

Introduction

> *"...Es geht mir vorwärts, aber langsam, und es kostet unendliche und äußerst langweilige Rechnungen..."*[1]
> *Sophus Lie, Letter to A. Mayer of April 4, 1874.*

Solving equations has been one of the most important driving forces in the history of mathematics. Particularly well known is the problem of solving algebraic equations of order higher than four and its eventual solution by Lagrange, Ruffini, Abel and Galois. The complete answer given by the latter author provided not only the solution for the problem at hand, but also established a new field in mathematics, the theory of groups, which in turn laid the foundation of modern algebra. A knowledgeable and detailed review of this subject and various other topics discussed below may be found in the book by Wussing [190], see also the *Historical Remarks* by Bourbaki [16].

Much less known is the fact that the theory of differential equations took a similar course in the second half of the 19th century. At that time, solving ordinary differential equations (ode's) had become one of the most important problems in applied mathematics, about 200 years after Leibniz and Newton introduced the concept of the derivative and the integral of a function. Numerous phenomena in the physical sciences were described by formulas involving differentiation and integration, and the need arose to determine the functional dependencies between the variables involved. In other words, the problem of solving a differential equation was born. In this book the phrase *solution of an ode* means an expression for the general solution in some function field, e.g., in terms of elementary or Liouvillian functions. The differential equation should vanish identically upon its substitution. If n is the order of the equation the general solution involving n constant parameters is searched for. It should not involve derivatives, infinite series or products. Furthermore, any numerical or graphical representation is excluded.

---

[1]Translation by the author: "...I proceed, but slowly, and it takes me infinite, extremely boring calculations..."

In the course of time, several *ad hoc* integration methods had been developed for special classes of ode's which occurred in the description of practical problems. It was quickly realized that a common feature of the various solution procedures consists in introducing new variables such that the given equation is transformed into an equation with known solution, or for which an integration method is available. Due to its importance, the special term *equivalence problem* was created for the task of recognizing whether two equations may be transformed into each other, and the entirety of equations sharing this property is called an *equivalence class*. In this general form, however, it is hardly justified to speak of a solution method because there is no indication of how to discover such a transformation in a concrete case, although infinitely many may exist. In particular, for nonlinear equations, efforts to find a solution along these lines very often terminate in the worst case possible: The attempts are abandoned without any evidence whether or not a solution in closed form does exist. It is therefore highly desirable to develop *algorithmic* solution schemes for the largest possible classes of equations.

The importance of equivalence problems had been realized before in algebraic geometry where the behavior of algebraic forms under the action of certain transformation groups, and in particular its *invariants* w.r.t. to these transformations, had been studied. The analogy with these methods turned out to be extremely successful in the realm of differential equations. In his article on invariants, Forsyth [47] expresses this as follows: *Similarity in properties of differential equations and of algebraic equations has long been of great value, both in the development of the theory and in the indication of methods of practical solution of the former equations.* Based on preceding work by Cockle [31], Laguerre [99], Brioschi [19] and Halphen [64], Forsyth applied this principle in particular for calculating invariants of linear differential equations under predefined classes of transformations.

This was essentially the state of affairs when Sophus Lie got interested in this problem in the sixties of the 19th century. The most important guide for him was Galois' theory for solving algebraic equations that had become widely known due to Liouville's efforts around 1850. The most significant recognition of Sophus Lie was that the transformation properties of an ode under certain groups of continuous transformations play a fundamental role in answering this question, very much like the permutations of the solutions of an algebraic equation furnish the key to understanding its solution behavior. This is best explained by his own words, which are taken from the first chapter of his book [113] on differential equations: [2]

---

[2]Translation by the author: "Previous investigations on ordinary differential equations as they may be found in the customary textbooks do not form a systematic entirety. Special integration theories have been developed, e.g., for homogeneous differential equations, for linear differential equations and other special forms of integrable differential equations. The mathematicians failed to observe, however, that the special theories may be subordinated to a general method. The foundation of this method is the concept of an *infinitesimal transformation* and closely related to it the concept of a *one-parameter group*."

Introduction

"Die älteren Untersuchungen über gewöhnliche Differentialgleichungen, wie man sie in den gebräuchlichen Lehrbüchern findet, bilden kein systematisches Ganzes. Man entwickelte specielle Integrationstheorien z. B. für homogene Differentialgleichungen, für die linearen Differentialgleichungen und andere specielle Formen von Differentialgleichungen. Es war aber den Mathematikern entgangen, dass diese speciellen Theorien sich unter eine allgemeine Methode unterordnen lassen. Das Fundament dieser Methode ist der Begriff der infinitesimalen Transformation und der damit aufs engste zusammenhängende Begriff der eingliedrigen Gruppe."

In order to provide an idea of how Lie acquired the intuition that finally led him to create his theory, and to understand the basic principle underlying it, a few simple but typical examples will be presented. Consider first the equation

$$y''(y + x) + y'(y' - 1) = 0 \tag{1.1}$$

from the collection of solved equations by Kamke [85], equation 6.133 in his enumeration. If new variables u and v are introduced by $x = u + v$, $y = u - v$ with $v \equiv v(u)$, equation (1.1) assumes the form

$$2v''u + v'^2 - 1 = 0.$$

On the other hand, the transformation $x = u + 2v$, $y = 3u - 4v$ yields

$$v''u - \tfrac{1}{2}v''v + \tfrac{2}{5}v'^3 - \tfrac{9}{10}v'^2 + \tfrac{1}{20}v' + \tfrac{3}{10} = 0.$$

Finally $x = \tfrac{1}{2}(v^2 + u)$, $y = \tfrac{1}{2}(v^2 - u)$ gives the particularly simple equation

$$v''v^3 + \tfrac{1}{4} = 0. \tag{1.2}$$

From this latter equation the solution is readily obtained in the form

$$C_1 v^2 = C_1^2 (u + C_2)^2 - \tfrac{1}{4}$$

by means of two quadratures as may easily be verified by substitution. [3] Thus finding among the infinity of possible variable transformations the particular one yielding (1.2) amounts to solving equation (1.1). Equation (1.2) has another remarkable property. Any of the following substitutions to new variables z and $w \equiv w(z)$,

$$u = -1 - \tfrac{1}{z}, \quad v = -\tfrac{w}{z}, \qquad u = -\tfrac{4z + 4}{8z + 7}, \quad v = -\tfrac{w}{8z + 7} \quad \text{or}$$

$$u = -\tfrac{9z - 9}{z - 10}, \quad v = -\tfrac{w}{z - 10}, \tag{1.3}$$

does *not change* its form, i.e., the result in all three cases is $w''w^3 + \tfrac{1}{4} = 0$. A transformation with this particular property is called a *symmetry* of the

---

[3] The details of this example may be found on page 323.

differential equation. The three substitutions (1.3) may be subsumed under the class of transformations of the form

$$u = \frac{(z+a)b^2}{1-(z+a)cb^2}, \qquad v = \frac{wb}{1-(z+a)cb^2} \qquad (1.4)$$

corresponding to the parameter values $(a, b, c) = (1, 1, 1), (1, 2, 2)$ and $(-1, 3, 1)$ respectively. Most remarkable, the entirety of transformations (1.4) forms a group whose elements are parametrized in terms of a, b and c. It turns out that invariance of (1.2) under the group of transformations (1.4) is a characteristic feature not only of eq. (1.2) but of all equations obtained from it by a variable transformation, i.e., for the full equivalence class determined by (1.2) which may be taken as a canonical representative. In general, a canonical form is not uniquely determined by the symmetry type. The allowed freedom may be used in order to obtain an especially simple solution. An invariance transformation of any member of the equivalence class determined by (1.2) is obtained from (1.3) or (1.4) if the same variable transformation is applied to it that leads to the transformed equation. The first transformation of (1.3) for example yields the symmetry

$$x = \frac{v}{(u-v)^2} - \frac{1}{2}, \quad y = \frac{u}{(u-v)^2} + \frac{1}{2}$$

of (1.1). The entirety of invariance transformations related by a variable change is called a *symmetry type*. In general a group of transformations may be a symmetry group for equations that are not equivalent to each other, e.g., if they have not the same order. The entirety of equations allowing the same symmetry type is called a *symmetry class*, it is the union of equivalence classes.

With the insight gained by examples similar to those just described, Lie developed a solution scheme for ode's that may be traced back to the following question: Are there any nontrivial transformations leaving the form of the given equation invariant? If the answer is affirmative like for eq. (1.2) above, they determine a canonical form of the given equation. If the solution of the canonical equation may be obtained and moreover it is possible to determine the transformation of the given equation to a canonical form, the original problem is solved.

Due to the crucial role of various kinds of transformations in Lie's theory of differential equations, a major part of his work is devoted to establishing a systematic theory of *groups of continuous transformations*. Although it was only a device for his main goal of solving ode's, the theory of *Lie groups* and the closely related *Lie algebras* as they have been named by Hermann Weyl became well-established fields in mathematics, independent from differential equations.

Contrary to that, Lie's theory for solving ordinary differential equations has been almost forgotten after his death. Most textbooks on differential equations do not even mention his name or the concept of a symmetry. For a long

Introduction 5

time the practical work of solving ode's has remained almost unchanged from the way it is described in the above quotation by Lie. Basically it consists of a set of heuristics that are applied one after another, usually accompanied by database like tools in the form of collections of solved examples as the one by Kamke [85], or the more recent collections by Polyanin [149] or Sachdev [157]. This fact certainly needs some explanation. The most important obstacle for applying his methods is the fact that they require huge amounts of analytical calculations that hardly may be performed by pencil and paper for almost any nontrivial example. The obvious remedy is to employ one of the computer algebra systems that have become available lately to carry them through.

There is another more fundamental reason. It is related to the question to what extent is his theory *constructive*, i.e., how much effort is needed to design algorithms from it that may be applied for solving problems? This is particularly important if the goal is a piece of computer algebra software that accepts an ode as input and returns the best possible answer according to the underlying theory. Lie himself did not attach a high value to this goal because he probably realized the practical difficulties due to the size of calculations.

After Lie had recognized that the symmetry of an ode is a fundamental concept that allows finding its solutions in closed form, he described several solution procedures based on it. On the one hand, if the infinitesimal generators for the symmetry group of a given ode may be determined explicitly, the canonical form and possibly the solution may be constructed from it. This is described in detail in Lie's book [113] on the subject; see also the books by Olver [140], Stephani [174] and Bluman [13]. In a second version the vector fields generating the symmetry transformations are also determined explicitly. The proper integration problem is reformulated and solved in terms of an associated linear partial differential equation (pde), a subject that Lie had considered in detail about ten years earlier. This procedure is also described in Lie's book quoted above. There is still another approach of solving equations with symmetries. Its symmetry type is determined by suitable manipuations of the determining equations of the symmetries *without* solving them explicitly. It is shown that *in principle* the transformations to canonical form may be obtained from it. Lie [109] outlined this proceeding in a series of articles in the *Archiv for Mathematik* but never came back to it. The reason probably is that it does not seem possible to set up the equations for the finite transformations in terms of the given ode to be solved.

In order to make this latter approach into a working method, two fundamental discoveries that were made within about twenty years after Lie's death turned out to be of fundamental importance. The first is Loewy's theory of linear ode's, Loewy [127], and its generalization to certain systems of linear pde's [105]. This was another effort to use the analogies between algebraic equations and differential equations, i.e., the reducibility of a linear ode and its representation in terms of irreducible equations of lowest degree. Loewy's main result is a theorem that guarantees the existence of a unique decomposition in terms of largest so-called *completely reducible factors*. The second is

Janet's theory of linear pde's, Janet [83], and in particular the canonical form he introduced for any such system which is known today as a *Janet basis*. This concept is closely related to Buchberger's *Gröbner basis* in polynomial ideal theory, and turned out to be fundamental in the realm of differential equations.

The importance of these results for Lie's theory originates from the fact that the symmetries of any differential equation are obtained from its so-called *determining system*, a linear homogeneous system of pde's. Representing it as a Janet basis allows one to identify its symmetry type from the coefficients. Furthermore, Loewy's and Janet's results allow to break down the solution algorithms for nonlinear ode's based on its symmetries uniquely and systematically into a small number of basic problems for which solution algorithms may be designed. At each step and in particular for the final result, the function field necessary for representing the answer is obtained.

Sophus Lie was not the only person who had the idea of generalizing Galois' results to differential equations. Almost at the same time as he developed his theories in Leipzig, Picard and Vessiot [146] in Paris initiated another effort with a similar objective. Their approach is usually referred to as *Picard-Vessiot-theory* or *differential Galois theory*. Their goal is more limited than Lie's because they consider only *linear* ode's. Very much like ordinary Galois theory deals with the field extensions generated by the solutions of univariate algebraic equations, differential Galois theory investigates the differential field extensions over a base field generated by the solutions of linear ode's. They obtain a theory that establishes a one-to-one correspondence between the structure of the differential Galois group of an equation and the function field generated by its solutions. Therefore the theory of Picard and Vessiot may very well claim to be the proper analogue of Galois' theory for linear ode's.

Although Lie aimed for a general and complete theory for solving nonlinear ode's as well, this goal has only partially been achieved. In Drach's [38] thesis that has been supervised by É. Picard and J. Tannery, the state of affairs is expressed as follows: *...Lie's théorie des groupes à l'intégration des équations n'est pas la véritable généralization de la méthode employée par Galois pour les équations algébriques.*[4] Drach's main objections are that Lie's theory is not complete in the following sense: An equation may have a symmetry group that cannot be employed for solving it. This holds for first order equations, and may occur for equations of higher order as well. On the other hand, Lie's theorems are not reciprocal, i.e., an equation may have a closed form solution yet may not have a nontrivial symmetry group. An example is the equation given by Gonzales-Lopez [54]

$$y''y - y'^2 - y'y^2(x^2 + 1) - 2y^3x = 0. \tag{1.5}$$

---

[4]Translation by the author: "...Lie's group theory for the integration of equations is not the true generalization of the method applied by Galois for algebraic equations."

Introduction 7

This equation has a trivial symmetry group, yet its general solution is

$$y = \frac{C_1^3}{(C_1 x + 1)^2 + C_1^2 + 1 + C_2 e^{C_1 x}}. \tag{1.6}$$

In other words, Lie's theory does not establish a unique relation between the existence of symmetries of an equation and its solution behavior.

In this context the following question is frequently asked: What is the relationship between Lie's theory and the theory of Picard and Vessiot? Although their point of departure is basically the same and group theory is an essential part of both theories, the answer is that there is virtually no connection between them. They are different approaches to the general problem of solving ordinary differential equations in closed form. Consequently, there arises a second question: What is the proper generalization of Galois' theory for solving *nonlinear* differential equations? There does not seem to exist a satisfactory answer at present.

Another important aspect is raised by the following question: To what extent do these theories support solving concrete problems? Although the theory of Picard and Vessiot is quite satisfactory from a theoretical point of view, it is only of limited value for obtaining explicit solutions. On the other hand, as it will turn out later on in this book, for solving nonlinear ordinary differential equations Lie's theory is a powerful tool. The solutions for the nonlinear second order equations in Chapter 6 of Kamke's collection, for example, are almost exclusively due to Lie symmetries. Furthermore, any equation obtained from them by a variable transformation may also be solved by applying Lie's theory, i.e., instead of a finite number of examples their full equivalence classes become amenable to the solution procedure.

Sometimes the objection is raised that most equations do not have nontrivial Lie symmetries and therefore Lie's theory does not apply. Although this is true, one should keep in mind that a similar argument applies to linear ode's and their differential Galois groups, and is also true for algebraic equations. The equations that actually occur in applications very often have a special structure that expresses itself in the existence of symmetries. It is of utmost importance to recognize these situations because it leads to a better understanding of the underlying problem, and to take advantage of them for the solution procedure. In this way Lie's theory may finally lead to a better understanding of the distinctive features that make an equation solvable, and may even help to raise the intuition that is necessary in order to establish a more complete theory.

As mentioned before, the proceeding described in Lie [109] is the basis for most of the solution algorithms for second and third order ode's in this book. In this approach, the symmetry type of an ode, being an invariant under point transformations, serves the purpose of narrowing down the possible equivalence classes to which it may belong. The totality of equations from a certain family sharing the same symmetry type is called a *symmetry class*; it

8

is a union of equivalence classes. In this form the close relation between the symmetry analysis and equivalence problems becomes particularly obvious. This point has been discussed in the preface of the book by Olver [141].

After the symmetry class of an equation and a possible canonical form are known, the function field of admitted equivalence transformations must be specified in order to obtain a well-defined problem. Furthermore, it should be possible to solve equivalence problems within this function field *constructively* and to design solution algorithms. In this book most ode's are polynomial in the derivatives with coefficients that are rational functions of the dependent and the independent variables. The smallest field containing the coefficients is called the *base field*. More general function fields are extensions of the base field, e.g., elementary or Liouvillian extensions. The overall scheme for solving second- or higher order ode's along these lines decomposes in three major steps.

▷ Determine the symmetry class of the given equation. This is achieved in terms of the Janet basis coefficients for the determining system of its symmetries; it proceeds completely in the base field.

▷ Transform the equation to a canonical form corresponding to its equivalence class which is contained in the symmetry class. To this end a system of linear or Riccati-like pde's will be set up for which solutions in well-defined function fields may be determined algorithmically.

▷ Solve the canonical form and determine the solution of the given equation from it.

In addition to developing solution schemes for individual equations, equations containing undetermined elements like parameters or functions depending on a specified set of indeterminates may be considered. Examples are linear equations, Riccati equations or Abel equations with coefficients that are undetermined functions of the independent variable. In other words, whole families of equations are considered that are identified by a certain structure. The problem then is to decompose the totality of equations with a given structure into symmetry or equivalence classes. The importance of this proceeding arises from the fact that many equations occuring in practical problems are specified in terms of its structure.

The scope of this book has been restricted deliberately to ordinary differential equations of low order. The aim is to obtain solution algorithms for well-defined classes of equations, leading to the best result including the possible answer that a solution of a certain type does *not* exist. Any extension of this scope would lead to a significant increase of the size of the book. The simplest pde's for a scalar function depending on two variables, or the simplest systems of ode's for two functions, require a classification of the finite Lie groups of three-space comprising several hundred entries. It remains an interesting

Introduction 9

and challenging problem to develop solution schemes for these cases along the lines described in this book.

The goals of this book are twofold. On the one hand, it intends to be a source of reference for people working actively in the field. To this end, the detailed discussion of many subjects and the extensive tabulations in the main text and especially in the appendices should prove to be an indispensible tool. Beyond that it might serve as a textbook if supplemented by some additional reading material, e.g., on the theory of ode's in general or on Lie groups and Lie algebras. Some bibliographic hints are given in the introductory remarks to the individual chapters.

The subsequent Chapter 2 provides the mathematical foundation for the rest of the book. It deals mainly with linear differential equations. For linear ode's the most important topic dealt with is Loewy's theory for the unique decomposition into completely reducible components. This is the basis for handling more general problems later on. For linear pde's a fairly complete description of Janet's work is given. Of particular importance are those systems of pde's that allow a finite-dimensional solution space. The dimension of this space is called the *order* of the system. This generalizes the notion of order of a linear ode. For values up to three, a complete classification of Janet bases is given. Loewy's decomposition of linear ode's is generalized to such systems. In Chapter 3, those results from the theory of continuous groups of a two-dimensional manifold $\mathbb{R}^2$ are presented that are relevant for the main subject of this book, i.e., solving ode's. The close relation between Lie's symmetry analysis and the equivalence problem is the subject of Chapter 4. The two subsequent Chapters are the core of the book. The subject of Chapter 5 is to identify the symmetry class to which a given quasilinear equation of order two or three belongs. To this end it is sufficient to know a Janet basis for its determining system, it is *not* necessary to determine the symmetry generators explicitly. The same is true for solving a given ode as is shown in Chapter 6. The crucial problem of transforming a given ode with a nontrivial symmetry to canonical form is achieved by solving a system of linear or Riccati-like pde's. As mentioned above, the symmetry class serves essentially the purpose of identifying a canonical form. Lie's second approach, i.e., to determine the Lie algebra of symmetry generators explicitly and to generate the canonical form of the differential equations from it, is the subject of Section 3 in Chapter 6.

Various topics are postponed to the appendices A through F. The solutions of selected exercises are given in Appendix A. A collection of useful formulae is listed in Appendix B. Various properties of ideals of monomials that are needed for Janet's algorithm are presented in Appendix C. In Appendix D for selected linear ode's from Kamke's collection the Loewy decomposition is given explicitly. In Appendix E the symmetries of the equations of second and third order of Kamke's collection are listed. They serve as a reference in many places of this book. Appendix F contains a description of the user interface of the software system ALLTYPES which is an important part of this book.

10

It provides a great many functions for carrying out the calculations that are necessary in order to apply the theory to practical problems. Most of these calculations are too voluminous to be performed by pencil and paper. This software may be accessed through the website www.alltypes.de.

The most important sources of information for this book were Lie's original writings on the subject, edited in seven volumes of the *Gesammelte Abhandlungen* by Friedrich Engel, and also Engels' *Anmerkungen* in the supplements to these volumes. Furthermore, there are the books written under Lie's guidance as quoted in the references at the end of this book. A lot of interesting background information may be found in the biography of Sophus Lie by Stubhaug [175].

The author would like to thank several colleagues and students for their helpful comments. In particular, Günter Czichowski, the late Eckehart Hotzel and several anonymous referees. The abiding encouragement by Carl-Adam Petri and the excellent working environment at the Fraunhofer Institute of Ulrich Trottenberg are also gratefully acknowledged.

Chapter 2

Linear Differential Equations

This chapter provides the fundamental algorithms for working with *linear* ordinary and partial differential equations that are the building blocks for the solution algorithms given later in this book. Systems of linear pde's are characterized in the first place by the number m of dependent and the number n of independent variables. Additional quantities of interest are the number N of equations, the order of the highest derivatives that may occur and the smallest function field in which the coefficients are contained, called the *base field*. Without further specification this is assumed to be the field of rational functions in the independent variables with rational number coefficients. The special case $m = n = N = 1$ corresponds to a single linear ode. The fundamental concepts described in this chapter are the *Loewy decomposition*, Loewy [127], and the *Janet basis*, Janet [83]. The latter term is chosen to honour the French mathematician Maurice Janet who described this concept and gave an algorithm to obtain it. After it had been forgotten for about fifty years, it was rediscovered [163] and utilized in various applications as described in this book. A good survey is also given in the article by Plesken and Robertz [148]. Janet bases are the differential counterpart of Gröbner bases that have been introduced by Bruno Buchberger and are a well-established tool in polynomial ideal theory and algebraic geometry now. The relevance of a Janet basis for the main subject of this book, i.e., solving ordinary differential equations or ode's for short, originates from the fact that the symmetries of any such equation are determined by a system of linear homogeneous pde's. General references for this chapter are the book by Ince [80] or the two volumes on linear equations by Schlesinger [160]. For Chapter 2.1 the first 150 pages of the book by van der Put and Singer [185], or the article by Buium and Cassidy [23] are highly recommended.

2.1 Linear Ordinary Differential Equations

The theory of linear ode's is important for several reasons. In the first place, many problems in connection with pde's may be traced back to certain standard problems involving ode's. Secondly, the theory of linear ode's is more complete than the theory of pde's and therefore serves as a guide for what kind of results may be expected for the latter. Many of the concepts and the methods introduced in this section will be needed later on. Symmetry properties of linear ode's will be discussed in detail in Chapter 5.5.

Solving Linear ODE's. A linear homogeneous ode for a differential indeterminate y, for which $y' = \dfrac{dy}{dx}$ and more generally $y^{(n)} = \dfrac{dy^{(n-1)}}{dx}$ is defined for any natural number n, will be written as

$$L(y) \equiv y^{(n)} + q_1 y^{(n-1)} + q_2 y^{(n-2)} + \ldots + q_{n-1} y' + q_n y = 0. \qquad (2.1)$$

If not specified otherwise the coefficients q_k, $k = 1, \ldots, n$, are rational functions in the independent variable x with rational number coefficients, i.e., $q_k \in \mathbb{Q}(x)$ which is called the *base field*.

Equation (2.1) allows the *trivial solution* $y \equiv 0$. The general solution contains n constants $C_1, \ldots, C_n$. Due to its linearity, these constants appear in the form $y = C_1 y_1 + \ldots + C_n y_n$ with the y_k linearly independent over the field of constants, e. g. $C_i \in \mathbb{Q}$ in this case. They form a so-called *fundamental system* and generate a n-dimensional vector space over the field of constants $\mathbb{Q}$. The coefficients q_k of (2.1) may be rationally expressed in terms of a fundamental system and its derivatives. To this end the *Wronskian*

$$W^{(n)}(y_1, \ldots, y_n) = \begin{vmatrix} y_1 & y_2 & \cdots y_n \\ y_1' & y_2' & \cdots y_n' \\ \cdots & \cdots & \cdots \\ y_1^{(n-1)} & y_2^{(n-1)} & \cdots y_n^{(n-1)} \end{vmatrix} \qquad (2.2)$$

is defined. It is different from zero iff the $y_k(x)$ are independent over the constants. More generally, the determinants

$$W_k^{(n)}(y_1, \ldots, y_n) = \frac{\partial}{\partial y^{(n-k)}} \begin{vmatrix} y & y_1 & y_2 & \cdots y_n \\ y' & y_1' & y_2' & \cdots y_n' \\ \cdots & \cdots & \cdots & \cdots \\ y^{(n)} & y_1^{(n)} & y_2^{(n)} & \cdots y_n^{(n)} \end{vmatrix} \qquad (2.3)$$

are defined for $k = 0, 1, \ldots, n$. There are the obvious relations $W_0^{(n)} = (-1)^{n-1} W^{(n)}$ and $\dfrac{dW^{(n)}}{dx} = (-1)^{n-1} W_1^{(n)}$. The determinant at the right hand side of (2.3) vanishes for $y = y_k$, $k = 1, \ldots, n$. Therefore, given a fundamental system $y_1, \ldots, y_n$, (2.3) is another way of writing the left hand side of (2.1). This yields the representation

$$q_k = (-1)^k \frac{W_k^{(n)}(y_1, \ldots, y_n)}{W^{(n)}(y_1, \ldots, y_n)} \qquad (2.4)$$

Linear Differential Equations

for the coefficients q_k, $k = 1, \ldots, n$ in terms of a fundamental system, i.e., (2.1) is uniquely determined by its solution space. For $k = 1$ there follows $W_1^{(n)} + q_1 W^{(n)} = W^{(n)'} + q_1 W^{(n)} = 0$, i.e., $W^{(n)} = \exp\left(-\int q_1 dx\right)$.

It is always possible to replace equation (2.1) by another equation over the same base field, its so called *rational normal form*

$$\bar{y}^{(n)} + \bar{q}_2 \bar{y}^{(n-2)} + \ldots + \bar{q}_{n-1} \bar{y}' + \bar{q}_n \bar{y} = 0,$$

by introducing a new dependent variable $\bar{y}(x)$ by $y = \exp\left(-\frac{1}{n}\int q_1(x)dx\right)\bar{y}$. More general transformations with the same effect are discussed in Chapter 4.1.

Solving an equation (2.1) amounts to determining n independent elements of a fundamental system, called a *full set of solutions*. In order to make this a rigorous concept, the function field containing the y_k's has to be specified. After that an algorithm has to be designed in order to obtain the solution explicitly.

The simplest problem is to ask for solutions in the base field. If it is $\mathbb{Q}(x)$, this amounts to searching for the rational solutions of (2.1). In general any rational solution has the form

$$y = C_1 y_1 + \ldots + C_k y_k \tag{2.5}$$

with $0 \leq k \leq n$, $y_k \in \mathbb{Q}(x)$, C_k constant and $y_1, \ldots, y_k$ linearly independent over constants. It may occur that all elements of a fundamental system are rational, i.e., $k = n$, or that the equation does not have a rational solution at all. The most general rational solution is the one with the highest possible value for k. If $k < n$ it is an important problem to find the minimal extension of the base field containing a full set of solutions.

Determining the most general solution in the base field occurs as a sub-problem of many other problems. It is therefore important that an efficient algorithm is available for this purpose. Such an algorithm is designed according to the following general principles that are valid for more general function fields as well. At first the problem has to be made finite, i.e., the possible solutions have to be parametrized such that a finite number of candidates is obtained. Secondly, a search procedure must be provided that identifies the desired answer among the candidates, or assures that a solution does not exist.

The following observation yields an important constraint for the possible form of any rational solution of (2.1). Assume a rational solution has a pole of order k at the finite position $x = x_0$. If it is substituted into (2.1), a term proportional to $\dfrac{1}{(x - x_0)^{k+n}}$ is generated from the highest derivative $y^{(n)}$. Furthermore, it is assumed that both the coefficients and the solution are uniquely represented as partial fractions. Then in order to make the left hand side vanish, at least one more term of order $k + n$ must occur. Because the lower derivatives of y cannot produce such a term, one of the coefficients must

14

contain a pole at x_0. As a consequence, the only possible poles of a rational solution of (2.1) are those that occur in the coefficients $q_i(x)$. Alternatively, if (2.1) is considered as an equation over $\mathbb{Q}[x]$, by multiplication with the least common multiple of the coefficient denominators, the possible poles are the zeros of the coefficient of the highest derivative $y^{(n)}$.

For any of these poles the coefficient of the highest order term must vanish. This leads to a nonlinear equation for the possible highest order whose largest solution in natural numbers is the desired bound. Similar arguments apply for the behavior at infinity. A polynomial ansatz with undetermined coefficients within these limits leads to a system of linear algebraic equations the solutions of which determine the rational solutions. The following algorithm is based on these considerations. More details may be found in Schwarz [162] and the references given there.

ALGORITHM 2.1   *RationalSolutionsLode(L(y))*. Given a linear homogeneous ode

$$ L(y) \equiv y^{(n)} + q_1 y^{(n-1)} + q_2 y^{(n-2)} + \ldots + q_{n-1} y' + q_n y = 0 $$

with $q_k(x) \in \mathbb{Q}(x)$, the maximal number of linearly independent rational solutions is returned.

$S1$: *Singularities and bounds.* Identify the possible positions of poles and determine an upper bound for any of them.

$S2$: *Solve linear system.* Set up the linear system for the undetermined coefficients for an ansatz within the limits found in $S1$ and determine its solution. If only the trivial solution exists, return an empty list $\{\}$.

$S3$: *Return result.* For each solution found in $S2$ generate a solution of $L(y)$ and return the list $\{y_1, \ldots, y_k\}$.

EXAMPLE 2.1   Consider the equation $y'' + \left(2 + \frac{1}{x}\right)y' - \frac{4}{x^2}y = 0$. It has no. 2.201 in Kamke's collection. For the order N of a singularity at infinity there follows immediately $N = 0$. For the order M of the only other pole at $x = 0$ there follows $M(M + 1) - M - 4 = 0$, i.e., $M = 2$. In step $S2$ the ansatz $\frac{a_2}{x^2} + \frac{a_1}{x} + a_0$ yields the system $3a_1 + 4a_2 = 0$, $2a_0 + a_1 = 0$ with the solution $a_1 = -\frac{4}{3}a_2$, $a_0 = \frac{2}{3}a_2$. With the normalization $a_2 = 1$ the only rational solution obtained in step $S3$ has the form $\frac{1}{x^2} - \frac{4}{3x} + \frac{2}{3}$, i.e., $k = 1$ in this case. The second element of a fundamental system is not rational, it will be obtained in Example 2.14 below by a different method. ⬜

Solving a linear *inhomogeneous* equation

$$ y^{(n)} + q_1 y^{(n-1)} + q_2 y^{(n-2)} + \ldots + q_{n-1} y' + q_n y = r \tag{2.6} $$

may be traced back to the corresponding homogeneous problem with $r = 0$ and integrations by the method of *variation of constants*. To this end, the C_i in the general solution $y = C_1 y_1 + \ldots + C_n y_n$ of the homogeneous equation

Linear Differential Equations 15

are considered as functions of x. Substituting this expression into (2.6) and imposing the constraints

$$C_1' y_1^{(k)} + C_2' y_2^{(k)} + \ldots + C_n' y_n^{(k)} = 0 \qquad (2.7)$$

for $k = 0, \ldots, n-2$ yields the additional condition

$$C_1' y_1^{(n-1)} + C_2' y_2^{(n-1)} + \ldots + C_n' y_n^{(n-1)} = r. \qquad (2.8)$$

The linear system (2.7), (2.8) for the C_k' has always a solution due to the non-vanishing determinant of its coefficient matrix which is the Wronskian $W^{(n)}$. Consequently, solving the inhomogeneous equation (2.6) requires only integrations if a fundamental system for the corresponding homogeneous problem is known.

EXAMPLE 2.2 If y_1 and y_2 is a fundamental system for $y'' + q_1 y' + q_2 y = 0$, the general solution of $y'' + q_1 y' + q_2 y = r$ may be written as

$$y = C_1 y_1 + C_2 y_2 + y_1 \int \frac{r y_2}{W} dx - y_2 \int \frac{r y_1}{W} dx \qquad (2.9)$$

where $W = y_1' y_2 - y_2' y_1$. Let the equation

$$y'' - \frac{4x}{2x-1} y' + \frac{4}{2x-1} y = \frac{e^{-2x}}{2x-1}$$

be given. A fundamental system for the homogeneous equation is $y_1 = x$, $y_2 = e^{2x}$. Therefore $r = \frac{e^{-2x}}{2x-1}$ and $W = (2x-1)e^{2x}$. Substituting this into the above equation leads to

$$y = C_1 e^{2x} + C_2 x + e^{2x} \int \frac{x e^{-4x}}{(2x-1)^2} dx - x \int \frac{e^{-2x}}{(2x-1)^2} dx. \qquad \square$$

Inhomogeneous problems will occur frequently later on if a linear homogeneous equation of third order is reducible but not completely reducible.

If a linear ode (2.1) does not have a full set of solutions in its base field $\mathcal{F}$, i.e., if $k < n$ in (2.5), the dimension of its solution space may increase if an enlarged function field is admitted for a fundamental system. This is achieved by adjoining solutions of algebraic or differential equations of order not higher than n to the base field. If θ is the element to be adjoined, the enlarged differential extension field is denoted by $\mathcal{F}\langle\theta\rangle$, extensions by more than a single element are denoted by $\mathcal{F}\langle\theta_1, \ldots, \theta_k\rangle$. Frequently the newly adjoined elements are required to obey special equations over the previously defined field. If these equations are algebraic or of first order, a special term is introduced for the corresponding extension.

DEFINITION 2.1 *(Liouvillian extensions).* *Let $\mathcal{F}$ be a differential field with derivation $\frac{d}{dx} \equiv \,'$. A simple Liouvillian extension $\mathcal{F}\langle\theta\rangle$ is obtained by adjoining an element θ such that one of the following holds.*

    i) Algebraic extension, θ satisfies a polynomial equation with coefficients in $\mathcal{F}$.

    ii) Extension by the exponential of an integral, $\theta' - a\theta = 0$, $a \in \mathcal{F}$.

    iii) Extension by an integral, $\theta' - a = 0$, $a \in \mathcal{F}$.

If this process is repeated and any finite number of elements $\theta_1, \ldots, \theta_k$ is adjoined, a Liouvillian extension $\mathcal{F}\langle\theta_1, \ldots, \theta_k\rangle$ is obtained.

The special case $\theta = e^x$, $\theta = \log x$ or θ algebraic is also called an *elementary extension*. If a full set of solutions is not obtained in a Liouvillian extension, solutions of higher order equations have to be adjoined to the base field which are called *Picard-Vessiot extensions*. They may be considered as the equivalent of the splitting field of an algebraic equation.

A systematic study of the possible field extensions is based on the concept of an irreducible linear ode, and the factorization of reducible equations into irreducible components. A basic prerequisite for this discussion is a theory for solving certain nonlinear equations due to Riccati that are discussed next.

Solving Riccati Equations. Any n-th order linear homogeneous ode (2.1) may be transformed into a nonlinear equation of order $n - 1$ for a function $z(x)$ by means of the variable change $y' = zy$. The higher order derivatives of y are then $y^{(\nu)} = \phi_\nu(z)y$ where

$$\phi_\nu(z) = \frac{d\phi_{\nu-1}(z)}{dx} + z\phi_{\nu-1}(z), \quad \nu \geq 1 \text{ and } \phi_0 = 1. \tag{2.10}$$

For $\nu \leq 4$ their explicit form is

$$\phi_0 = 1, \quad \phi_1 = z, \quad \phi_2 = z' + z^2, \quad \phi_3 = z'' + 3zz' + z^3,$$
$$\phi_4 = z''' + 3z'^2 + 4zz'' + 6z'z^2 + z^4.$$

Substitution into (2.1) yields the *Riccati equation associated to (2.1)*

$$\mathcal{R}_x^n(z) \equiv q_0\phi_n + q_1\phi_{n-1} + q_2\phi_{n-2} + \ldots + q_{n-1}\phi_1 + q_n\phi_0 = 0 \tag{2.11}$$

of order $n - 1$ for the unknown function $z(x)$; the $q_k(x)$ for $k = 1, \ldots, n$ are the same as in (2.1), $q_0 = 1$. For $n = 2$, 3 and 4 these equations are explicitly given by

$$z' + z^2 + q_1 z + q_2 = 0, \tag{2.12}$$
$$z'' + 3zz' + z^3 + q_1(z' + z^2) + q_2 z + q_3 = 0, \tag{2.13}$$
$$z''' + 4zz'' + 3z'^2 + 6z'z^2 + z^4 + q_1(z'' + 3zz' + z^3) + q_2(z' + z^2) + q_3 z + q_4 = 0. \tag{2.14}$$

Linear Differential Equations

Similar to linear equations, the existence of rational solutions of any Riccati equation is of fundamental interest. If the general solution of an equation of order n is rational it contains n constants. If this is not true, rational solutions involving fewer constants or no constants at all may exist. The latter solutions are called *special rational solutions.*

In order to make this more precise first it has to be made clear when two solutions are considered as essentially different. To this end the following concept of equivalence of rational functions is introduced. Two rational functions $p, q \in \mathbb{Q}(x)$ are called *equivalent* if there exists another rational function $r \in \mathbb{Q}(x)$ such that $p - q = \frac{r'}{r}$ holds, i.e., if p and q differ only by the logarithmic derivative of another rational function. It is easily seen that this defines an equivalence relation on $\mathbb{Q}(x)$. Applying this concept, the following estimate of essentially different rational solutions of a Riccati equation may be obtained.

LEMMA 2.1　*A generalized Riccati equation of order n has at most $n + 1$ pairwise non equivalent rational solutions.*

The proof of this result may be found in the article by Li and Schwarz [104]. For equations of order one and two a detailed description of all possible rational solutions is given next.

THEOREM 2.1　*If a first order Riccati equation $z' + z^2 + az + b = 0$ with $a, b \in \mathbb{Q}(x)$ has rational solutions, one of the following cases applies:*

i) *The general solution is rational and has the form*

$$z = \frac{r'}{r + C} + p \tag{2.15}$$

where $p, r \in \bar{\mathbb{Q}}(x)$, $\bar{\mathbb{Q}}$ is a suitable algebraic extension of $\mathbb{Q}$, and C is a constant.

ii) *There is only one, or there are two inequivalent special rational solutions.*

PROOF　Let y_1, y_2 be a fundamental system for the linear homogeneous equation $y'' + ay' + by = 0$ corresponding to the given Riccati equation $z' + z^2 + az + b = 0$. If both $\frac{y_1'}{y_1}$ and $\frac{y_2'}{y_2}$ are not rational, a rational solution does not exist. If $\frac{y_1'}{y_1}$ is rational but $\frac{y_2'}{y_2}$ is not, $\frac{y_1'}{y_1}$ is the only rational solution. If both $\frac{y_1'}{y_1}$ and $\frac{y_2'}{y_2}$ are rational but

$$\frac{C_1 y_1' + C_2 y_2'}{C_1 y_1 + C_2 y_2} \tag{2.16}$$

18

is not, $\frac{y_1'}{y_1}$ and $\frac{y_2'}{y_2}$ are two special rational solutions of the Riccati equation. This covers case ii). If $\frac{y_1'}{y_1}$, $\frac{y_2'}{y_2}$ and the ratio (2.16) are rational, it may be rewritten as

$$\frac{C_1 y_1' + C_2 y_2'}{C_1 y_1 + C_2 y_2} = \frac{C y_1'/y_1 + y_2'/y_1}{C + y_2/y_1} = \frac{(y_2/y_1)'}{C + y_2/y_1} + \frac{y_1'}{y_1} \tag{2.17}$$

with $C = \frac{C_1}{C_2}$. The assignments $p = \frac{y_1'}{y_1}$ and $r = \frac{y_2}{y_1}$ yield the expressions given for case i). $\qquad\square$

For first order Riccati equations there exists an extension of the preceding theorem for which there is no analogue if the order is higher than one.

COROLLARY 2.1   *The general first order Riccati equation $y' + py^2 + qy + r = 0$ may be transformed into normal form $z' + z^2 + az + b = 0$ by the variable change $y = \frac{z}{p}$ with the result $a = q - \frac{p'}{p}$, $b = pr$. The following three cases may be distinguished, depending on the number of special solutions of the normal form equation that are already known.*

i) *If a single special solution z_1 is known, the general solution is $z = z_1 + \frac{1}{w}$ where w is the general solution of $w' - (2z_1 + a)w = 1$.*

ii) *If two special solutions z_1 and z_2 are known, the general solution is $z = \frac{r'}{C + r} + z_1$ with $r = -\exp\left(\int (z_1 - z_2) dx\right)$.*

iii) *If three special solutions z_1, z_2 and z_3 are known, the general solution may be written as $\frac{z - z_2}{z - z_1} : \frac{z_3 - z_2}{z_3 - z_1} = C$ or*

$$z = \frac{(z_1 - z_2)\frac{z_1 - z_3}{z_2 - z_3}}{C - \frac{z_1 - z_3}{z_2 - z_3}} + z_1.$$

From these representations it follows in particular that two special rational solutions yield the general rational solution if z_1 and z_2 are equivalent, i.e., if the difference $z_1 - z_2$ is a logarithmic derivative. If three special rational solutions are known, the general solution is always rational.

The proof is obtained by substituting the various solutions into the given equation. It is instructive to consider various examples corresponding to the alternatives of the above results.

EXAMPLE 2.3   The equation considered in Example 2.5 below has the two special rational solutions $z = x$ and $z = x + \frac{2}{x}$. Because the difference of the two special solutions may be written as a logarithmic derivative in the form $\frac{2}{x} = \frac{(x^2)'}{x^2}$, the preceding discussion shows that the general solution must be rational. With $z_1 = x$ and $z_2 = x + \frac{2}{x}$, the same expression for the general solution is obtained as before. If $z_1 = x + \frac{2}{x}$ and $z_2 = x$ is chosen, its form is

$$z = \frac{\frac{2}{x^3}}{\bar{C} - \frac{1}{x^2}} + x + \frac{2}{x}.$$

The two constants are related by $C\bar{C} + 1 = 0$. □

EXAMPLE 2.4   The equation $z' + z^2 - \left(1 + \frac{1}{x}\right)z + \frac{1}{x} = 0$ has the two special rational solutions $z_1 = 1$, $z_2 = \frac{1}{x+1}$. The difference $z_1 - z_2 = \frac{x}{x+1}$ is not a logarithmic derivative. As a consequence, the general solution

$$z = \frac{xe^{-x}}{C - (x+1)e^{-x}} + 1$$

is not rational. □

THEOREM 2.2   *If a second order Riccati equation*

$$z'' + 3zz' + z^3 + a(z' + z^2) + bz + c = 0$$

with $a, b, c \in \mathbb{Q}(x)$ has rational solutions, one of the following cases applies.

i) *The general solution is rational and has the form*

$$z = \frac{C_2 u' + v'}{C_1 + C_2 u + v} + p \tag{2.18}$$

where $p, u, v \in \bar{\mathbb{Q}}(x)$, $\bar{\mathbb{Q}}$ is a suitable algebraic extension of $\mathbb{Q}$, and C_1 and C_2 are constants.

ii) *There is a single rational solution containing a constant; it has the form (2.15).*

iii) *There is a rational solution containing a single constant as in the preceding case, and in addition a single special rational solution.*

iv) *There is only a single one, or there are two or three special rational solutions that are pairwise inequivalent.*

The proof for the second order equation proceeds along similar lines. Therefore it is left as Exercise 2.6.

For Riccati equations of order higher than two the general structure of rational solutions has been described by Li and Schwarz [104]. Their result is given here without proof.

THEOREM 2.3   *The rational solutions of the Riccati equation (2.11) associated to (2.1) may be described as follows. Let $r_i \in \mathbb{Q}(x)$ and $P_i \equiv$*

20

$\{p_{i,1}, \ldots, p_{i,k}\}$, $p_{i,j} \in \mathbb{Q}[x]$, $C_{i,j}$ constants, the sets P_i linearly independent over constants for $i = 1, \ldots, m$ and $j = 1, \ldots, k$, and define

$$S_{P_i}^{r_i} \equiv \left\{ r_i + \frac{C_{i,1}p'_{i,1} + \ldots + C_{i,k}p'_{i,k}}{C_{i,1}p_{i,1} + \ldots + C_{i,k}p_{i,k}} \right\}.$$

The rational solutions of (2.11) are the disjoint union of $S_{P_i}^{r_i}$, $i = 1, \ldots, m$. Moreover, $\sum_{i=1}^{m} |P_i|$ is not greater than n.

Due to its connection to many problems treated later on, a constructive procedure for determining the rational solutions of a Riccati equation is needed. Similar to linear equations, the first step is to find the position of its singularities. For this purpose the explicit expression

$$\phi_\nu(z) = \sum_{k_0 + 2k_1 + \ldots + \nu k_{\nu-1} = \nu} \frac{\nu!}{k_0!k_1!\ldots k_{\nu-1}!} \left(\frac{z}{1}\right)^{k_0} \left(\frac{z'}{2!}\right)^{k_1} \cdots \left(\frac{z^{(\nu-1)}}{\nu!}\right)^{k_{\nu-1}}$$

(2.19)

for the functions $\phi_\nu(z)$ in (2.11) is useful. It follows from a formal analogy to the iterated chain rule of di Bruno [22]. For the subsequent discussion several of its properties are required, e. g. the order of a pole of ϕ_ν which it exhibits as a consequence of a pole of z. These properties are discussed in Exercise 2.13 and are taken for granted from now on without mentioning it.

Let x_0 be the position of a finite pole of order $M > 1$ in a possible solution $z(x)$. Due to (2.19) it generates exactly one term proportional to $\dfrac{1}{(x - x_0)^{\nu M}}$ in $\phi_\nu(z)$. In order to match it with some other term, the same pole must occur at least in a single coefficient $q_k(x)$ at the left hand side of equation (2.11). From this it follows that poles of order higher than one may occur only at the pole positions of the coefficients. The same is true for a pole of any order N at infinity. A bound for the order at any of these singularities is obtained by determining the growth for the various terms and looking for the largest integer where at least two terms have the same value. Due to the nonlinearity of the Riccati equation, the equations for the coefficients of the various singularities in a partial fraction expansion obey nonlinear algebraic equations that may have nonrational numbers as solutions. Therefore rational solutions of Riccati equations are searched in appropriate algebraic extensions $\bar{\mathbb{Q}}$ of the rational numbers.

These arguments do not apply for finite poles of order one. Any term in (2.19) generates a pole of order ν in this case. Nonleading terms originating from coefficient singularities may interfere with them. As a consequence, the finite first order poles are not confined to those poles occuring in the coefficients. It is a new problem to find their positions. According to Markoff [130] or Singer [169] one may proceed as follows. Let z be the desired solution of an nth order Riccati equation. Due to the fact that the residues of the additional first order poles are natural numbers, their contribution appears as a polynomial factor p in the solution of the corresponding linear equation. If the factor

Linear Differential Equations

originating from the higher order coefficient singularities is already known by the methods described above, a linear homogeneous equation for p may be obtained; and its possible polynomial solutions determine the additional first order poles of z. These polynomial solutions may be obtained by specializing the algorithm *RationalSolutionsLode*. The subsequent algorithm is designed accordingly.

ALGORITHM 2.2   *RationalSolutionsRiccati(R)*. Given a Riccati equation R of order n, return the rational solutions in $\bar{\mathbb{Q}}(x)$.

$S1$: *Higher order poles and bounds.* Determine the possible positions of poles of order higher than one that may occur in a solution of R and determine an upper bound for each of them.

$S2$: *Solve algebraic system.* Set up an algebraic system for the undetermined coefficients of an ansatz within the limits found in $S1$, determine the coefficients from it and construct the corresponding solutions of R. If a solution does not exist, return an empty list.

$S3$: *First order poles.* For each solution obtained in $S2$ generate the corresponding linear equation for a function p and determine its polynomial solutions.

$S4$: *Construct solution.* From each polynomial solution obtained in $S3$ and the corresponding solution obtained in $S2$ construct a solution of the given Riccati equation.

$S5$: *Return result.* Discard those solutions obtained in $S4$ that are subcases of some other solution, and return a list with the remaining ones.

EXAMPLE 2.5   Consider the equation $z' + z^2 - \left(2x + \frac{1}{x}\right)z + x^2 = 0$. A power x^N in a possible solution generates terms of order $N-1$, $2N$, $N+1$, $N-1$ and 2, i.e., $N = 1$ is the largest value where at least two terms match each other. A pole at zero of the form $\frac{1}{x^M}$ leads to singular terms of order $M+1$, $2M$, $M-1$, $M+1$ and $M-2$, therefore again $M = 1$ is the desired bound. The ansatz $z = ax + b + \frac{c}{x}$ yields the algebraic system

$$a^2 - 2a + 1 = 0, \ ab - b = 0, \ b^2 + 2ac - 2c = 0, \ 2bc - b = 0, \ c^2 - 2c = 0$$

with the two solutions $a = 1$, $b = c = 0$ and $a = 1$, $b = 0$, $c = 2$, i.e., two special rational solutions are $z = x$ and $z = x + \frac{2}{x}$. The corresponding equations for a polynomial factor are $p'' + \frac{3}{x}p' = 0$ and $p'' - \frac{1}{x}p' = 0$. The former equation does not have any polynomial solution while a fundamental system for the latter is $\{1, x^2\}$. This yields the general solution $z = \dfrac{2x}{x^2 + C} + x$.   ⬚

Gcrd, Lclm and Factorization of Linear ODE's.   Typical questions that arise when dealing with linear differential equations are the following: Is it possible to obtain some or all elements of a fundamental system for a given linear ode as solutions of equations that are *simpler* than the given one? Without further specification this means that only equations of the same type, i.e., linear homogeneous ode's over the same base field, but of lower order, are admitted. More special problems are to decide whether all solutions of a given equation are also solutions of some other equation, or whether two given equations have some solutions in common. Certainly one would like to decide these questions *without* determining the solutions explicitly. For algebraic equations there are well-known algorithms for answering these questions.

For the subsequent discussion a more algebraic language is appropriate. To this end, let $\mathbb{Q}(x)[D]$ with $D = \frac{d}{dx}$ be the ring of ordinary differential operators in the indeterminate x with rational function coefficients. Its elements have the form

$$A \equiv a_0 D^n + a_1 D^{n-1} + \ldots + a_{n-1} D + a_n$$

with $a_k \in \mathbb{Q}(x)$. If $a_0 \neq 0$ the order of A is n. The linear ode (2.1) is obtained by applying such an operator with $a_0 = 1$, $a_i = q_i$ for $i = 1, \ldots, n$, to a differential indeterminate y, i.e., by writing $Ay = 0$. Let

$$B \equiv b_m D^m + b_1 D^{m-1} + \ldots + b_{m-1} D + b_m$$

be another operator. The sum $A + B$ of two operators is defined by termwise addition in the obvious way. The product AB is defined by the prescription $Db_j(x) = b_j(x)D + b'_j(x)$ from which the general expression

$$a_i(x)D^i b_j(x)D^j = \sum_{k=0}^{i} \binom{i}{k} a_i(x) b_j^{(k)} D^{i-k+j}$$

follows. These definitions make the set of differential operators $\mathbb{Q}(x)[D]$ into a ring. In general the product AB is different from BA, i.e., this ring is *not commutative*. Consequently, one has to distinguish between left and right factors in a product. In general, for differential operators A, B and C, $C = A\,B$ implies $ker(B) \subset ker(C)$, i.e., the solution space of B is a subspace of the solution space of C. Usually differential operators or linear differential equations are assumed to be primitive, i.e., the coefficient of the highest derivative is unity.

For the problems mentioned above, the existence of a division algorithm is essential. It is defined in complete analogy to the corresponding divison in the ring of univariate algebraic polynomials. As a consequence, if A and B are operators of order n and m respectively with $m \leq n$, there are unique operators Q and R such that $A = QB + R$ where the order of R is strictly less than m. This guarantees that the ring $\mathbb{Q}(x)[D]$ is Noetherian as shown by Ore [143]. When $R = 0$, B is a right factor of A and Q is called *the exact quotient* of A and B.

EXAMPLE 2.6   Let $A = D^2 + \frac{1}{x}D - \frac{1}{4} - \frac{1}{2x}$ and $B = D - \frac{1}{2}$ be given. The complete division scheme for A and B is

$$\left(D^2 + \tfrac{1}{x}D - \tfrac{1}{4} - \tfrac{1}{2x}\right) : \left(D - \tfrac{1}{2}\right) = D + \tfrac{1}{2} + \tfrac{1}{x}$$
$$\underline{-\left(D^2 - \tfrac{1}{2}D\right)}$$
$$\left(\tfrac{1}{2} + \tfrac{1}{x}\right)D - \left(\tfrac{1}{2} + \tfrac{1}{x}\right)\tfrac{1}{2}$$
$$\underline{-\left(\tfrac{1}{2} + \tfrac{1}{x}\right)D + \left(\tfrac{1}{2} + \tfrac{1}{x}\right)\tfrac{1}{2}}$$
$$0$$

This implies $D^2 + \tfrac{1}{x}D - \tfrac{1}{4} - \tfrac{1}{2x} = \left(D + \tfrac{1}{2} + \tfrac{1}{x}\right)\left(D - \tfrac{1}{2}\right)$. In this particular case the remainder vanishes, i.e., the operator A may be written as a product with the right factor B. Consequently, the equation $y'' + \tfrac{1}{x}y' - \left(\tfrac{1}{4} + \tfrac{1}{2x}\right)y = 0$ has the solution $\exp\left(\tfrac{x}{2}\right)$ in common with the equation $y' - \tfrac{1}{2}y = 0$. □

EXAMPLE 2.7   Let $A = D^3 - \tfrac{3}{x}D^2 + \left(1 + \tfrac{8}{x^2}\right)D - \tfrac{2}{x} - \tfrac{10}{x^3}$ and $B = D - \tfrac{2}{x}$. Proceeding as above, the representation $A = QB + R$, with the operators $Q = D^2 - \tfrac{1}{x}D + 1 + \tfrac{2}{x^2}$ and $R = \tfrac{1}{x^3}$ is obtained. This is easily verified by expanding the right hand side, or by going through the steps of the division scheme. □

The division algorithm for differential operators allows to calculate the *greatest common right divisor* or $Gcrd(A,B)$ of two given operators A and B by the Euclidian algorithm, exactly like for univariate polynomials, by utilizing $Gcrd(A, B) = Gcrd(B, R)$ if $A = QB + R$. Because this definition applies division on the right, it is called greatest common *right* divisor.   In this book the *greatest common left divisor* which may be defined similarly will not occur. The solution space of the differential equation corresponding to the $Gcrd$ is the intersection of the solution spaces of the differential equations corresponding to its arguments.

EXAMPLE 2.8   Let $A = D^2 + \tfrac{1}{x}D - \tfrac{1}{4} - \tfrac{1}{2x}$ and $B = D^2 + \left(x - \tfrac{1}{2}\right)D - \tfrac{x}{2}$. A first division yields $A = QB + R$ with $Q = 1$ and $R = \left(\tfrac{1}{x} - x + \tfrac{1}{2}\right)\left(D - \tfrac{1}{2}\right)$. Dividing B by R yields a vanishing remainder, i.e., $Gcrd(A, B) = D - \tfrac{1}{2}$. This result means that the two second order linear ode's $Ay = 0$ and $By = 0$ have the solution $\exp\left(\tfrac{x}{2}\right)$ of $y' - \tfrac{1}{2}y = 0$ in common. □

Similarly the *least common left multiple* or $Lclm(A,B)$ for two differential operators A and B may be defined. It is the operator of lowest order that is divided exactly by either argument from the right. The equation corresponding to the $Lclm$ is the equation of lowest order with solution space the sum of its components. Its order is obviously at most the sum of the orders of the two arguments, and it is strictly less than that if the two arguments have a nontrivial $Gcrd$. Both the $Gcrd$ and the $Lclm$ are commutative and associative.

Given any two operators, its $Lclm$ may be determined as follows. Applying the preceding remarks, determine its order and generate an ansatz with undetermined coefficients. The conditions of divisibility generate a system of linear algebraic equations, the solution of which determines the desired $Lclm$.

EXAMPLE 2.9   Let A and B as in Example 2.8. Due to the first order $Gcrd$, the $Lclm$ is of order three. Dividing $D^3 + c_1 D^2 + c_2 D + c_3$ by A and B, and equating the coefficients of the resulting remainder to zero leads to a system of linear algebraic equations for c_1, c_2 and c_3. Substituting its solution into the third order ansatz finally yields

$$Lclm(A, B) = D^3 + \left(x + \frac{2}{x} - \frac{2x - \frac{1}{2}}{x^2 - \frac{1}{2}x - 1}\right)D^2$$

$$+ \left(\frac{3}{4} - \frac{1}{x} + \frac{\frac{1}{2}x - \frac{9}{4}}{x^2 - \frac{1}{2}x - 1}\right)D - \frac{1}{4}x - \frac{1}{2} + \frac{\frac{1}{4}x + 1}{x^2 - \frac{1}{2}x - 1}.$$

From this construction it follows that $\exp\left(\frac{x}{2}\right)$ is one solution of the corresponding third order ode, the two remaining ones are independent solutions of $Ay = 0$ and $By = 0$ respectively. ⬜

The general question of reducibility of a given operator or its corresponding equation, and how to obtain its factorization into irreducible components is considered next. An operator or an equation is called *irreducible* if a factorization into lower order operators over the same base field does not exist. In full generality this problem has been solved at the end of the 19th century by Beke [9] and Schlesinger [160], vol. II, it requires to solve the so-called associated equations, their results are given next.

THEOREM 2.4   *(Beke 1894, Schlesinger 1897) Let the nth order linear ode*

$$y^{(n)} + Q_1 y^{(n-1)} + \ldots + Q_n y = 0 \tag{2.20}$$

for $y(x)$ be given, and let

$$y^{(m)} + q_1 y^{(m-1)} + \ldots + q_m y = 0 \tag{2.21}$$

be a possible right factor with $1 \leq m \leq n - 1$, and $N \equiv \binom{n}{m}$. Further define $w_0 \equiv W^{(m)}(y_1, \ldots, y_m)$ to be the Wronskian of (2.21), and $w_k \equiv W_k^{(m)}$ as defined by (2.3) such that $q_k = (-1)^k \frac{w_k}{w_0}$; w_0 obeys the associated equation

$$w_0^{(\bar{N})} + r_1 w_0^{(\bar{N}-1)} + \ldots + r_{\bar{N}} w_0 = 0 \tag{2.22}$$

of order $\bar{N} \leq N$ with $r_i \in \mathbb{Q}\langle Q_1, \ldots, Q_n \rangle$ for all i. The w_k may be expressed as

$$w_k = l_{k, \bar{N}-1} w_0^{(\bar{N}-1)} + l_{k, \bar{N}-2} w_0^{(\bar{N}-2)} + \ldots + l_{k,0} w_0$$

with $l_{i,j} \in \mathbb{Q}\langle Q_1, \ldots, Q_n \rangle$ for all i and j. A factor of order m of (2.20) exists iff (2.22) has a hyperexponential solution, i.e., a solution such that $\frac{w_0'}{w_0} \in \mathbb{Q}(x)$.

Linear Differential Equations 25

The proof of this theorem and more details may be found in the above quoted references, see also Schwarz [162], Bronstein [21] and van Hoeij [73] where factorization algorithms are described.

EXAMPLE 2.10   Let $n = 3$ and $m = 2$, i.e., $y''' + Q_1 y'' + Q_2 y' + Q_3 y = 0$ and y_1, y_2 be a fundamental system of $y'' + q_1 y' + q_2 y = 0$. If this latter equation is a factor of the former, the determinants

$$w_0 = \begin{vmatrix} y_1 & y_2 \\ y_1' & y_2' \end{vmatrix}, \quad w_1 = \begin{vmatrix} y_1 & y_2 \\ y_1'' & y_2'' \end{vmatrix} \text{ and } w_2 = \begin{vmatrix} y_1' & y_2' \\ y_1'' & y_2'' \end{vmatrix}$$

obey $w_0' = w_1$, $w_1' = w_2 - Q_1 w_1 - Q_2 w_0$ and $w_2' = -Q_1 w_2 + Q_3 w_0$. Reordering this system in a *lex* order with $w_2 > w_1 > w_0$ yields

$$w_0''' + 2Q_1 w_0'' + (Q_1' + Q_1^2 + Q_2)w_0' + (Q_2' + Q_1 Q_2 - Q_3)w_0 = 0$$

and $w_1 = w_0'$, $w_2 = w_0'' + Q_1 w_0' + Q_2 w_0$, i.e., explicitly $q_1 = -\dfrac{w_0'}{w_0}$ and $q_2 = \dfrac{w_0''}{w_0} + Q_1 \dfrac{w_0'}{w_0} + Q_2$. □

In this book the general Beke-Schlesinger scheme is not considered further. Rather factorizations of second and third order equations are obtained by a direct method that leads to a more detailed description of the possible factorizations and the corresponding solutions. The subsequent lemma reduces these factorization problems to solving Riccati equations.

LEMMA 2.2   *Determining the right irreducible factors of a linear homogeneous ode up to order three with rational function coefficients amounts to finding rational solutions of Riccati equations.*

i) A second order ode $y'' + Ay' + By = 0$ has right factors of the form $y' + ay = 0$ with $A, B \in \mathbb{Q}(x)$ if and only if $a \in \mathbb{Q}(x)$ satisfies

$$a' - a^2 + Aa - B = 0.$$

ii) A third order ode $y''' + Ay'' + By' + Cy = 0$ with $A, B, C \in \mathbb{Q}(x)$ has a right factor of the form $y' + ay = 0$ if and only if $a \in \mathbb{Q}(x)$ satisfies

$$a'' - 3aa' + a^3 + A(a' - a^2) + Ba - C = 0.$$

It has a right factor of the form $y'' + by' + cy = 0$ if and only if $b, c \in \mathbb{Q}(x)$ are solutions of

$$a'' - 3aa' + a^3 + 2A(a' - a^2) + (A' + A^2 + B)a - B' - AB - C = 0,$$
$$b = -(a' - a^2 + Aa - B).$$

PROOF   A first order factor $y' + ay = 0$ implies

$$y'' + (a' - a^2)y = 0, \quad y''' + (a'' - 3aa' + a^3)y = 0.$$

26

Reduction w.r.t. to these constraints yields immediately the given conditions. A second order factor $y'' + ay' + by = 0$ implies

$$y''' + (a' - a^2 + b)y' + (b' - ab)y = 0.$$

Reduction of the third order equation leads to the two conditions

$$a' - a^2 + aA + b - B = 0, \quad b' - ab + bA - C = 0.$$

The former one may be solved for b. If it is substituted into the latter the two conditions of the theorem are obtained. ▯

EXAMPLE 2.11   Let the fourth order linear ode

$$L \equiv y^{(4)} - \left(6x + \frac{1}{x}\right)y''' + 12x^2 y'' - (9x^3 - 7x)y' + (2x^4 - 6x^2)y = 0 \quad (2.23)$$

be given. The corresponding third order Riccati equation has the two rational solutions $z = x$ and $z = 2x$. They yield the two first order factors

$$L_1 \equiv y' - xy \text{ and } L_2 \equiv y' - 2xy$$

with

$$Lclm(L_1, L_2) = y'' - \frac{3x^2 + 1}{x}y' + 2x^2 y.$$

Dividing it out from L yields

$$L_3 \equiv y'' - 3xy' + \left(x^2 + 3 - \frac{2}{x^2}\right)y.$$

This equation is irreducible. Therefore equation (2.23) has the representation $L = L_3 Lclm(L_1, L_2)$ in terms of largest completely reducible right factors. ▯

There is an important difference between the factorization of polynomials and the factorization of differential operators, i.e., the latter is not unique as the following example shows.

EXAMPLE 2.12   (Landau 1902) Consider $D^2 - \frac{2}{x}D + \frac{2}{x^2}$. Two possible factorizations are

$$\left(D - \frac{1}{x}\right)^2 \text{ and } \left(D - \frac{1}{x(1+x)}\right)\left(D - \frac{1+2x}{x(1+x)}\right).$$

More general, the factorization

$$\left(D - \frac{1}{x(1+ax)}\right)\left(D - \frac{1+2ax}{x(1+ax)}\right)$$

depends on a constant parameter a. ▯

Linear Differential Equations

EXAMPLE 2.13 Another aspect of the non-uniqueness of factorizations may be seen from the following representation of Landau's operator as least common multiple, e. g. in the form

$$Lclm\left(D - \frac{2}{x}, D - \frac{\frac{1}{2}}{x} - \frac{\frac{3}{2}}{x+2}\right) \text{ or } Lclm\left(D - \frac{1}{x} - \frac{1}{x+1}, D - \frac{1}{x} - \frac{1}{x-1}\right).$$

In general, the representation $Lclm\left(D - \frac{1}{x} - \frac{1}{C+x}\right)$ with $C = C_1$ and $C = C_2$, $C_1 \neq C_2$ is valid. This means that the least common left multiple of any pair of first order operators corresponding to two different values $C_1 \neq C_2$ yields Landau's operator. The explanation of these phenomena is provided by Loewy's results that are given next. $\square$

Loewy's Decomposition. The factorization of differential operators into irreducible factors is not unique, as Landau's example has shown. In order to obtain a unique decomposition of a differential operator or the corresponding linear differential equation in terms of lower order components, according to Loewy [127] or Ore [143] one has to proceed as follows. By applying a factorization algorithm, determine *all* irreducible right factors beginning with lowest order. The *Lclm* of these factors is called the *largest completely reducible part*, it is uniquely determined by the given operator. If its order equals n, the order of the differential operator, the operator is called *completely reducible* and the procedure terminates. If this is not the case, the *Lclm* is divided out and the same procedure is repeated with the exact quotient. After a finite number of steps the desired decomposition is obtained. By the above definition an irreducible operator is completely reducible. This proceeding is based on the following theorem by Loewy.

THEOREM 2.5 *(Loewy 1906) Let $D = \frac{d}{dx}$ a derivative and $a_i \in \mathcal{F}$ some differential field with derivative D. A differential operator*

$$L \equiv D^n + a_1 D^{n-1} + \ldots + a_{n-1}D + a_n$$

of order n may be written uniquely as the product of largest completely reducible factors $L_k^{(d_k)}$ of order d_k over the same base field in the form

$$L = L_m^{(d_m)} L_{m-1}^{(d_{m-1})} \ldots L_1^{(d_1)}$$

with $d_1 + \ldots + d_m = n$. The factors $L_k^{(d_k)}$ are unique.

The proof of this fundamental theorem may be found in the above quoted article by Loewy. In honor of Loewy the $L_j^{(k)}$ are called *Loewy factors*. The rightmost of them is simply called *Loewy factor*, it represents the largest completely reducible part.

In order to apply Theorem 2.5 for generating a fundamental system of the corresponding linear ode, the further decomposition into irreducible factors is required as is described next.

28

COROLLARY 2.2   *Any factor $L_k^{(d_k)}$, $k = 1, \ldots, m$ in Theorem 2.5 may be written as the least common left multiple of irreducible operators $l_{j_i}^{(e_i)}$ with $e_1 + \ldots + e_k = d_k$ in the form*

$$L_k^{(d_k)} = Lclm(l_{j_1}^{(e_1)}, l_{j_2}^{(e_2)}, \ldots, l_{j_k}^{(e_k)}).$$

This result provides a detailed description of the function spaces containing the solution of a reducible linear differential equation. A generalization of this result to certain systems of pde's will be given later in this chapter. More general field extensions are studied by differential Galois theory. Good introductions are the above quoted articles by Kolchin, Singer and Bronstein, and the lecture by Magid [128].

For a fixed value of the order of a linear ode there is a finite set of possibilities differing by the number and the order of factors, some of which may contain parameters. Each alternative is called a *type of Loewy decomposition*. The complete answer for the most important cases $n = 2$ and $n = 3$ is listed in the following corollaries.

COROLLARY 2.3   *For n=2 the possible types of Loewy decompositions are*

$$\mathcal{L}_1^2 : l^{(2)}, \quad \mathcal{L}_2^2 : l_1^{(1)}l_2^{(1)}, \quad \mathcal{L}_3^2 : Lclm(l_1^{(1)}, l_2^{(1)}), \quad \mathcal{L}_4^2 : Lclm(l^{(1)}(C))$$

where C is a constant. All types except $\mathcal{L}_2^2$ are completely reducible.

The proof follows immediately from Lemma 2.2 and the classification of solutions of Riccati equations given in Theorem 2.1. A factor containing a constant corresponds to a factorization that is not unique, any special value for C generates an irreducible factor. Because the originally given operator has order 2, two different special values must be chosen in order to represent it as $Lclm$.

EXAMPLE 2.14   The examples for the various types of Loewy decompositions of second order equations are taken from Kamke's collection, Chapter 2, they correspond to equations (2.162), (2.136), (2.201) and (2.146) respectively; as usual $D = \frac{d}{dx}$.

$$\mathcal{L}_1^2 : y'' + \tfrac{1}{x}y' + \left(1 - \tfrac{\nu^2}{x^2}\right)y = \left(D^2 + \tfrac{1}{x}D + 1 - \tfrac{\nu^2}{x^2}\right)y = 0,$$

$$\mathcal{L}_2^2 : y'' + \tfrac{1}{x}y' - \left(\tfrac{1}{4} + \tfrac{1}{2x}\right)y = \left(D + \tfrac{1}{2} + \tfrac{1}{x}\right)\left(D - \tfrac{1}{2}\right)y = 0,$$

$$\mathcal{L}_3^2 : y'' + \left(2 + \tfrac{1}{x}\right)y' - \tfrac{4}{x^2}y$$
$$= Lclm\left(D + \tfrac{2}{x} - \tfrac{2x-2}{x^2 - 2x + \tfrac{3}{2}}, \; D + 2 + \tfrac{2}{x} - \tfrac{1}{x + \tfrac{3}{2}}\right)y = 0,$$

$$\mathcal{L}_4^2 : y'' - \tfrac{6}{x^2}y = Lclm\left(D - \tfrac{5x^4}{x^5 + C} - \tfrac{2}{x}\right)y = 0. \qquad \Box$$

Linear Differential Equations 29

These examples show clearly how a Loewy decomposition makes the solution space in the various functions fields completely explicit. Due to the irreducibility of all factors this representation is unique, except for type $\mathcal{L}_4^2$ decompositions which allow linear transformations over constants. The same result for equations of order three is given next.

COROLLARY 2.4   *For n=3 the possible decompositions are*

$$\mathcal{L}_1^3 : l^{(3)}, \quad \mathcal{L}_2^3 : l^{(2)} l^{(1)}, \quad \mathcal{L}_3^3 : l_1^{(1)} l_2^{(1)} l_3^{(1)}, \quad \mathcal{L}_4^3 : Lclm(l_1^{(1)}, l_2^{(1)}) l_3^{(1)},$$

$$\mathcal{L}_5^3 : Lclm(l_1^{(1)}(C)) l_2^{(1)}, \quad \mathcal{L}_6^3 : l^{(1)} l^{(2)}, \quad \mathcal{L}_7^3 : l_1^{(1)} Lclm(l_2^{(1)}, l_3^{(1)}),$$

$$\mathcal{L}_8^3 : l_1^{(1)} Lclm(l_2^{(1)}(C)), \quad \mathcal{L}_9^3 : Lclm(l_1^{(1)}, l_2^{(1)}, l_3^{(1)}), \quad \mathcal{L}_{10}^3 : Lclm(l^{(2)}, l^{(1)}),$$

$$\mathcal{L}_{11}^3 : Lclm(l_1^{(1)}(C), l_2^{(1)}), \quad \mathcal{L}_{12}^3 : Lclm(l^{(1)}(C_1, C_2))$$

where C, C_1 and C_2 are constants. The types $\mathcal{L}_1^3$, and $\mathcal{L}_9^3$ through $\mathcal{L}_{12}^3$ are completely reducible. In the remaining cases the Loewy factor is of order less than three.

Whenever a constant C appears in a factor, two different special values may be chosen such that the corresponding $Lclm$ generates a second order operator. Similarly for the two constants in the last decomposition, three different pairs of special values may be chosen such that the $Lclm$ generates a third order operator.

EXAMPLE 2.15   Most of the examples of the various types of Loewy decompositions of third order equations are again taken from Kamke's collection, Chapter 3. The first three of them correspond to equations 3.6, 3.76 and 3.73 respectively.

$$\mathcal{L}_1^3 : y''' + 2xy' + y = 0,$$

$$\mathcal{L}_2^3 : y''' + \frac{1}{x^4} y'' - \frac{2}{x^6} y = \left(D^2 + \left(\frac{2}{x} + \frac{1}{x^4} \right) D + \frac{2}{x^5} \right) \left(D - \frac{2}{x} \right) y = 0,$$

$$\mathcal{L}_3^3 : y''' - \left(\frac{2}{x+1} + \frac{2}{x} \right) y'' + \left(\frac{6}{x} + \frac{4}{x^2} - \frac{6}{x+1} \right) y' + \left(\frac{8}{x} - \frac{8}{x^2} - \frac{4}{x^3} - \frac{8}{x+1} \right) y$$
$$= \left(D - \frac{2}{x+1} - \frac{1}{x} \right) \left(D - \frac{1}{x} \right) \left(D - \frac{2}{x} \right) y = 0.$$

There is no type $\mathcal{L}_4^3$ decomposition in Kamke's collection. An example is

$$\mathcal{L}_4^3 : y''' + \left(x - \frac{1}{x+1} - \frac{1}{x-1} + \frac{2}{x} \right) y'' + \left(1 + \frac{1}{x+1} - \frac{1}{x-1} \right) y'$$
$$= Lclm \left(D + x, D + \frac{1}{x} \right) Dy = 0.$$

The next four examples are equations 3.58, 3.45, 3.37 and 3.74 respectively

30

from Kamke's collection.

$$\mathcal{L}_5^3 : y''' + \left(\frac{1}{4} + \frac{7}{2x} - \frac{1}{4x^2}\right)y'' + \left(\frac{1}{x} + \frac{1}{x^2}\right)y' + \frac{1}{2x^2}y$$

$$= Lclm\left(D - \frac{1}{x+C} + \frac{2}{x}\right)\left(D + \frac{1}{4} - \frac{1}{2x} - \frac{1}{4x^2}\right)y = 0,$$

$$\mathcal{L}_6^3 : y''' + \frac{4}{x}y'' + \left(1 + \frac{2}{x^2}\right)y' + \frac{3}{x}y = \left(D + \frac{3}{x}\right)\left(D^2 + \frac{1}{x}D + 1\right)y = 0,$$

$$\mathcal{L}_7^3 : y''' - y'' - \left(\frac{1}{x-2} - \frac{1}{x}\right)y' + \left(\frac{1}{x-2} - \frac{1}{x}\right)y$$

$$= \left(D + \frac{1}{x-2} + \frac{1}{x}\right)Lclm\left(D - 1, D - \frac{2}{x}\right)y = 0,$$

$$\mathcal{L}_8^3 : y''' - \frac{1}{x}y'' + \frac{1}{x^2}y' = DLclm\left(D - \frac{2x}{x^2+C}\right)y = 0.$$

The decompositions of type $\mathcal{L}_9^3$ in Kamke's collection are more complicated. A simple example is

$$\mathcal{L}_9^3 : y''' + \left(x + 1 - \frac{2x}{x^2-2} - \frac{2x+1}{x^2+x+1}\right)y'' + \left(x - 2 - \frac{2x-1}{x^2+x+1}\right)y'$$

$$- \left(2 - \frac{2x}{x^2-2} - \frac{2}{x^2+x+1}\right)y = Lclm\left(D + 1, D + x, D - \frac{2}{x}\right)y = 0.$$

The next example is Kamke's equation 3.29.

$$\mathcal{L}_{10}^3 : y''' + \frac{3}{x}y'' + y = Lclm\left(D^2 - \left(1 - \frac{2}{x}\right)D + 1 - \frac{1}{x}, D + 1 + \frac{1}{x}\right)y = 0.$$

There is no type $\mathcal{L}_{11}^3$ decomposition in Kamke's collection. An example is

$$\mathcal{L}_{11}^3 : y''' - \left(1 - \frac{3}{x} + \frac{2x+1}{x^2+x-1}\right)y'' - \frac{x+3}{x^2+x-1}y'$$

$$- \left(\frac{3}{x} - \frac{3x+4}{x^2+x-1}\right)y = Lclm\left(D + \frac{1}{x} - \frac{2x}{x^2+C}, D - 1\right)y = 0.$$

Finally there is equation 3.71 from Kamke's collection.

$$\mathcal{L}_{12}^3 : y''' - \left(\frac{1}{x+3} + \frac{2}{x}\right)y'' + \left(\frac{4}{3x} + \frac{2}{x^2} - \frac{4}{3(x+3)}\right)y' + \left(\frac{2}{3x} - \frac{2}{x^2} - \frac{2}{3(x+3)}\right)y$$

$$= Lclm\left(D - \frac{3x^2 + 2C_1x + C_2}{x^3 + C_1x^2 + C_2x + C_2}\right)y = 0. \quad \square$$

In Appendix D a fairly complete list of Loewy decompositions of the linear equations listed in Chapters 2 and 3 of Kamke's collection is given.

Linear Differential Equations

EXAMPLE 2.16   The type $\mathcal{L}_6^3$ decomposition considered in the above example will be derived now by applying the procedure of Example 2.10. The coefficients of the given third order equation are $Q_1 = \frac{4}{x}$, $Q_2 = 1 + \frac{2}{x^2}$ and $Q_3 = \frac{3}{x}$. They lead to the associated equation

$$w_0''' + \frac{8}{x}w_0'' + \left(1 + \frac{14}{x^2}\right)w_0' + \left(\frac{1}{x} + \frac{4}{x^3}\right)w_0 = 0$$

with the solution $w_0 = \frac{1}{x}$. It yields $q_1 = \frac{1}{x}$ and $q_2 = 1$, i.e., the second order factor $y'' + \frac{1}{x}y' + y = 0$.   □

There remains to be discussed how the solution procedure for a linear ode with a nontrivial Loewy decomposition is simplified. The general principle is described in the following corollary the proof of which is obvious.

COROLLARY 2.5   Let a linear differential operator P of order n factor into $P = QR$ with R of order m and Q of order $n - m$. Further let $y_1, \ldots, y_m$ be a fundamental system for $R(y) = 0$, and $\bar{y}_1, \ldots, \bar{y}_{n-m}$ a fundamental system for $Q(y) = 0$. Then a fundamental system for $P(y) = 0$ is given by the union of $y_1, \ldots, y_m$ and special solutions of $R(y) = \bar{y}_i$ for $i = 1, \ldots, n - m$.

Consequently, solving an inhomogeneous equation of any order with known fundamental system for the homogeneous part requires only integrations. At first this result will be applied to reducible second order equations. With $D = \frac{d}{dx}$, $y \equiv y(x)$ and $a_i \equiv a_i(x)$, for the three nontrivial decompositions the following fundamental systems y_1 and y_2 are obtained. In decomposition $\mathcal{L}_3^2$ a_1 is not equivalent to a_2 as defined on page 17.

$\mathcal{L}_2^2 : (D + a_2)(D + a_1)y = 0,$

$$y_1 = \exp\left(-\int a_1 dx\right), \quad y_2 = y_1 \int \exp\left(\int (a_1 - a_2)dx\right)dx.$$

$\mathcal{L}_3^2 : Lclm(D + a_2, D + a_1)y = 0, \quad y_i = \exp\left(-\int a_i dx\right), \quad i = 1, 2,$

$\mathcal{L}_4^2 : Lclm(D + a(C))y = 0, \quad y_i = \exp\left(-\int a(C_i)dx\right), \quad i = 1, 2.$

There is an important difference between the last two cases. A fundamental system corresponding to a $\mathcal{L}_3^2$ decomposition is linearly independent over the base field, whereas this is not true for the last decomposition type $\mathcal{L}_4^2$. It is left as Exercise 2.5 to derive these fundamental systems and their properties. An example for each decomposition type is given next.

EXAMPLE 2.17   The decompositions of Example 2.14 are considered again. The first equation is the well-known Bessel equation with a fundamental system $J_\nu(x)$ and $I_\nu(x)$ in terms of series expansions. For the type $\mathcal{L}_2^2$ equation with $a_2 = \frac{1}{2} + \frac{1}{x}$ and $a_1 = -\frac{1}{2}$, a fundamental system is $y_1 = \exp\left(\frac{x}{2}\right)$ and

$y_2 = \exp\left(\frac{x}{2}\right) \int e^{-x} \frac{dx}{x}$. For the type $\mathcal{L}_3^2$ decomposition,

$$a_2 = \frac{2}{x} - \frac{4x - 4}{2x^2 - 4x + 3} \quad \text{and} \quad a_1 = 2 + \frac{2}{x} - \frac{2}{2x + 3}.$$

Two independent integrations yield $y_1 = \frac{2}{3} - \frac{4}{3x} + \frac{1}{x^2}$ and $y_2 = \frac{2}{x} + \frac{3}{x^2} e^{-2x}$. Finally in the last case $a(C) = -\frac{5x^4}{x^5 + C} + \frac{2}{x}$, integration yields $y = (x^5 + C)\frac{1}{x^2}$ from which the fundamental system $y_1 = x^3$ and $y_2 = \frac{1}{x^2}$ corresponding to $C = 0$ and $C \to \infty$ follows. $\quad\square$

For differential equations of order three there are four decompositions into first order factors with no constants involved. Fundamental systems for them may be obtained as follows.

$\mathcal{L}_3^3:\ (D + a_3)(D + a_2)(D + a_1)y = 0,\ \ y_1 = \exp\left(-\int a_1 dx\right),$

$\quad y_2 = y_1 \int \exp\left(\int(a_1 - a_2)dx\right)dx,\ y_3 = y_1 y_2 \int \exp\left(\int(a_2 - a_3)dx\right)dx.$

$\mathcal{L}_4^3:\ Lclm(D + a_3, D + a_2)(D + a_1)y = 0,\ \ y_1 = \exp\left(-\int a_1 dx\right),$

$\quad y_2 = y_1 \int \exp\left(\int(a_1 - a_2)dx\right)dx,\ y_3 = y_1 \int \exp\left(\int(a_1 - a_3)dx\right)dx.$

$\mathcal{L}_7^3:\ (D + a_3)Lclm(D + a_2, D + a_1)y = 0,\ \ y_i = \exp\left(-\int a_i dx\right),\ i = 1, 2,$

$\quad y_3 = y_1 \int \exp\left(\int(a_1 - a_3)dx\right)\frac{dx}{a_2 - a_1} - y_2 \int \exp\left(\int(a_2 - a_3)dx\right)\frac{dx}{a_2 - a_1}.$

$\mathcal{L}_9^3:\ Lclm(D + a_3, D + a_2, D + a_1)y = 0,\ \ y_i = \exp\left(-\int a_i dx\right),\ i = 1, 2, 3.$

The proof of these formulas is elementary, it is based on the above Corollary 2.5 and the expression for the solution of an inhomogeneous second order linear ode given in Example 2.2.

EXAMPLE 2.18 The type $\mathcal{L}_3^3$ decomposition considered in Example 2.15 has coefficients $a_1 = -\frac{2}{x}$, $a_2 = -\frac{1}{x}$ and $a_3 = \frac{1}{x} - \frac{2}{x+1}$. If they are substituted in the above expressions, the fundamental system $y_1 = x^2$, $y_2 = x^2 \log x$ and $y_3 = x + x^3 + x^2 \log(x)^2$ is obtained. $\quad\square$

EXAMPLE 2.19 The type $\mathcal{L}_7^3$ decomposition considered in Example 2.15 has coefficients $a_1 = -\frac{2}{x}$, $a_2 = -1$ and $a_3 = \frac{1}{x} + \frac{1}{x-2}$. If they are substituted in the above expressions, the fundamental system $y_1 = x^2$, $y_2 = e^x$ and

$$y_3 = \frac{x(x^2 - 2)}{4(x - 2)} + \frac{x^2}{4} \log \frac{x - 2}{x} + e^x \int e^{-x} \frac{dx}{(x - 2)^2}$$

is obtained. $\quad\square$

There are various decompositions into first order factors involving one or two constants. Their fundamental systems may be obtained analogous to the preceding cases. The difference is that there may be dependencies over the

Linear Differential Equations

base field or an extension of it. All three elements of a fundamental system corresponding to a decomposition of type $\mathcal{L}_{12}^3$ are dependent over the base field. Two elements are dependent over the base field for decompositions of type $\mathcal{L}_8^3$ and $\mathcal{L}_{11}^3$, and two elements of a type $\mathcal{L}_5^3$ decomposition are dependent over the simple field extension of the base field generated by its first order right factor.

There are two decomposition types involving second order irreducible factors.

$\mathcal{L}_2^3: (D^2 + a_3 D + a_2)(D + a_1)y = 0$. Let $\bar{y}_2$ and $\bar{y}_3$ be a fundamental system of the left factor. Then

$$y_1 = \exp\left(-\int a_1 dx\right), \quad y_2 = y_1 \int \frac{\bar{y}_2}{y_1} dx, \quad y_3 = y_1 \int \frac{\bar{y}_3}{y_1} dx.$$

$\mathcal{L}_6^3: (D + a_3)(D^2 + a_2 D + a_1)y = 0$. Let y_1 and y_2 be a fundamental system of the right factor, $W = y_1' y_2 - y_2' y_1$ its Wronskian. Then

$$y_3 = y_1 \int \exp\left(-\int a_3 dx\right) \frac{y_2}{W} dx - y_2 \int \exp\left(-\int a_3 dx\right) \frac{y_1}{W} dx.$$

EXAMPLE 2.20   Consider the following equation with the type $\mathcal{L}_2^3$ decomposition.

$$y''' + \frac{x^2 + 1}{x} y'' + \frac{4x^2 - 4}{x^2} y' + \frac{x^2 - 3}{x} y = \left(D^2 + \frac{1}{x} D + \frac{x^2 - 4}{x^2}\right)(D + x)y = 0.$$

The left factor corresponds to the Bessel equation for $n = 2$ with the fundamental system $J_2(x)$ and $Y_2(x)$. Applying the above formulas, the fundamental system

$$y_1 = \exp\left(-\tfrac{1}{2}x^2\right), \quad y_2 = y_1 \int J_2(x) \exp\left(\tfrac{1}{2}x^2\right) dx, \quad y_3 = y_1 \int Y_2(x) \exp\left(\tfrac{1}{2}x^2\right) dx$$

in terms of integrals over Bessel functions and exponentials is obtained.   ⬜

Symmetric Products and Symmetric Powers. The equations that are obeyed by the powerproducts of the elements of fundamental systems of given linear ode's are important tools for understanding the field extensions generated by the solutions of certain irreducible equations. They are called the *symmetric product* and the *symmetric power*. The discussion in this subsection is based on results by Singer and Ulmer [170, 172].

DEFINITION 2.2   *(Symmetric product, symmetric power) Let $M = 0$ and $N = 0$ be linear differential equations with fundamental systems $\{y_1, \ldots, y_m\}$ and $\{z_1, \ldots, z_n\}$ respectively. The equation $L = 0$ for which the products $\{y_1 z_1, \ldots, y_1 z_n, \ldots, y_m z_n\}$ generate the solution space is called the symmetric product and is written $L = M \circledS N$. If $M = N$, this is also called the symmetric power and written $M^{\circledS 2}$. The k-th power $M^{\circledS k}$ is defined correspondingly. Sometimes the same terminology is applied to the operators that generate the respective equations.*

34

Singer [170] has shown that if M has order n, the order of $M^{\circledS k}$ is not higher than $\binom{k+n-1}{n-1}$. If $n = 2$, the order of $M^{\circledS k}$ is $k + 1$.

The symmetric product is associative, i.e., there holds $(L_1 \circledS L_2) \circledS L_3 = L_1 \circledS (L_2 \circledS L_3)$. Whenever a linear ode of order three or more is the symmetric power of some linear second order ode, its solution is reduced to this latter equation.

EXAMPLE 2.21   Let $L_3 \equiv y''' - \frac{3}{x^2} y' + \frac{3}{x^3} y = 0$ with solutions $y_1 = \frac{1}{x}$, $y_2 = x$ and $y_3 = x^3$ and $L_2 \equiv z'' - \frac{3}{4x^2} z = 0$ with solutions $z_1 = \frac{1}{\sqrt{x}}$ and $z_2 = x\sqrt{x}$. Then $y_1 = z_1^2$, $y_2 = z_1 z_2$ and $y_3 = z_2^2$. Consequently, $L_3 = L_2^{\circledS 2}$. □

In order to be applicable for a solution procedure it is necessary to know *a priori* that a given linear ode is the symmetric power of some other equation. In Theorem 5.21 it will be seen that the Lie symmetries of a linear equation provide this information.

In some applications the symmetric powers of a given linear ode have to be determined. According to the above mentioned references one may proceed as follows. In order to construct $M \circledS N$, assume y and z are solutions of M and N respectively. Differentiate their product $w = yz$ such that

$$ w' = y'z + yz', \ \ w'' = y''z + 2y'z' + yz'', \dots, w^{(k)} = \sum_{j=0}^{k} \binom{k}{j} y^{(j)} z^{(k-j)} $$

and reduce the result w.r.t. M and N. In the resulting expressions there occur at most mn different terms $y^{(i)} z^{(j)}$ with $i < m$, $j < n$. Consequently, for some k these products are linearly dependent over constants, and a linear relation for the derivatives of w results, it is the desired symmetric product equation. Its coefficients are in the base field of the original equations.

EXAMPLE 2.22 *(Kovacic 1986)* Consider the equation $y'' + \left(\frac{3}{16x^2} - \frac{1}{x}\right) y = 0$. If y_1 and y_2 are two solutions one obtains for the product $z = y_1 y_2$ the following linear system for $y_1^{(i)} y_2^{(j)}$, $i, j = 0, 1$.

$$ y_1 y_2 = z, \ \ y_1' y_2 + y_1 y_2' = z', \ \ 2y_1' y_2' + \left(\frac{2}{x} - \frac{3}{8x^2}\right) y_1 y_2 = z'', $$
$$ \left(\frac{4}{x} - \frac{3}{4x^2}\right)(y_1' y_2 + y_1 y_2') - \left(\frac{2}{x^2} - \frac{3}{4x^3}\right) y_1 y_2 = z'''. $$

Transforming it to triangular form leads to the second symmetric power

$$ z''' - \left(\frac{4}{x} - \frac{3}{4x^2}\right) z' + \left(\frac{2}{x^2} - \frac{3}{4x^3}\right) z = 0. \qquad\qquad □ $$

Computing symmetric powers may become fairly complex if the value of the power increases and the coefficients of the given equation are complicated. Therefore an explicit list of symmetric powers of a second order equation $L \equiv y'' + ry = 0$ is given in Appendix B on page 379. An example is the second symmetric power $z''' + 4rz' + 2r'z = 0$ of $y'' + ry = 0$.

Linear Differential Equations

Galois Theory for Second Order Equations. The structure of function fields containing the solutions of a *reducible* linear second order ode is obvious from its Loewy decomposition $\mathcal{L}_2^2$, $\mathcal{L}_3^2$ or $\mathcal{L}_4^2$. A complete answer embracing also *irreducible* equations is provided by the theory of Picard and Vessiot which is also called *differential Galois theory*. For second order equations it is explicitly described by Singer and Ulmer [172]. In the differential case, the Galois groups are subgroups of the general linear group $GL_2(\mathbb{C})$ over the complex numbers $\mathbb{C}$. According to Kaplansky [88], page 41, the Galois group is a unimodular group contained in $SL_2(\mathbb{C})$ if the equation is in rational normal form $y'' + ry = 0$. In order to obtain the possible Galois groups of such equations the algebraic subgroups of $SL_2(\mathbb{C})$ must be known, they are described next.

THEOREM 2.6   *The algebraic subgroups of $SL_2(\mathbb{C})$ may be classified, up to conjugacy, as follows.*

i) The Borel group $B = \left\{ \begin{pmatrix} a & b \\ 0 & a^{-1} \end{pmatrix}, \ a, b \in \mathbb{C}, \ a \neq 0 \right\}.$

ii) For $m \in \mathbb{N}$, the groups $B_m = \left\{ \begin{pmatrix} a & b \\ 0 & a^{-1} \end{pmatrix}, \ a, b \in \mathbb{C}, \ a^m = 1 \right\}.$

iii) The group $Z = \left\{ \begin{pmatrix} a & 0 \\ 0 & a^{-1} \end{pmatrix}, \ a \in \mathbb{C}, \ a \neq 0 \right\}.$

iv) For $m \in \mathbb{N}$, the cyclic groups $Z_m = \left\{ \begin{pmatrix} a & 0 \\ 0 & a^{-1} \end{pmatrix}, \ a \in \mathbb{C}, \ a^m = 1 \right\}.$

v) The group $D = \left\{ \begin{pmatrix} a & 0 \\ 0 & a^{-1} \end{pmatrix}, \begin{pmatrix} 0 & 1 \\ -1 & 0 \end{pmatrix}, \ a \in \mathbb{C}, \ a \neq 0 \right\}.$

vi) For $m \in \mathbb{N}$, the dihedral groups

$$D_m = \left\{ \begin{pmatrix} a & 0 \\ 0 & a^{-1} \end{pmatrix}, \begin{pmatrix} 0 & 1 \\ -1 & 0 \end{pmatrix}, \ a \in \mathbb{C}, \ a^m = 1 \right\}.$$

vii) The tetrahedral group of order 24 which is generated by ($\theta^8 = 1$)

$$A_4^{SL_2} = \left\langle \begin{pmatrix} i & 0 \\ 0 & -i \end{pmatrix}, \begin{pmatrix} 0 & 1 \\ -1 & 0 \end{pmatrix}, \frac{1}{\sqrt{2}} \begin{pmatrix} \theta^7 & \theta^7 \\ \theta^5 & \theta \end{pmatrix} \right\rangle.$$

viii) The octahedral group of order 48 which is generated by ($\theta^8 = 1$)

$$S_4^{SL_2} = \left\langle \begin{pmatrix} i & 0 \\ 0 & -i \end{pmatrix}, \begin{pmatrix} 0 & 1 \\ -1 & 0 \end{pmatrix}, \begin{pmatrix} \theta & 0 \\ 0 & \theta^7 \end{pmatrix}, \frac{1}{\sqrt{2}} \begin{pmatrix} \theta^7 & \theta^7 \\ \theta^5 & \theta \end{pmatrix} \right\rangle.$$

36

ix) The icosahedral group of order 120 which is generated by $(\eta^5 = 1)$

$$A_5^{SL_2} = \left\langle -\begin{pmatrix} \eta^3 & 0 \\ 0 & \eta^2 \end{pmatrix}, \frac{1}{\eta^2 - \eta^3} \begin{pmatrix} \eta + \eta^4 & 1 \\ -1 & -\eta + \eta^4 \end{pmatrix} \right\rangle.$$

x) The full unimodular group $SL_2(\mathbb{C})$.

A good reference for this result, including an introduction to the literature, is the article by Martins [131]. At first they are applied to reducible equations.

THEOREM 2.7   *(Singer and Ulmer 1993) Let $L(y) \equiv y'' + ry = 0$ be a reducible second order ode with $r \in \mathbb{Q}(x)$, $L^{\circledS k}$ its $k-$th symmetric power for $k \geq 1$, and s_k the number of independent nontrivial rational solutions of $L^{\circledS k}$. The following Galois groups $\mathcal{G}(L)$ have to be distinguished.*

> i) $\mathcal{G}(L)$ *is conjugate to the Borel group B or one of its subgroups B_m if and only if there is a single nontrivial solution y such that $\frac{y'}{y} \in \mathbb{Q}(x)$. Furthermore, $\mathcal{G}(L)$ is conjugate to B_m if and only if $s_k = 0$ for $k = 1, \ldots, m - 1$ and $s_m = 1$.*

> ii) $\mathcal{G}(L)$ *is conjugate to Z or one of its finite subgroups Z_m if and only if there are two independent nontrivial solutions y_i such that $\frac{y_i'}{y_i} \in \mathbb{Q}(x)$ for $i = 1, 2$. There holds $y_1 y_2 \in \mathbb{Q}(x)$. Furthermore, $\mathcal{G}(L)$ is conjugate to Z_m if and only if for $k < m$ there holds $s_k = 0$ for k odd, $s_k = 1$ for k even and $s_m = 2$ for m odd or $s_m = 3$ for m even.*

The proof of this theorem may be found in the article of Singer and Ulmer [172] on page 24. The criteria for identifying B_m and Z_m are based on the tabulations given in Section 6.1 and 6.2 of [131]. The theorem covers the Loewy decompositions different from $\mathcal{L}_1^2$. The decomposition type $\mathcal{L}_2^2$ corresponds to case *i)*, the types $\mathcal{L}_3^2$ and $\mathcal{L}_4^2$ correspond to *case ii)*.

Applying this theorem, the following result for a reducible linear ode may always be obtained: If the criteria for a fixed value of m are fulfilled, the group is either B_m or Z_m. If not, the Galois group may be either of them for a value m greater than that for which the number of rational solutions of symmetric powers can be determined, or it is B or Z. In order to obtain a complete answer, a bound for the possible value of m is needed. Such a bound does not seem to be known at present. A more complete discussion of these questions may be found in the above quoted article by Martins [131]. For equations with type $\mathcal{L}_4^2$ Loewy decomposition a complete answer is always possible.

COROLLARY 2.6   *An equation with Galois group Z_1 or Z_2 has a type $\mathcal{L}_4^2$ Loewy decomposition.*

The proof of this corollary is left as Exercise 2.7.

$$\textit{Linear Differential Equations} \qquad 37$$

EXAMPLE 2.23 The equation

$$L(y) \equiv y'' + \left(\frac{3}{16x^2} + \frac{1}{4(x-1)^2} - \frac{1}{4x(x-1)} \right) y = 0$$

is discussed by Ulmer and Weil [182], page 192. Its type $\mathcal{L}_2^2$ Loewy decomposition $\left(D + \frac{1}{2(x-1)} + \frac{1}{4x} \right)\left(D - \frac{1}{2(x-1)} - \frac{1}{4x} \right) y = 0$ entails case $i)$ of the above theorem. A fundamental system is

$$y_1 = (x-1)^{1/2}x^{1/4}, \quad y_2 = (x-1)^{1/2}x^{1/4} \log \frac{\sqrt{x}-1}{\sqrt{x}+1}$$

with $\frac{y_1'}{y_1} = \frac{3x-1}{4x(x-1)} \in \mathbb{Q}(x)$ and $\frac{y_2'}{y_2}$ not in the base field. Because $y_1^4 \in \mathbb{Q}(x)$ the Galois group is B_4. ⬜

EXAMPLE 2.24 The equation $L(y) \equiv y'' + \frac{3}{16x^2}y = 0$ is discussed by Ulmer and Weil [182], page 182 and page 191. Its type $\mathcal{L}_3^2$ Loewy decomposition $Lclm\left(D - \frac{1}{4x}, D - \frac{3}{4x} \right) y = 0$ yields the fundamental system $y_1 = x^{1/4}$ and $y_2 = x^{3/4}$. There holds $y_1 y_2 = x$ and $y_1^4 = x$. Consequently, according to the above theorem, case $ii)$, the Galois group is the cyclic group of order 4.

$$\mathcal{G}(L) = \left\{ \begin{pmatrix} 1 & 0 \\ 0 & 1 \end{pmatrix}, \begin{pmatrix} -1 & 0 \\ 0 & -1 \end{pmatrix}, \begin{pmatrix} i & 0 \\ 0 & -i \end{pmatrix}, \begin{pmatrix} -i & 0 \\ 0 & i \end{pmatrix} \right\}. \qquad ⬜$$

The next example is interesting because its solution requires the extension of the field of constants.

EXAMPLE 2.25 The equation $L(y) \equiv y'' + \frac{7}{16x^2}y = 0$ with $\mathcal{L}_3^2$ Loewy decomposition $Lclm\left(D - \frac{2-\sqrt{-3}}{4x}, D - \frac{2+\sqrt{-3}}{4x} \right) y = 0$ is discussed by Ulmer and Weil [182], page 197. A fundamental system is $y_1 = x^{(2+i\sqrt{3})/4}$ and $y_2 = x^{(2-i\sqrt{3})/4}$. For $k \leq 48$, the symmetric powers $L^{\circledS k}$ have no rational solution for k odd, and a single rational solution for k even. By the above Theorem 2.7, case $ii)$, the conditions for Z_m are not satisfied for $m \leq 48$. Consequently, the Galois group is either Z_m with $m > 48$ or Z. ⬜

From the listing in Appendix D it is obvious that most second order linear ode's given there fall under the category covered by the above theorem. The Galois groups of irreducible second order equations with type $\mathcal{L}_1^2$ decomposition are discussed next. As far as possible, the conditions for the various types of Galois groups are expressed in terms of solutions of symmetric powers.

THEOREM 2.8 (Singer and Ulmer 1993, Ulmer and Weil 1996) Let $L(y) \equiv y'' + ry = 0$ be an irreducible second order ode with $r \in \mathbb{Q}(x)$. The following Galois groups $\mathcal{G}(L)$ have to be distinguished; s_k is the number of independent nontrivial rational solutions of $L^{\circledS k} = 0$.

i) $\mathcal{G}(L)$ is conjugate to D if and only if there holds $s_k = 0$ for $k \neq 0 \mod(4)$ and $s_k = 1$ for $k = 0 \mod(4)$ for all k. It is conjugate to D_m if and only if for $k < 2m$ the same conditions hold as in the previous case, $s_{2m} = 1$ for m even and $s_{2m+2} = 2$ for m odd.

ii) $\mathcal{G}(L)$ is conjugate to the tetrahedral group $A_4^{SL_2}$ if and only if $s_k = 0$ for $k \leq 5$, $s_6 = 1$.

iii) $\mathcal{G}(L)$ is conjugate to the octahedral group $S_4^{SL_2}$ if and only if $s_k = 0$ for $k \leq 7$, $s_8 = 1$.

iv) $\mathcal{G}(L)$ is conjugate to the icosahedral group $A_5^{SL_2}$ if and only if $s_k = 0$ for $k \leq 11$, $s_6 = 12$.

v) $\mathcal{G}(L)$ is the full group $SL_2(\mathbb{C})$ if none of the preceding cases holds.

These criteria allow determining a finite Galois group of an irreducible second order equation algorithmically. For the imprimitive groups D and D_m of case i) the same remarks on page 36 apply as for the groups of reducible equations covered by Theorem 2.7.

In order to obtain algebraic solutions of a second order equation explicitily, the following result of Ulmer and Weil [182] is useful.

THEOREM 2.9   (Ulmer and Weil 1996) Let $L(y) = y'' + r(x)y = 0$ be an irreducible second order ode with $r \in k$, k some differential field, and let $u \equiv \frac{y'}{y}$. Then the zeroes of $P(u) = u^m + \sum_{i=0}^{m-1} b_i u^i$ with $b_i \in k$ are solutions of the Riccati equation $u' + u^2 + r = 0$ if and only if the coefficient b_{m-1} is the negative logarithmic derivative of an exponential solution of $L^{\circledS m} = 0$. The coefficients b_i of $P(u)$ are determined from the system

$$b_m = 1, \quad b_{i-1} = \frac{1}{m+1-i}\left[(i+1)rb_{i+1} - b_i' + b_{m-1}b_i\right]$$

for $0 < i \leq m-1$. In particular for $m = 2$ there holds $b_0 = r - \frac{1}{2}(b_1' - b_1^2)$.

The proof may be found in the article by Ulmer and Weil [182]. The special case $m = 2$ is considered in Exercise 2.8.

EXAMPLE 2.26   The equation considered in Example 2.22 is irreducible whereas its second symmetric power is reducible. Consequently, its Galois group is imprimitive. The right factor $z' - \frac{1}{2x}z = 0$ of the second symmetric power yields the solution $z = \sqrt{x}$ from which $b_1 = \frac{1}{2x}$ and $b_0 = \frac{1}{16x^2} - \frac{1}{x}$ are obtained, i.e., $P(u) = u^2 - \frac{1}{2x}u + \frac{1}{16x^2} - \frac{1}{x}$ with solutions $u_{1,2} = \frac{1 \pm 4\sqrt{x}}{4x}$. They yield the fundamental system $y_{1,2} = x^{1/4}e^{\pm 2\sqrt{x}}$.   □

Linear Differential Equations 39

EXAMPLE 2.27   Equation 2.290 from Kamke's collection

$$y'' + \frac{27x}{27x^2 + 4}y' - \frac{3}{27x^2 + 4}y = 0$$

is irreducible with rational normal form $\bar{y}'' + \frac{1}{4}\frac{405x^2 - 264}{(27x^2 + 4)^2}\bar{y} = 0$. Its second symmetric power is reducible, its type $\mathcal{L}_{10}^3$ Loewy decomposition has a first order right factor $z' - \frac{27x}{27x^2 + 4}z = 0$. Consequently, its Galois group is imprimitive. By Theorem 2.9

$$P(u) = u^2 + \frac{27x}{27x^2 + 4}u + \frac{3}{4}\frac{135x^2 - 16}{27x^2 + 4}$$

is obtained. Its zeros $u_{1,2} = \frac{1}{2}(x^2 + \frac{4}{27})^{-1}\left(\frac{3}{2}x \pm \sqrt{x^2 + \frac{4}{27}}\right)$ yield the solutions $\bar{y}_{1,2} = (x^2 + \frac{4}{27})^{1/4}\left(x + \sqrt{x^2 + \frac{4}{27}}\right)^{\pm 1/3}$ of the equation in rational normal form, and finally the solutions $y_{1,2} = \left(x + \sqrt{x^2 + \frac{4}{27}}\right)^{\pm 1/3}$ of the original equation are obtained. □

EXAMPLE 2.28   Airy's equation $y'' - xy = 0$ is discussed by Singer and Ulmer [172]. Its 6th symmetric power

$$y^{(7)} - 56xy^{(5)} - 140y^{(4)} + 784x^2y''' + 2352y'' - 4(576x^3 - 295)y' - 3456x^2y = 0$$

is irreducible as shown by these authors. By Theorem 2.8, case v) the Galois group of Airy's equation is $SL_2(\mathbb{C})$. □

At this point the discussion of linear ode's is complete. For a second order equation a nontrivial Loewy decomposition provides the complete answer for its solutions in terms of integrations. This implies that the Galois group of the equation is a proper subgroup of the respective linear group. If the equation is irreducible there may be a finite Galois group leading to algebraic solutions. If the Galois group is $SL_2(\mathbb{C})$, there may be some interesting information on the solutions in terms of transcendents that have been introduced in the literature in terms of so-called *special functions* like e. g. Bessel functions, Legendre functions etc. To recognize them amounts to solving an equivalence problem which is considered in Section 4.1 on page 170. If none of these alternatives applies, its solutions have to be investigated numerically or by graphical methods or, if it is interesting enough for large classes of problems, new transcendents may be introduced. In principle these remarks may be generalized to linear ode's of any order.

The collection of solved ode's by Kamke [87] contains more than 400 linear equations of second order in Chapter C.2, and about 80 linear equation of third order in Chapter C.3. Most of them are listed in Appendix D of this book, and their Loewy decomposition is given if it is nontrivial, i.e., if it is not of type $\mathcal{L}_1^2$ or $\mathcal{L}_1^3$.

2.2 Janet's Algorithm

The theory of systems of linear homogeneous partial differential equations (pde's) is of interest in its own right, independent of its applications for finding symmetries and invariants of differential equations. Any such system may be transformed in infinitely many ways by adding linear combinations of its members or derivatives thereof without changing its solution space. In general it is a difficult question whether there exist nontrivial solutions at all, or what the degree of arbitrariness of the general solution is. It may consist of a finite set of constants, in which case there is a finite dimensional solution space over the constants. This is true for the systems considered in this book originating from symmetry analysis of ode's. On the other hand, the general solution may involve one or more functions depending on various numbers of arguments. The largest of these numbers is called the *differential type*, the number of functions depending on them is called the *typical differential dimension* by Kolchin [91], combined they have been baptized the *gauge* [62].

These questions were the starting point for Maurice Janet. He introduced a unique representation for these systems nowadays known as a *Janet basis*, similar to a Gröbner basis representation of a system of algebraic equations. There are several advantages to using a Janet basis representation for a system of linear homogeneous pde's. The degree of arbitrariness of its general solution, i.e., its gauge, is easily obtained from it. Due to the uniqueness of the coefficients of a Janet basis, additional invariants may be obtained providing further information on the solutions.

Both Gröbner bases and Janet bases have an important feature in common. Except in very simple cases it is virtually impossible to calculate any of them by pencil-and-paper, i.e., an efficient computer algebra implementation is crucial for handling concrete problems. A first example will make some of these points clear.

EXAMPLE 2.29   Consider the following linear homogeneous systems for two differential indeterminates z and w depending on x and y. The first system is defined by $f_i = 0$, $i = 1, \ldots, 5$ where

$$f_1 \equiv w_y + \frac{x}{2y(x^2 + y)} z_y - \frac{1}{y} w, \quad f_2 \equiv z_{xy} + \frac{y}{x} w_y + \frac{2y}{x} z_x,$$

$$f_3 \equiv w_{xy} - \frac{2x}{y} z_{xx} - \frac{x}{y^2} w_x,$$

$$f_4 \equiv w_{xy} + z_{xy} + \frac{1}{2y} w_y - \frac{1}{y} w_x + \frac{x}{y} z_y - \frac{1}{2y^2} w,$$

$$f_5 \equiv w_{yy} + z_{xy} - \frac{1}{y} w_y + \frac{1}{y^2} w.$$

(2.24)

Linear Differential Equations 41

And the second one is $g_i = 0$, $i = 1, \ldots, 4$ where

$$g_1 \equiv z_{yy} + \frac{1}{2y} z_y, \quad g_2 \equiv w_{xx} + 4y^2 w_y - 8y^2 z_x - 8yw,$$

$$g_3 \equiv w_{xy} - \frac{1}{2} z_{xx} - \frac{1}{2y} w_x - 6y^2 z_y, \tag{2.25}$$

$$g_4 \equiv w_{yy} - 2z_{xy} - \frac{1}{2y} w_y + \frac{1}{2y^2} w.$$

A priori, there does not seem to be any connection between systems (2.24) and (2.25). In particular it is not obvious whether there is any relation between the space of solutions they describe. The answer is easily obtained, however, if they are written as

$$f_1 = \frac{x}{2y(x^2 + y)} e_2 + e_4, \quad f_2 = \frac{\partial}{\partial x} e_2 + \frac{y}{x} e_4 + \frac{2y}{x} e_1,$$

$$f_3 = \frac{\partial}{\partial y} e_3 - \frac{2x}{y} \frac{\partial}{\partial x} e_1, \tag{2.26}$$

$$f_4 = \frac{\partial}{\partial x} e_2 + \frac{\partial}{\partial y} e_3 + \frac{1}{2y} e_4 - \frac{1}{y} e_3 + \frac{x}{y} e_2, \quad f_5 = \frac{\partial}{\partial x} e_2 + \frac{\partial}{\partial y} e_4$$

and

$$g_1 = \frac{\partial}{\partial y} e_2 + \frac{1}{2y} e_2, \quad g_2 = \frac{\partial}{\partial x} e_3 + 4y^2 e_4 - 8y^2 e_1,$$

$$g_3 = \frac{\partial}{\partial y} e_3 - \frac{1}{2} \frac{\partial}{\partial x} e_1 - 6y^2 e_2 - \frac{1}{4y} e_3, \tag{2.27}$$

$$g_4 = \frac{\partial}{\partial y} e_4 - 2 \frac{\partial}{\partial x} e_2 + \frac{1}{2y} e_4$$

where

$$e_1 \equiv z_x + \frac{1}{2y} w, \quad e_2 \equiv z_y, \quad e_3 \equiv w_x, \quad e_4 \equiv w_y - \frac{1}{y} w. \tag{2.28}$$

From this representation it is clear that both systems (2.24) and (2.25) allow all solutions of the system $e_1 = 0$, $e_2 = 0$, $e_3 = 0$, $e_4 = 0$. Explicitly, they are $z = C_1 + C_2 x$ and $w = -2C_2 y$, where C_1 and C_2 are the integration constants, i.e., they form a two-dimensional vector space over the constants. It will be seen later on that there are no more solutions of (2.24) and (2.25). ☐

Modules over Rings of Differential Operators. From the above example it is apparent that the object of primary interest is not an individual system of pde's, but the module that is generated from it over an appropriate ring of differential operators; they are called $\mathcal{D}$-modules. This situation is similar to commutative algebra where the ideals generated by sets of polynomials are closely related to the algebraic manifolds described by the corresponding set of algebraic equations. It turns out that the language of differential algebra applied in the books by Kaplansky [88] and Kolchin [91] is the proper setting for this discussion. Good introductions to $\mathcal{D}$-modules are the articles by Oaku and Shimoyama [139] and Quadrat [150]. A few basic results from these references will be given next.

An operator δ on a ring $\mathcal{R}$ is called a *derivation operator* if $\delta(a + b) = \delta(a) + \delta(b)$ and $\delta(ab) = \delta(a)b + a\delta(b)$ for all elements $a, b \in \mathcal{R}$. A ring with a single derivation operator is called an *ordinary differential ring*. If there is a finite set Δ of derivation operators such that $\delta\delta'(a) = \delta'\delta(a)$ for $a \in \mathcal{R}$, $\delta, \delta' \in \Delta$ the ring is called a *partial differential ring*. Analogous definitions apply for an *ordinary differential field* and a *partial differential field* if the ring $\mathcal{R}$ bis replaced by a field $\mathcal{F}$.

Let Θ denote the commutative semigroup generated by the elements of Δ. If it is written multiplicatively, its elements can be expressed in the form $\theta = \prod_{\delta \in \Delta} \delta^{e(\delta)}$, where each exponent $e(\delta)$ is a natural number; $e(\delta) = 0$ defines the identity. The sum $\sum_{\delta \in \Delta} e(\delta)$ is called the *order* of θ. The elements of Θ are called *derivative operators* or simply *derivatives*.

In this book most frequently rings of differential operators with derivation operators $\partial_i = \dfrac{\partial}{\partial x_i}$ and rational function coefficients will be applied, they are denoted by $\mathcal{D} = \mathbb{Q}(x_1, \ldots, x_n)[\partial_1, \ldots, \partial_n]$. Its elements have the form

$$d = \sum r_{i_1,\ldots,i_n}(x_1, \ldots, x_n)\partial_1^{i_1} \ldots \partial_n^{i_n} \tag{2.29}$$

where $r_{i_1,\ldots,i_n} \in \mathbb{Q}(x_1, \ldots, x_n)$, almost all $r_{i_1,\ldots,i_n}$ equal to zero. The ring $\mathcal{D}$ is noncommutative with commutation rules $\partial_i a = a\partial_i + \dfrac{\partial a}{\partial x_i}$ for $1 \le i \le n$ and $a \in \mathbb{Q}(x_1, \ldots, x_n)$.

An m-dimensional left vector module $\mathcal{D}^m$ over $\mathcal{D}$ has elements $(d_1, \ldots, d_m)$ with $d_i \in \mathcal{D}$ for all i. The sum of two elements of $\mathcal{D}^m$ is defined by componentwise addition, multiplication with ring elements d by $d(d_1, \ldots, d_m) = (dd_1, \ldots, dd_m)$.

The relation between the submodules of $\mathcal{D}^m$ and systems of linear pde's is established as follows. Let $(u^1, \ldots, u^m)^T$ be an m-dimensional column vector of differential indeterminates. Then the product

$$(d_1, \ldots, d_m)(u^1, \ldots, u^m)^T = d_1 u^1 + d_2 u^2 + \ldots + d_m u^m \tag{2.30}$$

defines a *linear differential polynomial* in the u's that may be considered as the left hand side of a partial differential equation; $u^1, \ldots, u^m$ are called the *dependent variables* or *functions*, depending on the *independent variables* $x_1, \ldots, x_n$.

A $N \times m$ matrix $\{c_{i,j}\}$, $i = 1, \ldots, N$, $j = 1, \ldots, m$, $c_{i,j} \in \mathcal{D}$, defines a system of N linear homogeneous pde's

$$c_{i,1} u^1 + \ldots + c_{i,m} u^m = 0, \quad i = 1, \ldots, N. \tag{2.31}$$

The ith equation of (2.31) corresponds to the vector

$$(c_{i,1}, c_{i,2}, \ldots, c_{i,m}) \in \mathcal{D}^m \text{ for } i = 1, \ldots, N. \tag{2.32}$$

This correspondence between the elements of $\mathcal{D}^m$, the differential polynomials (2.30) and its associated pde's(2.31) allows turning from one representation to

Linear Differential Equations

the other whenever it is appropriate. The module generated by these vectors determines the solutions of the system.

EXAMPLE 2.30   In Example 2.29 there holds $n = m = 2$. The elements

$$\left(\partial_y - \frac{1}{y}, \frac{x}{2y(x^2 + y)} \partial_y \right), \quad \left(\frac{y}{x} \partial_y, \partial_{xy} + \frac{2y}{x} \partial_x \right), \quad \left(\partial_{xy} - \frac{x}{y^2} \partial_x, \frac{2x}{y} \partial_x^2 \right),$$

$$\left(\partial_{xy} + \frac{1}{2y} \partial_y - \frac{1}{y} \partial_x - \frac{1}{2y^2}, \partial_{xy} + \frac{x}{y} \partial_y \right), \quad \left(\partial_y^2 - \frac{1}{y} \partial_y + \frac{1}{y^2}, \partial_{xy} \right)$$

generate the submodule of $\mathcal{D}^2$ corresponding to system (2.24). The same submodule is generated by

$$\left(\frac{1}{2y}, \partial_x \right), \quad (0, \partial_y), \quad (\partial_x, 0), \quad \left(\partial_y - \frac{1}{y}, 0 \right)$$

corresponding to system (2.28). The respective equations are obtained by applying the module elements to $\binom{w}{z}$ with w, z differential indeterminates with nonvanishing derivatives w.r.t. x and y. ⬜

Given any module $\mathcal{M} \subset \mathcal{D}^m$, there are two fundamental problems associated with it.

▷ *Membership in* $\mathcal{M}$. Establish a procedure for deciding membership in $\mathcal{M}$. This problem will be solved by constructing a Janet basis for $\mathcal{M}$.

▷ *Structure of* $\mathcal{M}$. Determine all maximal submodules $\mathcal{N}$ with the property $\mathcal{M} \subset \mathcal{N} \subseteq \mathcal{D}^m$.

Answering these questions will be the subject of the remaining part of this section.

Term Orders and Rankings.   Just like an ordering of monomials is a fundamental prerequisite for defining a Gröbner basis in commutative algebra, an ordering of partial derivatives is necessary for defining canonical forms for differential operators and differential equations. At first it will be given in the language of differential polynomials and pde's.

DEFINITION 2.3   *(Ranking) A ranking of derivatives is a total ordering such that for any derivatives δ, δ_1 and δ_2, and any derivation operator θ there holds $\delta \leq \theta\delta$, $\delta_1 \leq \delta_2 \Rightarrow \delta\delta_1 \leq \delta\delta_2$.*

A *term* $t_{i,j}$ in a differential polynomial (2.31) is a differential indeterminate u^α or a derivative of it, multiplied by an element of the base field. The left hand side of a linear homogeneous pde is a sum of terms. For such a pde, the ordering of partial derivatives defined above generates a unique order for its terms. A system of linear homogeneous pde's with terms $t_{i,j}$ is always arranged in the form

$$\boxed{t_{1,1}} > t_{1,2} > \ldots > t_{1,k_1}$$
$$\wedge$$
$$\boxed{t_{2,1}} > t_{2,2} > \ldots > t_{2,k_2}$$
$$\wedge$$
$$\vdots$$
$$\wedge$$
$$\boxed{t_{N,1}} > t_{N,2} > \ldots > t_{N,k_N}$$

such that the ordering relations are valid as indicated. Each line of this scheme corresponds to a differential polynomial or a differential equation of the system, its terms are arranged in decreasing order from left to right. In order to save space, sometimes several equations are arranged into a single line. In these cases, in any line the leading terms *increase* from left to right. The terms in the rectangular boxes are the *leading terms* containing the *leading derivative*, i.e., the leading term is that term preceding any other term in the respective differential polynomial or equation w.r.t. the given ordering. For any given system of pde's there is a finite number of term orderings leading to different arrangements of terms, a trivial upper bound being the number of permutations of its terms. In any term ordering there is a *lowest term* which is always one of the dependent variables corresponding to $e(\delta) = 0$. By Dickson's lemma [33], Chapter 2, $ 4, any decreasing sequence of derivatives must terminate.

In order to generate the above mentioned form of a system of pde's effectively it is advantageous to have a predicate to decide whether or not any pair of derivatives is in correct order. To this end, a *matrix of weights (matrice de côtes)* has been introduced by Riquier [153]. The subsequent discussion follows closely his original work and a more complete and systematic treatment of the subject by Thomas [176].

First of all a fixed enumeration $(u^1, \ldots, u^m)$ is chosen for the dependent and $(x_1, \ldots, x_n)$ for the independent variables. To any derivative of a function u^α a matrix I of weights is attached comprising a single column as follows.

$$\frac{\partial u^\alpha}{\partial x_1^{i_1} \partial x_2^{i_2} \ldots \partial x_n^{i_n}} \longleftrightarrow I \equiv (i_1, i_2, \ldots, i_n, 0, \ldots, 1, \ldots, 0)^T. \qquad (2.33)$$

The position of the element 1 is the $(n + \alpha)^{th}$ row. The function u^α itself corresponds to a matrix with $i_1 = i_2 = \ldots = i_n = 0$. For a partial derivative (2.33), $(i_1, \ldots, i_n)$ is called the *derivative vector*. Let J be the matrix of weights associated with some other derivative of a function u^β. Riquier's method of placing these derivatives in a definite order is based on the following rules. Let there be given a matrix M with $n + m$ columns corresponding to the independent and the dependent variables, and any number of rows. By definition the derivative represented by I is higher than the one represented

Linear Differential Equations 45

by J if the first nonzero element in the column $M(I-J)$ is positive, and vice versa if it is negative.

The particular form of the matrix M determines the type of ordering. Thomas [153] discussed the possible selections in full generality. He described the most general matrix leading to a given ordering and gave a canonical form for a single representative. For the applications in this book, basically three types of orderings are applied, a *lexicographic ordering*, a *graded lexicographic ordering* and a *graded reverse lexicographic ordering* abbreviated by *lex*, *grlex* and *grevlex* respectively.

In the lexicographic ordering, derivatives are ordered first by functions such that $u_i > u_j$ if $i < j$ in the arrangement $(u^1, \ldots, u^m)$. If the functions contained in two derivatives are identical to each other, the order of its partial derivatives is determined like in the lexicographic term order of two algebraic monomials, i.e., $x_i > x_j$ if $i < j$ in the arrangement $(x_1, \ldots, x_n)$. The power of an individual variable in the algebraic case corresponds to the order of the derivative w.r.t. this variable. The ordering according to these rules is achieved by the matrix

$$M_{lex} = \begin{pmatrix} 0\,0 \ldots 0 & m & m-1 & \ldots & 1 \\ 1\,0 \ldots 0\,0 & 0 & & \ldots & 0 \\ 0\,1 \ldots 0\,0 & 0 & & \ldots & 0 \\ \vdots\;\vdots & & \vdots & \ldots & \\ 0\,0 \ldots 1\,0 & 0 & & \ldots & 0 \end{pmatrix}$$

whereupon the lower left $n \times n$ corner is the n-dimensional unit matrix.

The graded lexicographic ordering *grlex* is obtained if the total orders of the two derivatives are compared first. If they are different from each other, the higher one precedes the other. If not, the above *lex* order is applied. The corresponding matrix M_{grlex} is obtained from M_{lex} if the first line $\{1\,1\,\ldots 1\,0\,0 \ldots 0\}$ is included, i.e., if it has the form

$$M_{grlex} = \begin{pmatrix} 1\,1 \ldots 1\,0 & 0 & & \ldots & 0 \\ 0\,0 \ldots 0 & m & m-1 & \ldots & 1 \\ 1\,0 \ldots 0\,0 & 0 & & \ldots & 0 \\ 0\,1 \ldots 0\,0 & 0 & & \ldots & 0 \\ \vdots\;\vdots & & \vdots & \ldots & \\ 0\,0 \ldots 1\,0 & 0 & & \ldots & 0 \end{pmatrix}.$$

Finally, the graded reverse lexicographic ordering *grevlex* at first compares the total order of the derivatives like in the *grlex* ordering. If they are equal to each other, they are ordered by the functions. If they are identical, derivative orders w.r.t. individual variables are compared, beginning with the last variable. The first pair with a different value decides the order, the derivative with the lower value is higher than the other one. The following matrix generates this ordering.

$$M_{grevlex} = \begin{pmatrix} 1 & 1 & \ldots & 1 & 0 & 0 & \ldots & 0 \\ 0 & 0 & \ldots & 0 & m & m-1 & \ldots & 1 \\ 0 & 0 & \ldots & -1 & 0 & 0 & \ldots & 0 \\ & \vdots & \vdots & & & \vdots & \ldots & \\ 0 & -1 & \ldots & 0 & 0 & 0 & \ldots & 0 \\ -1 & 0 & \ldots & 1 & 0 & 0 & \ldots & 0 \end{pmatrix}.$$

There are numerous variations of these orderings in addition to the obvious permutations of dependent and independent variables among themselves. In any lexicographic ordering, for example, the independent variables may be compared ahead of the dependent ones, or a combination of *lex* and *grlex* orderings may be applied.

The generation of orderings by weight matrices provides an easy algorithmic way for establishing it for any number of derivatives. The following algorithm *Higher?* establishes a reflexive ordering of partial derivatives if a matrix of weights is given. Because in a linear homogeneous system each term contains a unique derivative, it may also be applied for ordering the terms of its members.

ALGORITHM 2.3   *Higher?*(d_1, d_2, M). Given two derivatives d_1 and d_2, and a matrix M of weights, *true* is returned if d_2 does not precede d_1.

$S1$: *Equality?* If $d_1 = d_2$ return *true*.

$S2$: *Generate cotes.* Set up matrices I_1 and I_2 for d_1 and d_2 according to (2.33).

$S3$: *Apply M.* Generate the column $M(I_1 - I_2)$. If the first nonvanishing element is positive, return *true* and *false* otherwise.

EXAMPLE 2.31   Consider a problem with $m = n = 2$, comprising two functions (w, z) depending on (y, x). For the lexicographic orderings $w > z$, $y > x$ and derivatives not higher than two, (2.33) yields

$$\begin{matrix} z & z_x & z_y & z_{xx} & z_{xy} & z_{yy} \end{matrix}$$
$$\begin{pmatrix} 0 \\ 0 \\ 0 \\ 1 \end{pmatrix} \begin{pmatrix} 1 \\ 0 \\ 0 \\ 1 \end{pmatrix} \begin{pmatrix} 0 \\ 1 \\ 0 \\ 1 \end{pmatrix} \begin{pmatrix} 2 \\ 0 \\ 0 \\ 1 \end{pmatrix} \begin{pmatrix} 1 \\ 1 \\ 0 \\ 1 \end{pmatrix} \begin{pmatrix} 0 \\ 2 \\ 0 \\ 1 \end{pmatrix}$$

for the derivatives of z. The corresponding vectors for w are obtained if the last two elements $0, 1$ in each column are replaced by $1, 0$. Then the matrix

$$M = \begin{pmatrix} 1 & 1 & 0 & 0 \\ 0 & 0 & 2 & 1 \\ 1 & 0 & 0 & 0 \\ 0 & 1 & 0 & 0 \end{pmatrix}$$

generates the graded lexicographic ordering

$$w_{yy} > w_{xy} > w_{xx} > z_{yy} > z_{xy} > z_{xx} > w_y > w_x > z_y > z_x > w > z.$$

The matrix

$$M = \begin{pmatrix} 0\,0\,2\,1 \\ 1\,0\,0\,0 \\ 0\,1\,0\,0 \end{pmatrix}$$

generates the ordering

$$w_{yy} > w_{xy} > w_{xx} > w_y > w_x > w > z_{yy} > z_{xy} > z_{xx} > z_y > z_x > z$$

which is a lexicographic ordering. ⬜

The orderings described so far may be uniquely translated into module orderings for elements of $\mathcal{D}^m$ if the transformation rules described in the preceding subsection on pages 41-43 are applied.

EXAMPLE 2.32   The orderings of the preceding example are reconsidered. Applying the above rules, for $w > z$, $y > x$, $w = (1,0)$ and $z = (0,1)$ the graded lexicographic ordering leads to

$$(\partial_{yy},0) > (\partial_{xy},0) > (\partial_{xx},0) > (0,\partial_{yy}) > (0,\partial_{xy}) > (0,\partial_{xx}) >$$
$$(\partial_y,0) > (\partial_x,0) > (0,\partial_y) > (0,\partial_x) > (1,0) > (0,1)$$

for the elements of $\mathcal{D}^2$ of order not higher than two. Similarly the lexicographic ordering yields

$$(\partial_{yy},0) > (\partial_{xy},0) > (\partial_{xx},0) > (\partial_y,0) > (\partial_x,0) > (1,0) >$$
$$(0,\partial_{yy}) > (0,\partial_{xy}) > (0,\partial_{xx}) > (0,\partial_y) > (0,\partial_x) > (0,1). \quad ⬜$$

The latter order considered in this example is the POT order which has been introduced by Adams and Loustaunau [3] for modules over polynomial rings. Their TOP order does not have an exact counterpart among the orders discussed above.

Reduction and Autoreduction. The first fundamental operation of Janet's algorithm is the *reduction*. Given any pair of linear differential polynomials e_1 and e_2, it may occur that the leading derivative of e_2, or of a suitable derivative ∂e_2 of it, equals some derivative in e_1. This coincidence may be applied to remove the respective term from e_1 by an operation which is called a *reduction step*. To this end, multiply the appropriate derivative ∂e_2 by the coefficient of its leading derivative in the term containing it in e_1, and subtract it from e_1 multiplied by the leading coefficient of e_2. If e_2 is monic which is usually the case, this latter operation may be skipped, otherwise the result has to be made monic. In general, reductions may occur several times. However, due to the properties of the term orderings described above and the genuine lowering of terms in any reduction step, by Dickson's lemma the following algorithm always terminates after a finite number of iterations. After its completion no further reductions are possible. In this case e_1 is called *completely reduced*

48

with respect to e_2. For any linear differential polynomial e the abbreviations

$$Lder(e), \quad Lterm(e), \quad Lcoef(e), \quad Lfun(e)$$

are applied to denote the leading derivative, the leading term, the leading co-efficient or the leading function of e. $Coefficient(e,t)$ denotes the coefficient of the term t contained in e.

ALGORITHM 2.4 $Reduce(e_1, e_2)$. Given two monic linear differential poly-nomials e_1 and e_2, e_1 is returned completely reduced w.r.t. e_2.

$S1$: $Reduction\ possible?$ Search for a term t in e_1 the derivative of which may be obtained from the leading derivative of e_2 by applying some appropriate ∂. If none is found, return e_1.

$S2$: $Reduction\ step$. Set $e_1 := e_1 - Coefficient(e_1, t)\partial e_2$ and $e_1 := Monic\ e_1$, then goto $S1$.

EXAMPLE 2.33   Let $e_1 \equiv z_y - \frac{x^2}{y^2} z_x - \frac{x-y}{y^2} z$, $f_1 \equiv z_x + \frac{1}{x} z$ and $f_2 \equiv z_y + \frac{1}{y} z$. Reduction of e_1 w.r.t. f_1 yields in a single step $z_y + \frac{1}{y} z$, whereas reduction w.r.t. f_2 yields $-\frac{x^2}{y^2} z_x - \frac{x}{y^2} z$.   ☐

For some applications, a slightly more general algorithm for reduction is desirable, reducing a given equation w.r.t. to a system of pde's instead of a single one.

ALGORITHM 2.5 $Reduce(e, S)$. Given a linear differential polynomial e and a system of linear monic polynomials $S = \{f_1, f_2, \ldots\}$, e is returned reduced w.r.t. S.

$S1$: $Reduction\ possible?$ Search for a polynomial $f_i \in S$ w.r.t. which e may be reduced. If none is found, return e.

$S2$: $Reduction\ performed$. Set $e := Reduce(e, f_i)$ and goto to $S1$.

It is obvious that its action is achieved by repeated application of the preceding algorithm $Reduce$. Termination follows from the finiteness of the system S and the termination of the former algorithm. If a differential polynomial has been reduced w.r.t. to a given system, it is called in $normal\ form$ w.r.t. to this system. In general the result is $not\ unique$.

EXAMPLE 2.34   Let e_1, f_1 and f_2 be defined as in Example 2.33. If now e_1 is reduced w.r.t. to the system $\{f_1, f_2\}$, the result is zero, independent of the order in which the reductions are performed.   ☐

EXAMPLE 2.35   Consider f_4 in (2.24) and its reduction w.r.t. the system comprising f_1 and f_2. Reduction w.r.t. to f_1 is achieved in two steps. At first the leading derivative w_{xy} is removed by subtraction of $\frac{\partial f_1}{\partial x}$. After division

Linear Differential Equations 49

by the leading coefficient the result is

$$f_{4a} \equiv z_{xy} + \frac{x^2 + y}{2x^2 y - x + 2y^2} w_y$$

$$+ \frac{2x^5 + 4x^3 y + x^2 + 2xy^2 - y}{(2x^2 y - x + 2y^2)(x^2 + y)} z_y - \frac{x^2 + y}{(2x^2 y - x + 2y^2)y} w.$$

In the second step the term proportional to w_y is removed by subtraction of a proper multiple of f_1. This leads to

$$f_{4b} \equiv z_{xy} + \frac{4x^5 y + 8x^3 y^2 - x^3 + 2x^2 y + 4xy^3 - xy - 2y^2}{(2x^2 y - x + 2y^2)(x^2 + y)y} z_y.$$

The result f_{4b} may be further reduced w.r.t. to f_2. This yields

$$f_{4c} \equiv w_y - \frac{4x^6 y + 8x^4 y^2 - x^4 + 2x^3 y + 4x^2 y^3 - x^2 y - 2xy^2}{(2x^2 y - x + 2y^2)(x^2 + y)y^2} z_y + 2z_x.$$

Now there is again a reduction possible w.r.t. f_1, which yields the answer

$$f_{4d} \equiv z_y - \frac{4(2x^2 y - x + 2y^2)(x^2 + y)y^2}{(4x^5 y + 8x^3 y^2 - x^3 + 2x^2 y^2 + 2x^2 y + 4xy^3 - 2xy + 2y^3 - 2y^2)x} z_x$$

$$- \frac{(2x^2 y - x + 2y^2)(x^2 + y)y}{(4x^5 y + 8x^3 y^2 - x^3 + 2x^2 y^2 + 2x^2 y + 4xy^3 - 2xy + 2y^3 - 2y^2)x} w = 0.$$

At this point f_{4d} cannot be further reduced by any other equation. ⬚

For any given system, in general it is possible that some elements may be reduced w.r.t. the remaining ones. Janet's algorithm requires that at first all such reductions be performed. A system with the property that there is no reduction possible of any element w.r.t. to any other in the system is called an *autoreduced system*.

The following algorithm accepts any ordered system of linear differential polynomials as input and returns the corresponding autoreduced system by calling the algorithm $Reduce(e_i, e_j)$ repeatedly. These calls are organized as follows. The lowest element of the given system which may be applied for any reduction is selected and all reductions w.r.t. this element are performed. The resulting system is treated in the same way until no further reduction is possible in the full system, and the result is returned.

ALGORITHM 2.6 *AutoReduce(S)*. Given an ordered system of linear monic differential polynomials $S = \{e_1, e_2, \ldots\}$, a new system is returned, such that each element is completely reduced w.r.t. to any other member of the system.

$S1$: *Reduction possible?* Find the first element e_k in S w.r.t. which any reduction may be performed in $\{e_{k+1}, e_{k+2} \ldots\}$. If none is found, return the system S unchanged.

50

$S2$: *Perform reductions.* Set
$$S := \{e_1, e_2, \ldots e_k\} \cup \{Reduce(e_j, e_k) | j = k + 1, k + 2, \ldots\},$$
reorder S properly and return to $S1$.

Termination is assured due to the finiteness of the system and the fact that any call to *Reduce* in step $S2$ generates a result with at least a single term replaced by a lower one.

EXAMPLE 2.36   Again system (2.24) is considered. In the first iteration of *AutoReduce* the lowest element f_1 is applied for removing all partial derivatives of w w.r.t. y in the remaining elements with the result

$$w_y + \frac{x}{2x^2y + 2y^2} z_y - \frac{1}{y}w, \quad z_{xy} - \frac{1}{2x^2 + 2y} z_y + \frac{2y}{x} z_x + \frac{1}{x}w,$$

$$z_{xy} + \frac{4x^5y + 8x^3y^2 - x^3 + 2x^2y + 4xy^3 - xy - 2y^2}{4x^4y^2 - 2x^3y + 8x^2y^3 - 2xy^2 + 4y^4} z_y,$$

$$z_{xy} + (4x^2 + 4y)z_{xx} + \frac{2x^3 - 2x^2y + 2xy - 2y^2}{xy} w_x - \frac{x^2 - y}{x^3 + xy} z_y,$$

$$z_{yy} - \frac{2x^2y + 2y^2}{x} z_{xy} - \frac{x^2 + 2y}{x^2y + y^2} z_y.$$

The derivatives of z_{xy} are removed by means of the second element next.

$$w_y + \frac{x}{2x^2y + 2y^2} z_y - \frac{1}{y}w,$$

$$z_y - \frac{8x^4y^3 - 4x^3y^2 + 16x^2y^4 - 4xy^3 + 8y^5}{4x^6y + 8x^4y^2 - x^4 + 2x^3y^2 + 2x^3y + 4x^2y^3 - 2x^2y + 2xy^3 - 2xy^2} z_x$$

$$- \frac{4x^4y^2 - 2x^3y + 8x^2y^3 - 2xy^2 + 4y^4}{4x^6y + 8x^4y^2 - x^4 + 2x^3y^2 + 2x^3y + 4x^2y^3 - 2x^2y + 2xy^3 - 2xy^2} w,$$

$$z_{xx} + \frac{x - y}{2xy} w_x - \frac{2x^2 - x - 2y}{8x^5 + 16x^3y + 8xy^2} z_y - \frac{y}{2x^3 + 2xy} z_x - \frac{1}{4x^3 + 4xy}w,$$

$$z_{xy} - \frac{1}{2x^2 + 2y} z_y + \frac{2y}{x} z_x + \frac{1}{x}w,$$

$$z_{yy} - \frac{x^3 + x^2y^2 + 2xy + y^3}{x^3y + xy^2} z_y + \frac{4x^2y^2 + 4y^3}{x^2} z_x + \frac{2x^2y + 2y^2}{x^2}w.$$

The next six iterations generate systems that are too large to be reproduced here. After one more iteration, the lowest element becomes $z_x + \frac{1}{2y}w$, and an additional iteration generates the final result

$$z_x + \frac{1}{2y}w, \quad z_y, \quad w_x, \quad w_y + \frac{1}{y}w. \qquad \square$$

Completion and Integrability Conditions. Although autoreduction already removes a lot of arbitrariness from any sytem of differential polynomials, there are still some shortcomings which disqualify them from being considered as canonical form. In the first place, the outcome of autoreduction is highly non-unique as the result of Example 2.36 and its comparison with the autoreduced system (2.25) shows. Furthermore, differentiating in (2.25) g_2 w.r.t. y and g_3 w.r.t. x, and reducing the result w.r.t. to the remaining elements of the system, the following result is obtained.

$$\frac{\partial g_2}{\partial y} \equiv w_{xxy} + 2yw_y - 16yz_x - 10w,$$

$$\frac{\partial g_3}{\partial y} \equiv w_{xxy} - \frac{1}{2}z_{xxx} - 6y^2 z_{xy} - \frac{1}{4y}z_{xx} - \frac{1}{4y^2}w_x - 3yz_y.$$

Either of them may be considered as a representation of the leading derivative w_{xxy} in terms of its reductum which do *not coincide*. It is obvious that infinitely many instances like this may be generated by suitable differentiations. These apparent inconsistencies already hold the clue on how to proceed further, i.e., an autoreduced system must be supplemented by additional equations such that these inconsistencies disappear.

In order to obtain a finite problem, according to Janet [83] certain elements which make it into a *complete system* have to be included. To this end, at first the full system of pde's is partitioned into subsystems with the same function in the leading derivative. By definition, a system of linear differential polynomials of this kind is called *complete* if the derivative vectors of the leading terms in any subsystem form a complete set in the sense that it is defined in Appendix C for degree vectors of monomials. Because differentiation corresponds to multiplication by the respective variable, the subsequently described completion algorithm is a straightforward generalization of the completion process for systems of monomials that is described in Appendix C.

ALGORITHM 2.7   *CompleteSystem(S)*. Given an ordered autoreduced system of linear differential polynomials S with leading derivative vectors D, the complete system corresponding to S is returned.

$S1$: *Separate*. Represent S in the form $S := S_1 \cup S_2 \cup \ldots \cup S_k$ such that the elements in each S_i contain the same leading function with leading derivatives D_i.

$S2$: *Completion*. For each i set $\bar{D}_i := Complete(D_i)$.

$S3$: *Generate systems*. For each S_i generate the elements corresponding to a $\bar{D}_i$ and assign it to $\bar{S}_i$.

$S4$: *Generate complete system*.   Collect the $\bar{S}_i$, assign them to $\bar{S} := \{\bar{S}_1, \bar{S}_2, \ldots \bar{S}_k\}$ and return $Reorder(\bar{S})$

Termination of this process is assured by the finiteness of the completion algorithm $Complete(u)$ for systems of monomials described in Appendix C.

Consider any linear homogeneous system, choose two elements with the same leading function and solve both with respect to its leading term. By suitable cross differentiation it is always possible to obtain two new polynomials such that the derivatives on the left hand sides are identical to each other. Intuitively it is expected that the same should hold true for the right hand sides if all reductions w.r.t. to the remaining elements of the system are performed. This consideration justifies the following definition.

DEFINITION 2.4   *(Integrability condition). Let e_1 and e_2 be two elements of a system of linear differential polynomials. Let ∂ and ∂' be derivatives of minimal order that make their leading derivatives identical to each other. Then the constraint*

$$Lcoef(\partial'e_2)\partial e_1 - Lcoef(\partial e_1)\partial' e_2 = 0$$

is called an integrability condition for the system S.

The integrability conditions correspond to the S-pairs in a Gröbner basis calculation. A priori there are infinitely many of them in any given system of pde's. It is one of the achievements of Janet to identify a *finite* number of them such that the remaining ones follow. This is the subject of the next theorem.

THEOREM 2.10   *(Janet 1920) Let $S = \{e_1, e_2, \ldots\}$ be an autoreduced complete system of linear differential polynomials. In order to satisfy all integrability conditions of this system it is sufficient to satisfy*

$$\frac{\partial e_i}{\partial x_k} - \frac{\partial^{m_1+\ldots+m_p} e_j}{\partial x_{i_1}^{m_1} \ldots \partial x_{i_p}^{m_p}} = 0 \tag{2.34}$$

provided that the leading derivatives of the two expressions at the left hand side are identical to each other, x_k is a nonmultiplier for e_i and $x_{i_1}, \ldots, x_{i_p}$ are multipliers for e_j.

PROOF   In a complete system the leading derivatives of a function parametrize the classes of all principal derivatives. Derivatives w.r.t. the multiplier variables reproduce the derivatives of each class. Therefore in order to obtain any principal derivative in two different ways, at least a single derivation w.r.t. to a nonmultiplier variable must be involved. Let x_k be a nonmultiplier variable for e_i. The leading derivatives of the derived element must be contained in some other class. Let the leading derivative of e_j parametrize this class. Therefore it must be possible by deriving it w.r.t. to its multiplier variables a suitable number of times to generate the same leading derivative as in $\frac{\partial e_i}{\partial x_k}$. Let this requirement be fulfilled for (2.34). Due to the uniqueness of the representation of principal derivatives, any other element satisfying this condition has the form

$$\frac{\partial^{n_1+\ldots+n_p} e_j}{\partial x_{i_1}^{n_1} \ldots \partial x_{i_p}^{n_p}} \left(\frac{\partial e_i}{\partial x_k} - \frac{\partial^{m_1+\ldots+m_p} e_j}{\partial x_{i_1}^{m_1} \ldots \partial x_{i_p}^{m_p}} \right) = 0.$$

The above condition (2.34) assures their vanishing.   ⬜

Linear Differential Equations   53

It may be said somewhat loosely that a system of pde's for which all integrability conditions are satisfied contains all elements *explicitly* that are essential for determining the general solution. This property is of such a fundamental importance that a special term is introduced for it.

DEFINITION 2.5   *(Janet basis). For a given term ordering an autoreduced system of linear differential polynomials is called a Janet basis if all integrability conditions reduce to zero.*

An autoreduced system satisfying all integrability conditions is also called *coherent.* The following algorithm that accepts any system of linear differential polynomials and transforms it into a new system such that all integrability conditions are satisfied is due to Janet [83].

ALGORITHM 2.8   *JanetBasis(S).* Given a system of linear differential polynomials $S = \{e_1, e_2, \ldots\}$, the Janet basis corresponding to S is returned.

$S1$: *Autoreduction.* Set $S := Autoreduce(S)$.

$S2$: *Completion.* Set $S := CompleteSystem(S)$.

$S3$: *Find integrability conditions.* Find all pairs of leading terms v_i of e_i and v_j of e_j such that differentiation w.r.t. a nonmultiplier x_{i_k} and multipliers $x_{j_1}, \ldots, x_{j_l}$ leads to $\dfrac{\partial v_i}{\partial x_{i_k}} = \dfrac{\partial^{p_1 + \ldots + p_l} v_j}{\partial x_{j_1}^{p_1} \ldots \partial x_{j_l}^{p_l}}$ and determine the integrability conditions

$$c_{i,j} = Lcoef(e_j) \cdot \frac{\partial e_i}{\partial x_{i_k}} - Lcoef(e_i) \cdot \frac{\partial^{p_1 + \ldots + p_l} e_j}{\partial x_{j_1}^{p_1} \ldots \partial x_{j_l}^{p_l}}.$$

$S4$: *Reduce integrability conditions.* For all $c_{i,j}$ set
$$c_{i,j} := Reduce(c_{i,j}, S).$$

$S5$: *Termination?* If all $c_{i,j}$ are zero return S, otherwise make the assignment $S := S \cup \{c_{i,j} | c_{i,j} \neq 0\}$, reorder S properly and goto $S1$.

PROOF   Steps $S1$, $S2$ and $S4$ are finite due to the finiteness of the algorithms called there. Applying Theorem 2.10 to all subsystems corresponding to like leading functions leads to the integrability conditions in step $S3$. Upon reduction in step $S4$, the leading derivatives of any nonvanishing integrability condition are lower than any leading derivative in the current system. Therefore this process must terminate after a finite number of iterations.   ☐

The above definition of a Janet basis has the advantage of being constructive, i.e., it may be applied in straightforward manner to decide whether a given system of linear differential polynomials is a Janet basis and, if this is not the case, to transform it into one. The most important property of Janet bases is established in the next theorem. Sometimes this property of a Janet basis is used as the definition.

THEOREM 2.11   *Let J be a Janet basis generating a module $\mathcal{M} \subset \mathcal{D}^m$. Then an element $e \in \mathcal{D}^m$ is contained in $\mathcal{M}$ if and only if it may be reduced to zero by J.*

54

By means of this theorem it will be shown next that the result obtained from the algorithm *JanetBasis* is unique.

THEOREM 2.12   *For a given module $\mathcal{M}$ and a given term ordering there exists a unique Janet basis.*

PROOF   Assume for a given module $\mathcal{M}$ there are two different Janet bases $J = \{j_1, \ldots, j_k\}$ and $H = \{h_1, \ldots, h_l\}$. Let their elements be monic and ordered such that the leading terms increase with increasing index as explained on page 44. By Theorem 2.11 there holds

$$Reduce(j_i, H) = 0 \text{ for } i = 1, \ldots, k \text{ and } Reduce(h_i, J) = 0 \text{ for } i = 1, \ldots, l.$$

In the first place this requires that the leading terms of j_1 and h_1 are identical, because if e. g. $Lterm(j_1) < Lterm(h_1)$ holds, j_1 cannot be reduced by any element of H. Consequently, the leading term of $Reduce(j_1, h_1) = j_1 - h_1$ is lower than the common leading term of both j_1 and h_1 and cannot be reduced any further. By Theorem 2.11 it follows that $j_1 - h_1 = 0$, i.e., $j_1 = h_1$. Due to this equality, both j_1 and h_1 cannot take part in any further reduction. Therefore the same argument may be repeated with j_1 and h_1 omitted and so forth until the complete bases are covered.   ⬚

EXAMPLE 2.37   As a first example consider the system (2.25) from above. It is already autoreduced. Integrability conditions can only arise from the subsystem comprising g_2 with multipliers x and y, and g_3 and g_4 with multiplier y each. As it is already complete, there is no action in step $S2$. Their multipliers lead in step $S3$ to the integrability conditions

$$z_{xxx} + 8y^2 w_{yy} + \frac{1}{y} w_{xx} - 4y^2 z_{xy} - 32yz_x - 16w = 0, \ z_{xxy} - 4y^2 z_{yy} - 8yz_y = 0.$$

In step $S4$ their left-hand sides are reduced to

$$g_5 \equiv z_{xxx} + 8y^2 w_{yy} + \frac{1}{y} w_{xx} - 4y^2 z_{xy} - 32yz_x - 16w,$$

$$g_6 \equiv z_{xxy} - 4y^2 z_{yy} - 8yz_y.$$

Because the result does not vanish, the algorithm is repeated with the enlarged system comprising equations $g_1, \ldots, g_6$. The subsystem comprising g_1, g_5, g_6 is completed in step $S2$ to

$$z_{yy} + \frac{1}{2y} z_y, \ z_{xxx} + 12y^2 z_{xy} - 24yz_x - 12w,$$

$$z_{xxy} - 6yz_y = 0, \ z_{xyy} + \frac{1}{2y} z_{xy}.$$

The first element has multiplier variables x and y, for the remaining ones y is the only multiplier variable. They yield

$$z_{xxy} + 12y^2 z_{yy} + 12yz_y, \ z_{xyy} + \frac{1}{2y} z_{xy} - \frac{1}{y^2} w_y - \frac{2}{y^2} z_x.$$

Linear Differential Equations 55

In step $S4$ they are reduced to z_y, $w_y + 2z_x$. The enlarged system generated in $S5$ is autoreduced to

$$z_x + \frac{1}{2y}w, \quad z_y, \quad w_x, \quad w_y - \frac{1}{y}w$$

at the beginning of the third pass. Because the only two integrability conditions reduce to zero, this is the final answer. ▯

2.3 Properties of Janet Bases

From now on the emphasis will be shifted from linear differential polynomials to the systems of lpde's that are associated with them, and their solutions in suitable function fields. Furthermore, it is assumed frequently that the dimension of these solution spaces is finite, and that both the number of dependent and of independent variables is not greater than two. These constraints are a priori known to be true for the applications later in this book. It turns out that systems with these properties allow a complete classification; and explicit solution procedures may be designed for them.

Type and Order of a Janet Basis. It turns out that the leading derivatives are of extraordinary importance for any Janet basis. They are applied for defining the *type of a Janet basis*.

DEFINITION 2.6 *(Type of a Janet basis) Let $f = \{f_1, \ldots, f_p\}$ be a Janet basis for a fixed term order. Its leading derivatives $\{Lder(f_1), \ldots, Lder(f_p)\}$ define the type of the Janet basis f.*

The leading derivatives divide the totality of derivatives into two classes for which a special name is introduced.

DEFINITION 2.7 *(Parametric and principal derivatives. Rank). Those derivatives that may be obtained from the leading derivatives of a Janet basis by suitable differentiation are called principal, the remaining ones are called parametric. The number of parametric derivatives is called its rank.*

The distinctive feature of a Janet basis as compared to an arbitrary system of pde's may be expressed in terms of its parametric derivatives as follows: A Janet basis does not generate any relation between the parametric derivatives by differentiation and elimination, i.e., the parametric derivatives are independent of each other.

Another quantity of importance for any system of pde's is the degree of arbitrariness of its general solution. As a matter of fact, this question was the starting point for Janet's work. In general it may consist of any number of functions with varying numbers of arguments. A special case is represented by

56

those systems the general solution of which contains only a finite number of constants. Due to the linearity, their general solution is a linear combination of a fundamental system, i.e., it generates a finite dimensional vector space. A straightforward extension of this property which is true for linear ode's leads to the following definition.

DEFINITION 2.8 *(Order). If the general solution of a linear homogeneous system of pde's contains a finite number of constants, this number is called the order of the system. It is identical to the dimension of its solution space.*

The importance of a Janet basis originates from the fact that the dimension of the solution space is obvious from it due to the following result. Its proof may be found in the book by Kolchin [91], Chapter IV, Section 5.

THEOREM 2.13 *(Kolchin 1973) If the rank of a Janet basis is finite, then it is identical to its order.*

EXAMPLE 2.38 The type of the Janet basis generated in Example 2.37 is determined by $\{z_x, z_y, w_x, w_y\}$. The only parametric derivatives are z and w; therefore it allows a two-dimensional solution space as obtained previously. This information is available *without* determining the solution explicitly. □

THEOREM 2.14 *If the dimension of the solution space of a system of linear homogeneous pde's is finite, any Janet basis in lexicographic term ordering has a subsystem containing only the lowest dependent variable. This subsystem also has a finite-dimensional solution space. Its lowest equation is a linear ode in the lowest independent variable.*

PROOF Let the dependent variables be ordered as $z_1 > z_2 \ldots > z_m$, the independent variables as $x_1 > x_2 > \ldots > x_n$. The finite dimension of the solution space requires that the Janet basis contains equations with leading terms $\dfrac{\partial^{k_i} z_m}{\partial x_i^{k_i}}$ for $i = 1, \ldots, n$ and $k_i \geq 0$. This assures the finite-dimensional solution space for the lowest dependent variable. The applied term ordering guarantees that in these equations there does not occur a term containing a function z_i with $i < m$. Due to the ordering of the independent variables, the equation with leading term $\dfrac{\partial^{k_n} z_m}{\partial x_n^{k_n}}$ cannot contain a derivative w.r.t. to x_k and $k < n$, i.e., it is an ode w.r.t. to x_n. □

EXAMPLE 2.39 Consider the Janet basis

$$\left\{ z_{1,x} + \frac{1}{y-2} z_2 - \frac{1}{x} z_1, \ z_{1,y}, \ z_{2,x}, \ z_{2,y} - \frac{1}{y-2} z_2 \right\} \qquad (2.35)$$

in *grlex* order with $z_2 > z_1$ and $x > y$. There are four possible *lex* term orderings with the following result.

$$\left\{ z_{1,xx} - \frac{1}{x} z_{1,x} + \frac{1}{x^2} z_1, \ z_{1,y}, \ z_2 + (y-2) z_{1,x} - \frac{y-2}{x} z_1 \right\} \text{ for } z_2 > z_1, y > x.$$

$$\left\{z_{2,x},\ z_{2,y} - \frac{1}{y-2}z_2,\ z_{1,x} - \frac{1}{x}z_1 + \frac{1}{y-2}z_2, z_{1,y}\right\} \text{ for } z_1 > z_2, y > x.$$

$$\left\{z_{1,y},\ z_{1,xx} - \frac{1}{x}z_{1,x} + \frac{1}{x^2}z_1,\ z_2 + (y-2)z_{1,x} - \frac{y-2}{x}z_1\right\} \text{ for } z_2 > z_1, x > y$$

and finally

$$\left\{z_{2,y} - \frac{1}{y-2}z_2,\ z_{2,x},\ z_{1,y},\ z_{1,x} - \frac{1}{x}z_1 + \frac{1}{y-2}z_2\right\} \text{ for } z_1 > z_2, x > y.$$

In each of these Janet bases the two lowest equations generate a Janet basis for a single function. If $y > x$ the lowest equation is a linear ode in x over the base field $\mathbb{Q}(y)$ and vice versa. □

$\mathcal{J}_1^{(1,2)}$	$\mathcal{J}_{2,1}^{(1,2)}$	$\mathcal{J}_{2,2}^{(1,2)}$
$z_x \mid z$	$z_y \mid z_x, z$	$z_x \mid z$
$z_y \mid z$	$z_{xx} \mid z_x, z$	$z_{yy} \mid z_y, z$

$\mathcal{J}_{3,1}^{(1,2)}$	$\mathcal{J}_{3,2}^{(1,2)}$	$\mathcal{J}_{3,3}^{(1,2)}$
$z_y \mid z_x, z$	$z_{xx} \mid z_y, z_x, z$	$z_x \mid z$
$z_{xxx} \mid z_{xx}, z_x, z$	$z_{xy} \mid z_y, z_x, z$	$z_{yyy} \mid z_{yy}, z_y, z$
	$z_{yy} \mid z_y, z_x, z$	

TABLE 2.1: The types of Janet bases for $m = 1$, $n = 2$, i.e., for a single function $z(x, y)$ of order up to 3. The term ordering is *grlex* with $y > x$. Any type is uniquely characterized by its leading derivatives in the square boxes.

Classification for m=1, n=2. The simplest generalization of a linear homogeneous ode is a system of lpde's of finite order for a single function. For any fixed value of the order and a fixed term ordering there is a finite number of possible types of Janet bases. They are obtained by the following considerations. In the *grlex* term ordering with $y > x$ the derivatives of order not higher than three are

$$z_{yyy} > z_{xyy} > z_{xxy} > z_{xxx} > z_{yy} > z_{xy} > z_{xx} > z_y > z_x > z.$$

According to the definition of a parametric derivative, for an r-dimensional solution space, principal derivatives must be selected from this arrangement such that none of it may be obtained by differentiation from any of the remaining ones and that they allow r parametric derivatives. For $r = 1$, 2 or 3 the possible selections are as follows.

$$r = 1: \quad \{z_x, z_y\},$$
$$r = 2: \quad \{z_y, z_{xx}\}, \{z_x, z_{yy}\},$$
$$r = 3: \quad \{z_y, z_{xxx}\}, \{z_{xx}, z_{yx}, z_{yy}\}, \{z_x, z_{yyy}\}.$$

In Table 2.1 the structure of the corresponding Janet bases is indicated; and a unique notation $\mathcal{J}_{r,k}^{(m,n)}$ for any of them is introduced. Here m and n denote the number of dependent and independent variables respectively, r denotes the order, and k is a consecutive enumeration within this set. The symbol $\mathcal{J}^{(m,n)}$ without the lower indices denotes collectively all Janet basis types corresponding to the given values of m and n. Each line corresponds to an element of the Janet basis, its leading derivative is enclosed in a box, the parametric derivatives that may occur are listed to the right in decreasing order. In a concrete Janet basis they are multiplied by a coefficient from the base field.

The possible arrangements of parametric and leading derivatives may also be visualized geometrically. To this end, for any dependent variable its partial derivatives are assigned to a point of a rectangular lattice. Parametric derivatives are indicated by open dots, principal derivatives by heavy dots. To the six types of Janet bases defined in Table 2.1 there correspond the following graphs.

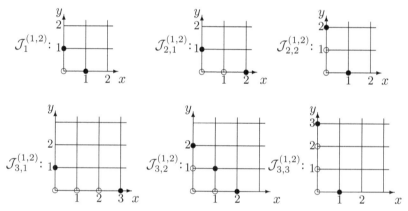

The classification of Janet bases given above provides a complete survey of all possible Janet bases for $m = 1$, $n = 2$ if the dimension of the solution space is not higher than three. If it is a priori known for a particular system that the dimension of its solution space is at most three, it is *guaranteed* that the resulting Janet basis type is included in this listing.

Next to the classification problem there occurs the following question: What are the most general coefficients for any given type such that the Janet basis property is assured? This is a new phenomenon that does not have a counterpart for linear ode's of the form (2.1). Its coefficients q_i may be chosen arbitrarily from the base field, yet the resulting equation has an n−dimensional solution space, only the function field in which a fundamental system is contained is determined by this choice. This is *not* true for Janet bases for systems of lpde's. The constraints for the coefficients guaranteeing the Janet basis property are extremely important, particularly if they are combined with

Linear Differential Equations

additional restrictions expressing further properties of the solutions as they will occur in later applications. Only if the former constraints are completely known, the latter ones may be identified as new features originating from the special problem.

The coefficient constraints for a given Janet basis type are essentially obtained in steps $S2$, $S3$ and $S4$ of the algorithm *JanetBasis* given above. Because the type of the Janet basis is fixed now, upon completion of step $S3$ the nonvanishing integrability conditions are *not* added to the system. Rather in order to preserve the type of the Janet basis, the coefficients of the various parametric derivatives are supposed to vanish identically. The conditions obtained in this way represent a system of pde's for the coefficient functions. If it is satisfied, the full system is guaranteed to be the most general Janet basis of the specified type. This is achieved by insertion of the additional step $S5a$ as follows.

$S5a$: *Separate Integrability Conditions.* Assign

$$C := \bigcup_{c_{i,j}} Coefficients(c_{i,j})$$

and return C.

In general the integrability conditions obtained in step $S5a$ are a system of nonlinear pde's that may comprise redundant equations. However, the following important property makes it amenable for further simplification.

LEMMA 2.3 *The integrability conditions obtained in step $S5a$ are quasilinear, i.e., they are linear in the highest derivatives.*

PROOF In step $S2$ and the first part of $S3$ only differentiations of the given linear homogeneous equations are performed, i.e., the resulting system is such that the reductum of any equation is linear and homogeneous in the coefficients. The same is true for the integrability conditions $c_{i,j}$ obtained in the second part of $S3$ because the leading terms cancel each other. For any given $c_{i,j}$ two kinds of reductions in step $S4$ may occur. In the first place there are those applying an equation e_k with an index different from i and j. They can never generate a quadratic term because there is no intersection between the coefficient variables of any equation involved. On the other hand, any reduction w.r.t. equation e_i or e_j involves only derivatives that are lower than those involved in forming the integrability condition. As a consequence any derivative of maximal order can only occur linearly. ▯

Applying the modified version of algorithm *JanetBasis* to the entries of Table 2.1 yields the following answer.

THEOREM 2.15 *For $m = 1$, $n = 2$ and order $r \leq 3$ the coherence conditions or integrability conditions (IC's for short) are explicitly given by the following constraints.*

$$\mathcal{J}_1^{(1,2)} : \{z_x + az, \ z_y + bz\}. \quad IC : a_y - b_x = 0.$$

$$\mathcal{J}_{2,1}^{(1,2)}: \qquad \{z_y + a_1 z_x + a_2 z, \; z_{xx} + b_1 z_x + b_2 z\}.$$

$$IC's: \quad \begin{aligned} a_{1,xx} - b_{1,x}a_1 - a_{1,x}b_1 + 2a_{2,x} - b_{1,y} &= 0, \\ a_{2,xx} + a_{2,x}b_1 - 2a_{1,x}b_2 - b_{2,x}a_1 - b_{2,y} &= 0. \end{aligned}$$

$$\mathcal{J}_{2,2}^{(1,2)}: \qquad \{z_x + a_1 z, \; z_{yy} + b_1 z_y + b_2 z\}.$$

$$IC's: \quad b_{1,x} - 2a_{1,y} = 0, \quad a_{1,yy} + a_{1,y}b_1 - b_{2,x} = 0.$$

$$\mathcal{J}_{3,1}^{(1,2)}: \qquad \{z_y + a_1 z_x + a_2 z, \; z_{xxx} + b_1 z_{xx} + b_2 z_x + b_3 z\}.$$

$$a_{1,xx} - \tfrac{1}{3}b_{1,y} - \tfrac{1}{3}b_{1,x}a_1 + a_{2,x} - \tfrac{1}{3}a_{1,x}b_1 = 0,$$

$$IC's: \quad \begin{aligned} a_{2,xxx} + a_{2,xx}b_1 - b_{3,y} - b_{3,x}a_1 + a_{2,x}b_2 - 3a_{1,x}b_3 &= 0, \\ b_{1,xy} + b_{1,xx}a_1 + 6a_{2,xx} - 3b_{2,y} - 3b_{2,x}a_1 + \tfrac{4}{3}b_{1,y}b_1 & \\ + 2b_{1,x}a_{1,x} + 2a_{2,x}b_1 - 6a_{1,x}b_2 + \tfrac{4}{3}b_1(a_1 b_1)_x &= 0. \end{aligned}$$

$$\mathcal{J}_{3,2}^{(1,2)}: \quad \begin{aligned} \{z_{xx} + a_1 z_y + a_2 z_x + a_3 z, \; z_{xy} + b_1 z_y + b_2 z_x + b_3 z, \\ z_{yy} + c_1 z_y + c_2 z_x + c_3 z\}. \end{aligned}$$

$$a_{1,y} - b_{1,x} + a_1 b_2 - a_1 c_1 - a_2 b_1 + a_3 + b_1^2 = 0,$$

$$a_{2,y} - b_{2,x} - a_1 c_2 + b_1 b_2 - b_3 = 0,$$

$$IC's: \quad a_{3,y} - b_{3,x} - a_1 c_3 - a_2 b_3 + a_3 b_2 + b_1 b_3 = 0,$$

$$b_{1,y} - c_{1,x} + a_1 c_2 - b_1 b_2 + b_3 = 0,$$

$$b_{2,y} - c_{2,x} + a_2 c_2 - b_1 c_2 - b_2^2 + b_2 c_1 - c_3 = 0,$$

$$b_{3,y} - c_{3,x} + a_3 c_2 - b_1 c_3 - b_2 b_3 + b_3 c_1 = 0.$$

$$\mathcal{J}_{3,3}^{(1,2)}: \qquad \{z_x + a_1 z, \; z_{yyy} + b_1 z_{yy} + b_2 z_y + b_3 z\}.$$

$$IC's: \quad \begin{aligned} b_{1,x} - 3a_{1,y} &= 0, \quad a_{1,yy} - \tfrac{1}{3}b_{2,x} + \tfrac{2}{3}a_{1,y}b_1 = 0, \\ b_{2,xy} - 3b_{3,x} + \tfrac{1}{3}b_{2,x}b_1 - 2b_{1,y}a_{1,y} + 3a_{1,y}b_2 - \tfrac{2}{3}a_{1,y}b_1^2 &= 0. \end{aligned}$$

PROOF   For the $\mathcal{J}_1^{(1,2)}$ Janet basis the only IC follows by subtracting

$$z_{xy} + (a_y - ab)z \quad \text{and} \quad z_{xy} + (b_x - ab)$$

from each other. These expressions are obtained by deriving the two elements of the Janet basis w.r.t. y and x respectively and reducing the result. The condition obtained assures the path independence of integrals involving the gradient vector field of z.

As a second example, the IC's for the type $\mathcal{J}_{2,1}^{(1,2)}$ Janet basis will be obtained. To this end, the first element has to be derived twice w.r.t. to the

Linear Differential Equations 61

multiplier x and all possible reductions have to be performed. The result is

$$z_{xy} + (a_{1,x} + a_2 - a_1b_1)z_x + (a_{2,x} - a_1b_2)z,$$

$$z_{xxy} + (a_{1,xx} + 2a_{2,x} - 2a_{1,x}b_1 - a_1b_{1,x} - a_1b_2 - a_2b_1 + a_1b_1^2)z_x$$

$$+(a_{2,xx} - 2a_{1,x}b_2 - a_1b_{2,x} - a_2b_2 + a_1b_1b_2)z.$$

The second element has to be derived w.r.t. to its nonmultiplier y. After all reductions are performed, the expression

$$z_{xxy} + (b_{1,y} - a_{1,x}b_1 - a_2b_1 - a_1b_2 + a_1b_1^2)z_x$$

$$+ (b_{2,y} - a_{2,x}b_1 - a_2b_2 + a_1b_1b_2)z$$

is obtained. If the two elements with the leading derivative z_{xxy} are subtracted from each other, the resulting expression must vanish identically. Because the parametric derivatives z_x and z are independent, its coefficients must vanish. After some simplification this yields the given IC's. The calculations for the remaining cases are similar. □

EXAMPLE 2.40   Consider the set of differential polynomials

$$\left\{z_{xx} - \frac{y}{x(x+y)}z_y + \frac{1}{x}z_x, \; z_{xy} + \frac{1}{x+y}z_y, \; z_{yy} + \frac{1}{x+y}z_y\right\}.$$

Its leading terms suggest that it might be a type $\mathcal{J}_{3,2}^{(1,2)}$ Janet basis. The nonvanishing coefficients

$$a_1 = -\frac{y}{x(x+y)}, \quad a_2 = \frac{1}{x} \quad \text{and} \quad b_1 = c_1 = \frac{1}{x+y}$$

satisfy the conditions given in the above theorem, i.e., the Janet basis property is assured *without* running the Janet basis algorithm. As is easily shown, a basis for its three-dimensional solution space is $\{1, \log x, \log(x+y)\}$.

In general, any change in the coefficients destroys the Janet basis property, e.g. if the sign of the last term in the first equation is reversed. Running this new system through the algorithm *JanetBasis* yields the type $\mathcal{J}_{2,1}^{(1,2)}$ Janet basis $\{z_y, z_{xx} - \frac{1}{x}z_x\}$. A basis for its two-dimensional solutions space is $\{1, x^2\}$. □

Classification for m=n=2. The second case to be discussed in full detail involves two dependent variables, they are denoted by z_1 and z_2. In *grlex* term ordering with $z_2 > z_1$ and $y > x$, the derivatives of order not higher than two are

$$\ldots z_{2,yy} > z_{2,yx} > z_{2,xx} > z_{1,yy} > z_{1,yx} > z_{1,xx} >$$

$$z_{2,y} > z_{2,x} > z_{1,y} > z_{1,x} > z_2 > z_1.$$

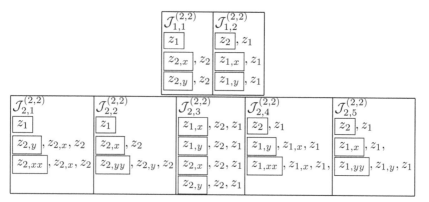

TABLE 2.2: The Janet bases of types $\mathcal{J}_{1,k}^{(2,2)}$ and $\mathcal{J}_{2,k}^{(2,2)}$ for linear homogeneous systems in $z_1(x,y)$ and $z_2(x,y)$. The term ordering is *grlex* with $z_2 > z_1$ and $y > x$. Any type is uniquely characterized by its leading derivatives in the square boxes.

For an r-dimensional solution space, principal derivatives must be selected from this arrangement such that none of it may be obtained by differentiation from any of the remaining ones and such that there are r parametric derivatives. For $r = 1$, 2 or 3 the possible selections are as follows.

$r = 1$: $\{z_2\}, \{z_1\}$

$r = 2$: $\{z_{2,x}, z_2\}, \{z_{2,y}, z_2\}, \{z_2, z_1\}, \{z_{1,y}, z_1\}, \{z_{1,x}, z_1\}$

$r = 3$: $\{z_{2,xx}, z_{2,x}, z_2\}, \{z_{2,y}, z_{2,x}, z_2\}, \{z_{2,yy}, z_{2,y}, z_2\}, \{z_{2,x}, z_2, z_1\},$
$\{z_{2,y}, z_2, z_1\}, \{z_{1,x}, z_2, z_1\}, \{z_{1,y}, z_2, z_1\}, \{z_{1,yy}, z_{1,y}, z_1\},$
$\{z_{1,y}, z_{1,x}, z_1\}, \{z_{1,xx}, z_{1,x}, z_1\}.$

In Tables 2.2 and 2.3 the structure of the corresponding Janet bases is indicated and a unique notation for any of them is introduced. Each line corresponds to an equation of the Janet basis, its leading derivative is enclosed in a box, the parametric derivatives that may occur in any equation are listed to the right of it in decreasing order. In a concrete Janet basis they are multiplied by a coefficient from the base field. Similar schemes may be set up for higher dimensional solution spaces or for values of m and n different from 2. The number of alternatives increases quickly if these values increase.

In the graphical representation there correspond to each Janet basis as many drawings as there are dependent variables. For the five possible types of Janet bases $\mathcal{J}_{2,k}^{(2,2)}$ these graphs are as follows.

Linear Differential Equations

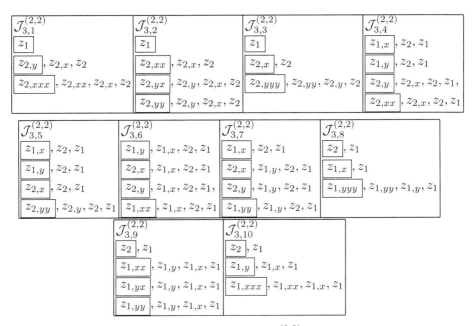

TABLE 2.3: The Janet bases of types $\mathcal{J}_{3,k}^{(2,2)}$, $k = 1, \ldots, 10$ for linear homogeneous systems in $z_1(x,y)$ and $z_2(x,y)$. The term ordering is *grlex* with $z_2 > z_1$ and $y > x$. Any type is uniquely characterized by its leading derivatives in the square boxes.

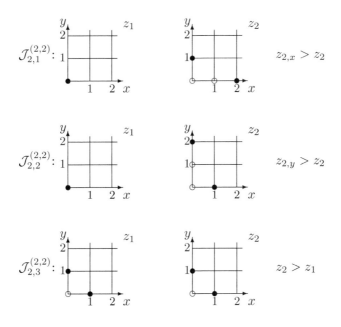

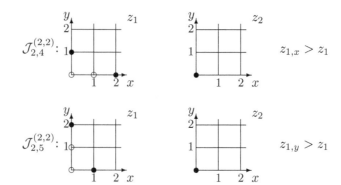

From this representation the distinctive properties of parametric and non-parametric derivatives are obvious.

The analogue of Theorem 2.15 for Janet bases of type $\mathcal{J}^{(2,2)}$ of order not greater than three is given next. Due to the size of the result it is split into two cases, at first order one and two Janet bases are considered, and the third order case thereafter.

THEOREM 2.16 *For $m = n = 2$ and order 1 or 2, the coherence conditions or integrability conditions (IC's for short) are explicitly given by the following constraints.*

$$\mathcal{J}_{1,1}^{(2,2)} : \{z_1,\ z_{2,x} + az_2,\ z_{2,y} + bz_2\}.\quad IC : a_y - b_x = 0.$$

$$\mathcal{J}_{1,2}^{(2,2)} : \{z_2 + az_1,\ z_{1,x} + bz_1,\ z_{1,y} + cz_1\}.\quad IC : b_y - c_x = 0.$$

$$\mathcal{J}_{2,1}^{(2,2)} : \{z_1,\ z_{2,y} + a_1 z_{2,x} + a_2 z_2,\ z_{2,xx} + b_1 z_{2,x} + b_2 z_2\}.$$

$IC's$:
$$a_{1,xx} - b_{1,y} - b_{1,x} a_1 + 2 a_{2,x} - a_{1,x} b_1 = 0,$$
$$a_{2,xx} - b_{2,y} - b_{2,x} a_1 + a_{2,x} b_1 - 2 a_{1,x} b_2 = 0.$$

$$\mathcal{J}_{2,2}^{(2,2)} : \{z_1,\ z_{2,x} + a_1 z_2,\ z_{2,yy} + b_1 z_{2,y} + b_2 z_2\}.$$
$IC's$: $b_{1,x} - 2 a_{1,y} = 0,\ a_{1,yy} - b_{2,x} + a_{1,y} b_1 = 0.$

$$\mathcal{J}_{2,3}^{(2,2)} : \begin{aligned}&\{z_{1,x} + a_1 z_2 + a_2 z_1,\ z_{1,y} + b_1 z_2 + b_2 z_1,\\ &\ z_{2,x} + c_1 z_2 + c_2 z_1,\ z_{2,y} + d_1 z_2 + d_2 z_1\}.\end{aligned}$$

$IC's$:
$$b_{1,x} - a_{1,y} + a_1 d_1 - b_1 c_1 - a_1 b_2 + a_2 b_1 = 0,$$
$$b_{2,x} - a_{2,y} + a_1 d_2 - b_1 c_2 = 0,$$
$$d_{1,x} - c_{1,y} - a_1 d_2 + b_1 c_2 = 0,$$
$$d_{2,x} - c_{2,y} + c_1 d_2 - a_2 d_2 - c_2 d_1 + b_2 c_2 = 0.$$

$$\mathcal{J}_{2,4}^{(2,2)} : \{z_2 + a_1 z_1,\ z_{1,y} + b_1 z_{1,x} + b_2 z_1,\ z_{1,xx} + c_1 z_{1,x} + c_2 z_1\}.$$
$IC's$:
$$b_{1,xx} - c_{1,y} - c_{1,x} b_1 + 2 b_{2,x} - b_{1,x} c_1 = 0,$$
$$b_{2,xx} - c_{2,y} - c_{2,x} b_1 + b_{2,x} c_1 - 2 b_{1,x} c_2 = 0.$$

$$\mathcal{J}_{2,5}^{(2,2)} : \{z_2 + a_1 z_1,\ z_{1,x} + b_1 z_1,\ z_{1,yy} + c_1 z_{1,y} + c_2 z_1\}.$$

$$IC's: \quad c_{1,x} - 2b_{1,y} = 0,\ b_{1,yy} - c_{2,x} + b_{1,y}c_1 = 0.$$

THEOREM 2.17   *For $m = n = 2$ and order 3 the coherence conditions or integrability conditions (IC's for short) are explicitly given by the following constraints.*

$$\mathcal{J}_{3,1}^{(2,2)} : \quad \{z_1, z_{2,y} + a_1 z_{2,x} + a_2 z_2,\ z_{2,xxx} + b_1 z_{2,xx} + b_2 z_{2,x} + b_3 z_2\}.$$

$$IC's: \quad \begin{aligned} & a_{1,xx} - \tfrac{1}{3}b_{1,y} - \tfrac{1}{3}b_{1,x}a_1 + a_{2,x} - \tfrac{1}{3}a_{1,x}b_1 = 0, \\ & b_{1,xy} + b_{1,xx}a_1 + 6a_{2,xx} - 3b_{2,y} - 3b_{2,x}a_1 + \tfrac{4}{3}b_{1,y}b_1 + 2b_{1,x}a_{1,x} \\ & \quad + \tfrac{4}{3}b_{1,x}a_1 b_1 + 2a_{2,x}b_1 - 6a_{1,x}b_2 + \tfrac{4}{3}a_{1,x}b_1^2 = 0, \\ & a_{2,xxx} + a_{2,xx}b_1 - b_{3,y} - b_{3,x}a_1 + a_{2,x}b_2 - 3a_{1,x}b_3 = 0. \end{aligned}$$

$$\mathcal{J}_{3,2}^{(2,2)} : \quad \begin{aligned} & \{z_1, z_{2,xx} + a_1 z_{2,y} + a_2 z_{2,x} + a_3 z_2,\ z_{2,xy} + b_1 z_{2,y} + b_2 z_{2,x} + b_3 z_2, \\ & \qquad\qquad z_{2,yy} + c_1 z_{2,y} + c_2 z_{2,x} + c_3 z_2\}. \end{aligned}$$

$$IC's: \quad \begin{aligned} & b_{1,x} - a_{1,y} + a_1 c_1 - a_1 b_2 - b_1^2 + a_2 b_1 - a_3 = 0, \\ & b_{2,x} - a_{2,y} + a_1 c_2 + b_3 - b_1 b_2 = 0, \\ & b_{3,x} - a_{3,y} + a_1 c_3 - b_1 b_3 + a_2 b_3 - a_3 b_2 = 0, \\ & c_{1,x} - b_{1,y} - a_1 c_2 - b_3 + b_1 b_2 = 0, \\ & c_{2,x} - b_{2,y} + c_3 + b_1 c_2 - a_2 c_2 - b_2 c_1 + b_2^2 = 0, \\ & c_{3,x} - b_{3,y} + b_1 c_3 - a_3 c_2 - b_3 c_1 + b_2 b_3 = 0. \end{aligned}$$

$$\mathcal{J}_{3,3}^{(2,2)} : \quad \{z_1,\ z_{2,x} + a_1 z_2,\ z_{2,yyy} + b_1 z_{2,yy} + b_2 z_{2,y} + b_3 z_2\}.$$

$$IC's: \quad \begin{aligned} & b_{1,x} - 3a_{1,y} = 0,\ a_{1,yy} - \tfrac{1}{3}b_{2,x} + \tfrac{2}{3}a_{1,y}b_1 = 0, \\ & b_{2,xy} - 3b_{3,x} + \tfrac{1}{3}b_{2,x}b_1 - 2a_{1,y}b_{1,y} + 3a_{1,y}b_2 - \tfrac{2}{3}a_{1,y}b_1^2 = 0. \end{aligned}$$

$$\mathcal{J}_{3,4}^{(2,2)} : \quad \begin{aligned} & \{z_{1,x} + a_1 z_2 + a_2 z_1,\ z_{1,y} + b_1 z_2 + b_2 z_1, \\ & \qquad z_{2,y} + c_1 z_{2,x} + c_2 z_2 + c_3 z_1,\ z_{2,xx} + d_1 z_{1,x} + d_2 z_2 + d_3 z_1\}. \end{aligned}$$

$$IC's: \quad \begin{aligned} & a_1 c_1 + b_1 = 0,\ b_{2,x} - a_{2,y} + a_1 c_3 = 0, \\ & b_{1,x} - a_{1,y} + a_1 c_2 - a_1 b_2 + a_2 b_1 = 0, \\ & c_{1,xx} - d_{1,y} - d_{1,x}c_1 + 2c_{2,x} - c_{1,x}d_1 a_1 c_3 = 0, \\ & c_{2,xx} - d_{2,y} - d_{2,x}c_1 - 2c_{3,x}a_1 + c_{2,x}d_1 - 2c_{1,x}d_2 \\ & \qquad\qquad - a_{1,x}c_3 - a_1 c_3 d_1 + a_1 a_2 c_3 = 0, \\ & c_{3,xx} - d_{3,y} - d_{3,x}c_1 + c_{3,x}d_1 - 2c_{3,x}a_2 - 2c_{1,x}d_3 - a_{2,x}c_3 \\ & \qquad\qquad - c_2 d_3 + a_2 c_1 d_3 + b_2 d_3 + c_3 d_2 - a_2 c_3 d_1 + a_2^2 c_3 = 0. \end{aligned}$$

$$\mathcal{J}_{3,5}^{(2,2)}: \quad \{z_{1,x} + a_1 z_2 + a_2 z_1,\ z_{1,y} + b_1 z_2 + b_2 z_1,$$
$$z_{2,x} + c_1 z_2 + c_2 z_1,\ z_{2,yy} + d_1 z_{2,y} + d_2 z_2 + d_3 z_1\}.$$

$$a_1 = 0,\ b_{1,x} - b_1 c_1 + a_2 b_1 = 0,$$
$$b_{2,x} - a_{2,y} - b_1 c_2 = 0,\ d_{1,x} - 2c_{1,y} + b_1 c_2 = 0,$$
$$IC's: \ c_{1,yy} - d_{2,x} - 2c_{2,y} b_1 + c_{1,y} d_1 - b_{1,y} c_2 - b_1 c_2 d_1 + b_1 b_2 c_2 = 0,$$
$$c_{2,yy} - d_{3,x} + c_{2,y} d_1 - 2c_{2,y} b_2 - b_{2,y} c_2$$
$$- c_1 d_3 + a_2 d_3 + c_2 d_2 - b_2 c_2 d_1 + c_2 b_2^2 = 0.$$

$$\mathcal{J}_{3,6}^{(2,2)}: \quad \{z_{1,y} + a_1 z_{1,x} + a_2 z_2 + a_3 z_1,\ z_{2,x} + b_1 z_{1,x} + b_2 z_2 + b_3 z_1,$$
$$z_{2,y} + c_1 z_{1,x} + c_2 z_2 + c_3 z_1,\ z_{1,xx} + d_1 z_{1,x} + d_2 z_2 + d_3 z_1\}.$$
$$c_{1,x} - b_{1,y} + a_{1,x} b_1 - c_1 d_1 - a_1 b_1 d_1 + c_3$$
$$- b_1 c_2 + b_2 c_1 + a_1 d_3 - b_1^2 a_2 + a_3 b_1 = 0,$$
$$c_{2,x} - b_{2,y} + a_{2,x} b_1 - c_1 d_2 - a_1 b_1 d_2 + a_2 d_3 - a_2 b_1 b_2 = 0,$$
$$c_{3,x} - d_{3,y} + a_{3,x} b_1 - c_1 d_3 - a_2 b_1 d_3$$
$$+ b_2 c_3 - c_2 d_3 - a_1 b_2 d_3 + a_3 d_3 = 0,$$
$$IC's: \ a_{1,xx} - d_{1,y} - d_{1,x} a_1 - b_{1,x} a_2 + 2a_{3,x} - 2a_{2,x} b_1$$
$$- a_{1,x} d_1 + c_1 d_1 + a_1 b_1 d_2 - a_2 d_3 + a_2 b_1 b_2 = 0,$$
$$a_{2,xx} - d_{2,y} - d_{2,x} a_1 - b_{2,x} a_2 + a_{2,x} d_1 - 2a_{2,x} b_2 - 2a_{1,x} d_2$$
$$+ a_2 d_3 + c_2 d_2 + a_1 b_2 d_2 + a_2 b_1 d_2 - a_3 d_2 - a_2 b_2 d_1 + b_1^2 a_2 = 0,$$
$$a_{3,xx} - d_{3,y} - d_{3,x} a_1 - d_{3,x} a_2 + a_{3,x} d_1 - 2a_{2,x} d_3$$
$$- 2a_{1,x} d_3 + a_2 b_1 d_3 + c_3 d_2 + a_1 d_2 d_3 - a_2 d_1 d_3 + a_2 b_2 d_3 = 0.$$

$$\mathcal{J}_{3,7}^{(2,2)}: \quad \{z_{1,x} + a_1 z_2 + a_2 z_1,\ z_{2,x} + b_1 z_{1,y} + b_2 z_2 + b_3 z_1,$$
$$z_{2,y} + c_1 z_{1,y} + c_2 z_2 + c_3 z_1,\ z_{1,yy} + d_1 z_{1,y} + d_2 z_2 + d_3 z_1\}.$$

$$c_{1,x} - b_{1,y} + b_1 d_1 - b_1 c_2 + c_1^2 a_1 - a_2 c_1 + b_2 c_1 - d_3 = 0,$$
$$c_{2,x} - a_{1,y} c_1 + b_1 d_2 - a_1 c_3 + a_1 c_1 c_2 - b_{2,y} = 0,$$
$$c_{3,x} - a_{2,y} c_1 + b_1 d_3 + a_1 c_1 c_3 - a_2 c_3 + b_2 c_3 - c_2 d_3 = 0,$$
$$IC's: \ d_{1,x} + c_{1,y} a_1 - 2a_{2,y} + 2a_{1,y} c_1 - b_1 d_2 + a_1 c_3 - a_1 c_1 c_2 = 0,$$
$$a_{1,yy} - d_{2,x} - c_{2,y} a_1 + a_{1,y} d_1 - 2a_{1,y} c_2$$
$$+ a_1 d_3 + a_1 c_1 d_2 - a_2 d_2 + b_2 d_2 - a_1 c_2 d_1 + c_2^2 a_1 = 0,$$
$$a_{2,yy} - d_{3,x} - c_{3,y} a_1 + a_{2,y} d_1 - 2a_{1,y} c_3$$
$$+ a_1 c_1 d_3 + d_2 d_3 - a_1 c_3 d_1 + a_1 c_2 c_3 = 0.$$

$$\mathcal{J}_{3,8}^{(2,2)}: \quad \{z_2 + a_1 z_1,\ z_{1,x} + b_1 z_1,\ z_{1,yyy} + c_1 z_{1,yy} + c_2 z_{1,y} + c_3 z_1\}.$$
$$IC's: \ \begin{array}{l} c_{1,x} - 3b_{1,y} = 0,\ b_{1,yy} - \frac{1}{3} c_{2,x} + \frac{2}{3} b_{1,y} c_1 = 0, \\ c_{2,xy} - 3c_{3,x} + \frac{1}{3} c_{3,x} c_1 - 2c_{1,y} b_{1,y} + 3b_{1,y} c_2 - \frac{2}{3} b_{1,y} c_1^2 = 0. \end{array}$$

$$\mathcal{J}_{3,9}^{(2,2)} : \quad \{z_2 + a_1 z_1, \; z_{1,xx} + b_1 z_{1,y} + b_2 z_{1,x} + b_3 z_1,$$
$$z_{1,xy} + c_1 z_{1,y} + c_2 z_{1,x} + c_3 z_1, \; z_{1,yy} + d_1 z_{1,y} + d_2 z_{1,x} + d_3 z_1\}.$$

$$IC's : \quad c_3 - a_1 c_2 = 0, \; d_3 - a_1 d_2 = 0, \; b_{2,y} - b_{1,y}a_1 - a_{1,y}b_1 = 0,$$
$$c_{1,x} - c_{1,y} + c_1 d_1 - c_1^2 - b_2 + a_1 b_1 = 0, \; d_{1,x} - c_{1,y} = 0.$$

$$\mathcal{J}_{3,10}^{(2,2)} : \quad \{z_2 + a_1 z_1, z_{1,y} + b_1 z_{1,x} + b_2 z_1, z_{1,xxx} + c_1 z_{1,xx} + c_2 z_{1,x} + c_3 z_1\}.$$

$$IC's : \quad b_{1,xx} - \tfrac{1}{3}c_{1,y} - \tfrac{1}{3}c_{1,x}b_1 + b_{2,x} - \tfrac{1}{3}b_{1,x}c_1 = 0,$$
$$c_{1,xy} + c_{1,xx}b_1 + 6b_{2,xx} - 3c_{2,y} - 3c_{2,x}b_1 + \tfrac{4}{3}c_{1,y}c_1 + 2b_{1,x}c_{1,x}$$
$$+ \tfrac{4}{3}c_{1,x}b_1c_1 + 2b_{2,x}c_1 - 6b_{1,x}c_2 + \tfrac{4}{3}b_{1,x}c_1^2 = 0,$$
$$b_{2,xxx} + b_{2,xx}c_1 - c_{3,y} - c_{3,x}b_1 + b_{2,x}c_2 - 3b_{1,x}c_3 = 0.$$

This explicit form provides a complete understanding of the features of a given system of linear homogeneous pde's. Of particular importance is the clear separation of all correlations for the coefficients due to the coherence conditions given in the above theorem, and additional properties that may be due to the structure of a specific problem. For example, the lowest equation in any Janet basis of type $\mathcal{J}_{3,5}^{(2,2)}$ has always the form $z_{1,x} + a_2 z_1 = 0$, the vanishing of the term proportional to z_2 is *not* a characteristic feature of a specific problem. The importance of this classification will become clear in the next chapter where Janet bases of these types occur as determining systems for certain symmetry groups.

The quasilinearity of the constraints generated in step $S5a$ (see page 59) allows application of the algorithm *JanetBasis* to the system C defined there with the following modification. The algorithm is interrupted if an equation is generated that is not quasilinear. It turns out that this does not occur for the types listed in Table 2.1, Table 2.2 or 2.3. The term ordering applied for the integrability conditions is always *grlex* with

$$a_1 < a_2 < \ldots < b_1 < b_2 < \ldots < c_1 < c_2 < \ldots$$

and $x < y$. In most cases the Janet basis property is either obtained upon proper reordering or by autoreduction.

EXAMPLE 2.41   A typical case is type $\mathcal{J}_{3,3}^{(2,2)}$ for which in the first place the integrability conditions

$$b_{1,x} - 3a_{1,y} = 0, \quad a_{1,yy} - \tfrac{1}{3}b_{2,x} + \tfrac{2}{3}a_{1,y}b_1 = 0,$$
$$a_{1,yyy} + a_{1,yy}b_1 - b_{3,x} + a_{1,y}b_2 = 0$$

are obtained. Autoreduction generates the Janet basis from it by replacing the third equation with

$$b_{2,xy} - 3b_{3,x} + \tfrac{1}{3}b_{2,x}b_1 - 2b_{1,y}a_{1,y} + 3a_{1,y}b_2 - \tfrac{2}{3}a_{1,y}b_2^2 = 0. \qquad \square$$

Syzygies of Janet Bases. The coherence conditions of Theorems 2.15, 2.16 and 2.17 imply certain relations between the elements of a Janet basis which are known as *syzygies*. They will be required later on in this chapter for computing quotients. Let the Janet basis be $f = \{f_1, \ldots, f_p\}$ where $f_i \in \mathcal{D}^m$, $\mathcal{D} = \mathbb{Q}(x, y)[\partial_x, \partial_y]$. Syzygies of f are relations of the form

$$d_{k,1} f_1 + \ldots + d_{k,p} f_p = 0$$

where $d_{k,i} \in \mathcal{D}$, $i = 1, \ldots p$, $k = 1, 2, \ldots$. The $(d_{k,1}, \ldots, d_{k,p})$ may be considered as elements of the module $\mathcal{D}^p$. The totality of syzygies generates a submodule.

EXAMPLE 2.42   Consider the Janet basis $\{f_1 \equiv z_x + az, f_2 \equiv z_y + bz\}$ with the constraint $a_y = b_x$. The coherence condition for $z_x - az - f_1 = 0$ and $z_y + bz - f_2 = 0$ yields $az_y + a_y z - f_{1,y} - bz_x - b_x z + f_{2,x} = 0$. Reduction w.r.t. to the given equations and some simplification yields the single syzygy $(\partial_y + b) f_1 - (\partial_x + a) f_2 = 0$. ⬜

EXAMPLE 2.43   Consider the Janet basis

$$\left\{ f_1 \equiv z_{xx} + \frac{4}{x} z_x + \frac{2}{x^2} z, \ f_2 \equiv z_{xy} + \frac{1}{x} z_y, \ f_3 \equiv z_{yy} + \frac{1}{y} z_y - \frac{x}{y^2} z_x - \frac{2}{y^2} z \right\}.$$

The integrability condition for $z_{xx} + \frac{4}{x} z_x + \frac{2}{x^2} z - f_1 = 0$ and $z_{xy} + \frac{1}{x} z_y - f_2 = 0$ yields upon reduction and simplification $f_{1,y} + f_{2,x} - \frac{3}{x} f_2 = 0$. Similarly from the last two elements $f_1 - \frac{y^2}{x} f_{2,y} - \frac{y}{x} f_2 + \frac{y^2}{x} f_{3,x} + \frac{y^2}{x^2} f_3 = 0$ is obtained. Autoreduction of these two equations yields the final answer

$$\left(\partial_{yy} + \frac{3}{y} \partial_y - \frac{x}{y^2} \partial_x - \frac{2}{y^2} \right) f_2 - \left(\partial_{xy} + \frac{1}{x} \partial_y + \frac{2}{y} \partial_x + \frac{2}{xy} \right) f_3 = 0,$$

$$f_1 - \left(\frac{y^2}{x} \partial_y + \frac{y}{x} \right) f_2 + \left(\frac{y^2}{x} \partial_x + \frac{y^2}{x^2} \right) f_3 = 0. \qquad ⬜$$

Gcrd and Lclm of Modules. The *greatest common right divisor (Gcrd)* or *sum* of two modules $I \equiv\ <f_1, \ldots, f_p>$ and $J \equiv\ <g_1, \ldots, g_q>$, f_i, $g_j \in \mathcal{D}^m$ for all i and j, $\mathcal{D} = Q(x, y)[\partial_x, \partial_y]$, is generated by the union of the generators of I and J (Cox et al. [34], page 191). The solution space of the equations corresponding to $Gcrd(I, J)$ is the intersection of the solution spaces of its arguments. From this definition the following algorithm is self-explanatory.

ALGORITHM 2.9 $Gcrd(I, J)$. Given two modules $I \equiv\ <f_1, \ldots, f_p>$ and $J \equiv\ <g_1, \ldots, g_q>$ in $\mathcal{D}^m$, the Janet basis for its greatest common divisor is returned.

$S1$: *Join input generators.* Set $h := \{f_1, \ldots, f_p\} \cup \{g_1, \ldots, g_q\}$

$$\textit{Linear Differential Equations} \qquad 69$$

$S2$: *Create Janet basis.* Transform h in a Janet basis and return the result.

For $m = 1$ the *Gcrd* of the ideals $I, J \in \mathcal{D}$ is obtained. For $m = n = 1$ the Euclidean algorithm for computing the *Gcrd* of two linear ode's is subsumed as a special case under the above algorithm.

EXAMPLE 2.44 Consider the ideals $I = < \partial_x + \frac{y}{x}\partial_y, \, \partial_{xxx} + \frac{3}{x}\partial_{xx} >$ and

$$J = < \partial_{xx} + \frac{1}{x}\partial_x - \frac{1}{x^2}, \, \partial_{xy} + \frac{1}{x}\partial_y + \frac{1}{y}\partial_x + \frac{1}{xy}, \, \partial_{yy} + \frac{1}{y}\partial_y - \frac{1}{y^2} > .$$

Applying the above algorithm, the greatest common right divisor is obtained as $Gcrd(I, J) = < \partial_x + \frac{y}{x}\partial_y, \, \partial_{yy} + \frac{1}{y}\partial_y - \frac{1}{y^2} >$. In terms of solution spaces this result may be understood as follows. For I a basis of the solution space is $\{1, \frac{x}{y}, \frac{y}{x}\}$, and for J a basis is $\{\frac{1}{xy}, \frac{x}{y}, \frac{y}{x}\}$. A basis for their two-dimensional intersection space is $\{\frac{x}{y}, \frac{y}{x}\}$, it is the solution space of $Gcrd(I, J)$. $\quad\square$

The *Lclm* of two modules I and J is the *left intersection module* of its arguments. The solution space of the equations corresponding to $Lclm(I, J)$ is the smallest space containing the solution spaces of its arguments. An algorithm for generating the *Lclm* requires a more detailed analysis. Properly adjusting the corresponding algorithm from commutative algebra, see for example Cox et al. [34], Proposition 3.11 on page 218 and the article by Grigoriev and Schwarz [61], Section 3, the following algorithm is obtained.

ALGORITHM 2.10 $Lclm(I, J)$. Given the two modules $I \equiv < f_1, \ldots, f_p >$ and $J \equiv < g_1, \ldots, g_q >$ in $\mathcal{D}^m$, the Janet basis for its least common left multiple is returned.

$S1$: *Auxiliary system.* Introducing two differential indeterminates w and z, generate the system $\{f_1 w, \ldots, f_p w, g_1 w - g_1 z, \ldots, g_q w - g_q z\}$.

$S2$: *Generate Janet basis.* For the system obtained in $S1$ generate a Janet basis in *lex* term order with $w > z$.

$S3$: *Return result.* From the Janet basis obtained in $S2$ collect elements depending on z only and return the respective operators generating the result.

EXAMPLE 2.45 Consider the two ideals

$$I = < \partial_y + \frac{1}{y}, \partial_x + \frac{1}{x} > \quad \text{and} \quad J = < \partial_y + \frac{1}{x+y}, \partial_x + \frac{1}{x+y} > .$$

In step $S1$ the system

$$w_y + \frac{1}{y}w, \; w_x + \frac{1}{x}w,$$

$$w_y + \frac{1}{x+y}w - z_y - \frac{1}{x+y}z, \; w_x + \frac{1}{x+y}w - z_x - \frac{1}{x+y}z$$

70

is generated. The Janet basis obtained in step $S2$ is

$$z_x - \frac{y^2}{x^2}z_y + \frac{x-y}{x^2}z, \quad z_{yy} + \frac{2x+4y}{xy+y^2}z_y + \frac{2}{xy+y^2}z,$$

$$w + \frac{x^2+xy}{y}z_x + \frac{x}{y}z.$$

Consequently, the intersection ideal is

$$Lclm(I, J) =< \partial_x - \frac{y^2}{x^2}\partial_y + \frac{x-y}{x^2}, \partial_{yy} + \frac{2x+4y}{xy+y^2}\partial_y + \frac{2}{xy+y^2} > . \quad \Box$$

In some places later in this book a shortened notation for the $Lclm$ is applied, e. g. $Lclm(z_y + az, z_x + bz) = Lclm(\partial_y + a, \partial_x + b)z$.

Exact Quotient and Relative Syzygies. For ordinary differential operators the exact quotient has been defined on page 22. In order to generate a full system of solutions for a system of linear pde's which is reducible but not completely reducible, a proper generalization of this exact quotient is required. It is given next.

DEFINITION 2.9 *(Exact quotient, relative syzygies) Let $I \equiv< f_1, \ldots, f_p >\in \mathcal{D}^m$ and $J \equiv< g_1, \ldots, g_q >\in \mathcal{D}^m$ be such that $I \subseteq J$. The exact quotient module is generated by*

$$\{(e_{i,1}, \ldots, e_{i,q}) \in \mathcal{D}^q | e_{i,1}g_1 + \ldots + e_{i,q}g_q = f_i, i = 1, \ldots, p\}.$$

The relative syzygies module $Syz(I, J)$ is generated by

$$\{h = (h_1, \ldots, h_q) \in \mathcal{D}^q | h_1 g_1 + \ldots + h_q g_q \in I\}.$$

If for two modules there holds $I \subseteq J$, J is also called a *divisor* or a *component* of I. Obviously the relative syzygies module extends the syzygy module of J, the latter is obtained for the special choice $I = 0$. The following algorithm generates the relative syzygies module as the sum of the exact quotient and the syzygies module.

ALGORITHM 2.11 *RelativeSyzygies(I, J).* Given two modules $I \subset J \subset \mathcal{D}^m$ by $I \equiv< f_1, \ldots, f_p >$ and $J \equiv< g_1, \ldots, g_q >$, the Janet basis for its relative syzygies module $Syz(I, J) \in \mathcal{D}^q$ is returned.

$S1$: *Exact quotient.* For $i = 1, \ldots, p$ determine $e_i = (e_{i,1}, \ldots, e_{i,q})$ from $e_{i,1}g_1 + \ldots + e_{i,q}g_q = f_i$.

$S2$: *Include syzygies.* Generate the syzygies $s_k = s_{k,1}g_1 + \ldots + s_{k,q}g_q$ for $k = 1, 2, \ldots$.

$S3$: *Return Janet basis.* Join the quotients q_i and the syzygies s_k, transform them into a Janet basis and return it.

EXAMPLE 2.46   Consider the ideals

$$I =< \partial_y - \frac{x^2}{y^2}\partial_x - \frac{x-y}{y^2}, \ \partial_{xx} + \frac{2}{x}\frac{2x+y}{x+y}\partial_x + \frac{2}{x(x+y)} >$$

and $J =< \partial_x + \frac{1}{x}, \partial_y + \frac{1}{y} >$. There holds $I \subset J$. Step $S1$ generates the matrix $\left\{ \left(-\frac{x^2}{y^2}, 1 \right), \left(\partial_x + \frac{3x+y}{x(x+y)}, 0 \right) \right\}$. In step $S2$ the single syzygy $\left(\partial_y + \frac{1}{y}, \partial_x + \frac{1}{x} \right)$ is obtained. Step $S3$ generates the Janet basis for the relative syzygies module

$$Syz(I,J) =< \left(-\frac{x^2}{y^2}, 1 \right), \left(\partial_x + \frac{3x+y}{x(x+y)}, 0 \right), \left(\partial_y - \frac{x-y}{y(x+y)}, 0 \right) >\subset \mathcal{D}^2.$$

☐

EXAMPLE 2.47   (Li, Schwarz and Tsarev [105]) Consider the ideal $I = < \partial_y - \frac{x}{y}\partial_x, \partial_{xx} - \frac{xy-1}{x}\partial_x - \frac{y}{x} >$ and its right factor $J =< \partial_x - y, \partial_y - x >$. Division in step $S1$ yields $\left\{ \left(-\frac{x}{y}, 1 \right), \left(\partial_x + \frac{1}{x}, 0 \right) \right\}$. There is a single syzygy $(-\partial_y + x, \partial_x - y)$. From these three generators step $S3$ generates a Janet basis for the relative syzygies module $Syz(I,J) =< (-\frac{x}{y}, 1), \ (\partial_x + \frac{1}{x}, 0), \ (\partial_y, 0) >$.

☐

EXAMPLE 2.48   Consider the ideal $I =< \partial_{xx} + \frac{1}{x}\partial_x, \partial_{xy}, \partial_{yy} + \frac{1}{y}\partial_y >$. There is a single maximal ideal $J =< \partial_x, \partial_y >$ containing it. It yields the relative syzygies module $Syz(I,J) =< (\partial_x + \frac{1}{x}, 0), (\partial_y, 0), (0, \partial_x), (0, \partial_y + \frac{1}{y}) >$. It may be represented as the intersection of two maximal modules of order 1, i.e., there holds

$$Syz(I,J) = Lclm\big(< (1,0), (0,\partial_x)(0, \partial_y + \frac{1}{y}) >,$$

$$< (\frac{x}{y}, 1), (\partial_x + \frac{1}{x}, 0), (\partial_y, 0) > \big). \quad ☐$$

Similar as for ordinary differential equations there remains to be discussed how lower order right factors of a module of partial differential operators simplify the solution procedure, compare the discussion on page 31. The following result is due to Grigoriev and Schwarz [62].

THEOREM 2.18   Let $I \equiv< f_1, \ldots, f_p >$ and $J \equiv< g_1, \ldots, g_q >$ be two submodules of $\mathcal{D}^m$ such that $I \subseteq J$, with solution space V_I and V_J of finite dimension d_I and d_J respectively. Let the relative syzygies module $Syz(I,J) \equiv K \subset \mathcal{D}^q$ be generated by $< h_1, \ldots, h_r >$, and $z = (z_1, \ldots, z_m)$ and $\bar{z} = (\bar{z}_1, \ldots, \bar{z}_q)$ be differential indeterminates. A basis for the solution space V_I may be constructed as follows.

i) Determine a basis of V_J from $g_1 z = 0, \ldots, g_q z = 0$, let its elements be $v_k = (v_{k,1}, \ldots, v_{k,m})$, $k = 1, \ldots, d_J$.

ii) *Determine a basis of V_K from $h_1\bar{z} = 0, \ldots, h_r\bar{z} = 0$, let its elements be*
$w_k = (w_{k,1}, \ldots, w_{k,q})$, $k = 1, \ldots, d_K = d_I - d_J$.

iii) *For each inhomogeneous system $g_1\bar{z} = w_{k,1}, \ldots, g_q\bar{z} = w_{k,q}$, $k = 1, \ldots, d_K$*
determine a special solution $\bar{v}_k = (\bar{v}_{k,1}, \ldots, \bar{v}_{k,m})$.

A basis of the solution space V_I is $\{v_1, \ldots, v_{d_J}, \bar{v}_1, \ldots, \bar{v}_{d_K}\}$.

EXAMPLE 2.49   In Example 2.46, the right factor J corresponds to the
system $z_x + \frac{1}{x}z = 0$, $z_y + \frac{1}{y}z = 0$, its solution yields the basis element
$v_1 = \frac{1}{xy}$. The relative syzygies lead to the system

$$z_2 - \frac{x^2}{y^2}z_1 = 0, \quad z_{1,x} + \frac{3x+y}{x(x+y)}z_1 = 0, \quad z_{1,y} - \frac{x-y}{y(x+y)}z_1 = 0,$$

it yields the basis element $z_1 = \frac{y}{x(x+y)^2}$, $z_2 = \frac{x}{y(x+y)^2}$. Substituting it
into the inhomogeneous system results in the special solution $\bar{v}_1 = \frac{1}{x+y}$.
Consequently, a basis for the originally given system is $\{\frac{1}{xy}, \frac{1}{x+y}\}$.   ⧠

EXAMPLE 2.50   In Example 2.47 the system of pde's corresponding to J
is $z_x - yz = 0$, $z_y - xz = 0$, it yields the basis $\{e^{xy}\}$ for V_J. The relative
syzygies module leads to the system $z_2 - \frac{x}{y}z_1 = 0$, $z_{1,x} + \frac{1}{x}z_1 = 0$ and
$z_{1,y} = 0$; its one-dimensional solution space is generated by $z_1 = \frac{1}{x}$, $z_2 = \frac{1}{y}$.
There is a single inhomogeneous system $z_x - yz = \frac{1}{x}$, $z_y - xz = \frac{1}{y}$ with
the special solution $\bar{v}_1 = e^{xy}Ei(-xy)$. The exponential integral is defined
by $Ei(ax) = \int e^{ax}\frac{dx}{x}$. A basis for the solution space of the originally given
system is $\{e^{xy}, e^{xy}Ei(-xy)\}$.   ⧠

EXAMPLE 2.51   In Example 2.48, to the first order right factor corresponds
the system $z_x = 0$, $z_y = 0$, it yields the basis $\{1\}$ vor V_J. The two arguments
of the $Lclm$ lead to the systems $\bar{z}_1 = 0$, $\bar{z}_{2,x} = 0$, $\bar{z}_{2,y} + \frac{1}{y}\bar{z}_2 = 0$ with the
solution $\bar{z}_{1,1} = 0$, $\bar{z}_{1,2} = 0$, and $\bar{z}_2 + \frac{x}{y}\bar{z}_1 = 0$, $\bar{z}_{1,x} + \frac{1}{x}\bar{z}_1 = 0$, $\bar{z}_{1,y} = 0$ with
the solution $\bar{z}_{2,1} = \frac{1}{x}$, $\bar{z}_{2,2} = -\frac{1}{y}$. Substituting them into the inhomogeneous
system yields the two special solutions $\log y$ and $\log \frac{x}{y}$. Therefore a basis for
the given system is $\{1, \log y, \log \frac{x}{y}\}$.   ⧠

For efficiency reasons it is advantageous to have closed forms available for
those relative syzygies that occur frequently in applications. In this way the
Janet basis calculation which occurs in step $S3$ of the above algorithm may
be avoided. Because these applications usually involve solving differential
equations, they are given in terms of differential polynomials.

Linear Differential Equations 73

LEMMA 2.4   *A Janet basis $\{z_y + A_1 z_x + A_2 z, z_{xx} + B_1 z_x + B_2 z\}$ of type $\mathcal{J}_{2,1}^{(1,2)}$ has a component $\{z_1 \equiv z_x + az, z_2 \equiv z_y + bz\}$ of type $\mathcal{J}_1^{(1,2)}$ iff there holds $b - A_2 + A_1 a = 0$,*

$$a_x - a^2 + B_1 a - B_2 = 0, \quad a_y + A_1 a^2 + (A_{1,x} - A_1 B_1)a - A_{2,x} + A_1 B_2 = 0.$$

The relative syzygies module is of type $\mathcal{J}_{1,2}^{(2,2)}$ and is explicitly given by

$$\{z_2 + A_1 z_1, z_{1,x} + (B_1 - a)z_1, z_{1,y} + (A_{1,x} - A_1 B_1 + 2aA_1 + b)z_1\}.$$

PROOF   Reduction of the type $\mathcal{J}_{2,1}^{(1,2)}$ Janet basis w.r.t. $z_1 \equiv z_x + az$ and $z_2 \equiv z_y + bz$ yields the first two generators of the quotient under the assumption that b is determined from the first constraint and the equation with leading term a_x is valid. The third element defining the quotient follows if in the integrability condition $a_y - b_x = 0$ the variable b is eliminated applying the first two constraints. In Example 2.42 it has been shown that the type $\mathcal{J}_1^{(1,2)}$ Janet basis obeys the syzygy $z_{1,y} - z_{2,x} + bz_1 - az_2 = 0$. Reducing it w.r.t. to the two previous relations, the last generator is obtained. $\quad\square$

EXAMPLE 2.52   Consider again the Janet bases from Example 2.46. By the above lemma, the coefficients

$$A_1 = -\frac{x^2}{y^2}, \quad A_2 = -\frac{x-y}{y^2}, \quad B_1 = \frac{2}{x}\frac{2x+y}{x+y} \quad \text{and} \quad B_2 = \frac{2}{x(x+y)}$$

yield immediately the Janet basis for the relative syzygies module

$$< \left(-\frac{x^2}{y^2}, 1\right), \left(\partial_x + \frac{3x+y}{x(x+y)}, 0\right), \left(\partial_y - \frac{x-y}{y(x+y)}, 0\right) > . \quad\square$$

LEMMA 2.5   *A Janet basis $\{z_x + A_1 z, z_{yy} + B_1 z_y + B_2 z\}$ of type $\mathcal{J}_{2,2}^{(1,2)}$ has a component $\{z_1 \equiv z_x + az, z_2 \equiv z_y + bz\}$ of type $\mathcal{J}_1^{(1,2)}$ iff there holds*

$$A_1 - a = 0, \quad b_y - b^2 - B_1 b - B_2 = 0, \quad b_x + a_y - B_{1,x} = 0.$$

Its relative syzygies module is of type $\mathcal{J}_{1,1}^{(2,2)}$ and is explicitly given by

$$\{z_1, z_{2,x} + az_2, z_{2,y} + (B_1 - b)z_2\}.$$

The proof is similar as for Lemma 2.4 and is therefore omitted.

EXAMPLE 2.53   Consider the Janet basis $\{z_x + \frac{1}{x}z, z_{yy} + \frac{1}{y}z_y - \frac{1}{y^2}z\}$. Dividing out $\{z_1 \equiv z_x + \frac{1}{x}z, z_2 \equiv z_y - \frac{1}{y}z\}$ yields $\{z_1, z_{2,x} + \frac{1}{x}z_2, z_{2,y} + \frac{2}{y}z_2\}$. The original system has the solutions $\{\frac{y}{x}, \frac{1}{xy}\}$ whereas $\frac{y}{x}$ is the only solution

74

of its first order component $\{z_1, z_2\}$. The second member is obtained by solving the quotient system with the result $z_1 = 0$, $z_2 = -\dfrac{2}{xy^2}$, and then the inhomogeneous system $z_x + \frac{1}{x}z = 0$, $z_y - \frac{1}{y}z = -\dfrac{2}{xy^2}$ with the result $\frac{1}{xy}$. $\square$

Explicit forms for the possible quotients of Janet bases of order three are given next without proof.

LEMMA 2.6 *If a Janet basis* $\{z_y + A_1 z_x + A_2 z, z_{xxx} + B_1 z_{xx} + B_2 z_x + B_3 z\}$ *of type* $\mathcal{J}_{3,1}^{(1,2)}$ *has a component* $\{z_1 \equiv z_x + az, z_2 \equiv z_y + bz\}$ *of type* $\mathcal{J}_1^{(1,2)}$, *its quotient is of type* $\mathcal{J}_{2,4}^{(2,2)}$ *and is explicitly given by*

$$\{z_2 + A_1 z_1, \, z_{1,y} + A_1 z_{1,x} + (A_{1,x} + A_2)z_1,$$
$$z_{1,xx} + (B_1 - a)z_{1,x} + (B_2 - aB_1 - 2a_x + a^2)z_1\}.$$

If it has a component $\{z_1 \equiv z_y + a_1 z_x + a_2 z, z_2 \equiv z_{xx} + b_1 z_x + b_2 z\}$ *of type* $\mathcal{J}_{2,1}^{(1,2)}$, *its quotient is of type* $\mathcal{J}_{1,1}^{(2,2)}$ *and is explicitly given by*

$$\{z_1, \, z_{2,x} + (B_1 - b_1)z_2, \, z_{2,y} + (2A_{1,x} - A_1 B_1 + b_1 A_1 + A_2)z_2\}.$$

EXAMPLE 2.54 Consider the Janet basis $\{z_y + \frac{x}{y}z_x, \, z_{xxx} + \frac{3}{x}z_{xx}\}$. Dividing out the type $\mathcal{J}_1^{(1,2)}$ component $\{z_x - \frac{1}{x}z, \, z_y + \frac{1}{y}z\}$ yields the type $\mathcal{J}_{1,1}^{(2,2)}$ Janet basis

$$\left\{z_2 + \frac{x}{y}z_1, \, z_{1,y} + \frac{x}{y}z_{1,x} + \frac{1}{y}z_1, \, z_{1,xx} + \frac{4}{x}z_{1,x} + \frac{2}{x^2}z_1\right\}. \qquad \square$$

LEMMA 2.7 *If a Janet basis*

$$\{z_{xx} + A_1 z_y + A_2 z_x + A_3 z, \, z_{xy} + B_1 z_y + B_2 z_x + B_3 z,$$
$$z_{yy} + C_1 z_y + C_2 z_x + C_3 z\}$$

of type $\mathcal{J}_{3,2}^{(1,2)}$ *has a component* $\{z_1 \equiv z_x + az, z_2 \equiv z_y + bz\}$ *of type* $\mathcal{J}_1^{(1,2)}$, *its quotient is of type* $\mathcal{J}_{2,3}^{(2,2)}$ *and is explicitly given by*

$$\{z_{1,x} + A_1 z_2 + (A_2 - a)z_1, \, z_{1,y} + (B_1 - a)z_2 + B_2 z_1,$$
$$z_{2,x} + B_1 z_2 + (B_2 - b)z_1, \, z_{2,y} + (C_1 - b)z_2 + C_2 z_1\}.$$

If it has a component $\{z_1 \equiv z_y + a_1 z_x + a_2 z, z_2 \equiv z_{xx} + b_1 z_x + b_2 z\}$ *of type* $\mathcal{J}_{2,1}^{(1,2)}$, *its quotient is of type* $\mathcal{J}_{1,2}^{(2,2)}$ *and is explicitly given by*

$$\{z_2 + A_1 z_1, \, z_{1,x} + (a_1 A_1 + B_1)z_1, \, z_{1,y} + (C_1 - 2a_1^2 A_1 - a_2)z_1\}.$$

Linear Differential Equations

If it has a component $\{z_1 \equiv z_x + a_1 z, z_2 \equiv z_{yy} + b_1 z_y + b_2 z\}$ of type $\mathcal{J}_{2,2}^{(1,2)}$, its quotient is of type $\mathcal{J}_{1,2}^{(2,2)}$ and is explicitly given by

$$\{z_2 + C_2 z_1, z_{1,x} + (A_2 - a_1)z_1, z_{1,y} + B_2 z_1\}.$$

EXAMPLE 2.55   Consider the type $\mathcal{J}_{3,2}^{(1,2)}$ Janet basis

$$\left\{ z_{xx} + \frac{4}{x} z_x + \frac{2}{x^2} z, \; z_{xy} + \frac{1}{x} z_y, \; z_{yy} + \frac{1}{y} z_y - \frac{x}{y^2} z_x - \frac{2}{y^2} z \right\}.$$

Dividing out the type $\mathcal{J}_1^{(1,2)}$ Janet basis $\{z_1 \equiv z_x + \frac{1}{x} z, \; z_2 \equiv z_y + \frac{1}{y} z\}$ yields the type $\mathcal{J}_{2,3}^{(2,2)}$ Janet basis

$$\left\{ z_{1,x} + \frac{3}{x} z_1, \; z_{1,y}, \; z_{2,x} + \frac{1}{x} z_2 - \frac{1}{y} z_1, \; z_{2,y} - \frac{x}{y^2} z_1 \right\}.$$

Dividing out the type $\mathcal{J}_{2,2}^{(1,2)}$ Janet basis $\{z_1 \equiv z_x + \frac{1}{x} z, \; z_2 \equiv z_{yy} + \frac{1}{y} z_y - \frac{1}{y^2} z\}$ yields the type $\mathcal{J}_{1,2}^{(2,2)}$ Janet basis $\{z_2 - \frac{x}{y^2} z_1, \; z_{1,x} + \frac{3}{x} z_1, \; z_{1,y}\}$. $\square$

LEMMA 2.8   If a Janet basis $\{z_x + A_1 z, z_{yyy} + B_1 z_{yy} + B_2 z_y + B_3 z\}$ of type $\mathcal{J}_{3,3}^{(1,2)}$ has a component $\{z_1 \equiv z_x + az, z_2 \equiv z_y + bz\}$ of type $\mathcal{J}_1^{(1,2)}$, its quotient is of type $\mathcal{J}_{2,2}^{(2,2)}$ and is explicitly given by

$$\{z_1, z_{2,x} + A_1 z_2, z_{2,yy} + (B_1 - b)z_{2,y} + (B_2 - bB_1 - 2b_y + b^2)z_2\}.$$

If it has a component $\{z_1 \equiv z_x + a_1 z, z_2 \equiv z_{yy} + b_1 z_y + b_2 z\}$ of type $\mathcal{J}_{2,2}^{(1,2)}$, its quotient is of type $\mathcal{J}_{1,1}^{(2,2)}$ and is explicitly given by

$$\{z_1, z_{2,x} + A_1 z_2, z_{2,y} + (B_1 - b_1)z_2\}.$$

Construction of Janet Bases. It is difficult to construct examples for the various types of Janet bases by prescribing the coefficients. The reason is that finding sets of coefficients satisfying the coherence conditions comes down to solving systems of nonlinear pde's for which solution algorithms are not known.

A different approach starts from the solutions. A basis for the solution space is given, and the problem is to generate a Janet basis with this solution space. For a single linear ode the answer has been given in (2.4), providing explicit expressions for the coefficients in terms of a fundamental system. For the two second order Janet bases of type $\mathcal{J}^{(1,2)}$ the answer is to be obtained in Exercises 2.9 and 2.10.

It should be emphasized, however, that proceeding in this way only special classes of Janet bases may be obtained, i.e., those that are completely

76

reducible with first order right factors. Due to the requirement that the Janet basis to be constructed has the rational function field of the independent variables as base field, the fundamental system has to be hyperexponential. Any other choice in general will lead to coefficients that are not in this base field. This problem exists already for linear ode's.

An algorithm will be described now which is based on the constructive proof of a theorem due to Lie [112], vol. I, page 183. In the special case where the dependency on the integration constants is linear, the desired system is also linear and homogeneous. Assume the following system of equations is given.

$$z_1 \equiv C_1 z_{1,1} + \ldots + C_r z_{1,r},$$

$$\vdots \qquad \qquad \vdots \qquad \qquad (2.36)$$

$$z_m \equiv C_1 z_{m,1} + \ldots + C_r z_{m,r}$$

where $z_{j,k}$, $j = 1, \ldots, m$, $k = 1, \ldots, r$ are given functions of x and y, $z_1, \ldots, z_m$ are undetermined functions of x and y, and the C_k are constants to be eliminated. To this end, by a suitable number of differentiations, an enlarged system of rank r is generated from which the C_k may be eliminated. Upon resubstitution, the desired system of pde's is obtained. The algorithm $ConstructJanetBasis$ takes expressions (2.36) as input and returns the corresponding Janet basis in the specified term ordering.

ALGORITHM 2.12 $ConstructJanetBasis(\{z_1, \ldots z_m\}, O)$. Given a system of expressions of the form (2.36), the Janet basis in the term ordering O is returned such that (2.36) is its general solution.

$S1$: *Set up linear system.* Determine d such that $(d+1)(d+2) > r$.

$S2$: *Rank?* Generate linear system (L) for $C_1, \ldots, C_r$ by differentiating w.r.t. x and y up to order d. It has the form with $j = 1, \ldots, m$.

$$C_1 z_{j,1} + \ldots + C_r z_{j,r} = z_j,$$
$$C_1 z_{j,1,x} + \ldots + C_r z_{j,r,x} = z_{j,x},$$
$$C_1 z_{j,1,y} + \ldots + C_r z_{j,r,y} = z_{j,y}, \qquad (L)$$
$$\vdots \qquad \vdots \qquad \qquad \vdots$$
$$C_1 z_{j,1,yy\ldots} + \ldots + C_r z_{j,r,yy\ldots} = z_{j,yy\ldots},$$

$S3$: *Rank?* If the rank of the coefficient matrix is less than r, set $d := d+1$ and goto $S2$.

$S4$: *Determine C's.* Take a subsystem of (L) of rank r and determine $C_1, \ldots, C_r$ from it.

$S5$: *Generate Janet basis.* Substitute the values for the C's obtained in step $S4$ into the remaining equations and generate a Janet basis. If its solution space has dimension r, return it, otherwise set $d := d+1$ and goto $S2$.

Linear Differential Equations 77

EXAMPLE 2.56   For $m = 1$, $r = 2$ let $z_1 = z = C_1(x^2 - y^2) + C_2xy$. In step $S1$, $d = 1$ is found. Step $S2$ generates the system of equations

$$C_1(x^2 - y^2) + C_2xy = z, \quad 2C_1x + C_2y = z_x, \quad -2C_1y + C_2x = z_y.$$

The rank of the coefficient matrix is two. In step $S4$ the last two equations are employed for determining C_1 and C_2 with the result

$$C_1 = \frac{x}{2(x^2 + y^2)}z_x - \frac{y}{2(x^2 + y^2)}z_y, \quad C_2 = \frac{y}{x^2 + y^2}z_x + \frac{x}{x^2 + y^2}z_y.$$

Substitution into the first equation yields a single first order equation for z. Therefore in step $S5$ the value of d is increased to 2 and the algorithm continues with step $S2$ where the system $2C_1 = z_{xx}$, $C_2 = z_{xy}$, $-2C_1 = z_{yy}$ is generated. Substituting C_1 and C_2 and joining the previously obtained first order equation leads in step $S5$ to the Janet basis

$$z_x + \frac{y}{x}z_y - \frac{2}{x}z = 0, \quad z_{yy} - \frac{2y}{x^2 + y^2}z_y + \frac{2}{x^2 + y^2}z = 0 \qquad (2.37)$$

which is the final answer.   ⬚

2.4   Solving Partial Differential Equations

Until now the goal has been to obtain information on the solutions of linear pde's without determining them explicitly. Furthermore, it has been shown how given pde's may be replaced by simpler ones, e. g. by decomposing them into lower order equations. A common feature of these efforts was that the necessary calculations took their course within the base field of the originally given equations. In this section, the goal is to determine solutions in closed form. This may require to enlarge the function field. Various cases corresponding to special values of n and m, the number of independent and dependent variables respectively, and the number N of equations will be considered. Except in the last subsection, all equations are of first order.

Linear Equations for a Single Function. First order systems containing a single dependent variable and any number of independent variables, i.e., $m = 1$ and n arbitrary, have been studied extensively in the literature. The discussion in this section follows closely the *Encyklopädie* article by von Weber [187], or Chapter II in Goursat's book [57], see also Kamke [87], page 45 ff. At first, $N = 1$ is assumed. Let the function $z(x_1, \ldots, x_n)$ be determined by the homogeneous equation

$$Az \equiv a_1\frac{\partial z}{\partial x_1} + a_2\frac{\partial z}{\partial x_2} + \ldots + a_n\frac{\partial z}{\partial x_n} = 0. \qquad (2.38)$$

78

In particular there is no term proportional to z, and $a_k \equiv a_k(x_1, \ldots, x_n)$ belongs to $\mathbb{Q}(x_1, x_2, \ldots, x_n)$ or an extension of it. A *fundamental system* or *integral basis* for this equation is a set of $n-1$ independent functions

$$\psi_1(x_1, \ldots, x_n), \ldots, \psi_{n-1}(x_1, \ldots, x_n)$$

such that $A\psi_k = 0$ for $k = 1, \ldots, n-1$. The following theorem reduces the solution of (2.38) to solving a system of ode's. Its proof may be found in the literature quoted above.

THEOREM 2.19   *The general solution of equation (2.38) may be described as follows.*

i) *A fundamental system for (2.38) is given by n-1 first integrals of the so-called characteristic system*

$$\frac{dx_1}{dt} = a_1(x_1, \ldots, x_n), \ldots, \frac{dx_n}{dt} = a_n(x_1, \ldots, x_n).$$

ii) *The general solution of (2.38) is given by an undetermined function $\Psi(\psi_1, \ldots, \psi_{n-1})$ of the first integrals.*

It should be noted, however, that there is no guarantee that the fundamental system may always be explicitly determined. The reason is that solving the characteristic system amounts to solving general first order ode's for which an algorithm does not exist.

There is a remarkable generalization of the above theorem to inhomogeneous *quasilinear* equations by Kamke [86] II, §3, no. 15.

THEOREM 2.20   *A solution of the quasilinear first order equation*

$$a_1 \frac{\partial z}{\partial x_1} + a_2 \frac{\partial z}{\partial x_2} + \ldots + a_n \frac{\partial z}{\partial x_n} = r \tag{2.39}$$

where $a_k \equiv a_k(x_1, \ldots, x_n, z)$ for $k = 1, \ldots, n$ and $r \equiv r(x_1, \ldots, x_n, z)$ may be obtained as follows. Determine any integral $w = \psi(x_1, \ldots, x_n, z)$ of

$$a_1 \frac{\partial w}{\partial x_1} + a_2 \frac{\partial w}{\partial x_2} + \ldots + a_n \frac{\partial w}{\partial x_n} + r \frac{\partial w}{\partial z} = 0$$

by Theorem 2.19 and solve $\psi(x_1, \ldots, x_n, z) = C$ for z; $C \neq 0$ is a constant.

The special case $n = 2$ of the above theorem will occur frequently later on in this book. Therefore it is described in more detail in the following corollary.

COROLLARY 2.7   *The general solution of the linear partial differential equation*

$$z_x + a_1 z_y + a_2 z + a_3 = 0 \tag{2.40}$$

with $a_k \equiv a_k(x, y) \in \mathbb{Q}(x, y)$, $k = 1, \ldots, 3$ may be described as follows. Let $\phi(x, y) = const$ be the first integral obtained from the solution of the first order ode $\frac{dy}{dx} = a_1(x, y)$. Let $\bar{y} = \phi(x, y)$ and assume that the inverse $y = \bar{\phi}(x, \bar{y})$ exists. Then the general solution of (2.40) is

$$z(x, y) = \left(\Phi(\phi) - \int a_3(x, \bar{\phi}) \exp\left(\int a_2(x, \bar{\phi}) dx \right) dx \right) \exp\left(- \int a_2(x, \bar{\phi}) dx \right) \Big|_{\bar{y} = \phi} \tag{2.41}$$

with Φ an undetermined function of its argument.

PROOF   It is based on the procedure described in Kamke [87], Section 4.2. A homogeneous equation corresponding to (2.40) for a new function $w(x, y, z)$ is $w_x + a_1 w_y - (a_2 z + a_3) w_z = 0$. A first integral ϕ is obtained from $\frac{dy}{dx} = a_1$. If it is applied for introducing a new variable $\bar{y}$, the equation $\bar{w}_x - (a_2 z + a_3)|_{y=\bar{\phi}} \bar{w}_z = 0$ for $\bar{w}$ is obtained with the first integral

$$\psi = z \exp \int a_2(x, \bar{\phi}) dx + \int a_3(x, \bar{\phi}) \exp\left(\int a_2(x, \bar{\phi}) dx \right) dx.$$

Consequently, the general solution $w = \Psi(\phi, \psi)$ is obtained with an undetermined function Ψ. Resubstituting $\bar{y} = \phi(x, y)$ and solving for z yields (2.41). ▯

This corollary is particularly useful if a special solution of the homogeneous equation is already known, then solving the inhomogeneous equation is reduced to an integration.

EXAMPLE 2.57   Let $2xz_x + 3yz_y = 1$ be the given equation. With the notation of Corollary 2.7, $a_1 = \frac{3y}{2x}$, $a_2 = 0$ and $a_3 = -\frac{1}{2x}$. The first order equation $\frac{dy}{dx} = \frac{3y}{2x}$ leads to $\phi(x, y) = \frac{y^2}{x^3}$. From (2.41) the general solution is obtained in the form $z(x, y) = \Phi\left(\frac{y^2}{x^3} \right) - \frac{1}{2} \log x$ with Φ an undetermined function of its argument. ▯

Now, let $N > 1$. It is interesting to see how several concepts like integrability conditions and reductions have evolved in the course of time for this particular case. The subsequent theorem shows how the *Jacobian normal form* (Kamke [87], Chapter 6.2) is obtained naturally as a Janet basis.

THEOREM 2.21   *Let a first order system for a function $z(x_1, \ldots, x_n)$ be given as*

$$a_{1,1} \frac{\partial z}{\partial x_1} + a_{1,2} \frac{\partial z}{\partial x_2} + \ldots + a_{1,n} \frac{\partial z}{\partial x_n} = 0,$$

$$a_{2,1} \frac{\partial z}{\partial x_1} + a_{2,2} \frac{\partial z}{\partial x_2} + \ldots + a_{2,n} \frac{\partial z}{\partial x_n} = 0,$$

$$\cdots \qquad\qquad \cdots \tag{2.42}$$

$$a_{k,1} \frac{\partial z}{\partial x_1} + a_{k,2} \frac{\partial z}{\partial x_2} + \ldots + a_{k,n} \frac{\partial z}{\partial x_n} = 0.$$

80

The coefficients $a_{i,j}(x_1, \ldots, x_n)$ belong to some function field $\mathcal{F}$. For the variable ordering $x_1 > x_2 > \ldots > x_n$, the Janet basis for this system is identical to the Jacobian normal form, possibly after a new enumeration of the x_i,

$$\frac{\partial z}{\partial x_r} + b_{r,r+1}\frac{\partial z}{\partial x_{r+1}} + \ldots + b_{r,n}\frac{\partial z}{\partial x_n} = 0,$$

$$\frac{\partial z}{\partial x_{r-1}} + b_{r-1,r+1}\frac{\partial z}{\partial x_{r+1}} + \ldots + b_{r-1,n}\frac{\partial z}{\partial x_n} = 0,$$

$$\ldots \tag{2.43}$$

$$\frac{\partial z}{\partial x_1} + b_{1,r+1}\frac{\partial z}{\partial x_{r+1}} + \ldots + b_{1,n}\frac{\partial z}{\partial x_n} = 0$$

with $1 \leq r \leq n$ and $b_{i,j} \in \mathcal{F}$.

PROOF   By algebraic reduction of the system (2.42), the following autoreduced form may always be achieved, possibly after a new enumeration of the independent variables.

$$\frac{\partial z}{\partial x_p} + \bar{a}_{p,p+1}\frac{\partial z}{\partial x_{p+1}} + \ldots + \bar{a}_{p,n}\frac{\partial z}{\partial x_n} = 0,$$

$$\frac{\partial z}{\partial x_{p-1}} + \bar{a}_{p-1,p+1}\frac{\partial z}{\partial x_{p+1}} + \ldots + \bar{a}_{p-1,n}\frac{\partial z}{\partial x_n} = 0,$$

$$\ldots \qquad \ldots \tag{2.44}$$

$$\frac{\partial z}{\partial x_1} + \bar{a}_{1,p+1}\frac{\partial z}{\partial x_{p+1}} + \ldots + \bar{a}_{1,n}\frac{\partial z}{\partial x_n} = 0$$

with $1 \leq p \leq n$. The integrability conditions have the form

$$\frac{\partial}{\partial x_j}\left(\frac{\partial z}{\partial x_i} + \bar{a}_{i,p+1}\frac{\partial z}{\partial x_{p+1}} + \ldots + \bar{a}_{i,n}\frac{\partial z}{\partial x_n}\right)$$
$$- \frac{\partial}{\partial x_i}\left(\frac{\partial z}{\partial x_j} + \bar{a}_{j,p+1}\frac{\partial z}{\partial x_{p+1}} + \ldots + \bar{a}_{j,n}\frac{\partial z}{\partial x_n}\right) = 0$$

for $1 \leq i < j \leq p$. All second order terms in these expressions cancel each other identically. The remaining first order equations are reduced w.r.t. to (2.44). If the result is always zero, (2.44) is the desired Janet basis. Otherwise those that do not vanish are added to the system (2.44) and the process is repeated. The results described in the preceding section guarantee that this process terminates after a finite number of iterations. ☐

EXAMPLE 2.58   The following system of pde's

$$z_{x_2} + (x_4 x_3 - x_2)z_{x_3} + (x_4 x_3 x_1 - x_2 x_1 + x_2)z_{x_4} = 0,$$

$$z_{x_1} + (x_4 + x_2 - 3x_1)z_{x_3} + (x_4 x_1 + x_3 + x_2 x_1)z_{x_4} = 0$$

for a function $z \equiv z(x_1, x_2, x_3, x_4)$ has been discussed by Imschenetzky [79]; as usual $z_{x_i} \equiv \frac{\partial z}{\partial x_i}$. The Janet basis for $x_1 > x_2 > x_3 > x_4$ is

$$\left\{z_{x_3} + x_1 z_{x_4}, \ z_{x_2} + x_2 z_{x_4}, \ z_{x_1} + (x_3 + 3x_1^2)z_{x_4}\right\}$$

Linear Differential Equations

and for $x_4 > x_3 > x_2 > x_1$ it is

$$\left\{ z_{x_2} - \frac{x_2}{x_3 + 3x_1^2} z_{x_1}, \quad z_{x_3} - \frac{x_1}{x_3 + 3x_1^2} z_{x_1}, \quad z_{x_4} + \frac{1}{x_3 + 3x_1^2} z_{x_1} \right\}.$$

From each of these Janet bases it is obvious that the general solution depends on an undetermined function of a single argument. Explicitly it is given by $\phi(x_4 - x_1 x_3 - \frac{1}{2}x_2^2 - x_1^3)$. $\square$

Linear Equations with Symmetries. In some applications it is advantageous to transform a linear equation for a function $f(x_1, x_1, \ldots, x_n)$

$$Af \equiv a_1 \frac{\partial f}{\partial x_1} + a_2 \frac{\partial f}{\partial x_2} + \ldots + a_n \frac{\partial f}{\partial x_n} = 0 \tag{2.45}$$

to a new set of variables $y_1, \ldots, y_n$. Let them be defined by $y_k \equiv \psi_k(x_1, \ldots, x_n)$ with the inversions $x_k \equiv \phi_k(y_1, \ldots, y_n)$. In these variables (2.45) becomes

$$B_1 \frac{\partial g}{\partial y_1} + B_2 \frac{\partial g}{\partial y_2} + \ldots + B_n \frac{\partial g}{\partial y_n} = 0 \tag{2.46}$$

where $B_k = A\psi_k|_{x_1 = \phi_1, \ldots, x_n = \phi_n}$ for $1 \leq k \leq n$ and $g \equiv f(\phi_1, \ldots, \phi_n)$. The proof of this expression is considered in Exercise 2.21.

EXAMPLE 2.59   Let $A = x\partial_x + y\partial_y$ by given and new variables u and v be introduced by $x = uv$, $y = \frac{v}{u}$ with the inverse $u = \sqrt{\frac{x}{y}}$, $v = \sqrt{xy}$. Applying the above rule yields the transformed operator $v\partial_v$. $\square$

Of particular interest are those nontrivial transformations leaving the equation $Af = 0$ unchanged. Due to its importance a special term is introduced for them.

DEFINITION 2.10   *(Symmetry of a linear pde) If a linear homogeneous pde (2.45) is invariant w.r.t. a variable transformation, it is called a symmetry of (2.45). This is also expressed by saying that the transformation is admitted by (2.45).*

It may occur that the totality of transformations admitted by a given equation (2.45) make up a group depending on one or more parameters. For this concept the reader is referred to Section 3.1. A convenient criterion for the invariance w.r.t. a group may be given in terms of its infinitesimal generators as follows.

THEOREM 2.22   *A linear homogeneous pde (2.45) is invariant under the group with the infinitesimal generator $X = \xi_1 \frac{\partial}{\partial x_1} + \ldots + \xi_n \frac{\partial}{\partial x_n}$ if and only if there holds $[A, X] = \rho(x_1, \ldots, x_n)A$.*

The commutator $[A, X]$ is defined by equation (3.10) in the next chapter. The solution procedure for an equation (2.45) varies with the type of the

82

symmetry group that it admits. A detailed discussion may be found in the original work of Lie [108], § 10. Two special cases that are of particular importance later on in this book are described next. The first one deals with a *solvable* symmetry group, for its definition see page 118.

THEOREM 2.23 *(Lie 1877) Let the linear homogeneous pde to be solved be given in the form*

$$Af \equiv a_1 \frac{\partial f}{\partial x_1} + a_2 \frac{\partial f}{\partial x_2} + \ldots + a_n \frac{\partial f}{\partial x_n} = 0.$$

Assume that it admits

$$X_k \equiv \xi_{k,1} \frac{\partial}{\partial x_1} + \xi_{k,2} \frac{\partial}{\partial x_2} + \ldots + \xi_{k,n} \frac{\partial}{\partial x_n}$$

for $k = 1, \ldots, n - 1$, *and that the relations* $[X_i, X_k] = \sum_{\rho=1}^{k-1} c_{i,k}^\rho X_\rho$ *for* $i < k$ *and* $i = 1, \ldots, n - 1$, $k = 2, \ldots, n - 1$, $c_{i,k}^\rho$ *constant are valid. Then a fundamental system may be obtained as follows. Let the determinant* Δ *be defined by*

$$\Delta = \begin{vmatrix} a_1 & a_2 & \cdots & a_n \\ \xi_{1,1} & \xi_{1,2} & \cdots & \xi_{1,n} \\ \vdots & & & \vdots \\ \xi_{n-1,1} & \xi_{n-1,2} & \cdots & \xi_{n-1,n} \end{vmatrix} \neq 0.$$

Then an integral Φ_1 *of* $Af = 0$ *is*

$$\Phi_1 = \int \frac{1}{\Delta} \begin{vmatrix} dx_1 & dx_2 & \cdots & dx_n \\ a_1 & a_2 & \cdots & a_n \\ \xi_{1,1} & \xi_{1,2} & \cdots & \xi_{1,n} \\ \vdots & & & \vdots \\ \xi_{n-2,1} & \xi_{n-2,2} & \cdots & \xi_{n-2,n} \end{vmatrix}. \tag{2.47}$$

If new variables $\bar{x}_1 = \Phi_1$, $\bar{x}_k = x_k$ *for* $k = 2, \ldots, n$ *are introduced, derivatives w.r.t. to* $\bar{x}_1$ *do not occur any more,* $\bar{A}f = 0$ *and* $\bar{X}_k f = 0$ *for* $k = 1, \ldots, n - 3$ *are a complete system allowing* $\bar{X}_{n-2}$. *This problem is of the same type as the original one with the number of variables diminished by one for which the same procedure may be repeated until* $n - 1$ *integrals are obtained.*

This theorem will be employed in Chapter 7 for solving ode's with symmetries. In order to apply it, the knowledge of a sufficient number of symmetries of a certain kind is required. Because there is no algorithmic method for determining them, this knowledge must come from the application from which the equation $Af = 0$ originates. There are other methods for taking advantage of symmetries for solving first order pde's or systems of pde's. A few cases are described in the subsequent examples.

$$\text{Linear Differential Equations} \qquad 83$$

EXAMPLE 2.60   Let $A \equiv \partial_x + \frac{y}{z}\partial_y + \frac{y}{z}\partial_z$, $X_1 \equiv \partial_x$, $X_2 \equiv x\partial_x + y\partial_y + z\partial_z$. Due to $[X_1, A] = 0$, $[X_2, A] = -A$, $[X_1, X_2] = X_1$ the above theorem can be applied. In the first step $\Delta = \frac{y}{x}(y - z)$ and the integral for (2.47) is

$$\int \frac{z}{y}\frac{1}{y - z}\left(\frac{y}{z}dy - \frac{y}{z}dz\right) = 2\log(y - z).$$

Therefore $\Phi_1 = y - z$ may be chosen. Introducing new variables $\bar{x} = x$, $\bar{y} = y - z$, $\bar{z} = z$ yields $\bar{A} = \bar{x}\partial_{\bar{x}}$ and $\bar{X}_1 = \partial_{\bar{x}}$. Integration and resubstitution gives $\Phi_2 = x - z + (y - z)\log y$.   □

THEOREM 2.24   (Lie 1874) Let the linear homogeneous pde $Af = 0$ as in the preceding theorem allow q infinitesimal generators $B_1, \ldots, B_q$. Assume that there are exactly m relations of the form

$$B_k = \beta^k_{m+1}B_{m+1} + \ldots + \beta^k_q B_q + \alpha A, \quad k = 1, \ldots, m.$$

Then the coefficients β^k_j obey $A\beta^k_j = 0$, i.e., they are integrals of the given linear pde $Af = 0$.

The proof may be found in Lie[107], §8, Theorem VII. A simple application is given next.

EXAMPLE 2.61   Let $A = \partial_x + z\partial_y - \frac{z(z^2 + 1)}{2x}\partial_z$. It allows the three symmetry generators $U_1 = \partial_y$, $U_2 = x\partial_x + y\partial_y$ and $U_3 = xy\partial_x + \frac{1}{2}y^2\partial_y - xz^2\partial_z$ because $[A, U_1] = 0$, $[A, U_2] = A$, $[A, U_3] = (xz + y)A$. The relation

$$U_3 + \left(\frac{2xz}{z^2 + 1} - y\right)U_2 + \left(\frac{2xz(xz - y)}{z^2 + 1} + \frac{y^2}{2}\right)U_1 - \frac{2x^2 z}{z^2 + 1}A = 0$$

yields the integrals $\Phi_1 = \frac{2xz}{z^2 + 1} - y$ and $\Phi_2 = \frac{2xz(xz - y)}{z^2 + 1} + \frac{y^2}{2}$.   □

Linear Systems for m=1, n=2 and N=2. In this subsection various kinds of linear pde's are considered that will occur frequently in the applications later on in this book. Usually its solution may be expressed in terms of path integrals in the $x - y$–plane. If $p \equiv p(x, y)$ and $q \equiv q(x, y)$ are two functions of x and y obeying $p_y = q_x$, it is defined by

$$\oint pdx + qdy = \int_{x_0}^x p(\xi, y_0)d\xi + \int_{y_0}^y q(x, \eta)d\eta = \int_{y_0}^y q(x_0, \eta)d\eta + \int_{x_0}^x p(\xi, y)d\xi. \tag{2.48}$$

The lower integration limit of the first integral contributes the integration constant. As a rule, any integration problem will be considered as solved as soon as it has been reduced to a form that is covered by the lemmata of this subsection. A straightforward generalization of a linear homogeneous ode is considered first.

84

LEMMA 2.9   *The general solution of the homogeneous system*

$$z_x + az = 0, \qquad z_y + bz = 0$$

where $a \equiv a(x,y)$, $b \equiv b(x,y)$ *and* $a_y = b_x$ *is* $z = \exp\left(-\oint a\,dx + b\,dy\right)$.

EXAMPLE 2.62   Consider $z_x + xy^2 z = 0$ and $z_y + x^2 yz = 0$. The integral (2.48) yields

$$\oint xy^2\,dx + x^2 y\,dy = \int \xi y_0^2\,d\xi + \int x^2 \eta\,d\eta$$

$$= \tfrac{1}{2}y_0^2(x^2 - x_0^2) + \tfrac{1}{2}x^2(y^2 - y_0^2) = \tfrac{1}{2}x^2 y^2 + C$$

where C is a constant, and finally $z = C\exp\left(-\tfrac{1}{2}x^2 y^2\right)$.   ☐

LEMMA 2.10   *The general solution of the inhomogeneous system*

$$z_x + az = p, \qquad z_y + bz = q$$

where a, b, p *and* q *depend on* x *and* y *and obey the integrability conditions* $a_y = b_x$ *and* $p_y - q_x = aq - bp$ *may be written in the form*

$$z = z_0\left(\oint \frac{p\,dx + q\,dy}{z_0} + C\right) \text{ where } z_0 = \exp\left(-\oint a\,dx + b\,dy\right)$$

is the solution of the homogeneous system with $p = q = 0$ *and* C *is a constant.*

PROOF   Assume that the desired solution has the form $z = C(x,y)z_0$ with z_0 a solution of the homogeneous system. Differentiation and substitution into the given system leads to $C_x = \frac{p}{z_0}$ and $C_y = \frac{q}{z_0}$ from which the given result is obvious.   ☐

EXAMPLE 2.63   Consider the system $z_x + yz = y$, $z_x + xz = x$. By Lemma 2.9 the homogeneous system has the solution $z_0 = Ce^{-xy}$. Substituting it into the integral of the preceding lemma leads to

$$z = e^{-xy}\left(\oint e^{xy}(y\,dx + x\,dy) + C\right) = Ce^{-xy} + 1.$$ ☐

A special second order system corresponding to Janet basis type $\mathcal{J}_{2,1}^{(1,2)}$ that will occur frequently later on is treated next.

LEMMA 2.11   *Let the function* $z(x,y)$ *be determined by the system of equations* $z_y + az_x = 0$, $z_{xx} + cz_x = 0$ *with* $(a_x - ac)_x = c_y$. *The nonconstant member of a fundamental system is given by*

$$z = \oint \exp\left(-\oint c\,dx + (a_x - ac)\,dy\right)(dx - a\,dy).$$

Linear Differential Equations 85

The proof of this lemma is the subject of Exercise 2.27.

EXAMPLE 2.64   An example of such a system is

$$z_y + \frac{x}{(x-1)y} z_x = 0, \quad z_{xx} - \frac{x^2 - 2x + 2}{x(x-1)} z_x = 0.$$

Applying the formulas given above yields for the nonconstant member of a fundamental system $z = \frac{1}{xy} e^x$. In this particular case the same answer may be obtained from the component $\left\{ z_x - \frac{x-1}{x} z, z_y + \frac{1}{y} z \right\}$ because the given system is completely reducible. In general, however, this is not true and the expression in the lemma must be applied. ⬜

Quasilinear Systems for m=1, n=2 and N≥2. Systems comprising more than two equations are the next topic. Moreover they do not need to be linear in the dependent variable any more.

LEMMA 2.12   *Let a system of first order quasilinear equations*

$$a_i z_x + b_i z_y = r_i, \quad 2 \le i \le N$$

for $z \equiv z(x, y)$ be given where $a_i \equiv a_i(x, y)$, $b_i \equiv b_i(x, y)$ and $r_i \equiv r_i(x, y, z)$, $a_1 b_2 - a_2 b_1 \ne 0$. Then z obeys

$$z_x = -\frac{b_1 r_2 - b_2 r_1}{a_1 b_2 - a_2 b_1}, \quad z_y = \frac{a_1 r_2 - a_2 r_1}{a_1 b_2 - a_2 b_1},$$

$$(a_1 b_2 - a_2 b_1) r_j + (a_2 b_j - a_j b_2) r_1 - (a_1 b_j - a_j b_1) r_2 = 0 \quad for \quad j \ge 3. \tag{2.49}$$

PROOF   The expressions for z_x and z_y follow by algebraic elimination from the first two equations. The remaining conditions either assure the coherence of the full system, or they reduce the originally given differential equations to an elimination problem. ⬜

Depending on the form of the right hand sides r_i, various alternatives have to be distinguished. If the r_i do not contain z, or if only r_1 and r_2 depend linearly on z and the remaining conditions are identically satisfied, the first two equations lead to a representation of z as a path integral by Lemma 2.9 or 2.10. If any r_i contains z, and if the conditions for $j \ge 3$ given in the above lemma may be solved for z, the general solution does not contain a constant, and the first two equations reduce to consistency conditions. If they are violated the full system is inconsistent. Systems of this kind occur in the next chapter when groups acting on a two-dimensional manifold are transformed into canonical form.

Partial Riccati-Like Systems: A Single Function. The decomposition of Janet bases into irreducible components has been shown to come down to

86

solving various kinds of partial differential equations. Their structure suggests to call them partial Riccati-like equations. In this section, those systems occuring in the decomposition of Janet bases of type $\mathcal{J}^{(1,2)}$ of order not higher than three will be considered.

First of all a few definitions are introduced. Two rational functions $f, g \in \mathbb{Q}(x, y)$ are defined to be an *integrable pair* (f, g) if $\partial_y f = \partial_x g$. Two integrable pairs (f, g) and (p, q) are said to be *equivalent*, denoted by " $\sim$ ", if there exists a nonzero $h \in \mathbb{Q}(x, y)$ such that $f - p = \frac{\partial_x h}{h}$ and $g - q = \frac{\partial_y h}{h}$; $\sim$ defines an equivalence relation on the set of integrable pairs.

Let a and b belong to $\mathbb{Q}(x, y)$. If they are not an integrable pair it may be possible that there exist two polynomials $p, q \in \mathbb{Q}[x, y]$ such that $(a + \frac{\partial_x p}{p}, b + \frac{\partial_y q}{q})$ is an integrable pair. In the above mentioned article by Li and Schwarz [104] an algorithm is described that allows determining a pair (p, q) for any given (a, b) if it does exist.

An element a of some larger function field comprising $\mathbb{Q}(x, y)$ is said to be a *hyperexponential* if both $\frac{\partial_x a}{a}$ and $\frac{\partial_y a}{a}$ belong to $\mathbb{Q}(x, y)$. If a is a hyperexponential, $(\frac{\partial_x a}{a}, \frac{\partial_y a}{a})$ is an integrable pair. Conversely, for an integrable pair (f, g), the hyperexponential $\exp{(\oint f dx + g dy)}$ may be constructed. If (a, b) and (p, q) are two integrable pairs, then $(f, g) \sim (p, q)$ if and only if the ratio of their hyperexponentials is in $\mathbb{Q}(x, y)$, possibly multiplied by some constant.

The simplest Riccati-like system involving a single unknown function is the subject of the next theorem.

THEOREM 2.25   *The first order Riccati-like system of pde's*

$$e_1 \equiv z_x + A_1 z^2 + A_2 z + A_3 = 0, \quad e_2 \equiv z_y + B_1 z^2 + B_2 z + B_3 = 0 \quad (2.50)$$

is coherent and its general solution depends on a single constant if and only if its coefficients satisfy the constraints

$$A_{1,y} - B_{1,x} - A_1 B_2 + A_2 B_1 = 0,$$
$$A_{2,y} - B_{2,x} - 2A_1 B_3 + 2B_1 A_3 = 0, \quad (2.51)$$
$$A_{3,y} - B_{3,x} - A_2 B_3 + A_3 B_2 = 0.$$

Let $A_k, B_k \in \mathbb{Q}(x, y)$, $A_1 \neq 0$, $B_1 \neq 0$ and system (2.51) be satsified. If (2.50) has a rational solution, one of the following alternatives applies.

i) *The general solution is rational and has the form*

$$\frac{1}{A_1} \frac{r_x}{r + C} + p = \frac{1}{B_1} \frac{r_y}{r + C} + p \quad (2.52)$$

   where $p, r \in \mathbb{Q}(x, y)$ and C is the integration constant.

ii) *There is a single one, or there are two inequivalent special rational solutions.*

Linear Differential Equations 87

PROOF The coefficient constraints are the integrability conditions for system (2.50). Substituing $z = \frac{u}{w}$ into the given equations yields

$$u_x + \frac{u(A_1 u - w_x)}{w} + A_2 u + A_3 w = 0, \quad u_y + \frac{u(B_1 u - w_y)}{w} + B_2 u + B_3 w = 0.$$

In order to obtain a linear system for u and w, the constraints

$$w_x - A_1 u + P w = 0, \quad w_y - B_1 u + Q w = 0 \tag{2.53}$$

are necessary where $P, Q \in \mathbb{Q}(x, y)$ still have to be determined. Substituting these conditions into the above system for u and w yields

$$u_x + (A_2 + P) u + A_3 w = 0, \quad u_y + (B_2 + Q) u + B_3 w = 0. \tag{2.54}$$

The consistency of the above substitution requires that the system comprising the equations (2.53) and (2.54) be a Janet basis. This is assured if

$$P_y - Q_x + A_1 B_3 - A_3 B_1 = P_y - Q_x + \tfrac{1}{2}(A_{2,y} - B_{2,x}) = 0.$$

The second form is obtained by applying (2.51). A possible choice for P and Q is $P = -\frac{1}{2} A_2$, $Q = -\frac{1}{2} B_2$ respectively. It leads to the linear system

$$u_x + \tfrac{1}{2} A_2 u + A_3 w = 0, \quad u_y + \tfrac{1}{2} B_2 u + B_3 w = 0,$$

$$w_x - A_1 u - \tfrac{1}{2} A_2 w = 0, \quad w_y - B_1 u - \tfrac{1}{2} B_2 w = 0.$$

Let $u = C_1 u_1 + C_2 u_2$, $w = C_1 w_1 + C_2 w_2$ be its general solution. If the ratio $\frac{u}{w}$ is rational, it may be rewritten as

$$z = \frac{u}{w} = \frac{C \dfrac{u_1}{w_1} + \dfrac{u_2}{w_2}}{C + \dfrac{w_2}{w_1}} = \frac{\dfrac{u_2}{w_1} - \dfrac{u_1}{w_1} \dfrac{w_2}{w_1}}{C + \dfrac{w_2}{w_1}} + \frac{u_1}{w_1}.$$

Applying

$$\left(\frac{w_2}{w_1}\right)_x = \frac{w_{2,x}}{w_1} - \frac{w_1 w_{1,x}}{w_1^2} = A_1\left(\frac{u_2}{w_2} - \frac{u_1 w_2}{w_1^2}\right),$$

$$\left(\frac{w_2}{w_1}\right)_y = \frac{w_{2,y}}{w_1} - \frac{w_1 w_{1,y}}{w_1^2} = B_1\left(\frac{u_2}{w_2} - \frac{u_1 w_2}{w_1^2}\right)$$

the assignment $r = \frac{w_2}{w_1}$ yields the representation of case i). On the other hand, if $\frac{u_1}{w_1}$ is rational and $\frac{u_2}{w_1}$ is not or vice versa, (2.50) has exactly one rational solution. If neither are rational, there is none. If both are rational but the ratio $\frac{u}{w}$ is not, there are exactly two rational solutions. These cases cover alternative ii). $\square$

Based on this theorem, an algorithm for determining rational solutions of systems (2.50) may be designed as follows. The solution procedure is initiated with the first equation. If its general solution is rational, it contains a

88

constant w.r.t. x and has the general form $\dfrac{r_x}{r + C(y)} + p$ with $r, p \in \mathbb{Q}(x, y)$.
Substituting this expression into the second equation yields the first order
equation

$$r_x C_y - (p_y + B_1 p^2 + B_2 p + B_3) C^2$$
$$- \left[r_{xy} + 2p_y r + 2B_1 p(r_x + pr) + B_2(r_x + 2pr) + 2B_3 r \right] C$$
$$- (r_{xy} + p_y r)r + r_x r_y - B_1(r_x + pr)^2 - B_2(r_x + pr)r - B_3 r^2 = 0$$
$$(2.55)$$

for the y–dependence of C. Because the dependence on x is completely
explicit, it may be decomposed w.r.t. powers of x. The coefficient of C^2
has the form of the second equation of (2.50) for p. Therefore, depending
on whether or not p satisfies this equation, this leads to a system of linear
or Riccati ode's. If its general solution is rational, the same is true for the
full system. Otherwise special solutions may exist leading to special rational
solutions of the full system. If the first equation does not allow a general
rational solution, special rational solutions may exist. Those also fullfilling
the second equation are special solutions of the full system. The following
algorithm is organized according to these steps.

ALGORITHM 2.13   *RationalSolutionPartialRiccati1* (e_1, e_2). Given two partial Riccati-like equations (2.50), its general rational solution, or the maximal
number of special rational solutions are returned, or *failed* if none exists.
$S1$: *Rational solutions of first equation.* Determine the rational solutions
of e_1. If there are none return *failed*.
$S2$: *Special rational solutions.* If the rational solutions obtained in $S1$
are special, substitute them into e_2 an return those that satisfy it.
$S3$: *Solve equation for C.* Substitute p and r as obtained in $S1$ into
(2.55) and solve the resulting system of ode's for C. If it does not have
any rational solutions return *failed*, otherwise substitute them into z
and return the result.

EXAMPLE 2.65   Consider the equations

$$z_x + z^2 + \frac{2xz}{x^2 - \dfrac{2}{y}} - \frac{\frac{2}{y}}{\left(x^2 - \dfrac{2}{y}\right)^2} = 0, \quad z_y + \frac{x}{y}z^2 + \left(\frac{2}{y - \dfrac{2}{x^2}} - \frac{1}{y}\right)z = 0.$$

Its coefficients obey the coherence conditions (2.51). In step $S1$ of the above
algorithm the general solution of the first equation is obtained in the form

$$z\big(x, y, \bar{C}(y)\big) = \frac{1}{\bar{C}(y) + x} - \frac{xy}{x^2 y - 2}.$$

In step $S3$ the single equation $\bar{C}_y + \bar{C} = 0$ for the y-dependence of $\bar{C}$ is
obtained. Its solution $\bar{C} = \dfrac{C}{y}$ with C a constant yields the general rational
solution $z = \dfrac{y}{xy + C} - \dfrac{xy}{x^2 y - 2}$ of the full system.   □

Linear Differential Equations

The system (2.50) is invariant w.r.t. exchange of x and y. Therefore it makes no difference if the algorithm is started by solving the equation determining the y-derivative, this is the subject of Exercise 2.25.

The special case $B_1 = 0$ of (2.50), i.e., a Riccati equation for the x-dependence and a linear equation for the y-dependence is of some interest in later applications. In Exercise 2.27 it is shown that the solution to any such system may be written as

$$z = \frac{1}{2A_1}\left(\frac{A_{1,x}}{A_1} - A_2\right) + \frac{1}{A_1}C \tag{2.56}$$

where $C \equiv C(x)$ satisfies the Riccati equation

$$C_x + C^2 + \frac{1}{2}\left(\frac{A_{1,x}}{A_1} - A_2\right)_x + A_1 A_3 - \frac{1}{4}\left(\frac{A_{1,x}}{A_1} - A_2\right)^2 = 0. \tag{2.57}$$

The terms not containing C are *independent of y*. This property makes it into a genuine ode problem.

A system closely related to the preceding one is

$$z_{xx} + z_x^2 + A_1 z_x + A_2 = 0, \quad z_y + B_1 z_x + B_2 = 0 \tag{2.58}$$

which occurs in the analysis of several symmetry classes later on in this book. The coherence conditions for its coefficients are

$$B_{1,xx} + 2B_{2,x} - (A_1 B_1)_x = A_{1,y},$$
$$B_{2,xx} + A_1 B_{2,x} - A_2 B_{1,x} - (A_2 B_1)_x = A_{2,y}.$$

The first equation (2.58) is a first order Riccati equation for z_x. If its general solution is rational, z is obtained from the system

$$z_x = \frac{r_x}{C + r} + p, \quad z_y = -\frac{B_1 r_x}{C + r} - B_1 p - B_2 \tag{2.59}$$

with $C \equiv C(y)$. The integrability condition for its right hand sides leads to the first order ode

$$r_x C_y - ((B_1 p + B_2)_x + p_y)C^2$$
$$- \left[2r((B_1 p + B_2)_x + p_y) + (B_1 r_x + r_y)_x\right]C \tag{2.60}$$
$$+ r_x r_y - (B_1 r_x + r_y)_x r + B_1 r_x^2 = 0$$

for the y-dependence of C. If $(B_1 p + B_2)_x + p_y = 0$ it is a linear equation, otherwise a Riccati equation. Substituting the solution for C into (2.59), the solution for z is obtained by Lemma 2.10. If the first equation (2.58) has only special rational solutions, they are substituted into the second one and z is again obtained by Lemma 2.10.

Partial Riccati-Like Systems: Two Functions. The next Riccati-like system to be considered contains two unknown functions z_1 and z_2, depending

90

on x and y. Originally it occured in Lie's investigation of certain second order ode's. As will be seen below, the decomposition of various types of Janet bases leads to systems of this kind as well. In order to make the algorithms to be designed more generally applicable, the following class of systems will be considered.

$$
\begin{aligned}
e_1 &\equiv \bar{z}_{1,x} + a_0 \bar{z}_1^2 + a_1 \bar{z}_1 + a_2 \bar{z}_2 + a_3 = 0, \\
e_2 &\equiv \bar{z}_{1,y} + b_0 \bar{z}_1 \bar{z}_2 + b_1 \bar{z}_1 + b_2 \bar{z}_2 + b_3 = 0, \\
e_3 &\equiv \bar{z}_{2,x} + c_0 \bar{z}_1 \bar{z}_2 + c_1 \bar{z}_1 + c_2 \bar{z}_2 + c_3 = 0, \\
e_4 &\equiv \bar{z}_{2,y} + d_0 \bar{z}_2^2 + d_1 \bar{z}_1 + d_2 \bar{z}_2 + d_3 = 0.
\end{aligned}
\tag{2.61}
$$

If $a_0 d_0 \neq 0$ is assumed, this system is coherent if either of the following sets of constraints is satisfied.

$$
\begin{aligned}
b_0 - d_0 = 0, \quad a_0 - c_0 = 0, \quad d_{0,x} + b_2 c_0 - c_2 d_0 = 0, \\
c_{3,y} - d_{3,x} + a_3 d_1 - b_3 c_1 - c_2 d_3 + c_3 d_2 = 0, \\
c_{2,y} - d_{2,x} + a_2 d_1 - b_2 c_1 - b_3 c_0 + 2 c_3 d_0 = 0, \\
c_{1,y} - d_{1,x} + a_1 d_1 - b_1 c_1 - c_0 d_3 + c_1 d_2 - c_2 d_1 = 0, \\
c_{0,y} - b_1 c_0 + c_1 d_0 = 0, \\
a_{3,y} - b_{3,x} - a_1 b_3 - a_2 d_3 + a_3 b_1 + b_2 c_3 = 0, \\
a_{2,y} - b_{2,x} - a_1 b_2 + a_2 b_1 - a_2 d_2 + a_3 d_0 + b_2 c_2 = 0, \\
a_{1,y} - b_{1,x} - a_2 d_1 + b_2 c_1 - 2 b_3 c_0 + c_3 d_0 = 0
\end{aligned}
\tag{2.62}
$$

or

$$
\begin{aligned}
d_1 = 0, \quad c_1 = 0, \quad c_0 = 0, \quad b_2 = 0, \quad b_0 = 0, \quad a_2 = 0, \\
d_{0,x} - c_2 d_0 = 0, \quad c_{3,y} - d_{3,x} - c_2 d_3 + c_3 d_2 = 0, \quad c_{2,y} - d_{2,x} + 2 c_3 d_0 = 0, \\
a_{3,y} - b_{3,x} - a_1 b_3 + a_3 b_1 = 0, \quad a_{1,y} - b_{1,x} - 2 a_0 b_3 = 0, \quad a_{0,y} - a_0 b_1 = 0.
\end{aligned}
\tag{2.63}
$$

If the first alternative (2.62) applies, new dependent variables z_1 and z_2 are introduced by

$$
\begin{aligned}
\bar{z}_1 &= \frac{1}{a_0} z_1 - \frac{1}{3 a_0} \left(a_1 + c_2 - \frac{\partial_x (a_0 d_0)}{a_0 d_0} \right), \\
\bar{z}_2 &= \frac{1}{d_0} z_2 - \frac{1}{3 d_0} \left(b_1 + d_2 - \frac{\partial_y (a_0 d_0)}{a_0 d_0} \right).
\end{aligned}
\tag{2.64}
$$

The system (2.61) corresponding to conditions (2.62) is transformed into the normalized coherent system

$$
\begin{aligned}
e_1 &\equiv z_{1,x} + z_1^2 + A_1 z_2 + A_2 z_1 + A_3 = 0, \\
e_2 &\equiv z_{1,y} + z_1 z_2 + B_1 z_2 + B_2 z_1 + B_3 = 0, \\
e_3 &\equiv z_{2,x} + z_1 z_2 + B_1 z_2 + B_2 z_1 + B_3 = 0, \\
e_4 &\equiv z_{2,y} + z_2^2 + D_1 z_2 + D_2 z_1 + D_3 = 0.
\end{aligned}
\tag{2.65}
$$

The new coefficients $A_1, \ldots, D_3$ are nonlinear expressions of the old ones $a_0, \ldots, d_3$. Due to the coherence they satisfy

$$
\begin{aligned}
A_{1,y} - B_{1,x} - A_2 B_1 + A_1 B_2 - A_1 D_1 + A_3 + B_1^2 &= 0, \\
A_{2,y} - B_{2,x} - A_1 D_2 + B_1 B_2 - B_3 &= 0, \\
A_{3,y} - B_{3,x} - A_2 B_3 - A_1 D_3 + A_3 B_2 + B_1 B_3 &= 0 \\
B_{1,y} - D_{1,x} + A_1 D_2 - B_1 B_2 + B_3 &= 0, \\
B_{2,y} - D_{2,x} + A_2 D_2 - B_2^2 + B_2 D_1 - B_1 D_2 - D_3 &= 0, \\
B_{3,y} - D_{3,x} + A_3 D_2 - B_2 B_3 - B_1 D_3 + B_3 D_1 &= 0.
\end{aligned}
\tag{2.66}
$$

EXAMPLE 2.66   Consider the equations

$$
\bar{z}_{1,x} + \bar{z}_1^2 + \frac{3y^2}{x^3} \bar{z}_1 + \frac{y^3}{x^4} \bar{z}_2 - \frac{3x^2 y^2 - 3y^4}{x^6} = 0, \quad \bar{z}_{1,y} + \bar{z}_1 \bar{z}_2 + \frac{2x^2 y - y^3}{x^5} = 0,
$$

$$
\bar{z}_{2,x} + \bar{z}_1 \bar{z}_2 - \frac{2x^2 y + y^3}{x^5} = 0, \quad \bar{z}_{2,y} + \bar{z}_2^2 + \frac{1}{x} \bar{z}_1 + \frac{3y}{x^2} \bar{z}_2 + \frac{x^2 + 3y^2}{x^4} = 0.
$$

Its coefficients satisfy the conditions (2.62), but $\bar{z}_{1,y} \neq \bar{z}_{2,x}$. Introducing new variables by $\bar{z}_1 = z_1 - \frac{y^2}{x^3}$, $\bar{z}_2 = z_2 - \frac{y}{x^2}$ yields

$$
z_{1,x} + z_1^2 + \frac{y^2}{x^3} z_1 + \frac{y^3}{x^4} z_2 = 0 \quad z_{1,y} + z_1 z_2 - \frac{y}{x^2} z_1 - \frac{y^2}{x^3} z_2 = 0,
$$

$$
z_{2,x} + z_1 z_2 - \frac{y}{x^2} z_1 - \frac{y^2}{x^3} z_2 = 0, \quad z_{2,y} + z_2^2 + \frac{1}{x} z_1 + \frac{y}{x^2} z_2 = 0,
$$

i.e., the normal form (2.65). $\qquad\qquad$ ⬚

If the second alternative (2.63) applies, new dependent variables z_1 and z_2 are introduced by

$$
\bar{z}_1 = \frac{1}{a_0} z_1, \quad \bar{z}_2 = \frac{1}{a_0} z_2.
$$

In this case the system (2.61) corresponding to conditions (2.63) is transformed into

$$
z_{1,x} + z_1^2 + A_1 z_1 + A_2 = 0, \quad z_{1,y} + B = 0,
$$

$$
z_{2,x} + z_2^2 + C = 0, \quad z_{2,y} + z_2^2 + D_1 z_2 + D_2 = 0,
$$

i.e., it decouples into two individual systems of the type (2.50) for z_1 and z_2 and may be treated by the methods of the preceding subsection.

The subsequent theorem describes the structure of the various types of rational solutions of system (2.65).

THEOREM 2.26   *If the coherent system (2.65) has a rational solution, one of the following alternatives applies with $r, s, p, q \in \mathbb{Q}(x, y)$ and $p_y = q_x$.*

92

i) *The general solution is rational and contains two constants. It may be written in the form*

$$z_1 = \frac{C_2 r_x + s_x}{C_1 + C_2 r + s} + p, \quad z_2 = \frac{C_2 r_y + s_y}{C_1 + C_2 r + s} + q.$$

ii) *There is a rational solution containing a single constant, it may be written in the form*

$$z_1 = \frac{r_x}{C + r} + p, \quad z_2 = \frac{r_y}{C + r} + q.$$

iii) *There is a solution as described in the preceding case, and in addition there is a special rational solution not equivalent to it.*

iv) *There is a single one, or there are two or three special rational solutions which are pairwise inequivalent.*

PROOF   The coefficient constraints are the integrability conditions for the system (2.65). Substituting $z_1 = \frac{u}{w}$, $z_2 = \frac{v}{w}$ yields

$$u_x + \frac{u(u - w_x)}{w} + A_1 u + A_2 v + A_3 w = 0,$$
$$u_y + \frac{u(v - w_y)}{w} + B_1 u + B_2 v + B_3 w = 0,$$
$$v_x + \frac{v(u - w_x)}{w} + B_1 u + B_2 v + B_3 w = 0,$$
$$v_y + \frac{v(v - w_y)}{w} + D_1 u + D_2 v + D_3 w = 0.$$

In order to obtain a linear system for u, v and w, the constraints

$$w_x - u + Pw = 0, \quad w_y - v + Qw = 0 \tag{2.67}$$

are necessary where $P, Q \in \mathbb{Q}(x, y)$ still have to be determined. Substituting these conditions into the above system yields

$$u_x + (A_1 + P)u + A_2 v + A_3 w = 0, \quad u_y + (B_1 + Q)u + B_2 v + B_3 w = 0,$$
$$v_x + B_1 u + (B_2 + P)v + B_3 w = 0, \quad v_y + D_1 u + (D_2 + Q)v + D_3 w = 0. \tag{2.68}$$

The consistency of the substitutions for z_1 and z_2 requires that the system comprising the equations (2.68) and (2.67) be a Janet basis. This is assured if there holds

$$P_y - Q_x + B_3 - B_4 = P_y - Q_x + \tfrac{1}{3}(A_{1,y} - B_{1,x} + B_{2,y} - D_{2,x}) = 0.$$

The second form has been obtained by applying equations (2.66). These conditions are satisfied if $P = -\tfrac{1}{3}(A_1 + B_2)$ and $Q = -\tfrac{1}{3}(B_1 + D_2)$ are

Linear Differential Equations

chosen. It yields the linear system

$$u_x + (\tfrac{2}{3}A_1 - \tfrac{1}{3}B_2)u + A_2 v + A_3 w = 0,$$
$$u_y + (\tfrac{2}{3}B_1 - \tfrac{1}{3}D_2)u + B_2 v + B_3 w = 0,$$
$$v_x + B_1 u - (\tfrac{1}{3}A_1 - \tfrac{2}{3}B_2)v + B_3 w = 0, \qquad (2.69)$$
$$v_y + D_1 u - (\tfrac{1}{3}B_1 - \tfrac{2}{3}D_2)v + D_3 w = 0,$$
$$w_x - u - \tfrac{1}{3}(A_1 + B_2)w = 0, \quad w_y - v - \tfrac{1}{3}(B_1 + D_2)w = 0$$

that is equivalent to (2.61). Let its general solution be

$$u = C_1 u_1 + C_2 u_2 + C_3 u_3, \quad v = C_1 v_1 + C_2 v_2 + C_3 v_3, \quad w = C_1 w_1 + C_2 w_2 + C_3 w_3$$

where $w_k \neq 0$ for all k. The corresponding solutions of (2.61) are

$$z_1 = \frac{u}{w} = \frac{C_1 u_1 + C_2 u_2 + C_3 u_3}{C_1 w_1 + C_2 w_2 + C_3 w_3}, \quad z_2 = \frac{v}{w} = \frac{C_1 v_1 + C_2 v_2 + C_3 v_3}{C_1 w_1 + C_2 w_2 + C_3 w_3}.$$

If exactly one or two of the ratios $\frac{u_k}{w_k}$ and $\frac{v_k}{w_k}$ are rational and the remaining ones are not, system (2.61) has the corresponding number of rational solutions. If all three ratios are rational but $\frac{u}{w}$ or $\frac{v}{w}$ are not, there are exactly three rational solutions. These alternatives combined make up case iv).

If not both $\frac{u}{w}$ and $\frac{v}{w}$ are rational it may occur that they become rational if one of the integration constants vanishes, e. g. C_3. This yields

$$z_1 = \frac{C_1 u_1 + C_2 u_2}{C_1 w_1 + C_2 w_2} = \frac{\dfrac{u_2}{w_1} - u_1 \dfrac{w_2}{w_1^2}}{C + \dfrac{w_2}{w_1}} + \frac{u_1}{w_1}, \quad z_2 = \frac{C_1 v_1 + C_2 v_2}{C_1 w_1 + C_2 w_2} = \frac{\dfrac{v_2}{w_1} - v_1 \dfrac{w_2}{w_1^2}}{C + \dfrac{w_2}{w_1}} + \frac{v_1}{w_1}.$$

Applying the relations

$$\left(\frac{w_k}{w_1}\right)_x = \frac{w_{k,x}}{w_1} - \frac{w_k w_{1,x}}{w_1^2} = \frac{u_k}{w_1} - u_1 \frac{w_k}{w_1^2},$$

$$\left(\frac{w_k}{w_1}\right)_y = \frac{w_{k,y}}{w_1} - \frac{w_k w_{1,y}}{w_1^2} = \frac{v_k}{w_1} - v_1 \frac{w_k}{w_1^2}$$

which are a consequence of the last two equations of (2.69), the solutions of case ii) are obtained. If in addition the ratios $\frac{u_3}{w_3}$ and $\frac{v_3}{w_3}$ are rational, this is case iii).

Assume now that $\frac{u}{w}$ and $\frac{v}{w}$ both are rational. These quotients may be written as

$$z_1 = \frac{u}{w} = \frac{C_1 \dfrac{u_1}{w_1} + C_2 \dfrac{u_2}{w_1} + \dfrac{u_3}{w_1}}{C_1 + C_2 \dfrac{w_2}{w_1} + \dfrac{w_3}{w_1}} = \frac{C_2 \left(\dfrac{u_2}{w_1} - u_1 \dfrac{w_2}{w_1^2}\right) + \dfrac{u_3}{w_1} - u_1 \dfrac{w_3}{w_1^2}}{C_1 + C_2 \dfrac{w_2}{w_1} + \dfrac{w_3}{w_1}} + \frac{u_1}{w_1},$$

$$z_2 = \frac{u}{w} = \frac{C_1\dfrac{v_1}{w_1} + C_2\dfrac{v_2}{w_1} + \dfrac{v_3}{w_1}}{C_1 + C_2\dfrac{w_2}{w_1} + \dfrac{w_3}{w_1}} = \frac{C_2\left(\dfrac{v_2}{w_1} - v_1\dfrac{w_2}{w_1^2}\right) + \dfrac{v_3}{w_1} - v_1\dfrac{w_3}{w_1^2}}{C_1 + C_2\dfrac{w_2}{w_1} + \dfrac{w_3}{w_1}} + \frac{v_1}{w_1}$$

where C_1 and C_2 have been appropriately redefined. Applying again the above expressions for the derivatives of $\frac{w_k}{w_1}$, the representation of case i) follow. $\quad\square$

In order to determine the rational solutions of system (2.65) a more general problem will be considered first. Let u and v be two functions of x and y. Furthermore, let the x-dependence of u be determined by a Riccati ode in x of order n, and the y-dependence of v by a Riccati ode in y of order m, both $m, n \geq 1$. In addition $u_y = v_x$ is required, i.e., the full Riccati-like system is

$$R_x^n(u) = 0, \quad R_y^m(v) = 0, \quad u_y - v_x = 0. \tag{2.70}$$

The structure of its rational solutions is described next, it is the analogue of Theorem 2.3 for pde's in two dependent and two independent variables.

THEOREM 2.27   *Let $r_i, s_i \in \mathbb{Q}(x, y)$ and $P_i = \{p_{i,1}, \ldots, p_{i,k}\}$, $p_{i,j} \in \mathbb{Q}[x, y]$, $C_{i,j} \in \bar{\mathbb{Q}}$, the sets P_i linearly independent over $\mathbb{Q}$, $j = 1, \ldots, k$, $i = 1, \ldots, m$ and define*

$$S_{P_i}^{r_i, s_i} \equiv \{r_i + \frac{\partial_x(\sum_{j=1}^k C_{i,j}p_{i,j})}{\sum_{j=1}^k C_{i,j}p_{i,j}}, \, s_i + \frac{\partial_y(\sum_{j=1}^k C_{i,j}p_{i,j})}{\sum_{j=1}^k C_{i,j}p_{i,j}}\}.$$

The rational solutions of (2.70) are the disjoint union of $S_{P_i}^{r_i, s_i}$, $i = 1, \ldots, m$ for some natural numbers k and m.

The proof of this result may be found in the article by Li and Schwarz [104].

The main steps for designing an algorithm that determines the rational solutions as described by the above theorem will be given next. At first the two given Riccati ode's are solved by applying the algorithm *RationalSolutionRiccati* on page 21. If rational solutions do exist, it returns expressions of the form given in Theorem 2.3. The $C_{i,j}$ are constants w.r.t. x or y now. Their dependence on y and x respectively has to be determined such that the structure described in Theorem 2.27 is achieved.

The last equation of the system (2.70) requires that its solutions be integrable pairs. In particular this is true for the terms r_i and s_i occuring in $S_{P_i}^{r_i, s_i}$ because the terms involving the sums over the $p_{i,j}$ are already integrable pairs by themselves. If (r_i, s_i) is not an integrable pair, but two polynomials have been found that make them into an integrable pair, the obvious relation

$$\frac{\partial_x(\sum C_{i,j}hp_{i,j})}{\sum C_{i,j}hp_{i,j}} = \frac{\partial_x(\sum C_{i,j}p_{i,j})}{\sum C_{i,j}p_{i,j}} + \frac{\partial_x h}{h} \tag{2.71}$$

which is valid for any $h \in \mathbb{Q}[x, y]$ may be applied to move over the logarithmic derivative of h to r_i and s_i without changing the full expression $S_{P_i}^{r_i, s_i}$. If for

Linear Differential Equations 95

a given value of i it is not possible to make (r_i, s_i) into an integrable pair by this procedure, the respective $S_{P_i}^{r_i,s_i}$ has to be discarded.

In order to make the integrable pairs obtained in this way into a genuine solution of the full system (2.65), the y- and the x- dependence of the constants w.r.t. x and y respectively have to be determined in a final step. Due to the fact that the x-dependence is completely explicit in the solution of the Riccati equation in x, a bound may be obtained for the polynomials in x representing the constants in y occuring in the solution of the Riccati equation in y, and vice versa. This leads to a finite linear problem over the constants, again details may be found in [104]. These steps are organized in terms of the subsequent algorithm.

ALGORITHM 2.14 $RationalSolutionPartialRiccati(R_x^n, R_y^m)$. Given a partial Riccati-like system of the form (2.70), its rational solutions are returned, or *failed* if none exists.

$S1$: *Solve Riccati ode's.* Determine the rational solutions of R_x^n and R_y^m by applying the algorithm *RationalSolutionRiccati*. If none exists neither for R_x^n nor for R_y^m, return *failed*.

$S2$: *Generate solution candidates.* From the solutions obtained in $S1$ form all possible sets $S_{P_i}^{r_i,s_i}$ and retain those that can be made into an integrable pair. If none contains a constant, return *failed*.

$S3$: *Generate solutions.* For each integrable pair obtained in step $S2$, solve the linear problem for determining the integration constants. If a consistent solution exists, construct a solution from it. Return the set of all solutions or *failed* if none has been found.

EXAMPLE 2.67   Consider the Riccati-like system

$$R_x^1(a) \equiv a_x + a^2 - \frac{2}{3x}a + \frac{2}{3x^2} = 0, \quad R_y^1(b) \equiv b_y + b^2 + \frac{4}{3y}b = 0, \quad (2.72)$$

supplemented by $a_y - b_x = 0$. In step $S1$, the special rational solutions $\frac{2}{3x}$ and $\frac{1}{x}$ are obtained for a, and $-\frac{1}{3y}$ and 0 for b. They yield the integrable pairs $(a, b) = \left(\frac{2}{3x}, -\frac{1}{3y}\right), \left(\frac{2}{3x}, 0\right), \left(\frac{1}{x}, -\frac{1}{3y}\right)$ and $\left(\frac{1}{x}, 0\right)$ in step $S2$. Because they do not contain any undetermined elements, they are returned as result. $\square$

EXAMPLE 2.68   Let the Riccati-like system

$$R_x^2(a) \equiv a_{xx} + 3a_x a + a^3 = 0 \text{ and } R_y^1(b) \equiv b_y + b^2 + \frac{2}{y-1}b = 0$$

supplemented by $a_y - b_x = 0$ be given. In step $S1$ the rational solutions

$$a = \frac{2C_1(y)x + C_2(y)}{C_1(y)x^2 + C_2(y)x + C_3(y)} \text{ and } b = \frac{C_4(x)}{C_4(x)y + C_5(x)} - \frac{1}{y-1}$$

are obtained. The single pair made up from them is left unchanged in step $S2$ because the parts not depending on the $C's$, 0 and $-\frac{1}{y-1}$, are an integrable pair. In step $S3$ the bounds 2 and 1 for the x- and the y-dependencies respectively lead to a linear system over the constants from which the final answer is obtained in the form

$$a = \frac{1}{x+y+C}, \quad b = \frac{1}{x+y+C} - \frac{1}{y-1}.$$

$\square$

Rational and Hyperexponential Solutions of Linear PDE's. In this subsection the equations under consideration are linear and may be of order higher than one. The fundamental constraint now is that their solution space be finite-dimensional. This may easily be decided if a Janet basis is generated first. Therefore it will be assumed from now on that the systems under consideration are already in Janet basis form.

The general system comprises m functions $z_1 > \ldots > z_m$ depending on n independent variables $x_1 > \ldots > x_n$ in *lex* order, with coefficients from the base field $\mathbb{Q}(x_1, \ldots, x_n)$. Similar as for linear ode's, the solution algorithm depends crucially on the function field in which the desired solutions are searched for. At first solutions in the base field are considered.

Two important simplifications may be performed according to Theorem 2.14. First of all, starting from a system involving m functions, it allows to generate m subsystems for a single function z_k, $k = 1, \ldots, m$, each of which has a finite-dimensional solution space as well. Secondly, any such subsystem involving a single unknown function, contains as lowest equation a linear ode in the lowest variable x_n. Its rational solutions may be determined by *RationalSolutionsLode* on page 14. As a result, the single unknown function may be expressed as a rational function of this lowest independent variable with coefficients depending on the higher variables. Upon substitution into the remaining equations of the system, a linear system for these coefficients is obtained. It is of the same type as the original system with the number of independent variables diminished by one. Therefore, repeating this procedure n times, all dependencies for the coefficients are removed. In a final step a linear algebra problem has to be solved in order to make the solutions for the m subsystems into a genuine solution of the originally given system for all m functions. The following algorithm is based on this scheme.

ALGORITHM 2.15   *RationalSolutionLinearSystem(L)*. Given a linear homogeneous system L for m functions $z_1, \ldots, z_m$, depending on $x_1, \ldots, x_n$ with coefficients from $\mathbb{Q}(x_1, \ldots, x_n)$, with finite-dimensional solution space. Its rational solutions are returned, or *failed* if none exists.

$S1$: *Systems for a single function.* Generate a system for each individual function $z_1, \ldots, z_m$ by taking it in turn as lowest dependent variable in *lex* order.

Linear Differential Equations 97

$S2$: *Solve linear ode.* For each system generated in $S1$, determine the general rational solution of the linear ode for the lowest independent variable, and store it on an auxiliary list. If all of them are trivial, return *failed*.

$S3$: *Generate systems for coefficients.* Substitute the rational solutions obtained in step $S2$ into the remaining equations of the respective system. If the coefficients still carry a dependency with them, go to $S1$.

$S4$: *Solve linear algebraic system.* Substitute the relations obtained in $S2$ and $S3$ into the original system L. Solve the resulting algebraic system, construct a solution from it and return the result.

The following example applies this algorithm, it originates from the symmetry analysis of equation 6.91 from Kamke's collection.

EXAMPLE 2.69   Consider the system

$$z_{2,x} - \frac{2y}{3x^2} z_1 + \frac{2}{3x} z_2 = 0, \quad z_{2,y} - \frac{2}{3x} z_1 - \frac{1}{3y} z_2 = 0,$$
$$z_{1,x} - \frac{4}{3x} z_1 + \frac{1}{3y} z_2 = 0, \quad z_{1,y} - \frac{1}{3y} z_1 + \frac{x}{3y^2} z_2 = 0. \tag{2.73}$$

In step $S1$ the two systems for z_1 and z_2 are generated.

$$z_{1,xx} - \frac{2}{3x} z_{1,x} + \frac{2}{3x^2} z_1 = 0, \quad z_{1,y} - \frac{x}{y} z_{1,x} + \frac{1}{y} z_1 = 0. \tag{2.74}$$

$$z_{2,xx} + \frac{4}{3x} z_{2,x} = 0, \quad z_{2,y} - \frac{x}{y} z_{2,x} - \frac{1}{y} z_1 = 0. \tag{2.75}$$

In step $S2$, the lowest equations for z_1 and z_2 yield the rational solutions $z_1 = C_1(y)x$ and $z_2 = C_2(y)$. Substituting them into the remaining equation of the respective system in $S3$ yields $C_{1,y} = 0$ and $C_{2,y} - \frac{1}{y}C_2 = 0$. Due to their dependencies, the algorithm returns to $S1$. There is nothing to do here because only a single equation is left in either case. Step $S2$ generates the solutions $C_1(y) = C_3$ and $C_2 = C_4 y$ with C_3, C_4 constants. There is no action in step $S3$. The linear algebraic system in step $S4$ reduces to the single equation $C_3 - C_4 = 0$. So the final answer is $z_1 = Cx$, $z_2 = Cy$, it generates a one-dimensional solution space. Because the general solution of (2.73) allows a two-dimensional solution space, this result shows that the general solution is not rational, i.e., there must be an additional basis element which is not rational, it will be determined in the next example. ⬚

The last problem to be considered in this chapter is to find hyperexponential solutions of linear homogeneous systems for $z_1, \ldots, z_m$. Here the number of independent variables is restricted to two, usually x and y. This is sufficient for the applications in the symmetry analysis in later chapters. The discussion follows closely the article by Li and Schwarz [104]. At first $m = 1$ is assumed, i.e., a system for a single function z depending on x and y with coefficients from

98

$\mathbb{Q}(x, y)$ is considered. By Theorem 2.14 two linear ode's w.r.t. x and y may be generated. For both of them the associated Riccati equation for a function $u \equiv \frac{z_x}{z}$ and $v \equiv \frac{z_y}{z}$ is constructed as explained on page 16. These equations are supplemented by the obvious integrability condition $u_y = v_x$ which makes u and v into an integrable pair. This subsystem is of type (2.70), its rational solutions may be determined by the algorithm *RationalSolutionPartialRiccati*. They correspond to hyperexponential solutions of the original linear system. The remaining equations of the system for z generate linear relations between the constants such that the solution space of the full system is generated. The following algorithm applies these steps.

ALGORITHM 2.16   *HyperexponentialSolutionLinearSystem(L)*. Given a linear homogeneous system L for a function z, depending on x and y, with a finite-dimensional solution space, its general hyperexponential solution is returned, or *failed* if none exists.

$S1$: *Generate Riccati-like system.* Generate two Janet bases in *lex* order from L with $x > y$ and $y > x$. From the two lowest equations generate associated Riccati equations in x and y respectivey.

$S2$: *Solve Riccati-like system.* Determine the rational solutions of the Riccati-like system corresponding to the equations obtained in step $S1$ by algorithm *RationalSolutionPartialRiccati*. If none is found, return *failed*.

$S3$: *Generate systems for coefficients.* From the solutions obtained in step $S2$ construct the hyperexponential solutions, substitute them into the remaining equations of the linear system and generate the linear system for the coefficients.

$S4$: *Solve linear algebraic system.* Solve the linear algebraic system obtained in step $S3$, construct a solution from it and return the result.

If in a system m functions $z_1, \dots, z_m$ are involved, at first m subsystems for each variable z_k, $k = 1, \dots, m$ are generated. To each of them the above algorithm is applied, and in a final step an additional linear system for the coefficients has to be solved in order to generate the solution of the full system.

EXAMPLE 2.70   In order to find the hyperexponential solutions of (2.73), system (2.74) for z_1 will be considered first. In step $S1$, from the two lowest equations

$$z_{1,xx} - \frac{2}{3x} z_{1,x} + \frac{2}{3x^2} z_1 = 0, \quad z_{1,yy} + \frac{4}{3y} z_{1,y} = 0 \qquad (2.76)$$

the Riccati equations (2.72) are generated. Its rational solutions have been determined in Example 2.67. The four integrable pairs obtained there yields the general solution

$$z = C_1 \frac{x^{2/3}}{y^{1/3}} + C_2 x^{2/3} + C_3 \frac{x}{y^{1/3}} + C_4 x.$$

Substitution of this expression into (2.74) leads to the relations $C_1 = C_4 = 0$, $C_2 - C_3 = 0$; consequently, the full two-dimensional solution space of (2.74) is generated by x and $\dfrac{x^{2/3}}{y^{1/3}}$.

It turns out that system (2.75) leads to two equations for z_2 which may be obtained from (2.76) by exchange of x and y, i.e., two basis elements for its solution space are y and $\dfrac{y^{2/3}}{x^{1/3}}$. Finally, if a general linear combination of these basis elements is substituted into (2.73), a linear system for the coefficients is obtained, the solution of which yields the final answer

$$z_1 = C_1 x + C_2 \frac{x^{2/3}}{y^{1/3}}, \quad z_2 = C_1 y + 2C_2 \frac{y^{2/3}}{x^{1/3}}.$$

This is the full solution space of system (2.73). $\quad\square$

Decomposing Janet Bases. The hyperexponential solutions discussed in the preceding section correspond to first order right factors. For Janet bases of order d three or higher, right factors of any order less than d may occur. A complete understanding of the solution space determined by such a Janet basis is only possible if these right factors may be found. To this end the analogue of the Loewy decomposition of linear ode's is required which in turn requires a generalized Beke-Schlesinger scheme for these pde's. For the applications later in this book the general case is not needed. Therefore only a few results for a certain third order Janet basis which occurs frequently in the symmetry analysis will be described. They are based to a large extent on a recent article by Li et al. [105].

For a nontrivial exact quotient q of two linear ode's l_1 and l_2 to exist, i.e., for a relation $l_1 = q l_2$ to be valid, it is necessary that the order of l_2 be strictly less than that of l_1. For Janet bases the situation is more complicated. The partial order defined in the subsequent lemma describes the possible candidates for exact division for the Janet basis types $\mathcal{J}^{(1,2)}$ and order not higher than three. The following shortened notation is introduced for describing these order relations:

$$a \prec \{b, c, d, \dots\} \quad \text{means} \quad a \prec b, \ a \prec c, \ a \prec d \dots.$$

LEMMA 2.13   *For Janet bases of type $\mathcal{J}^{(1,2)}$ and order not higher than three, the relations*

$$\mathcal{J}_1^{(1,2)} \prec \{\mathcal{J}_{2,1}^{(1,2)}, \ \mathcal{J}_{2,2}^{(1,2)}\},$$

$$\mathcal{J}_{2,1}^{(1,2)} \prec \{\mathcal{J}_{3,1}^{(1,2)}, \ \mathcal{J}_{3,2}^{(1,2)}\}, \ \mathcal{J}_{2,2}^{(1,2)} \prec \{\mathcal{J}_{3,2}^{(1,2)}, \ \mathcal{J}_{3,3}^{(1,2)}\}$$

define a partial order that may be represented by the diagram

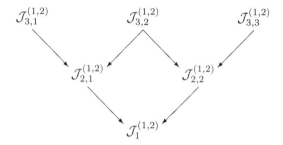

The exact quotient of two Janet bases e and f of these types may exist only if f is lower than e in this partial order.

PROOF   The partial order property is obvious. The leading derivative z_y in the type $\mathcal{J}_{3,1}^{(1,2)}$ Janet basis cannot be reduced by any element of a type $\mathcal{J}_{2,2}^{(1,2)}$ Janet basis, therefore the latter can never divide the former. The same is true for the leading derivative z_x of a type $\mathcal{J}_{3,3}^{(1,2)}$ Janet basis and a type $\mathcal{J}_{2,1}^{(1,2)}$ Janet basis. Similar arguments apply for the remaining cases.   □

The Janet basis of type $\mathcal{J}_{3,2}^{(1,2)}$ in the upper line of the above diagram occurs frequently in the symmetry analysis of ode's. Therefore the complete answer for its decomposition involving all possible Janet bases of lower order is given explicitly.

THEOREM 2.28   *A Janet basis*

$$\{z_{xx} + A_1 z_y + A_2 z_x + A_3 z,\ z_{xy} + B_1 z_y + B_2 z_x + B_3 z,\ z_{yy} + C_1 z_y + C_2 z_x + C_3 z\}$$

of type $\mathcal{J}_{3,2}^{(1,2)}$ may have the following components.

i) *A type $\mathcal{J}_1^{(1,2)}$ component $\{z_x + az,\ z_y + bz\}$ iff its coefficients a and b are rational solutions of*

$$a_x - a^2 + A_2 a + A_1 b - A_3 = 0,\ a_y - ab + B_2 a + B_1 b - B_3 = 0,$$
$$b_x - ab + B_1 b + B_2 a - B_3 = 0,\ b_y - b^2 + C_1 b + C_2 a - C_3 = 0.$$
(2.77)

ii) *A type $\mathcal{J}_{2,1}^{(1,2)}$ component $\{z_y + a_1 z_x + a_2 z,\ z_{xx} + b_1 z_x + b_2 z\}$ with*

$$a_1 = -\frac{w_1}{w_0},\ a_2 = \frac{w_2}{w_0},\ b_1 = A_1 \frac{w_1}{w_0} + A_2,\ b_2 = -A_1 \frac{w_2}{w_0} + A_3,\quad (2.78)$$

$$w_1 = -\frac{1}{A_1} w_{0,x} - \frac{A_2}{A_1} w_0,\ w_2 = -w_{0,y} - \frac{B_1}{A_1} w_{0,x} - \left(B_2 - \frac{A_2}{A_1} B_1\right) w_0$$
(2.79)

if $A_1 \neq 0$ and w_0 is a hyperexponential solution of a coherent third order system of type $\{w_{0,xx},\ w_{0,xy},\ w_{0,yy}\}$.

Linear Differential Equations

iii) *A component* $\{z_x + a_1 z, z_{yy} + b_1 z_y + b_2 z\}$ *of type* $\mathcal{J}_{2,2}^{(1,2)}$ *if* $A_1 = 0$. *Then the coefficients are* $a_1 = B_1$, $b_1 = C_1$ *and* $b_2 = C_3 - B_1 C_2$.

PROOF   The coefficient constraints for case i) and case iii) are obtained by direct reduction. The system (2.77) has been discussed in detail on page 90.

For case ii) let z_1 and z_2 be a fundamental system for a possible factor of type $\mathcal{J}_{2,1}^{(1,2)}$ and define

$$
w_0 = \begin{vmatrix} z_{1,x} & z_{2,x} \\ z_1 & z_2 \end{vmatrix}, \quad w_1 = \begin{vmatrix} z_{1,y} & z_{2,y} \\ z_1 & z_2 \end{vmatrix}, \quad w_2 = \begin{vmatrix} z_{1,y} & z_{2,y} \\ z_{1,x} & z_{2,x} \end{vmatrix}. \tag{2.80}
$$

The expressions (2.78) for the coefficients are obtained by substituting z_1 and z_2 into the type $\mathcal{J}_{2,1}^{(1,2)}$ system, using the given type $\mathcal{J}_{3,2}^{(1,2)}$ system for reduction and solving the respective linear algebraic systems. By differentiation and reduction w.r.t. to the given type $\mathcal{J}_{3,2}^{(1,2)}$ Janet basis the following first order system for the determinants w_0, w_1 and w_2 is obtained.

$$
w_{0,x} + A_1 w_1 + A_2 w_0 = 0, \quad w_{0,y} + w_2 + B_1 w_1 + B_2 w_0 = 0,
$$
$$
w_{1,x} - w_2 + B_1 w_1 + B_2 w_0 = 0, \quad w_{1,y} + C_1 w_1 + C_2 w_0 = 0,
$$
$$
w_{2,x} + (A_2 + B_1) w_2 + A_3 w_1 - B_3 w_0 = 0, \tag{2.81}
$$
$$
w_{2,y} + (B_2 + C_1) w_2 + B_3 w_1 - C_3 w_0 = 0.
$$

The coherence follows from the integrability conditions given in Theorem 2.15. The expressions (2.79) for w_1 and w_2 are obtained from the first two equations. If they are applied for elimination, due to the coherence the resulting system for w_0 may be transformed into a third order type $\mathcal{J}_{3,2}^{(1,2)}$ Janet basis in *grlex* term ordering with $y > x$. The details are discussed in Exercise 2.12. As a result the type $\mathcal{J}_{3,2}^{(1,2)}$ system for w_0 is obtained explicitly for general coefficients $A_1, A_2, \ldots, C_3$.   □

Although the system for w_0 obtained above appears quite cumbersome, it is useful for implementing decomposition algorithms because it is obtained without any Janet basis calculation.

EXAMPLE 2.71   If the type $\mathcal{J}_{3,2}^{(1,2)}$ Janet basis considered in Example 2.55 has a type $\mathcal{J}_1^{(1,2)}$ component, according to (2.77) its coefficients are solutions of

$$
a_x - a^2 + \tfrac{4}{x} a - \frac{2}{x^2} = 0, \quad a_y - ab + \tfrac{1}{x} b = 0,
$$
$$
b_x - ab + \tfrac{1}{x} b = 0, \quad b_y - b^2 + \tfrac{1}{y} b - \frac{x}{y^2} a + \frac{2}{y^2} = 0. \tag{2.82}
$$

The transformation (2.64) which is $a = -\bar{a} + \dfrac{5}{3x}$, $b = -\bar{b} + \dfrac{1}{3y}$ in this case

102

leads to

$$\bar{a}_x + \bar{a}^2 + \frac{2}{3x}\bar{a} - \frac{2}{9x^2} = 0, \quad \bar{a}_y + \bar{a}\bar{b} - \frac{1}{3y}\bar{a} - \frac{2}{3x}\bar{b} + \frac{2}{9xy} = 0,$$

$$\bar{b}_x + \bar{a}\bar{b} - \frac{1}{3y}\bar{a} - \frac{2}{3x}\bar{b} + \frac{2}{9xy} = 0, \quad \bar{b}_y + \bar{b}^2 - \frac{x}{y^2}\bar{a} + \frac{1}{3y}\bar{b} - \frac{2}{9y^2} = 0 \tag{2.83}$$

with the general solution

$$\bar{a} = \frac{C_2\frac{1}{y} + y}{C_1 + C_2\frac{x}{y} + xy} - \frac{1}{3x}, \qquad \bar{b} = -\frac{C_2\frac{x}{y^2} - x}{C_1 + C_2\frac{x}{y} + xy} + \frac{1}{3y}.$$

The special choice $C_1 = 0$, $C_2 \longrightarrow \infty$ yields the solution $\bar{a} = -\frac{2}{3x}$, $\bar{b} = \frac{2}{3y}$ and finally $a = \frac{1}{x}$, $b = \frac{1}{y}$. Due to the fact that the general solution of the Riccati-like system for a and b is rational, the originally given type $\mathcal{J}_{3,2}^{(1,2)}$ Janet basis is completely reducible. It is left as an exercise to determine the remaining components of its Loewy factor. $\Box$

Exercises

EXERCISE 2.1   In the following system of algebraic equations

$$x^3 + a_4 x^2 + a_3 y + a_2 x + a_1 = 0,$$
$$xy + b_4 x^2 + b_3 y + b_2 x + b_1 = 0,$$
$$y^2 + c_4 x^2 + c_3 y + c_2 x + c_1 = 0$$

x and y are considered as indeterminates, the a_i as parameters. Find the constraints for the parameters a_k, b_k and c_k such that the left hand sides of the above equations are a Gröbner basis in $grlex$ ordering with $y > x$.

EXERCISE 2.2   Let a third order ode $y''' + py' + \frac{1}{2}p'y = 0$ with $p \equiv p(x)$ be given. Show that a fundamental system $y_i, i = 1, \ldots, 3$ of the form $y_1 = z_1^2$, $y_2 = z_1 z_2$, $y_3 = z_2^2$ exists where $z_{1,2}$ is a fundamental system of $z'' + \frac{1}{4}pz = 0$. Later on it will be seen that an analogous property holds for the $n - th$ order linear homogeneous equation with suitable constraints for its coefficients and is a consequence of its symmetry structure.

EXERCISE 2.3   Generalize Example 2.2 to a third order lode $y''' + q_1 y'' + q_2 y' + q_3 y = r$ and apply the result to the special equation $y''' = r$.

EXERCISE 2.4   Let a second order linear homogeneous ode be given. If it is completely reducible it may be represented in the form

$$y'' + Py' + Qy = Lclm(D + p, D + q)y = 0 \text{ with } p \neq q.$$

Linear Differential Equations 103

Show that P and Q have the representation

$$P = p + q - \frac{p' - q'}{p - q}, \quad Q = pq - \frac{p'q - q'p}{p - q}.$$

Give an explicit expression for the general solution of the inhomogeneous equation $y'' + Py' + Q = R$.

EXERCISE 2.5   Determine the solutions corresponding to the various types of nontrivial Loewy decompositions of second order equations as given after Corollary 2.5.

EXERCISE 2.6   Prove Theorem 2.1 for second order Riccati equations.

EXERCISE 2.7   Prove Corollary 2.6.

EXERCISE 2.8   Prove Theorem 2.9 for $m = 2$.

EXERCISE 2.9   Express the coefficients of the type $\mathcal{J}_{2,1}^{(1,2)}$ Janet basis $\{z_y + a_1 z_x + a_2 z, z_{xx} + b_1 z_x + b_2 z\}$ in terms of a fundamental system $\{z_1, z_2\}$.

EXERCISE 2.10   The same problem for the type $\mathcal{J}_{2,2}^{(1,2)}$ Janet basis $\{z_x + a_1 z, z_{yy} + b_1 z_y + b_2 z\}$ with fundamental system $\{z_1, z_2\}$.

EXERCISE 2.11   Determine by direct reduction a system of pde's for the coefficients of a type $\mathcal{J}_{2,1}^{(1,2)}$ component for a type $\mathcal{J}_{3,2}^{(1,2)}$ Janet basis and compare the result with Theorem 2.28.

EXERCISE 2.12   Determine the system for w_0 of case $ii)$ in Theorem 2.28 explicitly.

EXERCISE 2.13   Determine the order of a pole in function (2.19) if $q(x)$ has a pole of order M at $x = x_0$. Distinguish the cases $M > 1$ and $M = 1$.

EXERCISE 2.14   Assume the type $\mathcal{J}_{2,1}^{(1,2)}$ Janet basis

$$z_y + A_1 z_x + A_2 z = 0, \ z_{xx} + B_1 z_x + B_2 z = 0$$

has the first order Loewy factor $z_x + az = 0, \ z_y + bz = 0$. Express the general solution in terms of integrals.

EXERCISE 2.15   The same problem for the type $\mathcal{J}_{2,3}^{(2,2)}$ Janet basis

$$z_{1,x} + A_1 z_2 + A_2 z_1 = 0, \ z_{1,y} + B_1 z_2 + B_2 z_1 = 0,$$
$$z_{2,x} + C_1 z_2 + C_2 z_1 = 0, \ z_{2,y} + D_1 z_2 + D_2 z_1 = 0$$

with the first order Loewy factor $z_2 + az_1 = 0, \ z_{1,x} + bz_1 = 0, \ z_{1,y} + cz_1 = 0$.

EXERCISE 2.16   Let $z_x + a_i z = 0, \ z_y + b_i z = 0, \ i = 1, 2$ be two first order systems. Determine its $Lclm$.

EXERCISE 2.17   Apply the results of Example 2.47 to obtain a basis for the solution space of the corresponding system of pde's.

104

EXERCISE 2.18   Let the system $z_y + a_1 z_x + a_2 z = v$, $z_{xx} + b_1 z_x + b_2 z = w$ of inhomogeneous pde's for $z(x, y)$ be given where $a_1, a_2, b_1, b_2, v, w \in \mathbb{Q}(x, y)$ and the left hand sides form a type $\mathcal{J}_{2,1}^{(1,2)}$ Janet basis. Express a special solution in terms of an integral if a fundamental system z_1 and z_2 of the corresponding homogeneous system is known.

EXERCISE 2.19   The same problem for $z_x + a_1 z = v$, $z_{yy} + b_1 z_y + b_2 z = w$ where the left hand sides form a type $\mathcal{J}_{2,2}^{(1,2)}$ Janet basis.

EXERCISE 2.20   The same problem for $z_y + a_1 z_x + a_2 z = v$, $z_{xxx} + b_1 z_{xx} + b_2 z_x + b_3 z = w$ where the left hand sides form a type $\mathcal{J}_{3,1}^{(1,2)}$ Janet basis and z_1, z_2 and z_3 are a fundamental system of the corresponding homogeneous system.

EXERCISE 2.21   Show how (2.46) is obtained from (2.45).

EXERCISE 2.22   Apply Theorem 2.20 to solve the equation $z_y + a z_x = r$ for $z(x, y)$ where $a \equiv a(x, y)$, $r = b(x, y)z + c(x, y)$.

EXERCISE 2.23   The equation $\dfrac{\partial I}{\partial x} + A\dfrac{\partial I}{\partial y} + 2I\dfrac{\partial A}{\partial y} + \dfrac{\partial^3 A}{\partial y^3} = 0$ has been considered by Drach [38], page 364. Use the result of the preceding exercise to describe its solution.

EXERCISE 2.24   Prove Theorem 2.24.

EXERCISE 2.25   Determine the general rational solution of the system considered in Example 2.65 by solving the second equation first.

EXERCISE 2.26   Assume a special solution z_1 for system (2.50) is known. Show that a second solution may be obtained by solving a linear system.

EXERCISE 2.27   Prove the representation (2.56) for the rational solutions of $z_x + A_1 z^2 + A_2 z + A_3 = 0$, $z_y + B_2 z + B_3 = 0$.

EXERCISE 2.28   Prove Lemma 2.11.

Chapter 3

Lie Transformation Groups

This chapter describes the basic facts of Lie's theory of continuous groups in a form that is suited for the applications to differential equations in later parts of this book. Originally Lie considered this only as an auxiliary subject for his main goal, i.e., solving differential equations in closed form by analogy with Galois' theory of algebraic equations. After recognizing the fundamental importance of these groups for this latter problem he developed a fairly complete theory for them. Its original objective, solving differential equations, was almost completely forgotten, most textbooks on differential equations do not even mention it at all. Later on the theory of continuous groups became a field of independent interest under the name *Lie groups*. The same is true for the algebraic objects introduced by Killing that were baptized *Lie algebras* by Hermann Weyl. They were obtained by abstraction from the commutators of vector fields occurring in Lie's theory.

In the first Section 3.1, basic concepts of the theory of continuous groups are put together for later reference, usually without proof. In Section 3.2 below, some properties of abstract Lie algebras are discussed. Section 3.3 applies the previous results to transformation groups of a two-dimensional manifold because these groups are most important for later parts of the book. Classifications of low-dimensional Lie algebras and transformation groups of a two-dimensional manifold are given in Section 3.4. The subject of Section 3.5 are so called *Lie systems*. These are special systems of linear homogeneous pde's with the additional feature that their solutions define the coefficients of infinitesimal generators of Lie transformation groups. These results are applied later in the symmetry analysis of differential equations.

3.1 Lie Groups and Transformation Groups

This section follows closely the article by Gorbatsevich and Onishchik in vol. 20 of the *Encyclopaedia of Mathematical Sciences* [43], part I; see also Chapter 1 of the book by Olver [140], or Chapters 5, 6 and 7 of Ibragimov's book [78].

Lie Groups. The main ingredients for this section are the algebraic concept of a group and the differential-geometric notion of a smooth manifold. The term *smooth* constrains the overlap functions of any coordinate chart to be C^∞ functions. These concepts are merged by the following definition which is the foundation for the subject of this chapter.

DEFINITION 3.1   *(Lie group) A set G is called a Lie group if there is given a structure on G satisfying the following three axioms.*

i) G is a group.

ii) G is a smooth manifold.

iii) The group operations $G \times G \to G$, $(x, y) \to xy$ and $G \to G$, $x \to x^{-1}$ are smooth functions.

If r is the dimension of the manifold, the group is called an *r-parameter Lie group* and is denoted by G_r.

For the applications in this book usually one is not interested in the full Lie group, but only in that part which is connected to the identity element. Starting with any global Lie group, a *local Lie group* may be constructed by taking a coordinate chart for the group manifold containing the identity, and retaining only those elements corresponding to parameter values contained in this identity neighborhood. From now on, the term Lie group will always be used in this sense, i.e., meaning the local group just described.

Let G and H be Lie groups. A smooth map $f : G \to H$ is a *homomorphism* if it is simultaneously a homomorphism of abstract groups and a smooth map of the group manifolds. A homomorphism is called an *isomorphism* if there exists a smooth inverse $f^{-1} : H \to G$. Two Lie groups are called *isomorphic* if an isomorphism between them exists. Due to the restriction to local groups, it is sufficient that these maps be defined for a neighborhood of the identities.

A subset H of a Lie group G is called a *Lie subgroup* if H is a subgroup of the abstract group G and a submanifold of the manifold G. If H is even a normal subgroup of the abstract group G, it is called a *Lie normal subgroup*. A Lie group G is called *simple* if the only normal Lie subgroups are the identity and G itself. The commutator group $G^{(1)} = G' = [G, G]$ is the group generated by all commutators $aba^{-1}b^{-1}$ with $a, b \in G$. The higher commutator groups are defined recursively by $G^{(\nu+1)} = G^{(\nu)'}$ for $\nu = 1, 2, \ldots$. A Lie group G_r is called *integrable* or *solvable* if its series of commutator groups $G^{(\nu)}$ terminates after a finite number of steps with the identity.

A Lie group is called *semi-simple* if it does not contain a solvable normal subgroup. Any Lie group has a maximal solvable normal subroup.

Let G_r be a r-parameter Lie group and $a, b \in G_r$ be two elements with the representation $a = (a_1, \ldots, a_r)$ and $b = (b_1, \ldots, b_r)$ in a coordinate chart. For the coordinates of the product $c = ab$ there holds

$$c_j = \phi_j(a_1, \ldots, a_r, b_1, \ldots, b_r) \quad \text{for} \quad j = 1, \ldots, r \tag{3.1}$$

Lie Transformation Groups

where ϕ_j are smooth functions of all arguments. Assume further now that the coordinate chart is such that the identity is represented by $e = (0, \ldots, 0)$. Then the lowest terms of (3.1) are

$$c_j = a_j + b_j + \tfrac{1}{2} \sum_{k,l} \gamma_{kl}^j a_k b_l + \ldots \tag{3.2}$$

with constants γ_{kl}^j. The omitted terms are of order three or higher with respect to a_k and b_l. The numbers

$$c_{kl}^j = \gamma_{kl}^j - \gamma_{lk}^j \tag{3.3}$$

are anti-symmetric in k and l, they are called the *structure constants* of the Lie group.

Lie Transformation Groups. Let M be a n-dimensional smooth manifold and G a Lie group. An *action T of the group G* on M is a smooth mapping

$$T : G \times M \to M, \quad T(g, x) \equiv gx \to \bar{x} \tag{3.4}$$

with the following properties:

$$T(e, x) = x, \quad T(a, T(b, x)) = T(ab, x) \tag{3.5}$$

for any $x \in M$, $g, a, b \in G$, $e \in G$ the unit element. Then G is called a *Lie transformation group* of the manifold M. For a r-parameter group G_r, in local coordinates the action (3.4) has the special form

$$\bar{x}_i = f_i(x_1, \ldots, x_n, a_1, \ldots, a_r) \tag{3.6}$$

with smooth functions f_i, $i = 1, \ldots, n$. In this form Lie originally introduced his continuous groups. Specializations of (3.6) are the basis for the symmetry analysis in later parts of this book.

An additional assumption is that the number r of parameters is minimal, i.e., that it is not possible to introduce a smaller number of parameters generating the same set of transformations. If this is true they are called *essential parameters*. It is proved in Lie [112], vol. I, pages 13-14, that this is guaranteed if there do not exist equations of the form

$$\sum_{k=1}^{r} \chi_k(a_1, \ldots, a_r) \frac{\partial f_i(x_1, \ldots, x_n, a_1, \ldots, a_r)}{\partial a_k} = 0$$

for $i = 1, \ldots, n$. Furthermore, it is shown there how a constructive criterion may be obtained from these requirements. From now on it is always assumed that group parameters are essential without explicitly mentioning it.

The elements $g \in G$ leaving a given element $x \in M$ invariant form the *stabilizer* of x. If T is an action of a group G on M, one can define an equivalence relation on M by

$$x \sim y \Longleftrightarrow x = gy \text{ for some } g \in G.$$

The equivalence classes are called the *orbits* of the action. Every point of $x \in M$ is contained in a unique orbit. The totality of orbits generate an *invariant decomposition* or simply *decomposition* of the manifold.

The action of a transformation group G on M is called *transitive* if for arbitrary $x, \bar{x} \in M$ there is a group element $g \in G$ such that $\bar{x} = gx$, otherwise it is called *intransitive*. If the action is transitive there is only a single orbit of if for $x_1, \ldots, x_k \in M$, $x_i \neq x_j$ for $i \neq j$, there is a $g \in G$ such that $\bar{x}_i = gx_i$, but the same is not true for $k + 1$, i.e., there are k orbits.

A transitive transformation group G of a manifold M is called *imprimitive* if M is the union of submanifolds that are permuted by the group. These submanifolds are called *systems of imprimitivity* of G. A group that is not imprimitive is called *primitive*.

Let G_r be a r-parameter Lie group with coordinates $a = (a_1, \ldots, a_r)$ and such that $e = (0, \ldots, 0)$. Defining the quantities

$$\xi_{k,i} \equiv \frac{\partial f_i}{\partial a_k}\bigg|_{a_1=\ldots=a_r=0} \tag{3.7}$$

for $k = 1, \ldots, r$ and $i = 1, \ldots, n$, expanding (3.6) at the identity and omitting terms of order higher than one in the group parameters there follows

$$\bar{x}_i = x_i + \xi_{1,i}a_1 + \ldots + \xi_{r,i}a_r + \ldots. \tag{3.8}$$

The vector fields

$$X_k = \xi_{k,1}\frac{\partial}{\partial x_1} + \ldots + \xi_{k,n}\frac{\partial}{\partial x_n}, \quad k = 1, \ldots, r, \tag{3.9}$$

are called the *infinitesimal generators* of the Lie transformation group. They are linearly independent over constants and obey the *commutation relations*

$$[X_k, X_l] = X_k X_l - X_l X_k = \sum_{j=1}^{r} c_{kl}^j X_j. \tag{3.10}$$

Mathematical objects obeying these relations have been baptized a *Lie algebra*. They occur in various branches of mathematics and form an independent subfield of algebra. Some of these algebraic properties are discussed in the next Section 3.2. If $X = [X_k, X_l]$ is the commutator of the two generators X_k and X_l, its coefficients may be expressed as

$$\xi_i = \sum_{j=1}^{n} \left(\xi_{k,i}\frac{\partial \xi_{l,i}}{\partial x_j} - \xi_{l,i}\frac{\partial \xi_{k,i}}{\partial x_j} \right) \quad \text{for } i = 1, \ldots, n. \tag{3.11}$$

As usual, if q generators $X_1, X_2, \ldots, X_q$ do not obey a relation

$$\lambda_1 X_1 + \ldots + \lambda_q X_q \equiv 0 \tag{3.12}$$

Lie Transformation Groups

109

with $\lambda_1, \ldots, \lambda_q$ constants, not all zero, they are said to be *linearly independent* over constants or simply *independent*. A generator X_k that may be expressed in the form

$$X_k = \lambda_1 X_1 + \ldots + \lambda_q X_q$$

is said to be *dependent on* $X_1, \ldots, X_q$. If q generators $X_1, X_2, \ldots, X_q$ do not obey a relation

$$\phi_1(x_1, \ldots, x_n)X_1 + \ldots + \phi_q(x_1, \ldots, x_n)X_q \equiv 0 \qquad (3.13)$$

with not all $\phi_i(x_1, \ldots, x_n)$ identically zero they are said to be *unconnected*. Any generator X_k that may be expressed in the form

$$X_k = \phi_1(x_1, \ldots, x_n)X_1 + \ldots + \phi_q(x_1, \ldots, x_n)X_q$$

is said to be *connected with* $X_1, \ldots, X_q$. Unconnected generators are obviously independent, but independent generators are not necessarily unconnected. For a transformation group of a n-dimensional manifold there cannot be more than n unconnected generators, though there may be any number of independent generators. If a set of unconnected generators is such that the commutator of any pair is connected with them, they are said to form a *complete system*.

The following rules for manipulating expressions involving a commutator follow immediately from the definition. Let $\phi(x_1, \ldots, x_n)$ and $\psi(x_1, \ldots, x_n)$ be two functions of $x_1, \ldots, x_n$, and X, X_1 and X_2 arbitrary vector fields. There holds

$$X(\phi + \psi) = X\phi + X\psi, \qquad X(\phi\psi) = \psi X\phi + \phi X\psi,$$

$$[\phi X_1, X_2] = \phi[X_1, X_2] - X_2 \phi X_1, \qquad [X_1, \phi X_2] = \phi[X_1, X_2] + X_1 \phi X_2.$$

Let $X_1, \ldots, X_r$ be the infinitesimal generators of an r-parameter Lie transformation group G with group parameters $a_1, \ldots, a_r$. The vector fields

$$Z_i = b_{i1}(a_1, \ldots, a_r)\frac{\partial}{\partial e_1} + \ldots + b_{ir}(a_1, \ldots, a_r)\frac{\partial}{\partial e_r}$$

where $b_{ik} = \sum_{j=1}^{r} c_{ik}^j e_j$ for $i = 1, \ldots, r$ and $k = 1, \ldots, r$ are the generators of a linear homogeneous transformation group of an r-dimensional manifold with coordinates $(e_1, \ldots, e_r)$. It is called the *adjoint group* of G.

Lie's Fundamental Theorems. Lie himself summarized several important results of his theory in terms of three fundamental theorems that are given next.

THEOREM 3.1   *(Lie's First Theorem) If the equations (3.6) define an r-parameter group, then the $\bar{x}_i$, considered as functions of $x_1, \ldots, x_n$ and $a_1, \ldots, a_r$, satisfy a system of pde's of the special form*

$$\frac{\partial \bar{x}_i}{\partial a_k} = \sum_{j=1}^{r} \psi_{j,k}(a_1, \ldots, a_r)\xi_{j,i}(\bar{x}_1, \ldots, \bar{x}_n)$$

110

for $k = 1, \ldots, r$. The determinant of the $\psi_{j,k}$ does not vanish identically. On the other hand, if functions $f_i(x_1, \ldots, x_n, a_1, \ldots, a_r)$ satisfy a system of differential equations of this form, they define a transformation group.

The proof of this theorem may be found in Lie [112], vol. I, Kapitel 2, pages 27-34 and vol. III, Kapitel 25, pages 545-564.

The importance of the infinitesimal generators of a group arises from the fact that a group, due to the restriction to local groups as explained above, is uniquely determined by them. This relation is established in the next theorem.

THEOREM 3.2   *(Lie's Second Theorem) Any r-parameter Lie group (3.6) defines r infinitesimal generators (3.9) satisfying $r(r-1)$ relations of the form (3.10). Conversely, r independent generators (3.9) satisfying the relations (3.10) define a Lie group.*

The proof of this theorem may be found in Lie [112], vol. III, Kapitel 25, pages 575-590. It is based on the system

$$\frac{d\bar{x}_i}{dt} = \lambda_1 \xi_{1,i}(\bar{x}_1, \ldots, \bar{x}_n) + \ldots + \lambda_r \xi_{r,i}(\bar{x}_1, \ldots, \bar{x}_n)$$

for the general one-parameter group contained in the r-parameter group (3.6); the $\lambda_1, \ldots, \lambda_r$ are constants. Together with the initial conditions $\bar{x}_i = x_i$ for $t = 0$, $i = 1, \ldots, n$, they determine the one-parameter groups uniquely. If for any fixed k, $1 \leq j \leq r$, $\lambda_k = a_k$, $\lambda_j = 0$ for $j \neq k$ are chosen, the system

$$\frac{d\bar{x}_i}{da_k} = \xi_{k,i}(\bar{x}_1, \ldots, \bar{x}_n) \tag{3.14}$$

for the one-parameter group corresponding to the generator X_k is obtained. Whenever the phrase *"the group of the X_k"* for a system of infinitesimal generators is used, it is based on Lie's second theorem and the restriction to the local group determined by them.

THEOREM 3.3   *(Lie's Third Theorem) If a system of constants c^j_{kl} satisfies the relations $c^j_{kl} + c^j_{lk} = 0$ and*

$$\sum_{s=1}^{r} (c^j_{ks} c^s_{li} + c^j_{ls} c^s_{ik} + c^j_{is} c^s_{kl}) = 0$$

for $i, j, k, l = 1, \ldots, r$, there exists a Lie group with these structure constants.

The proof of this theorem may be found in Lie [112], vol. III, pages 597 ff.

Similarity of Transformation Groups. If any transformation group is given by relations (3.6), an apparently different group may be obtained by introducing new coordinates of both the underlying manifold M and the group manifold G. Two groups related to each other in this way are not essentially

Lie Transformation Groups

different. A special term for this relation has been introduced by Lie [112], vol. I, page 24.

DEFINITION 3.2 *(Similarity of transformation groups) Let two r-parameter transformation groups be given by*

$$\bar{x}_i = f_i(x_1, \ldots, x_n, a_1, \ldots, a_r) \text{ and } \bar{y}_i = g_i(y_1, \ldots, y_n, b_1, \ldots, b_r).$$

If there exist diffeomorphisms $x_i = \varphi_i(y_1, \ldots, y_n)$ *for* $i = 1, \ldots, n$ *and* $a_k = \psi(b_1, \ldots, b_r)$ *for* $k = 1, \ldots, r$ *such that the two definitions become identical, the groups are called similar.*

The consequences of a simultaneous change of both coordinates are discussed in detail by Lie [112], vol. I, Kapitel 2 and by Eisenhart [41], §7. The transformation of the infinitesimal generators under a coordinate change of M occurs frequently in later applications. Let the two coordinates $(x_1, \ldots, x_n)$ and $(y_1, \ldots, y_n)$ be related by $x_k = \phi_k(y_1, \ldots, y_n)$ and $y_k = \psi_k(x_1, \ldots, x_n)$. Defining $\eta_i = X_k \psi_i(x_1, \ldots, x_n) \Big|_{x_1 = \phi_1, \ldots, x_n = \phi_n}$ for $i = 1, \ldots, n$, the transformed generator of (3.9) is

$$Y_k = \eta_1(y_1, \ldots, y_n) \frac{\partial}{\partial y_1} + \eta_n(y_1, \ldots, y_n) \frac{\partial}{\partial y_n}.$$

Of particular importance is the question of how the commutator of two vector fields is transformed under a change of coordinates. The answer is given in the subsequent lemma due to Lie [112], part I, page 84.

LEMMA 3.1 *If two vector fields* X_k *and* X_j *in coordinates* $(x_1, \ldots, x_n)$ *are transformed into* Y_1 *and* Y_2 *in coordinates* $(y_1, \ldots, y_n)$, *its commutator* $[X_k, X_j]$ *is transformed into* $[Y_k, Y_j]$.

As a consequence of this lemma the structure constants of a Lie algebra of vector fields are invariant under a coordinate change of the transformed manifold M. This is important because it decouples the algebraic properties of Lie algebras generated by vector fields from its appearance depending on the actual coordinates of M.

Given any two transformation groups the problem arises how to decide similarity and, in case the answer is affirmative, how to actually determine the transformation functions between them. The following theorem due to Lie essentially provides the answer.

THEOREM 3.4 *(Lie 1888). In order that two groups with infinitesimal generators* $X_1, \ldots, X_r$ *in coordinates* $(x_1, \ldots, x_n)$ *and* $Y_1, \ldots, Y_r$ *in coordinates* $(y_1, \ldots, y_n)$ *respectively be similar, the following conditions are necessary and sufficient.*

1. *The two groups must have the same structure, i.e., if the relations* $[X_i, X_j] = \sum_{k=1}^{r} c_{ij}^k X_k$ *are valid it must be possible to choose suitable*

112

linear combinations $\bar{Y}_i = \sum_{j=1}^r \gamma_{ij} Y_j$ with constant γ's such that the $\bar{Y}_i$ satisfy relations $[\bar{Y}_i, \bar{Y}_j] = \sum_{k=1}^r c_{ij}^k \bar{Y}_k$ with the same structure constants as for the X_i.

2. *If $X_1, \ldots, X_k$ are unconnected whereas $X_{k+i} = \sum_{j=1}^k \phi_{i,j}(x_1, \ldots, x_n) X_j$ for $k \leq r$ and $i = 1, \ldots, r - k$, the corresponding generators $\bar{Y}_1, \ldots, \bar{Y}_k$ are also unconnected whereas the relations $\bar{Y}_{k+i} = \sum_{j=1}^k \psi_{i,j}(y_1, \ldots, y_n) \bar{Y}_j$ are valid such that the equations $\phi_{i,j}(x_1, \ldots, x_n) = \psi_{i,j}(y_1, \ldots, y_n)$ are consistent. In particular they should not generate a relation among $x_1, \ldots, x_n$ on the one hand or $y_1, \ldots, y_n$ on the other.*

The proof may be found in Lie [112], part I, pages 353-355. It shows that the question of similarity of two groups may be divided into two independent parts. This separation is based on Lemma 3.1. In the first place, in order to be similar the infinitesimal generators must generate isomorphic Lie algebras. This is a purely algebraic problem that will be discussed in detail in the Section 3.2. Only if both Lie algebras are isomorphic and a basis transformation has been found such that their structure constants are identical, the second part dealing exclusively with the coordinate change of the transformed manifold applies.

A special kind of similarity is involved in defining the so-called *canonical variables* for a one-parameter group. Let the group equations in coordinates $(x_1, \ldots, x_n)$ be $\bar{x}_i = f_i(x_1, \ldots, x_n, a)$ with the single group parameter a. There exist always new coordinates $(u_1, \ldots, u_n)$ such that the group equations are $\bar{u}_i = u_i$ for $i = 1, \ldots, n - 1$ and $\bar{u}_n = u_n + b$, i.e., any one-parameter group is similar to a group of translations in a single coordinate. In these latter variables the infinitesimal generator is $\dfrac{\partial}{\partial u_n}$. In Section 3.3 finding a transformation to canonical form for groups acting in the plane will be discussed. Furthermore, canonical forms for groups involving more than a single parameter will be discussed.

Defining Equations of a Transformation Group. Starting with the infinitesimal generators of a group, a third representation in terms of the *defining equations*, as Lie [112], vol. I, page 185 called them, may be obtained. By definition, these equations are such that their general solution generates the vector space of the coefficients of the given generators. This representation is particularly important if the groups to be considered occur as symmetry groups of differential equations because they are usually defined in terms of systems of pde's.

Let the infinitesimal generators X_k of a group be given in the form (3.9). The general generator may be written as

$$\xi_1 \frac{\partial}{\partial x_1} + \ldots + \xi_n \frac{\partial}{\partial x_n} = \sum_{k=1}^r e_k \left(\xi_{k,1} \frac{\partial}{\partial x_1} + \ldots + \xi_{k,n} \frac{\partial}{\partial x_n} \right) \qquad (3.15)$$

with constants e_k. This defines the ξ_i by means of $\xi_i = \sum_{k=1}^{r} e_k \xi_{k,i}(x_1,\ldots,x_n)$. The $\xi_{k,i}$ are given functions of $x_1,\ldots,x_n$. It has been shown by Lie [112], vol. I, § 48, page 181 that for the functions ξ_i a system of linear pde's always may be constructed such that they form a fundamental system. Due to its importance a special name has been given to it.

DEFINITION 3.3 *(Defining equations of a group). The coefficients $\xi_1,\ldots,\xi_n$ of the general infinitesimal generator of a Lie transformation group defined by (3.15) satisfy a system of linear homogeneous pde's. They are called the defining equations of the group.*

Altogether there are three means now to describe a Lie transformation group. The finite equations (3.6), the infinitesimal generators (3.9) and the defining equations just described. Due to the restriction to local Lie groups they are essentially equivalent. The infinitesimal generators are obtained from the finite equations by differentiation. The converse requires solving linear pde's by Lie's second theorem. The same is true for determining the infinitesimal generators if the defining equations are given. The relation between these representations of a group and how to turn from one to another is illustrated by the following diagram. The operations corresponding to the downward arrows at the left can always be performed algorithmically. Contrary to that, there is no generally valid algorithm for the operations at the right corresponding to the upward arrows.

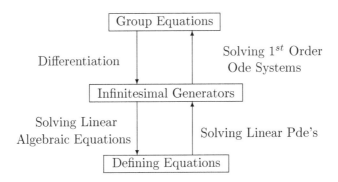

Invariants and Differential Invariants. A function $F(x_1,\ldots,x_n)$ is called an *invariant* of the r-parameter Lie transformation group G_r defined by (3.6) if there holds
$$F(\bar{x}_1,\ldots,\bar{x}_n) = F(x_1,\ldots,x_n).$$
Invariants of this kind are also called *absolute invariants* in order to distinguish them from another kind of invariants to be defined in Chapter 4. A criterion for invariance may be given in terms of the infinitesimal generators of the group defined by (3.9). A function $F(x_1,\ldots,x_n)$ is an invariant of the group if and only if there holds $X_k F(x_1,\ldots,x_n) = 0$ for $k = 1,\ldots,r$. If F is an invariant of a group, then also $\Phi(F)$ with Φ an undetermined function

114

is an invariant. If a group allows more than a single invariant, only the functionally independent invariants are relevant. If the generic rank q for the group generated by $X_1, \ldots, X_r$ is defined by

$$
rank \left\{
\begin{array}{c}
\xi_{1,1}, \ \xi_{1,2} \ \xi_{1,n} \\
\xi_{2,1}, \ \xi_{2,2} \ \xi_{2,n} \\
\vdots \quad \vdots \\
\xi_{r,1}, \ \xi_{r,2} \ \xi_{r,n}
\end{array}
\right\} \equiv q
$$

there are $n - q$ independent invariants. If $q = n$ there are no invariants at all and the group is transitive. The proof may be found in Lie [112], vol. I, Kapitel 13 or in the book by Eisenhart [41], §21.

The concept of a differential invariant emerges from the fact that a group acts on a manifold the coordinate variables of which may carry some dependencies with them. Whenever this occurs, the vector fields generating the group may be *prolonged* up to a predetermined order of the respective derivatives. These prolonged generators of the original group are considered as vector fields acting on a higher-dimensional manifold to which the above results may be applied. If different dependencies are assumed, there are different prolongations and consequently diverse types of differential invariants. In the remaining part of this subsection summation over repeated indices is always understood.

DEFINITION 3.4 *(Prolongation of a vector field) Let the vector field* $U = \xi_i(x, u)\partial_{x_i} + \eta^\alpha(x, u)\partial_{u^\alpha}$ *be the infinitesimal generator of a finite transformation group on a manifold with coordinates* $x = (x_1, \ldots, x_n)$ *and* $u = (u^1, \ldots, u^m)$ *with dependencies* $u^\alpha \equiv u^\alpha(x_1, \ldots, x_n)$ *for* $\alpha = 1, \ldots, m$. *Partial derivatives are denoted by* $u^\alpha_{i_1, \ldots, i_n} \equiv \dfrac{\partial^{i_1 + \ldots + i_n} u^\alpha}{\partial x_1^{i_1} \ldots x_n^{i_n}}$. *By definition the k-th prolongation of* U *is*

$$
U^{(k)} = U + \zeta^\alpha_i \frac{\partial}{\partial u^\alpha_i} + \ldots + \zeta^\alpha_{i_1, \ldots, i_k} \frac{\partial}{\partial u^\alpha_{i_1, \ldots, i_k}}.
$$

The functions $\zeta^\alpha_{i_1, \ldots, i_k}$ *are recursively defined by* $\zeta^\alpha_i = D_i(\eta^\alpha) - u^\alpha_s D_i(\xi_s)$ *and*

$$
\zeta^\alpha_{i_1, \ldots, i_k} = D_{i_k}(\zeta^\alpha_{i_1, \ldots, i_{k-1}}) - u^\alpha_{i_1, \ldots, i_{k-1} s} D_{i_k}(\xi_s).
$$

D_i *is the operator of total differentiation w.r.t.* x_i.

$$
D_i = \frac{\partial}{\partial x_i} + u^\alpha_i \frac{\partial}{\partial u^\alpha} + u^\alpha_{k,i} \frac{\partial}{\partial u^\alpha_k} + u^\alpha_{k,l,i} \frac{\partial}{\partial u^\alpha_{k,l}} \cdots
$$

If the vector fields generating a group are prolonged there arises the question as to what the algebraic relations between the prolonged vector fields are. The subsequent theorem provides the simple answer that the structure constants for the prolonged generators remain unchanged.

THEOREM 3.5   *Let $X_1, \ldots, X_r$ be vector fields generating a group acting on a manifold of arbitrary dimension obeying $[X_i, X_j] = \sum_{k=1}^{r} c_{ij}^k X_k$. Furthermore, let there be some sort of dependencies declared between these variables. The prolongations $X_i^{(l)}$ of any order obey the same commutation relations as the X's, i.e., there holds $[X_i^{(l)}, X_j^{(l)}] = \sum_{k=1}^{r} c_{ij}^k X_k^{(l)}$.*

The proof may be found in Lie [112], vol. I, §130 or Eisenhart [41], §27.

The next theorem due to Lie [114], page 760, assures the existence of an infinite number of differential invariants. Furthermore, it shows that only a finite number of them has to be determined by integrating a complete system of pde's.

THEOREM 3.6   *(Lie 1893) For any continuous group, acting on a manifold in any number of variables carrying certain dependencies, there exists an infinite number of differential invariants. They may be obtained by differentiation from a finite subset of them, it is called a full system of invariants.*

The first part of this result is obvious. By differentiating a sufficient number of times the number of variables increases until the condition for the existence of invariants mentioned before is obeyed.

3.2   Algebraic Properties of Vector Fields

While Lie developed his theory of continuous groups and of integrating differential equations in Leipzig, the high school teacher Wilhelm Killing, apparently not knowing Lie's work for a long time, developed a theory of certain algebraic structures that turned out to comprise the commutators of the infinitesimal generators occurring in Lie's theory as special cases. On Lie's suggestion, Elie Cartan in Paris took up the subject in his thesis, correcting and completing Killing's results, and continued for many years working in this field to become one of its most important contributors. Those topics of the resulting theory that will be applied later on are summarized in this section. In addition to Lie's publications, good references are the thesis of Cartan [25] and the book by Jacobson [82]. For algorithmic questions the article by Rand [151] or the thesis of de Graaf [37] is recommended. A comprehensive treatment including a survey of the more recent literature may be found in the three volumes of the Springer *Encyclopaedia* [43], entitled *Lie Groups and Lie Algebras*. A lot of interesting background information is covered in parts II and III of the book by Hawkins [68].

116

General Properties of Lie Algebras. By abstraction from its geometric origin, the algebraic objects obeying commutation relations (3.10) are studied as a subject of interest of its own right, they are called *Lie algebras*. First of all they have to be defined.

DEFINITION 3.5 *(Lie algebra). A Lie algebra $\mathcal{L}$ is a vector space over a field F equipped with a binary operation, called commutator, satisfying the following constraints for any $U, V, W \in \mathcal{L}$ and $\alpha, \beta \in F$.*
   i) $[U, V] = -[V, U]$ *(anti-commutativity)*
   ii) $[\alpha U + \beta V, W] = \alpha[U, W] + \beta[V, W]$ *(bilinearity)*
   iii) $[U, [V, W]] + [V, [W, U]] + [W, [U, V]] = 0$ *(Jacobi identity)*

In this book only finite dimensional Lie algebras are considered; and it is always assumed that the field F has characteristic zero. If not stated otherwise explicitly it will be the complex number field $\mathbb{C}$.

A Lie algebra is usually defined in terms of the commutators of basis elements in the form (3.10). The numerical values of the structure constants c_{ij}^k depend on the basis of the Lie algebra. If two Lie algebras are such that by a suitable change of the basis their structure constants become identical they are considered to be not essentialy different, clearly they are isomorphic. As usual in algebra, this notion of isomorphism establishes an equivalence relation between Lie algebras. As a consequence, a classification of Lie algebras may be performed by associating them to a canonical form which is given by a special basis. For algebras of low dimension canonical forms are given in Section 3.4 on page 135 ff.

Lie [112] has shown that deciding isomorphy of Lie algebras is a purely algebraic problem. To this end a suitable basis transformation has to be found such that in the new basis the commutators coincide with a canonical one. Let a basis of an r-dimensional Lie algebra be $\bar{U}_1, \ldots, \bar{U}_r$ with $[\bar{U}_i, \bar{U}_j] = \sum \bar{c}_{ij}^k \bar{U}_k$. The desired canonical basis elements $U_1, \ldots, U_r$ with commutation relations $[U_i, U_j] = \sum c_{ij}^k U_k$ are expressed in terms of the given ones as $U_i = \sum \alpha_{ij} \bar{U}_j$ with undetermined coefficients $\alpha_{ij} \in F$. Substituting these expressions into the canonical commutation relations leads to the system of algebraic equations

$$\sum_{l,m=1}^{r} \bar{c}_{lm}^n \alpha_{il} \alpha_{jm} = \sum_{k=1}^{r} c_{ij}^k \alpha_{kn}, \quad 1 \le i < j \le r, \ 1 \le n \le r \quad (3.16)$$

with the constraint $det\{\alpha_{ij}\} \neq 0$. This system has been given already by Lie [112], vol. I, page 290, equations (3). Its solution yields the desired transformation. In general it is not unique. It has been discussed in detail by Gerdt [52] applying Gröbner basis methods for determining its solution.

EXAMPLE 3.1   The vector fields considered in Example 3.3 below will be shown to generate a Lie algebra of type $l_{3,2}$. In order to determine the transformation to canonical generators explicitly, the new basis elements are written as $U_i = \alpha_{i1} \partial_x + \alpha_{i2} \partial_y + \alpha_{i3}(y \partial_x - x \partial_y)$. The commutation relations of $l_{3,2}$

Lie Transformation Groups   117

require that the coefficients satisfy

$$\alpha_{12}\alpha_{23} - \alpha_{13}\alpha_{22} = 0, \quad \alpha_{11}\alpha_{23} - \alpha_{13}\alpha_{21} = 0,$$
$$\alpha_{11}\alpha_{33} - \alpha_{13}\alpha_{31} + \alpha_{12} = 0, \quad \alpha_{12}\alpha_{33} - \alpha_{13}\alpha_{32} - \alpha_{11} = 0,$$
$$\alpha_{21}\alpha_{33} - \alpha_{23}\alpha_{31} - \alpha_{22} = 0, \quad \alpha_{22}\alpha_{33} - \alpha_{23}\alpha_{32} + \alpha_{21} = 0$$

and $\alpha_{13} = \alpha_{23} = 0$. Its general solution is $\alpha_{13} = \alpha_{23} = 0$, $\alpha_{33} = i$, $\alpha_{12} = -i\alpha_{11}$ and $\alpha_{22} = i\alpha_{21}$. The coefficients α_{11}, α_{21}, α_{31} and α_{32} remain undetermined. Choosing $\alpha_{11} = \alpha_{21} = 1$, $\alpha_{31} = \alpha_{32} = 0$ the final answer $U_1 = \partial_x - i\partial_y$, $U_2 = \partial_x + i\partial_y$ and $U_3 = i(y\partial_x - x\partial_y)$ is obtained. ⬜

Although a priori any Lie algebra with equal dimension is a candidate for the canonical form of any given Lie algebra, it is more efficient to narrow down the possible types in advance by calculating several invariants. This point has been emphasized by de Graaf [37], Section 7.1. The following example is taken from the above mentioned publication by Gerdt [52]. The characteristic polynomial which occurs there is explained below, see eq. (3.17).

EXAMPLE 3.2   Consider the Lie algebra of vector fields $v_1 = \partial_x$, $v_2 = \partial_t$, $v_3 = t\partial_x + \partial_u$, $v_4 = x\partial_x + 3t\partial_t - 2u\partial_u$ with nonvanishing commutators $[v_1, v_4] = [v_2, v_3] = v_1$, $[v_2, v_4] = 3v_2$ and $[v_3, v_4] = -2v_3$. According to definition (3.17) its characteristic polynomial $\omega(\omega - E_4)(\omega - 3E_4)(\omega + 2E_4)$ is squarefree and decomposes into linear factors. Consequently, the Lie algebra $\mathfrak{l}_{4,5}$ defined on page 136 is identified as the proper type. ⬜

A sub-vectorspace $\mathcal{M} \subset \mathcal{L}$ is a *subalgebra* if the commutator of any two elements from $\mathcal{M}$ is again in $\mathcal{M}$. This is written $[\mathcal{M}, \mathcal{M}] \subset \mathcal{M}$. An *invariant subalgebra* or *ideal* is such that any commutator with elements from $\mathcal{L}$ is in $\mathcal{M}$, i.e., $[\mathcal{M}, \mathcal{L}] \subset \mathcal{M}$. The *normalizer* of $\mathcal{M}$ is the set of all elements $l \in \mathcal{L}$ such that $[l, \mathcal{M}] \in \mathcal{M}$. The normalizer of $\mathcal{M}$ is a subalgebra. If $\mathcal{S}$ is a subset of $\mathcal{L}$ the *centralizer* of $\mathcal{S}$ is the set of all elements $c \in \mathcal{L}$ such that $[c, s] = 0$ for all $s \in \mathcal{S}$. The centralizer of $\mathcal{L}$ is called the *center*.

Fundamental problems are to classify all types of composition leading to nonisomorphic Lie algebras, and to determine simpler substructures within a given Lie algebra, i.e., to determine its invariant subalgebras. The *characteristic polynomial*

$$\Delta(\omega) \equiv |\sum E_i c_{ij}^k - \delta_{jk}\omega| =$$
$$\omega^r - \psi_1(E)\omega^{r-1} + \psi_2(E)\omega^{r-2} + \ldots + (-1)^{r-1}\psi_{r-1}(E)\omega$$

(3.17)

or correspondingly the *characteristic equation* $\Delta(\omega) = 0$ where $E_1, \ldots, E_r$ are r indeterminates turns out to be an important tool for dealing with these problems. It has been introduced by Killing [89] in order to answer the question in which two-dimensional subalgebras a given element is contained. The number of independent coefficients of the characteristic polynomial is called the *rank of the Lie algebra*. Important properties of the characteristic polynomial are given next.

118

THEOREM 3.7   *Let $\Delta(\omega)$ and the $\psi_k(E_1, \ldots, E_r)$ be defined by (3.17).*

   i) *The coefficient $\psi_k(E_1, \ldots, E_r)$ is a homogeneous polynomial of degree k in $E_1, \ldots, E_r$.*

  ii) *If new basis elements are introduced by $\bar{U}_i = \sum \alpha_{i,j} U_j$, the ψ_k are changed according to $\bar{\psi}_k(\bar{E}_1, \ldots, \bar{E}_r) = \psi_k(\sum \alpha_{i,1} E_i, \ldots, \sum \alpha_{i,r} E_i)$.*

 iii) *The coefficients $\psi_k(E_1, \ldots, E_r)$ satisfy the differential equations*

$$\sum_{\sigma,\tau=1}^{r} c_{\rho\sigma}^{\tau} E_\sigma \frac{\partial \psi}{\partial E_\tau} = 0,$$

   i.e. they are invariants of the adjoint group.

 iv) *The rank of a Lie algebra is not higher than the multiplicity of the root $\omega = 0$ of the characteristic equation.*

The proof may be found in Cartan's Thesis [25], Chapitre II. Property *ii)* shows that the structure of the coefficients ψ_k is invariant under a basis change of the corresponding Lie algebra. This property makes the characteristic polynomial into an important tool for determining the type of a Lie algebra as it is shown in the following example.

EXAMPLE 3.3   The Lie algebra generated by the vector fields ∂_x, ∂_y and $y\partial_x - x\partial_y$ has the nonvanishing structure constants $c_{32}^1 = -1$ and $c_{33}^2 = 1$. They yield the characteristic polynomial $(\omega^2 + E_3^2)\omega$. By Theorem 3.7 the Lie algebra is of type $l_{3,2}$ with $c = -1$ in the classification given in Section 3.4. This follows from the fact that it is the only algebra which has a third order and a first order term in its characteristic polynomial, and the dimension of its derived algebra (see below) has dimension 2.   ⬜

This example is of particular importance because it applies exactly the kind of reasoning that is used in Chapter 5 for identifying the symmetry types of differential equations.

Solvable Lie Algebras. The so-called *integrable* or *solvable groups* and the corresponding Lie algebras play an important role in the symmetry analysis of differential equations. This term is closely connected with their application for solving linear pde's as described in Theorem 2.23. The special structure of the commutators of the vector fields admitted by the given linear pde in this theorem is generalized as follows. The *derived Lie algebra* $\mathcal{L}' \equiv \mathcal{L}^{(1)}$ is generated by all commutators of $\mathcal{L}$, i.e., $\mathcal{L}' = [\mathcal{L}, \mathcal{L}]$. The *derived series* is the sequence $\mathcal{L}, \mathcal{L}', \ldots, \mathcal{L}^{(k)}$ where $\mathcal{L}^{(j)} = [\mathcal{L}^{(j-1)}, \mathcal{L}^{(j-1)}]$.

DEFINITION 3.6   *(Solvable Lie algebra) A Lie algebra of dimension r is called integrable or solvable if $\mathcal{L}^{(k)} = 0$ for some positive integer $k \leq r$.*

Lie Transformation Groups 119

The following theorem due to Lie and Engel provides a convenient means for identifying solvable Lie algebras if they are given by basis vectors and structure constants. Its proof may be found in Cartan's Thesis [25], *Théorème IV* on page 45.

THEOREM 3.8   *Let $\mathcal{L}$ be an r-dimensional Lie algebra with structure constants c_{ij}^k. Either of the following two conditions is necessary and sufficient for $\mathcal{L}$ to be solvable.*

i) The derived algebra of $\mathcal{L}$ has rank zero.

ii) For $i, j, k = 1, \ldots, r$ there holds $\sum_{\lambda,\mu,\nu=1}^{r} (c_{i\lambda}^{\mu} c_{j\mu}^{\nu} c_{k\nu}^{\lambda} - c_{i\mu}^{\lambda} c_{j\nu}^{\mu} c_{k\lambda}^{\nu}) = 0$.

EXAMPLE 3.4   Consider the three- and four-dimensional algebras listed on page 135. The first and the second derived algebras of $\mathfrak{l}_{3,1}$ and $\mathfrak{l}_{4,1}$ have dimension 3, therefore by definition they are not solvable. The second or the third derived algebras of the remaining algebras $\mathfrak{l}_{3,2}$ to $\mathfrak{l}_{3,6}$ and $\mathfrak{l}_{4,2}$ to $\mathfrak{l}_{4,17}$ have zero dimension, consequently they are solvable.   ☐

Important properties of solvable algebras are described next, they may be found in Cartan's Thesis [25] on pages 45-48.

THEOREM 3.9   *Solvable Lie algebras have the following important properties.*

i) Any solvable Lie algebra of dimension r contains an invariant subalgebra of dimension 1, which is contained in an invariant subalgebra of dimension 2 etc., i.e., r independent elements $U_1, U_2, \ldots, U_r$ may be found such that

$$[U_i, U_k] = \sum_{j=1}^{j=i} c_{i,k}^{j} U_j \quad for \quad i = 1, \ldots, r, \quad k = i+1, \ldots, r.$$

ii) The characteristic polynomial of a solvable algebra decomposes into linear factors, i.e.,

$$\Delta(\omega) = -\omega \prod_{i=1}^{i=r-1} \left(\sum_{\rho=i-1}^{\rho=r} E_\rho c_{\rho i}^{i} - \omega \right).$$

Semi-simple and simple Lie algebras.   If an algebra $\mathcal{L}$ is not solvable it may have invariant subalgebras that are solvable. There is always a unique largest solvable ideal which contains every solvable ideal, it is called the *radical* $\mathcal{R}$. If $\mathcal{R} = \mathcal{L}$ the algebra is solvable, if $\mathcal{R} = 0$ the algebra does not have any nontrivial solvable ideal and the following definition applies.

DEFINITION 3.7   *(Semi-simple and simple) If a nonsolvable Lie algebra does not have any nontrivial solvable invariant subalgebra it is called semi-simple. If it does not have any invariant subalgebra at all it is called simple.*

120

From the above discussion it follows that determining the radical of any Lie algebra is an important step for identifying its structure. The following result which is due to Cartan [25], Chapter III, Section 5, gives a constructive answer to this problem.

THEOREM 3.10   *(Cartan 1894) If $\mathcal{L}$ is a Lie algebra of dimension r with structure constants c_{ij}^k and characteristic polynomial (3.17), its radical is determined by a linear combination of the basis elements the coefficients of which obey the constraints $\sum_{\rho=1}^r c_{ij}^\rho \frac{\partial \psi_2(E)}{\partial E_\rho} = 0.$*

Because ψ_2 is a quadratic form, the constraints given in this theorem generate a system of linear homogeneous equations in the E_i, its solution determines the radical. The next theorem provides a constructive means for deciding semi-simplicity.

THEOREM 3.11   *A Lie algebra is semi-simple if and only if the discriminant of the coefficient $\psi_2(E)$ of the characteristic polynomial is different from zero.*

One of the major achievements of Killing and Cartan was to provide an explicit and complete classification of simple complex Lie algebras. It comprises four series of so-called classical algebras A_n, B_n, C_n and D_n for $n \geq 1$, 2, 3 or 4 respectively, and the exceptional algebras E_6, E_7, E_8, F_4 and G_2. The subindex denotes the rank of the respective algebra. Details may be found in the literature quoted at the beginning of this section. The importance of this classification becomes obvious from the following theorem which is also due to Killing and Cartan.

THEOREM 3.12   *Any semi-simple Lie algebra is the direct sum of invariant simple subalgebras.*

The decomposition ensured by this theorem is unique. The simple components may be transformed into canonical form. In order to obtain this representation explicitly for any given semi-simple Lie algebra suitable algorithms are required. A lot of information on these algorithmic problems may be found in the thesis of de Graaf [37] and the references given there; see also the article by Cartan [26] on this subject.

EXAMPLE 3.5   *(Cartan 1896)* Let a six-dimensional Lie algebra given by the vector fields

$$U_1 = \partial_x + x(x\partial_x + y\partial_y + z\partial_z), \ U_2 = \partial_y + x(x\partial_x + y\partial_y + z\partial_z),$$
$$U_3 = \partial_z + x(x\partial_x + y\partial_y + z\partial_z), \tag{3.18}$$
$$U_4 = z\partial_y - y\partial_x, \ U_5 = x\partial_z - z\partial_z, \ U_6 = y\partial_x - x\partial_y.$$

The coefficient $\psi_2(E) = E_1^2 + E_2^2 + E_3^2 + E_4^2 + E_5^2 + E_6^2$ has nonvanishing discriminant, consequently by Theorem 3.11 the Lie algebra is semi-simple.

It is easy to see that $\bar{U}_1 = U_1 + U_4$, $\bar{U}_2 = U_2 + U_5$, $\bar{U}_3 = U_3 + U_6$ and $\bar{U}_4 = U_1 - U_4$, $\bar{U}_5 = U_2 - U_5$, $\bar{U}_6 = U_3 - U_6$ generate two type $l_{3,1}$ algebras with $[\bar{U}_i, \bar{U}_{i+3}] = 0$ for $i = 1, 2, 3$, i.e., the vector fields (3.18) generate a semi-simple algebra which is the direct sum of two three-dimensional simple algebras of type $l_{3,1}$. □

The Levi Decomposition. A Lie algebra $\mathcal{L}$ having a nontrivial proper radical may be represented as a semidirect sum $\mathcal{R} \rtimes \mathcal{S}$ where $\mathcal{R}$ is the radical and $\mathcal{S}$ is semi-simple. The radical is unique and basis independent, the complement $\mathcal{S}$ is ismorphic to $\mathcal{L}/\mathcal{R}$. The formulation given in the subsequent theorem is taken from Rand [151].

THEOREM 3.13   *(Rand 1988) Let $\mathcal{L}$ be a Lie algebra with basis $l_1, \ldots, l_r$, radical $\mathcal{R}$ and Levi decomposition $\mathcal{R} \rtimes \mathcal{S}$. There is an algorithm that finds a new basis $r_1, \ldots, r_\rho, s_1, \ldots, s_\sigma$ with $\rho + \sigma = r$ such that $\mathcal{R}$ is generated by the r_i, $\mathcal{S}$ by the s_k and the commutation relations $[\mathcal{R}, \mathcal{R}] \subset \mathcal{R}$, $[\mathcal{S}, \mathcal{S}] = \mathcal{S}$ and $[\mathcal{R}, \mathcal{S}] \subseteq \mathcal{R}$ are valid.*

The proof of Rand [151] provides a constructive method for obtaining such a basis explicitly. It is not given here because the examples that are relevant later on in this book are simple enough such that it may be obtained by inspection, and the general method is not required.

EXAMPLE 3.6   Consider the Lie algebra $\mathbf{g}_{23}(r = 3)$ defined on page 139. It is generated by the vector fields

$$U_1 = \partial_x,\ U_2 = \partial_y,\ U_3 = x\partial_x,\ x\partial_y,\ U_4 = x\partial_y,\ U_5 = y\partial_y,$$
$$U_6 = x^2\partial_y,\ U_7 = x^2\partial_x + 2xy\partial_y$$

from which the nonvanishing structure constants $c_{13}^1 = 1$, $c_{14}^2 = 1$, $c_{17}^3 = 2$, $c_{17}^5 = 2$, $c_{25}^2 = 1$, $c_{27}^4 = 2$, $c_{36}^6 = 2$, $c_{37}^7 = 1$, $c_{45}^4 = 1$, $c_{47}^6 = 1$, $c_{56}^6 = -1$ and the coefficient $\psi_2 = 8E_1E_7 + 3E_5^2 - 6E_3E_5 + E_3^2$ are obtained. Applying Theorem 3.10 yields the constraints $E_1 = E_3 = E_7 = 0$, i.e., the radical is generated by U_2, U_4, U_5 and U_6. It is easy to see that U_1, $U_3 + U_5$ and U_7 generate a simple three-dimensional algebra. Consequently, the Levi decomposition has the form $l_{4,6}(a = b = 1) \rtimes l_{3,1}$ or, its equivalent in terms of vector fields

$$\{\partial_y, x\partial_y, x^2\partial_y, y\partial_y\} \rtimes \{\partial_x, x\partial_x + y\partial_y, x^2\partial_x + 2xy\partial_y\}. \qquad □$$

3.3   Group Actions in the Plane

For the main subject of this book Lie transformation groups acting on a two-dimensional manifold are most important. If not stated otherwise this will

122

be the complex plane $\mathbb{C}$. Therefore the results described in the two preceding sections are specialized to this case and illustrated by numerous examples, usually taken from the collection in Section 3.4. Of particular importance for later applications are constructive methods for dealing with these groups and its Lie algebras.

Finite Groups and Their Infinitesimal Generators. The action of an r parameter Lie transformation group in coordinates x and y is written as

$$\bar{x} = f(x, y, a_1, \ldots, a_r), \quad \bar{y} = g(x, y, a_1, \ldots, a_r) \tag{3.19}$$

with $f(x, y, 0, \ldots, 0) = x$ and $g(x, y, 0, \ldots, 0) = y$. If there is only a single parameter, it is denoted by a without a subindex. Conforming with (3.7), the coefficients of the infinitesimal generators $U_k = \xi_k(x, y)\partial_x + \eta_k(x, y)\partial_y$ are defined by

$$\xi_k(x, y) \equiv \frac{\partial f}{\partial a_k}\Big|_{a_1 = \ldots = a_r = 0} \quad \text{and} \quad \eta_k(x, y) \equiv \frac{\partial g}{\partial a_k}\Big|_{a_1 = \ldots = a_r = 0}$$

for $k = 1, \ldots, r$. According to Lie's First Theorem 3.1 there holds

$$\frac{\partial \bar{x}}{\partial a_k} = \sum_{j=1}^{r} \varphi_{j,k}(a_1, \ldots, a_r)\xi_j(\bar{x}, \bar{y}), \quad \frac{\partial \bar{y}}{\partial a_k} = \sum_{j=1}^{r} \psi_{j,k}(a_1, \ldots, a_r)\eta_j(\bar{x}, \bar{y}).$$

$$\tag{3.20}$$

EXAMPLE 3.7   The rotations of the plane are parametrized by

$$\bar{x} = x \cos \alpha - y \sin \alpha, \quad \bar{y} = x \sin \alpha + y \cos \alpha \tag{3.21}$$

with $0 \leq \alpha < 2\pi$. An additional second transformation of this type with parameter β

$$\bar{\bar{x}} = \bar{x} \cos \beta - \bar{y} \sin \beta, \quad \bar{\bar{y}} = \bar{x} \sin \beta + \bar{y} \cos \beta \tag{3.22}$$

may be combined into a single one corresponding to a parameter value $\gamma = \alpha + \beta$, the reverse operation of (3.21) corresponds to $-\alpha$, and no action to $\alpha = 0$. The infinitesimal generator is $U = -y\partial_x + x\partial_y$. The group equations (3.20) are $\frac{d\bar{x}}{d\alpha} = -\bar{y}$ and $\frac{d\bar{y}}{d\alpha} = \bar{x}$. A Janet basis for the defining equations in $grlex$, $\eta > \xi$, $y > x$ term order is $\{\eta + \frac{x}{y}\xi = 0, \xi_x = 0, \xi_y + \frac{1}{y}\xi = 0\}$.   ▯

EXAMPLE 3.8   Assume that the points with coordinates (x, y) are transformed into $(\bar{x}, \bar{y})$ according to $\bar{x} = x + a_1$ and $\bar{y} = y + a_2$. A second transformation of the same type with parameters b_1 and b_2 yields $\bar{\bar{x}} = \bar{x} + b_1$ and $\bar{\bar{y}} = \bar{y} + b_2$. Obviously the same effect may be achieved by a single transformation with parameter values $c_1 = a_1 + b_1$, $c_2 = a_2 + b_2$. The group equations (3.20) are $\frac{\partial \bar{x}}{\partial a_1} = 1$, $\frac{\partial \bar{x}}{\partial a_2} = 0$, $\frac{\partial \bar{y}}{\partial a_1} = 0$ and $\frac{\partial \bar{y}}{\partial a_2} = 1$. For the infinitesimal generators there follows $U_1 = \partial_x$ and $U_2 = \partial_y$ with $[U_1, U_2] = 0$, i.e., they generate the Lie algebra $l_{2,2}$ defined on page 135.   ▯

Lie Transformation Groups 123

In general a pair of substitution functions f and g does *not* define a group as the following example shows.

EXAMPLE 3.9   The transformations

$$\bar{x} = (a+1)x, \quad \bar{y} = y+a \quad \text{and} \quad \bar{x} = (b+1)\bar{x}, \quad \bar{\bar{y}} = \bar{y}+b \quad (3.23)$$

have the combined action

$$\bar{\bar{x}} = (a+1)(b+1)x, \quad \bar{\bar{y}} = y+a+b.$$

In this case it is obvious from the finite group equations that there does not exist a parameter c that may be expressed in the form $c = \phi(a,b)$. □

EXAMPLE 3.10   An example of a group depending on three parameters is generated by the motions in the plane. If two subsequent motions are defined by

$$\bar{x} = x\cos\alpha - y\sin\alpha + a, \quad \bar{y} = x\sin\alpha + y\cos\alpha + b,$$
$$\bar{\bar{x}} = \bar{x}\cos\beta - \bar{y}\sin\beta + c, \quad \bar{\bar{y}} = \bar{x}\sin\beta + \bar{y}\cos\beta + d$$

a simple calculation shows that the combined motion is of the same form and is generated by the parameter values

$$\alpha + \beta, \quad a\cos\beta + b\sin\beta + c, \quad a\sin\beta + b\cos\beta + d$$

for the rotation angle and the two translations. The infinitesimal generators corresponding to the parameters a, b and α respectively are $U_1 = \partial_x$, $U_2 = \partial_y$ and $U_3 = y\partial_x - x\partial_y$ with the nonvanishing commutators $[U_1, U_3] = -U_2$ and $[U_2, U_3] = U_1$. □

EXAMPLE 3.11   Consider the equations $\bar{x} = xe^{a_3} + a_1$, $\bar{y} = ye^{a_3} + a_2$. By differentiation there follows $\dfrac{\partial\bar{x}}{\partial a_1} = 1$, $\dfrac{\partial\bar{x}}{\partial a_3} = xe^{a_3}$, $\dfrac{\partial\bar{y}}{\partial a_2} = 1$ and $\dfrac{\partial\bar{y}}{\partial a_3} = ye^{a_3}$ for the nonvanishing derivatives. Consequently, corresponding to (3.20)

$$\phi = \psi = \begin{pmatrix} 1 & 0 & 0 \\ 0 & 1 & 0 \\ 0 & 0 & e^{a_3} \end{pmatrix}$$

and $\xi_1 = 1$, $\xi_2 = 0$, $\xi_3 = x$, $\eta_1 = 0$, $\eta_2 = 1$ and $\eta_3 = y$. They satisfy (3.20), i.e. the above expressions for $\bar{x}$ and $\bar{y}$ define a three parameter group. Substituting $a_2 = a_3 = 0$ yields $\bar{x} = x + a_1$, $\bar{y} = y$ and $U_1 = \partial_x$, substituting $a_1 = a_3 = 0$ yields $\bar{x} = x$, $\bar{y} = y + a_2$ and $U_2 = \partial_y$, finally substituting $a_1 = a_2 = 0$ yields $\bar{x} = xe^{a_3}$, $\bar{y} = ye^{a_3}$ and $U_3 = x\partial_x + y\partial_y$ with the commutators $[U_1, U_2] = 0$, $[U_1, U_3] = U_1$ and $[U_2, U_3] = U_2$, i.e., they generate a Lie algebra of type $l_{3,2}(c = 1)$. □

The finite transformations corresponding to an infinitesimal generator $U = \xi(x,y)\partial_x + \eta(x,y)\partial_y$ may be obtained from Lie's equations (3.14). In

124

coordinates x and y they have the form

$$\frac{d\bar{x}}{da} = \xi(\bar{x}, \bar{y}), \quad \frac{d\bar{y}}{da} = \eta(\bar{x}, \bar{y}). \tag{3.24}$$

This is a system of ode's for the two functions $\bar{x}(x, y, a)$ and $\bar{y}(x, y, a)$ w.r.t. to the group parameter a, complemented by the initial conditions $\bar{x}(a = 0) = x$, $\bar{y}(a = 0) = y$. The coordinates of the manifold M appear only as parameters. Although a solution for such a system always exists, in general it cannot be obtained in closed form. This is possible only if $\xi(\bar{x}, \bar{y})$ and $\eta(\bar{x}, \bar{y})$ have a special structure. The following examples illustrate these possibilites.

EXAMPLE 3.12   Let $U = x^2\partial_x + xy\partial_y$, i.e., $\frac{d\bar{x}}{da} = \bar{x}^2$, $\frac{d\bar{y}}{da} = \bar{x}\bar{y}$. Integration of the first equation yields $\bar{x} = -\dfrac{1}{a + \phi(x, y)}$ where ϕ is an undetermined function of its arguments. The requirement $\bar{x}(a = 0) = x$ leads to $\phi(x, y) = -\frac{1}{x}$ and $\bar{x} = \dfrac{x}{1 - ax}$. Substituting this value into the second equation yields $\frac{d\bar{y}}{da} = \dfrac{x\bar{y}}{1 - ax}$ from which there follows $\bar{y} = \dfrac{\psi(x, y)}{ax - 1}$ with ψ again an undetermined function of x and y. The condition $\bar{y}(a = 0) = y$ leads to $\psi(x, y) = -y$ and finally $\bar{y} = \dfrac{y}{1 - ax}$. $\square$

EXAMPLE 3.13   Let $U = (x + y)\partial_x$, i.e., $\frac{d\bar{x}}{da} = \bar{x} + \bar{y}$, $\frac{d\bar{y}}{da} = 0$. The last equation leads immediately to $\bar{y} = y$. Substitution into the first equation yields the linear first order equation $\frac{d\bar{x}}{da} - \bar{x} = y$ with the general solution $\bar{x} = e^a(ye^a + \phi(x, y))$. The initial condition $\bar{x}(a = 0) = x$ leads to $\phi(x, y) = x - y$, and the final answer $\bar{x} = e^a x + e^a(e^a - 1)y$. $\square$

EXAMPLE 3.14   Let $U = (y^2 + 1)\partial_x + (x^2 + 1)\partial_y$, i.e., $\frac{d\bar{x}}{da} = \bar{y}^2 + 1$, $\frac{d\bar{y}}{da} = \bar{x}^2 + 1$. Eliminating $\bar{x}$, the second order equation $\bar{y}''^2 - 4(\bar{y}' - 1)(\bar{y}^2 + 1) = 0$ is obtained for which a closed-form solution cannot be found. $\square$

For a selection of frequently occurring generators the corresponding finite transformation is given in Table 3.1.

In order to find the finite transformations of an r-parameter group if the transformations corresponding to the infinitesimal generators are known, a method described in Lie [114], Kapitel 7, §3 will be applied.

LEMMA 3.2   (Lie 1893) Let r one-parameter groups be determined by the transformation functions $f_i(x, y, a_i)$ and $g_i(x, y, a_i)$ for $i = 1, \ldots, r$. Then the general transformation of an r-parameter group is obtained recursively by

$$\begin{aligned} \bar{x} &= f_r(f_{r-1}(\ldots f_2(f_1(x, y, a_1), a_2), \ldots, a_{r-1}), a_r), \\ \bar{y} &= g_r(g_{r-1}(\ldots g_2(g_1(x, y, a_1), a_2), \ldots, a_{r-1}), a_r). \end{aligned} \tag{3.25}$$

EXAMPLE 3.15   Let a three-parameter group be determined by the infinitesimal generators $U_1 = \partial_x$, $U_2 = \partial_y$, $U_3 = x\partial_x + y\partial_y$. By means of the

Lie Transformation Groups

Generator	Transformation	Generator	Transformation
∂_x	$\bar{x} = x + a,\ \bar{y} = y$	∂_y	$\bar{x} = x,\ \bar{y} = y + a$
$x\partial_x$	$\bar{x} = xe^a,\ \bar{y} = y$	$y\partial_x$	$\bar{x} = x + ay,\ \bar{y} = y$
$x\partial_y$	$\bar{x} = x,\ \bar{y} = y + ax$	$y\partial_y$	$\bar{x} = x,\ \bar{y} = ye^a$
$y\partial_x - x\partial_y$	$\bar{x} = x\cos a + y\sin a,$ $\bar{y} = -x\sin a + y\cos a$	$y\partial_x + x\partial_y$	$\bar{x} = x\cosh a + y\sinh a,$ $\bar{y} = x\sinh a + y\cosh a$
$x^2\partial_x + xy\partial_y$	$\bar{x} = \dfrac{x}{1 - ax},\ \bar{y} = \dfrac{y}{1 - ax}$	$xy\partial_x + y^2\partial_y$	$\bar{x} = \dfrac{x}{1 - ay},\ \bar{y} = \dfrac{y}{1 - ay}$
$x^2\partial_x + 2xy\partial_y$	$\bar{x} = \dfrac{x}{1 - ax},\ \bar{y} = \dfrac{y}{(1 - ax)^2}$	$(x + y)\partial_x$	$\bar{x} = e^a x + (e^a - 1)y,\ \bar{y} = y$
$x\partial_x + (x + y)\partial_y$	$\bar{x} = e^a x,\ \bar{y} = e^a(ax + y)$	$x^2\partial_x + y^2\partial_y$	$\bar{x} = \dfrac{x}{1 - ax},\ \bar{y} = \dfrac{y}{1 - ay}$
$e^x\partial_y$	$\bar{x} = x,\ \bar{y} = y + ae^x$	$mx\partial_x + ny\partial_y$	$\bar{x} = xe^{ma},\ \bar{y} = ye^{na}$
$y^n\partial_x$	$\bar{x} = x + ay^n,\ \bar{y} = y$	$x^n\partial_y$	$\bar{x} = x,\ \bar{y} = y + ax^n$

TABLE 3.1: The finite transformations for a selection of frequently occurring infinitesimal generators are listed; a is the group parameter, the value $a = 0$ corresponds to the identity.

finite transformations given in Table 3.1 one gets $\bar{x} = x + a_1$, $\bar{y} = y$ for U_1, $\bar{\bar{x}} = \bar{x}$, $\bar{\bar{y}} = \bar{y} + a_2$ from U_2 and $\bar{\bar{\bar{x}}} = e^{a_3}\bar{\bar{x}}$, $\bar{\bar{\bar{y}}} = e^{a_3}\bar{\bar{y}}$ from U_3. Combining the three transformations and making an obvious change of variables, the finite transformations of the three-parameter group are obtained in the form

$$\bar{x} = (x + a_1)e^{a_3}, \quad \bar{y} = (y + a_2)e^{a_3}.$$

The final result must be independent of the order in which the infinitesimal generators are applied. Taking for example $U_1 = x\partial_x + y\partial_y$, $U_2 = \partial_x$ and $U_3 = \partial_y$ one obtains by a similar calculation

$$\bar{x} = e^{b_1}x + b_2, \quad \bar{y} = e^{b_1}y + b_3.$$

This is the same group with a different set of parameters. They are related by the equations $b_1 = a_3$, $b_2 = a_1 e^{a_3}$, $b_3 = a_2 e^{a_3}$. $\quad\square$

On page 112 a third representation for a Lie transformation group in terms of the so-called defining equations has been introduced. Solving these linear homogeneous systems of pde's is the main issue of Chapter 2. The reverse process, i.e., setting up the defining equations from a given set of infinitesimal generators, has also been discussed there on page 75. The algorithm *ConstructJanetBasis* has been designed for this purpose. The following simple example may be performed by pencil and paper, yet it shows all basic steps.

EXAMPLE 3.16　Let a group be given in terms of three vector fields by $\{\partial_x, \partial_y, y\partial_x - x\partial_y\}$, consequently $\xi = c_1 + c_3 y$ and $\eta = c_2 - c_3 x$. Differentiation leads to $\xi_x = 0$, $\xi_y = c_3$, $\eta_x = -c_3$ and $\eta_y = 0$. Eliminating c_3 there follows $\xi_x = 0$, $\xi_y + \eta_x = 0$, $\eta_y = 0$. It is not a Janet basis. In a *grlex* term ordering with $\eta > \xi$, $y > x$ the integrability condition $\xi_{yy} = 0$ is obtained. It has to be added to this system, i.e., the desired Janet basis is $\{\xi_x, \eta_x + \xi_y, \eta_y, \xi_{yy}\}$. ∎

The importance of the defining equations arises from the fact that the Lie symmetries of an ode are always obtained in this form in the first place, therefore these equations are called *determining system* in this context. In Section 3.5 it will be shown how many important properties of Lie groups may be obtained from these defining equations without solving them.

Geometric Properties. Let $\bar{x} = f(x, y, a)$ and $\bar{y} = g(x, y, a)$ be the finite equations of a one-parameter group with coordinates (x, y) and infinitesimal generator $U = \xi(x, y)\partial_x + \eta(x, y)\partial_y$. Under the action of this group an arbitrarily chosen point describes a curve the points of which are parametrized by a. Any two points on an individual curve may be transformed into each other by a suitable group element, they are *equivalent* w.r.t. this group. If they are located on different curves such a transformation does not exist. These curves, generated by the action of a one-parameter group on an arbitrary point of the plane, are called its *path curve*. They are special cases of the orbits defined on page 108. The path curves generate an invariant decomposition of the plane. By definition, the path curves corresponding to U obey equations (3.24). According to Theorem 2.19 these ode's with first integral $\varphi(x, y) = const$ are equivalent to the pde $U\varphi = 0$. It shows that the orbits of a one-parameter group of the plane are essentially determined by the invariant of the pde corresponding to its generator.

EXAMPLE 3.17 Consider the one-parameter group $\bar{x} = \frac{x}{1 - ax}$ and $\bar{y} = \frac{y}{1 - ax}$ with generator $U = x^2\partial_x + xy\partial_y$. The invariant of U is $\phi = \frac{y}{x}$. Consequently, its orbits are $\frac{y}{x} = C$, i.e., the straight lines through the origin.

∎

EXAMPLE 3.18 The group considered in the preceding example had the invariant $\phi = \frac{y}{x}$. Substituting the transformed variables $\bar{x}$ and $\bar{y}$ yields $\frac{\bar{y}}{\bar{x}} = \frac{y}{1 - ax} \frac{1 - ax}{x} = \frac{y}{x}$, i.e., the invariance of ϕ is explicit. ∎

EXAMPLE 3.19 The invariant of the one-parameter group with generator $U = ax\partial_x + by\partial_y$ is determined by $\frac{1}{a}\frac{dx}{x} = \frac{1}{b}\frac{dy}{y}$ with the result $\phi = \frac{y^a}{x^b}$.

∎

If the finite transformations of the group are known, deciding transitivity comes down to eliminating the group parameters $a_1, \ldots, a_r$ from equations (3.19) for given (x, y) and $(\bar{x}, \bar{y})$. Consider for example the group $\bar{x} = x + a_1$, $\bar{y} = y + a_2$. If $(\bar{x}, \bar{y})$ is considered to be chosen at will, the parameter values $a_1 = \bar{x} - x$, $a_2 = \bar{y} - y$ will perform the desired transformation. On the other hand, if the group is $\bar{x} = x$, $\bar{y} = y + a_1 x + a_2$, such a transformation does not

Lie Transformation Groups

exist because the transformations of the group leave x unchanged and a point with $\bar{x} \neq x$ can never be reached. The above question may be generalized to any number of points sharing this property.

EXAMPLE 3.20   The two-parameter groups $\mathbf{g}_4$ and $\mathbf{g}_{15}$ defined on page 138 are intransitive because the first coordinate is transformed as $\bar{x} = x$, i.e., if $\bar{x}$ is chosen different from x there is no group element that brings about the transformation. Group $\mathbf{g}_{25}$ has finite transformations $\bar{x} = e^{a_2}, \bar{y} = (y+a_1)e^{a_2}$ from which the expressions $a_1 = \dfrac{x\bar{y} - \bar{x}y}{x}$ and $a_2 = \log\dfrac{\bar{x}}{x}$ for the group parameters are obtained, i.e., the group is transitive. The same is true for the group $\mathbf{g}_{26}$ that has been considered before. □

EXAMPLE 3.21   Consider the group $\mathbf{g}_6$ on page 138 with finite transformations $\bar{x} = (x + a_1)e^{a_3}, \bar{y} = (y + a_2)e^{a_4}$. Let two different points $P_1 = (x_1, y_1)$ and $P_2(x_2, y_2)$ be arbitrarily given. Applying a transformation of the group to them yields $\bar{x}_i = (x_i + a_1)e^{a_3}, \bar{y}_i = (y_i + a_2)e^{a_4}$ for $i = 1, 2$. The four group parameters may be eliminated with the result

$$a_1 = \frac{x_2\bar{x}_1 - x_1\bar{x}_2}{\bar{x}_2 - \bar{x}_1}, \quad a_2 = \frac{y_2\bar{y}_1 - y_1\bar{y}_2}{\bar{y}_2 - \bar{y}_1}, \quad e^{a_3} = \frac{\bar{x}_1 - \bar{x}_2}{x_1 - x_2}, \quad e^{a_4} = \frac{\bar{y}_1 - \bar{y}_2}{y_1 - y_2}.$$

This means, for any given pairs P_1, P_2 and $\bar{P}_1 = (\bar{x}_1, \bar{y}_1)$, $\bar{P}_2 = (\bar{x}_2, \bar{y}_2)$ a transformation of $\mathbf{g}_6$ may be determined moving P_i to $\bar{P}_i$ for $i = 1, 2$. The same procedure for three arbitrarily chosen points leads to an inconsistent set of equations. Consequently, $\mathbf{g}_6$ is 2-fold transitive. □

If the finite transformations of the group are not known, or if their expressions are rather complicated, the criterion for transitivity in terms of the infinitesimal generators of the group given on page 114 may be applied. It allows deciding transitivity for all groups that are listed on page 138 from its vector field coefficients with the following answer. Transitive groups: All primitive groups $\mathbf{g}_1$, $\mathbf{g}_2$ and $\mathbf{g}_3$, the groups $\mathbf{g}_5$, $\mathbf{g}_6$, $\mathbf{g}_7$ and $\mathbf{g}_9$ to $\mathbf{g}_{12}$ allowing two systems of imprimitivity, the groups $\mathbf{g}_{13}$, $\mathbf{g}_{14}$ and $\mathbf{g}_{17}$ to $\mathbf{g}_{23}$ with a single system of imprimitivity and finally the groups $\mathbf{g}_{24}$, $\mathbf{g}_{25}$ and $\mathbf{g}_{26}$ allowing a one-parameter family of systems of imprimitivity. Intransitive groups: $\mathbf{g}_4$, $\mathbf{g}_8$, $\mathbf{g}_{15}$, $\mathbf{g}_{16}$ and $\mathbf{g}_{27}$.

The possible types of transitivity of the transformation groups of the plane are completely described by the following result due to Lie [112], part I, page 632.

THEOREM 3.14   *A finite continuous group of a two-dimensional manifold is at most four-fold transitive. The only group with maximal transitivity is the eight-parameter projective group* $\mathbf{g}_3$.

This discussion shows that a group on a two-dimensional manifold may allow a single invariant, and this occurs if and only if all generators are multiples of a single one. In particular this is true for any single-parameter group.

128

The transitive two-parameter group $\bar{x} = x + a_1$, $\bar{y} = y + a_2$ considered previously transforms any line $y = \alpha x + const.$ into $\bar{y} = \alpha \bar{x} + const'.$ with the same value of α, i.e., the straight lines with fixed slope α are permuted among themselves by the group. According to the definition given on page 108 they form a system of imprimitivity.

The number of systems of imprimitivity is an important distinguishing feature of any transformation group. For a Lie transformation group of the plane the complete answer has been obtained by Lie [114], page 342, as follows.

THEOREM 3.15   *A Lie transformation group of the plane is either primitive, or it has exactly one or two systems of imprimitivity, or there is a system of imprimitivity depending on a parameter or a function of one variable.*

In general it is difficult to identify a system of imprimitivity by means of the finite transformations of a group. A criterion for primitivity in terms of the infinitesimal generators of the group turns out to be more convenient. It is obtained by the following considerations. A system of imprimitivity in the plane may be regarded as a set of curves depending on a parameter which in turn may be described as the solution set of a first order ode. If an equation with this property does exist, the group under consideration must be a symmetry of it. Analytically this is assured (see Chapter 5) if for the generic rank of the prolonged coefficient matrix there holds

$$rank \left\{ \begin{matrix} \xi_1, & \eta_1, & \zeta_1 \\ \xi_2, & \eta_2, & \zeta_2 \\ & \vdots & \vdots \\ \xi_r, & \eta_r, & \zeta_r \end{matrix} \right\} = 3$$

where $\zeta_k = \eta_{k,x} + y'(\eta_{k,y} - \xi_{k,x}) - y'^2 \xi_{k,y}$, Lie [114], page 208. The $\zeta's$ are discussed in detail in Chapter 5.

In later applications the following provides a particularly well-suited criterion in terms of the Janet basis of a group (Schwarz [164], see also Neumer [135]).

THEOREM 3.16   *Let the infinitesimal generators of a symmetry group of the $x - y$–plane be of the form $U = \xi \partial_x + \eta \partial_y$ where the coefficients solve the corresponding determining system. In order for a system of imprimitivity different from $x = const.$ or $y = const.$ to exist, an equation*

$$\eta_y \omega + \eta_x - \eta \omega_y - \xi_y \omega^2 - \xi_x \omega - \xi \omega_x = 0 \tag{3.26}$$

with a suitable $\omega \equiv \omega(x,y)$ must be valid as a consequence of the determining system of this symmetry group. If $x = const.$ or $y = const.$ is a system of imprimitivity, (3.26) reduces to $\xi_y = 0$ or $\eta_x = 0$ respectively.

PROOF   Let the desired systems of imprimitivity be described by the first order ode $y' = \omega(x,y)$. By definition, the solutions of this first order equation are permuted by the original symmetry group. Therefore it must allow it as

Lie Transformation Groups
129

a symmetry group. Analytically this is expressed in terms of the requirement $U^{(1)}(y' - \omega) = 0$ on the solution set of $y' = \omega$ which is equivalent to (3.26), see Theorem 5.2; $\omega = 0$ corresponds to $y' = 0$ or $y = const.$, by exchange of x and y the condition for $x = const.$ is obtained. $\quad\square$

For Lie the number and type of systems of imprimitivity was an important tool in order to obtain a classification of the transformation groups allowed by various manifolds. The classification of the groups of $\mathbb{C}^2$ given later in this chapter is organized by its systems of imprimitivity. The following examples are taken from this listing on page 138.

EXAMPLE 3.22   The group $\mathbf{g}_1$ has the Janet basis $\{\eta_y + \xi_x, \xi_{xx}, \xi_{xy}, \xi_{yy}, \eta_{xx}\}$ for its defining equations. Reducing the condition (3.26) w.r.t. to it yields the equations $\omega = 0$, $\omega^2 = 0$, $\omega_x = 0$, $\omega_y = 0$ and $1 = 0$ which is obviously inconsistent. Therefore $\mathbf{g}_1$ is primitive. $\quad\square$

EXAMPLE 3.23   The group $\mathbf{g}_7$ has the Janet basis $\{\xi_y, \eta_x, \eta_y - c\xi_x, \xi_{xx}\}$ for its defining equations. From the first two equations $x = const.$ and $y = const.$ are obvious systems of imprimitivity according to the above lemma. The same group occurs in Example 5.16 as symmetry group. Reducing (3.26) w.r.t. to the second Janet basis given there leads to the equations $\omega_x = 0$, $\omega_y = 0$ and $(\omega - 6)(\omega + 1) = 0$ from which the two systems of imprimitivity $y - 6x = const$ and $y + x = const$ are obtained. $\quad\square$

EXAMPLE 3.24   The group $\mathbf{g}_{13}$ has the Janet basis $\{\xi_x - \frac{2}{y}\eta, \xi_y, \eta_y - \frac{1}{y}\eta, \eta_{xx}\}$ for its defining equations. The single system of imprimitivity $x = const$ is obvious from the second equation, reduction of (3.26) w.r.t. to this Janet basis leads to an inconsistent system. In Example 5.14 the same group type generates the Janet basis in some transformed variables. Reduction of (3.26) w.r.t. the Janet basis given there leads to the system

$$\omega_x + \frac{1}{x+y}\omega^2 - \frac{2}{x+y}\omega + \frac{1}{x+y} = 0, \quad \omega_y + \frac{1}{x+y}\omega^2 - \frac{2}{x+y}\omega + \frac{1}{x+y} = 0$$

and $(\omega - 1)^2 = 0$ with the unique solution $\omega = 1$, i.e., the single system of imprimitivity is $y - x = const.$ $\quad\square$

Similarity. Transformation to Canonical Form. The appearance of any Lie transformation group may be changed in two ways: Either by a change of the group parameters, leading essentially to a new basis of the Lie algebra generated by the infinitesimal generators, or by changing the coordinates of the transformed manifold, leaving the commutation relations between the infinitesimal generators unchanged according to Lemma 3.1, but changing the form of the generators as has been shown on page 111 for a n-dimensional manifold. Due to its importance for the rest of this chapter, it is given here explicitly for a two-dimensional manifold.

LEMMA 3.3   *Let $U = \xi(x, y)\partial_x + \eta(x, y)\partial_y$ be a vector field in the plane*

130

with coordinates x and y. Furthermore, let new coordinates u and v be defined by $u = \sigma(x, y)$ and $v = \rho(x, y)$ with the inverse $x = \phi(u, v)$ and $y = \psi(u, v)$. Then the coefficients of the transformed vector field $V = \alpha(u, v)\partial_u + \beta(u, v)\partial_v$ are

$$\alpha(u, v) = U\sigma\big|_{x=\phi, y=\psi}, \qquad \beta(u, v) = U\rho\big|_{x=\phi, y=\psi}.$$

A constructive method for deciding similarity of two transformation groups defined in terms of their infinitesimal generators has been given in Theorem 2. It is illustrated next by a simple example.

EXAMPLE 3.25   Consider the generators $U_1 = \partial_u$, $U_2 = \partial_v$, $U_3 = u\partial_u + v\partial_v$ in coordinates u and v with commutators $[U_1, U_2] = 0$, $[U_1, U_3] = U_1$ and $[U_2, U_3] = U_2$. They form a Lie algebra of type $l_{3,2}(c = 1)$. On the other hand, there are the generators $V_1 = \frac{1}{y}\partial_x + \frac{1}{x}\partial_y$, $V_2 = -\frac{x^2}{y}\partial_x + (x+y)\partial_y$ and $V_3 = \frac{1}{y}\partial_x + \left(y + \frac{1}{x}\right)\partial_y$ in coordinates x and y with commutators $[V_1, V_2] = V_1$, $[V_1, V_3] = V_1$ and $[V_2, V_3] = V_2 - V_3$. The linear combinations $\bar{V}_1 = \frac{1}{2}V_1$, $\bar{V}_2 = \frac{1}{2}(V_1 + V_2 - V_3)$ and $\bar{V}_3 = V_3 - V_1$ obey the same commutation as the generators U_i. Explicitly they are

$$\bar{V}_1 = \frac{1}{2y}\partial_x + \frac{1}{2x}\partial_y, \quad \bar{V}_2 = -\frac{x^2}{2y}\partial_x + \frac{x}{2}\partial_y, \quad \bar{V}_3 = y\partial_y.$$

The relations $U_3 = uU_1 + vU_2$ and $V_3 = xyV_1 + \frac{y}{x}V_2$ lead to the coordinate transformation $u = xy$ and $v = \frac{y}{x}$.   ▯

A special kind of similarity problem consists of the following question: Given the infinitesimal generators of a transformation group in some set of coordinates, is this group similar to another one which is also given in terms of its infinitesimal generators? This latter group is considered to be a canonical form chosen from a classification of groups like those described in Section 3.5. If similarity is assured, the transformation functions between the two coordinate sets may be determined.

For a one-parameter group there are no commutation relations. The canonical form of the group is the translation in one of the coordinates as explained on page 112. A constructive solution to this problem is given next.

LEMMA 3.4   *Let $\xi(x, y)\partial_x + \eta(x, y)\partial_y$ be a vector field with coordinates x and y, and ∂_v its canonical form in coordinates $u = \sigma(x, y)$, $v = \rho(x, y)$.*

i) If $\xi \neq 0$ and $\eta \neq 0$, the transformation functions are

$$\sigma(x, y) = \Phi(\phi(x, y)), \quad \rho(x, y) = \Psi(\phi(x, y)) + \int \frac{dx}{\xi(x, y = \bar{\phi})}\Big|_{\bar{y}=\phi}$$

where $\phi(x, y) = const$ is the first integral of $\frac{dy}{dx} = \frac{\eta}{\xi}$, $\bar{y} = \phi(x, y)$ with the inverse $y = \bar{\phi}(x, \bar{y})$.

Lie Transformation Groups

131

ii) If $\xi = 0$, $\eta \neq 0$ there holds $\sigma = \Phi(x)$, $\rho = \Psi(x) + \int \dfrac{dy}{\eta(x,y)}$.

iii) If $\xi \neq 0$, $\eta = 0$ there holds $\sigma = \Phi(y)$, $\rho = \Psi(y) + \int \dfrac{dx}{\xi(x,y)}$. In all cases Φ and Ψ are undetermined functions of its argument.

PROOF   According to Lemma 3.3, because $\alpha = 1$ and $\beta = 0$ now, σ and ρ satisfy the system $\xi\sigma_x + \eta\sigma_y = 0$ and $\xi\rho_x + \eta\rho_y = 1$. The solution of this system is obtained from Corollary 2.7 on page 78. $\quad\square$

This lemma shows that in general it is not guaranteed that the canonical form ∂_v for an arbitrary vector field may be obtained in closed form. The crucial step is to solve the first order differential equation $\dfrac{dy}{dx} = \dfrac{\eta}{\xi}$. This is only possible if the coefficients $\xi(x,y)$ and $\eta(x,y)$ are sufficiently simple. This remark will be important in Chapter 5 when symmetries of first order ode's are searched for.

EXAMPLE 3.26   Let $2x\partial_x + 3y\partial_y$ be the given vector field. The equation for σ is $2x\sigma_x + 3y\sigma_y = 0$. Its invariant $\dfrac{y^2}{x^3}$ yields $\sigma = \Phi\left(\dfrac{y^2}{x^3}\right)$ with Φ an undetermined function. The equation for ρ is $2x\rho_x + 3y\rho_y = 1$, it has been considered in Example 2.57. If $\Phi = 0$ is chosen in the solution for z given there, $\rho = -\frac{1}{2}\log x$ is obtained. Consequently, canonical variables are $u = \dfrac{y^2}{x^3}$, $v = -\frac{1}{2}\log x$ or $x = e^{-2v}$, $y = \sqrt{u}e^{-3v}$. $\quad\square$

Given an infinitesimal generator, the corresponding finite transformations may be obtained also by transforming it to canonical form first, and then substituting the backtransformation into the corresponding translation. In the next example this is shown for the group considered previously in Example 3.12.

EXAMPLE 3.27   Let the infinitesimal generator be $U = x^2\partial_x + xy\partial_y$. The transformation functions $\sigma(x,y)$ and $\rho(x,y)$ satisfy $x^2\sigma_x + xy\sigma_y = 0$ and $x^2\rho_x + xy\rho_y = 1$. By Corollary 2.7, the former equation leads to $\dfrac{dy}{y} = \dfrac{dx}{x}$ with the special solution $\phi \equiv \dfrac{y}{x} = C$. Introducing $\bar{x} = \phi$ as new variable, the integral for ρ yields

$$\rho = \int \frac{dy}{xy} = \bar{x}\int \frac{dy}{y^2} = -\frac{\bar{x}}{y} = -\frac{1}{x}.$$

The integration constant has been omitted. Substituting $u = \dfrac{y}{x}$ and $v = -\dfrac{1}{x}$ into the canonical transformation finally leads to

$$\frac{\bar{y}}{\bar{x}} = \frac{y}{x}, \quad -\frac{1}{\bar{x}} = -\frac{1}{x} + a \quad \text{or} \quad \bar{x} = \frac{x}{1+ax}, \quad \bar{y} = \frac{y}{1+ax}. \quad\square$$

Generator	$\sigma(x,y)$	$\rho(x,y)$	$\phi(u,v)$	$\psi(u,v)$
$x\partial_x$	y	$\log x$	e^v	u
$y\partial_x$	y	$\frac{x}{y}$	uv	u
$x\partial_x + y\partial_y$	$\frac{y}{x}$	$\log x$	e^v	ue^v
$x\partial_x - y\partial_y$	xy	$-\log y$	ue^v	e^{-v}
$ax\partial_x + by\partial_y$	$\frac{y^a}{x^b}$	$\frac{1}{b}\log y$	$u^{-1/b}e^{av}$	e^{bv}
$-y\partial_x + x\partial_y$	$\sqrt{x^2+y^2}$	$\arctan\frac{y}{x}$	$u\cos v$	$u\sin v$

TABLE 3.2: For a selection of vector fields $\xi(x,y)\partial_x + \eta(x,y)\partial_y$ the transformation functions $x = \phi(u,v)$, $y = \psi(u,v)$ to canonical form ∂_v, and the inverse functions $u = \sigma(x,y)$, $v = \rho(x,y)$ are given.

It is useful to have the transformation functions to canonical form for a selection of vector fields occurring frequently in applications easily available. They are listed in Table 3.2.

For groups with more than a single parameter, certain normal forms have been introduced by Lie. Let the generators be given as

$$U_i = \xi_i(x,y)\partial_x + \eta_i(x,y)\partial_y \text{ for } i = 1,\dots,r. \tag{3.27}$$

The determinants $\Delta_{ij} = \xi_i\eta_j - \xi_j\eta_i$ and especially $\Delta_{12} \equiv \Delta$ will be used frequently. The canonical coordinates are u and v. The solution to the transformation problem is based on the expressions for the coefficients of the transformed generators given in Lemma 3.3, and the solution procedure for quasilinear systems of pde's described in Lemma 2.12 of the preceding chapter. The operations required are determined above all by the dimension r of the group. For $r = 2$ the complete answer is given next.

LEMMA 3.5   Let $\xi_i\partial_x + \eta_i\partial_y$, $i = 1,2$ be the generators of a two-parameter group satisfying the proper commutator for one of the groups below. The two transformation functions $u = \sigma(x,y)$ and $v = \rho(x,y)$ to canonical variables are determined by the following systems of equations where $\Delta = \xi_1\eta_2 - \xi_2\eta_1$.

Group $\mathbf{g}_4$, $\{\partial_v, v\partial_v\}$: Constraint $\Delta = 0$.

$$\frac{dx}{\xi_1} = \frac{dy}{\eta_1}, \text{ solution } \phi(x,y) = C, \ \sigma \equiv \sigma(\phi(x,y)), \ \rho = \frac{\xi_2}{\xi_1}.$$

Group $\mathbf{g}_{15}, r = 2$, $\{\partial_v, u\partial_v\}$: Constraint $\Delta = 0$.

$$\sigma = \frac{\xi_2}{\xi_1}, \quad \rho = \int \frac{dy}{\eta_1(\sigma(x,y),y)}.$$

Group $\mathbf{g}_{26}$, $\{\partial_u, \partial_v\}$: Constraint $\Delta \neq 0$.

$$\sigma_x = \frac{\eta_2}{\Delta}, \ \sigma_y = -\frac{\xi_2}{\Delta}, \ \rho_x = -\frac{\eta_1}{\Delta}, \ \rho_y = \frac{\xi_1}{\Delta}.$$

Lie Transformation Groups

Group $\mathbf{g}_{25}$, $\{\partial_u, u\partial_u + v\partial_v\}$: *Constraint* $\Delta \neq 0$.

$$\sigma_x + \frac{\eta_1}{\Delta}\sigma = \frac{\eta_2}{\Delta}, \ \sigma_y - \frac{\xi_1}{\Delta}\sigma = -\frac{\xi_2}{\Delta}, \ \rho_x + \frac{\eta_1}{\Delta}\rho = 0, \ \rho_y - \frac{\xi_1}{\Delta}\rho = 0.$$

PROOF   For the group $\mathbf{g}_4$, Lemma 3.3 leads to the system

$$\xi_1\sigma_x + \eta_1\sigma_y = 0, \ \xi_1\rho_x + \eta_1\rho_y = 1, \ \xi_2\sigma_x + \eta_2\sigma_y = 0, \ \xi_2\rho_x + \eta_2\rho_y = \rho.$$

The existence of a nontrivial solution for σ requires the constraint $\Delta = 0$. If it is satisfied, σ is obtained from the first equation. The expression for ρ follows from the consistency of the two equations for ρ. For the second group $\mathbf{g}_{15}$ the same system is obtained except that ρ at the right hand side of the last equation is replaced by σ. Again $\Delta = 0$ is required. The consistency for the equations determining ρ_x and ρ_y now leads to the expressions for σ; ρ has to be determined from $\xi_1\rho_x + \eta_1\rho_y = 1$. The corresponding homogeneous equation has σ as its solution which is already known. Consequently, the integral for ρ follows by Lemma 3.4. For $\mathbf{g}_{26}$ the system for σ and ρ is

$$\xi_1\sigma_x + \eta_1\sigma_y = 1, \ \xi_1\rho_x + \eta_1\rho_y = 0, \ \xi_2\sigma_x + \eta_2\sigma_y = 0, \ \xi_1\rho_x + \eta_2\rho_y = 1,$$

and for $\mathbf{g}_{25}$

$$\xi_1\sigma_x + \eta_1\sigma_y = 1, \ \xi_1\rho_x + \eta_1\rho_y = 0, \ \xi_2\sigma_x + \eta_2\sigma_y = \sigma, \ \xi_2\rho_x + \eta_2\rho_y = \rho.$$

In either case algebraic elimination of the first partial derivatives yields the equations given above. $\quad\square$

EXAMPLE 3.28   Consider the vector fields $U_1 = \partial_x + \partial_y$, $U_2 = y\partial_x + x\partial_y$ satisfying $[U_1, U_2] = U_1$ generating a group of type $\mathbf{g}_{25}$. According to the above lemma, the transformation functions σ and ρ obey

$$\sigma_x + \frac{1}{x-y}\sigma = \frac{x}{x-y}, \ \sigma_y - \frac{1}{x-y}\sigma = -\frac{y}{x-y},$$

$$\rho_x + \frac{1}{x-y}\rho = 0, \ \rho_y - \frac{1}{x-y}\rho = 0.$$

Its solution is obtained by Lemma 2.9 and Lemma 2.10 with the result $\sigma = \frac{C_1}{x-y} + \frac{x+y}{2}$ and $\rho = \frac{C_2}{x-y}$. Choosing $C_1 = C_2 = 1$ yields $u \equiv \sigma(x, y) = \frac{1}{x-y} + \frac{x+y}{2}$ and $v \equiv \rho(x, y) = \frac{1}{x-y}$, and the inverse transformations $x \equiv \phi(u, v) = u - v + \frac{1}{2v}$ and $y \equiv \psi(u, v) = u - v - \frac{1}{2v}$. $\quad\square$

Differential Invariants. In the plane there is only a single dependency possible between the coordinates if no additional variables are introduced. Let the coordinates again be x and y and assume that y depends on x, i.e.,

134

$y \equiv y(x)$. This case covers ordinary differential equations in a single dependent variable. The next two examples deal with invariants of this type.

EXAMPLE 3.29   Consider the vector fields $U_1 = \partial_x$, $U_2 = 2x\partial_x + y\partial_y$ and $U_3 = x^2\partial_x + xy\partial_y$ obeying $[U_1, U_2] = 2U_1$, $[U_1, U_3] = U_2$ and $[U_2, U_3] = 2U_3$. If $y \equiv y(x)$ is assumed, its second prolongations are

$$U_1^{(2)} = U_1, \quad U_2^{(2)} = U_2 - y'\partial_{y'} - 3y''\partial_{y''}, \quad U_3^{(2)} = U_3 - (xy' - y)\partial_{y'} - 3xy''\partial_{y''}.$$

A simple calculation shows that the first two commutators of the prolonged vector fields remain unchanged. The last one is obtained from

$$U_2^{(2)}U_3^{(2)} = 4x^2\partial_x + 3xy\partial_y - (xy' - y)\partial_{y'} + 3xy''\partial_{y''}$$

and

$$U_3^{(2)}U_2^{(2)} = 2x^2\partial_x + xy\partial_y + (xy' - y)\partial_{y'} + 9xy''\partial_{y''}$$

with the result $U_2^{(2)}U_3^{(2)} - U_3^{(2)}U_2^{(2)} = 2U_3^{(2)}$.   ⬜

EXAMPLE 3.30   Consider the four-parameter group generated by ∂_x, ∂_y, $x\partial_y$ and $x\partial_x + y\partial_y$. If it is assumed that $y \equiv y(x)$, the fifth prolongations are ∂_x, ∂_y, $x\partial_y + \partial_{y'}$ and $x\partial_x + y\partial_y - y''\partial_{y''} - 2y'''\partial_{y'''} - 3y^{(4)}\partial_{y^{(4)}} - 4y^{(5)}\partial_{y^{(5)}}$. By the methods explained in Chapter 2 the invariants $\Phi_1 = \dfrac{y''^2}{y'''}$, $\Phi_2 = \dfrac{y''^3}{y^{(4)}}$ and $\Phi_3 = \dfrac{y''^4}{y^{(5)}}$ are obtained. The invariants Φ_1 and Φ_2 are a full system.   ⬜

More general types of differential invariants for transformation groups in the plane may be obtained if dependencies on variables different from the coordinates are assumed. In the next example the group is the same as in the preceding one, the coordinates are now v and w, they are assumed to depend on x and y.

EXAMPLE 3.31   Consider the four-parameter group generated by ∂_v, ∂_w, $v\partial_w$ and $v\partial_v + w\partial_w$. It is assumed now that both v and w depend on x and y. These latter variables are not transformed by the group. The second prolongations w.r.t. x and y are

$$\partial_v, \quad \partial_w, \quad v\partial_w + v_x\partial_{w_x} + v_y\partial_{w_y} + v_{xx}\partial_{w_{xx}} + v_{xy}\partial_{w_{xy}} + v_{yy}\partial_{w_{yy}},$$
$$v\partial_v + w\partial_w + v_x\partial_{v_x} + w_x\partial_{w_x} + v_y\partial_{v_y} + w_y\partial_{w_y}$$
$$+ v_{xx}\partial_{v_{xx}} + w_{xx}\partial_{w_{xx}} + v_{xy}\partial_{v_{xy}} + w_{xy}\partial_{w_{xy}} + v_{yy}\partial_{v_{yy}} + w_{yy}\partial_{w_{yy}}.$$

There are two first order invariants $\Phi_1 = \dfrac{v_y}{v_x}$ and $\Phi_2 = \dfrac{v_x w_y - v_y w_x}{v_x^2}$, and three second order invariants are

$$\Phi_6 = \frac{v_{xx}w_x - v_x w_{xx}}{v_x^2}, \quad \Phi_7 = \frac{v_{yy}w_y - v_y w_{yy}}{v_y^2}, \quad \Phi_8 = \frac{v_{xy}w_x - v_x w_{xy}}{v_x^2}.$$ ⬜

3.4 Classification of Lie Algebras and Lie Groups

Lie's classification of groups of the complex plane and of the complex Lie algebras of low dimension is described in this section. It is not explained how Lie arrived at these classifications, only references to the literature are given.

Lie Algebras of Low Dimensions. For dimension not higher than four the listing of complex Lie algebras is complete. In some cases there occur parameters. They have to be constrained suitably in order to avoid duplication. A few algebras of higher dimension are also given because they are needed for the symmetry analysis later on. In addition the following general rule is applied: Whenever an essential algebraic property like for example the dimension of the derived algebra changes for a special value of a parameter, the algebra corresponding to these parameter values is listed as a separate entry. More details may be found in Lie's original work [112], vol. III, or in vol. 41, Chapter 7 of the *Encyclopedia of Mathematical Sciences*.

Each entry of the subsequent listing is a representative of a full isomorphism class. In order to identify a given Lie algebra, membership to a particular isomorphism class has to be decided. Secondly, the transformation to canonical form may be determined.

Various properties of Lie algebras may be read off easily from this listing. For example, all two-dimensional algebras, all three-dimensional algebras except $l_{3,1}$ and all four-dimensional algebras except $l_{4,1}$ are solvable; $l_{3,1}$ is simple; $l_{4,1}$ has the Levi decomposition $l_1 \ltimes l_{3,1}$ where the radical is generated by U_4 and the simple three-dimensional algebra by the remaining generators. In the lisiting below only nonvanishing commutators are given.

One-dimensional algebras

l_1: *commutative*, Characteristic equation: $\omega = 0$.

Two-dimensional algebras

$l_{2,1}$: $[U_1, U_2] = U_1$, $dim\ l'_{2,1} = 1$, $dim\ l''_{2,1} = 0$.
Characteristic equation: $\omega^2 - E_2\omega = \omega(\omega - E_2) = 0$.
$l_{2,2}$: *commutative*. Characteristic equation: $\omega^2 = 0$.

Three-dimensional algebras

$l_{3,1}$: $[U_1, U_2] = U_1$, $[U_1, U_3] = 2U_2$, $[U_2, U_3] = U_3$, $dim\ l'_{3,1} = dim\ l''_{3,1} = 3$.
Characteristic equation: $\omega^3 + (4E_3E_1 - E_2^2)\omega = \omega(\omega^2 + 4E_3E_1 - E_2^2) = 0$.
$l_{3,2}(c)$: $[U_1, U_3] = U_1$, $[U_2, U_3] = cU_2, c \neq 0$, $dim\ l'_{3,2} = 2$, $dim\ l''_{3,2} = 0$.
Characteristic equation: $\omega^3 - (c+1)E_3\omega^2 + cE_3^2\omega = \omega(\omega - E_3)(\omega - cE_3) = 0$.
$l_{3,3}$: $[U_1, U_3] = U_1$, $[U_2, U_3] = U_1 + U_2$, $dim\ l'_{3,3} = 2$, $dim\ l''_{3,3} = 0$.
Characteristic equation: $\omega^3 - 2E_3\omega^2 + E_3^2\omega = \omega(\omega - E_3)^2 = 0$.
$l_{3,4}$: $[U_1, U_3] = U_1$, $dim\ l'_{3,4} = 1$, $dim\ l''_{3,4} = 0$.
Characteristic equation: $\omega^3 - E_3\omega^2 = \omega^2(\omega - E_3) = 0$.
$l_{3,5}$: $[U_2, U_3] = U_1$, $dim\ l'_{3,5} = 1$, $dim\ l''_{3,5} = 0$.
Characteristic equation: $\omega^3 = 0$.
$l_{3,6}$: *commutative*. Characteristic equation: $\omega^3 = 0$.

136

Four-dimensional algebras

$l_{4,1}$: $[U_1, U_2] = U_1$, $[U_1, U_3] = 2U_2$, $[U_2, U_3] = U_3$, $dim\, l'_{4,1} = dim\, l''_{4,1} = 3$.
Characteristic equation: $\omega^4 + (4E_3E_1 - E_2^2)\omega^2 = (\omega^2 + 4E_3E_1 - E_2^2)\omega^2 = 0$.

$l_{4,2}$: $[U_1, U_4] = U_1$, $[U_2, U_4] = U_1 + U_2$, $[U_3, U_4] = U_2 + U_3$,
$dim\, l'_{4,2} = 3$, $dim\, l''_{4,2} = 0$.
Characteristic equation: $\omega^4 - 3E_4\omega^3 + 3E_4^2\omega^2 - E_4^3\omega = \omega(\omega - E_4)^3 = 0$.

$l_{4,3}(c)$: $[U_1, U_4] = cU_1$, $[U_2, U_4] = (c+1)U_2$, $[U_3, U_4] = U_1 + cU_3$,
$dim\, l'_{4,3} = 3$, $dim\, l''_{4,3} = 0$. Characteristic equation:
$$\omega^4 - (3c+1)E_4\omega^3 + c(3c+2)E_4^2\omega^2 - c^2(c+1)E_4^3\omega$$
$$= \omega(\omega - cE_4)^2(\omega - (c+1)E_4) = 0.$$

$l_{4,4}$: $[U_1, U_4] = 2U_1$, $[U_2, U_3] = U_1$, $[U_2, U_4] = U_2$, $[U_3, U_4] = U_2 + U_3$,
$dim\, l'_{4,4} = 3$, $dim\, l''_{4,4} = 1$, $dim\, l'''_{4,4} = 0$. Characteristic equation:
$\omega^4 - 4E_4\omega^3 + 5E_4^2\omega^2 - 2E_4^3\omega = -\omega(\omega - 2E_4)(\omega - E_4)^2 = 0$.

$l_{4,5}(c)$: $[U_1, U_4] = cU_1$, $[U_2, U_3] = U_1$, $[U_2, U_4] = U_2$, $[U_3, U_4] = (c-1)U_3$,
$c \neq 1$, $dim\, l'_{4,5} = 3$, $dim\, l''_{4,5} = 1$, $dim\, l'''_{4,4} = 0$. Characteristic equation:
$$\omega^4 - 2cE_4\omega^3 + (c^2 + c - 1)E_4^2\omega^2 - c(c-1)E_4^3\omega$$
$$= \omega(\omega - E_4)(\omega - cE_4)(\omega - (c-1)E_4)) = 0.$$

$l_{4,6}(a, b)$: $[U_1, U_4] = U_1$, $[U_2, U_4] = aU_2$, $[U_3, U_4] = bU_3$, $ab \neq 0$,
$dim\, l'_{4,6} = 3$, $dim\, l''_{4,6} = 0$.
Characteristic equation: $\omega^4 - 2E_2\omega^3 + E_4^2\omega^2 = \omega^2(\omega - E_4)^2 = 0$.

$l_{4,7}$: $[U_1, U_4] = U_1$, $[U_2, U_4] = U_2$, $[U_3, U_4] = U_2 + U_3$,
$dim\, l'_{4,7} = 2$, $dim\, l''_{4,7} = 0$.
Characteristic equation: $\omega^4 - 3E_4\omega^3 + 3E_4^2\omega^2 - E_4^3\omega = \omega(\omega - E_4)^3 = 0$.

$l_{4,8}$: $[U_1, U_2] = U_2$, $[U_1, U_3] = U_2 + U_3$, $dim\, l'_{4,8} = 2$, $dim\, l''_{4,8} = 0$.
Characteristic equation: $\omega^4 + 2E_1\omega^3 + E_1^2\omega^2 = \omega^2(\omega + E_1)^2 = 0$.

$l_{4,9}(a)$: $[U_1, U_2] = U_2$, $[U_1, U_3] = aU_3$, $a \neq 0$, $dim\, l'_{4,9} = 2$, $dim\, l''_{4,9} = 0$.
Characteristic equation:
$$\omega^4 - (a+1)E_4\omega^3 + aE_4^2\omega^2 = \omega^2(\omega - E_4)(\omega - aE_4) = 0.$$

$l_{4,10}$: $[U_1, U_2] = U_2$, $[U_3, U_4] = U_4$, $dim\, l'_{4,10} = 2$, $dim\, l''_{4,10} = 0$.
Characteristic equation:
$$\omega^4 + (E_3 + e_1)\omega^3 + E_3E_1\omega^2 = \omega^2(\omega + E_1)(\omega + E_3) = 0.$$

$l_{4,11}$: $[U_2, U_4] = U_2$, $[U_3, U_4] = U_1$, $dim\, l'_{4,11} = 2$, $dim\, l''_{4,11} = 0$.
Characteristic equation: $\omega^4 - E_4\omega^3 = \omega^3(\omega - E_4) = 0$.

$l_{4,12}$: $[U_1, U_4] = U_1$, $[U_2, U_3] = U_1$, $[U_2, U_4] = U_2$,
$dim\, l'_{4,12} = 2$, $dim\, l''_{4,12} = 0$.
Characteristic equation: $\omega^4 - 2E_4\omega^3 + E_4^2\omega^2 = \omega^2(\omega - E_4)^2 = 0$.

$l_{4,13}$: $[U_1, U_4] = U_2$, $[U_3, U_4] = U_1$, $dim\, l'_{4,13} = 2$, $dim\, l''_{4,13} = 0$.
Characteristic equation: $\omega^4 = 0$.

$l_{4,14}$: $[U_1, U_4] = U_1$, $[U_2, U_4] = U_2$, $dim\, l'_{4,14} = 2$, $dim\, l''_{4,14} = 0$.
Characteristic equation: $\omega^4 - 2E_2\omega^3 + E_4^2\omega^2 = \omega^2(\omega - E_4)^2 = 0$.

$l_{4,15}$: $[U_1, U_4] = U_1$, $dim\, l'_{4,15} = 1$, $dim\, l''_{4,15} = 0$.
Characteristic equation: $\omega^4 - E_4\omega^3 = \omega^3(\omega - E_4)$.

$l_{4,16}$: $[U_1, U_2] = U_3$, $dim\, l'_{4,16} = 1$, $dim\, l''_{4,16} = 0$.
Characteristic equation: $\omega^4 = 0$.

Lie Transformation Groups

$l_{4,17}$: *commutative*. Characteristic equation: $\omega^4 = 0$.

In order to faciliate the comparison with the classification as given by Lie [112], vol. III, §136 and §137, in the subsequent table the Lie algebras $l_{4,i}$ and the corresponding equation number of Lie are given *without* the parameter constraints.

i in $l_{4,i}$	1	2	3	4	5	6	7	10	12	13	16	17
Lie's enumeration	58	70	68	63	62	67	64	65	69	71	72	73

Algebra $l_{4,8}$ is obtained from $l_{4,3}$ for the parameter value $c = -1$, the permutations U_2 and U_4, U_1 and U_2 and the replacement U_2 by $-U_2$. Algebra $l_{4,9}$ is obtained from $l_{4,6}$ for the parameter value $b = 0$. Algebra $l_{4,11}$ is obtained from $l_{4,3}$ for the parameter value $c = 0$. Algebra $l_{4,14}$ is obtained from $l_{4,6}$ for the parameter values $a = 1, b = 0$. Algebra $l_{4,15}$ is obtained from $l_{4,6}$ for the parameter values $a = b = 0$.

Some algebras with more than four dimensions

l_8: $[U_1, U_3] = U_1, [U_1, U_5] = U_2, [U_1, U_7] = U_4, [U_1, U_8] = U_6 + 2U_3$,
$[U_2, U_4] = U_1, [U_2, U_6] = U_2, [U_2, U_7] = 2U_6 + U_3, [U_2, U_8] = U_5$,
$[U_3, U_4] = -U_4, [U_3, U_5] = U_5, [U_3, U_8] = U_8, [U_4, U_5] = U_6 - U_3$,
$[U_4, U_8] = U_7, [U_5, U_6] = U_5, [U_6, U_7] = U_7$.

Characteristic equation: The two leading coefficients of (3.17) are

$$\psi_1 = -2E_6, \quad \psi_2 = 3E_2E_7 + 6E_1E_8 + E_6^2 + E_6E_3 - 2E_3^2.$$

Finite Transformation Groups. These are the most important groups occurring in this book because any group of Lie symmetries of an ode must be similar to one of them. As it will turn out later on, however, not all of them do actually occur as a symmetry group of an ode of low order. These questions will be considered in Chapter 5. The subsequent classifications correspond to a manifold $\mathbb{C}^2$ with coordinates x and y as given by Lie [112], vol. III, Kapitel 3, pages 28-78. In the real plane $\mathbb{R}^2$ the classification of primitive groups is different, it is given in the same book on pages 360-392. A more complete discussion may be found in Gonzales-Lopez [54], and in Chapter 6 of vol. I of [43]. A partial ordering for these groups has been given by Krause [96], page 260.

There are certain special classes of groups that have played an important role for the development of the subject, the so-called *linear groups*. They are subgroups of the group $GL(V)$ of linear transformations of a finite-dimensional vector space over the field $\mathbb{R}$ or $\mathbb{C}$. On the one hand they were the first examples of such groups investigated by Lie. Secondly, their structure turned out to be fundamental in the sense that other groups may be composed out of them by standard operations like products or semidirect products. These classes are described first.

138

i) The $n(n+2)$-parameter projective group

$$\bar{x}_k = \frac{a_{k,1}x_1 + a_{k,2}x_2 + \ldots + a_{k,n}x_n + a_{k,n+1}x_{n+1}}{a_{n+1,1}x_1 + a_{n+1,2}x_2 + \ldots + a_{n+1,n}x_{n+1}}$$

with generators ∂_{x_i}, $x_i\partial_{x_k}$, $x_i\sum x_j\partial_{x_j}$, $i,j,k = 1,\ldots,n$.

ii) The $n(n+1)$-parameter general linear group

$$\bar{x}_k = a_{k,1}x_1 + a_{k,2}x_2 + \ldots + a_{k,n}x_n + a_{k,n+1}$$

with generators ∂_{x_i}, $x_i\partial_{x_k}$.

iii) The n^2-parameter general linear homogeneous group

$$\bar{x}_k = a_{k,1}x_1 + a_{k,2}x_2 + \ldots + a_{k,n}x_n$$

with generators $x_i\partial_{x_k}$, $i,k = 1,\ldots,n$.

iv) The $n(n+1)-1$-parameter special linear group of those transformations in *ii)* with determinant equal to 1. Its generators are ∂_{x_i}, $x_i\partial_{x_k}$ and $x_i\partial_{x_i} - x_k\partial_{x_k}$, $i,k = 1,\ldots,n$, $i \neq k$.

v) The $n^2 - 1$-parameter special linear homogeneous group of those transformations in *iii)* with determinant equal to 1. Its generators are $x_i\partial_{x_k}$, $x_i\partial_{x_i} - x_k\partial_{x_k}$, $i,k = 1,\ldots,n$, $i \neq k$.

A systematic notation for the groups defined in *ii)*, *iii)*, *iv)* and *v)* is GL_n, SL_n, GLH_n and SLH_n. However, as van der Waerden [186], page 5, has pointed out, it became popular to use the notation GL_n and gl_n, SL_n and sl_n for the homogeneous groups *iii)* and *v)* that will be applied in this book. For $n = 2$, using the variables x and y, their explicit form is

$$gl_2 \equiv \{x\partial_x, y\partial_x, x\partial_y, y\partial_y\}, \quad \text{type } 1_{4,1},$$
$$sl_2 \equiv \{x\partial_y, y\partial_x, x\partial_x - y\partial_y\}, \quad \text{type } 1_{3,1}.$$

The problem of determining a basis for gl_2 and sl_2 such that the commutator table given in the above listing applies is considered in Exercise 3.11.

Each entry of the subsequent listing is a representative of a *group type*, i.e., the totality of all groups that are similar to this entry by a diffeomorphism. In order to identify any given group, its type has to be found. After that the transformation to the representative in this list may be determined.

Primitive groups

$\mathbf{g}_1$: $\{\partial_x, \partial_y, x\partial_y, x\partial_x - y\partial_y, y\partial_x\}$.
$\mathbf{g}_2$: $\{\partial_x, \partial_y, x\partial_y, y\partial_y, x\partial_x, y\partial_x\}$.
$\mathbf{g}_3$: $\{\partial_x, \partial_y, x\partial_y, y\partial_y, x\partial_x, y\partial_x, x^2\partial_x + xy\partial_y, xy\partial_x + y^2\partial_y\}$.

Groups with two systems of imprimitivity $x = const$ and $y = const$.

$\mathbf{g}_4$: $\{\partial_y, y\partial_y\}$.

$\mathbf{g}_5$: $\{\partial_x, \partial_y, y\partial_y\}$.

$\mathbf{g}_6$: $\{\partial_x, \partial_y, y\partial_y, x\partial_x\}$.

$\mathbf{g}_7$: $\{\partial_x, \partial_y, x\partial_x + cy\partial_y\}$ with $c \neq 0, 1$.

$\mathbf{g}_8$: $\{\partial_y, y\partial_y, y^2\partial_y\}$.

$\mathbf{g}_9$: $\{\partial_x, \partial_y, y\partial_y, y^2\partial_y\}$.

$\mathbf{g}_{10}$: $\{\partial_x + \partial_y, x\partial_x + y\partial_y, x^2\partial_x + y^2\partial_y\}$.

$\mathbf{g}_{11}$: $\{\partial_x, \partial_y, y\partial_y, x\partial_x, y^2\partial_y\}$.

$\mathbf{g}_{12}$: $\{\partial_x, \partial_y, y\partial_y, x\partial_x, y^2\partial_y, x^2\partial_x\}$.

Groups with the system of imprimitivity $x = const$

$\mathbf{g}_{13}$: $\{\partial_x, 2x\partial_x + y\partial_y, x^2\partial_x + xy\partial_y\}$.

$\mathbf{g}_{14}$: $\{y\partial_y, \partial_x, x\partial_x, x^2\partial_x + xy\partial_y\}$.

$\mathbf{g}_{15}$: $\{\phi_1(x)\partial_y, \phi_2(x)\partial_y, \dots, \phi_r(x)\partial_y\}$ with $r \geq 2$. Size: $r \geq 2$.

$\mathbf{g}_{16}$: $\{\phi_1(x)\partial_y, \phi_2(x)\partial_y, \dots, \phi_r(x)\partial_y, y\partial_y\}$ with $r \geq 2$. Size: $r + 1 \geq 3$.

$\mathbf{g}_{17}$: $\{e^{\alpha_k x}\partial_y, xe^{\alpha_k x}\partial_y, \dots, x^{\rho_k}e^{\alpha_k x}\partial_y, \partial_x\}$ with $l \geq 1, 1 \leq k \leq l$,
$\sum \rho_k + l \geq 2$, $\alpha_1(\alpha_1 - 1) = 0$. Size: $\sum \rho_k + l + 1 \geq 3$.

$\mathbf{g}_{18}$: $\{e^{\alpha_k x}\partial_y, xe^{\alpha_k x}\partial_y, \dots, x^{\rho_k}e^{\alpha_k x}\partial_y, y\partial_y, \partial_x\}$ with $l \geq 1, 1 \leq k \leq l$,
$l + \sum \rho_k \geq 2$, $\alpha_1 = 0, \alpha_2 = 1$. Size: $\sum \rho_k + l + 2 \geq 4$.

$\mathbf{g}_{19}$: $\{\partial_y, x\partial_y, \dots, x^{r-1}\partial_y, \partial_x, x\partial_x + cy\partial_y\}$ with $r \geq 2$. Size: $r + 2 \geq 4$.

$\mathbf{g}_{20}$: $\{\partial_y, x\partial_y, \dots, x^{r-1}\partial_y, \partial_x, x\partial_x + (ry + x^r)\partial_y\}$ with $r \geq 1$. Size: $r + 2 \geq 3$.

$\mathbf{g}_{21}$: $\{\partial_y, x\partial_y, \dots, x^{r-1}\partial_y, y\partial_y, \partial_x, x\partial_x\}$ with $r \geq 2$. Size: $r + 3 \geq 5$.

$\mathbf{g}_{22}$: $\{\partial_y, x\partial_y, \dots, x^{r-1}\partial_y, \partial_x, 2x\partial_x + (r-1)y\partial_y, x^2\partial_x + (r-1)xy\partial_y\}$
with $r \geq 2$. Size: $r + 3 \geq 5$.

$\mathbf{g}_{23}$: $\{\partial_y, x\partial_y, \dots, x^{r-1}\partial_y, y\partial_y, \partial_x, x\partial_x, x^2\partial_x + (r-1)xy\partial_y\}$ with $r \geq 2$.
Size: $r + 4 \geq 6$.

Groups with the system of imprimitivity $y = \alpha x + const$

$\mathbf{g}_{24}$: $\{\partial_x, \partial_y, x\partial_x + y\partial_y\}$.

$\mathbf{g}_{25}$: $\{\partial_y, x\partial_x + y\partial_y\}$.

$\mathbf{g}_{26}$: $\{\partial_x, \partial_y\}$.

Groups with the system of imprimitivity $y = \phi(x) + const$

$\mathbf{g}_{27}$: $\{\partial_y\}$.

This listing of the finite groups of the complex plane follows most closely the one given in Lie [112], vol. III. For the indicated range of parameters the listing is complete without any duplication. In earlier publications Lie gave some of these groups in slightly different form, compare the discussion in Lie [112], vol. III, page 76. Group $\mathbf{g}_7$ with $c \neq 1$ is listed separately from group $\mathbf{g}_{24}$ because the value $c = 1$ in the latter group generates a different system of imprimitivity than for $c \neq 1$, and for the value $c = 0$ it is similar to $\mathbf{g}_5$.

A further distinction made by Lie concerns the *multiplicity* of a system of imprimitivity. It generates the additional distinctions as follows. For group $\mathbf{g}_{22}$ with $r = 3$ the system $x = const.$ counts twofold (last line on page 71), for $r = 2$ and $r \geq 4$ it counts as a single system (first item after the heading on page 72: *b) Die invariante Schar zählt doppelt*). For group $\mathbf{g}_{20}$ and $r = 1$

140

the system is counted twofold (last line on page 72), all remaining cases $r > 1$ are single counted (item six from top of page 72). Finally the imprimitivity system for group $\mathbf{g}_{19}$ with $c = 1$ counts twofold (second line from bottom of page 72) and only single for the remaining cases $c \neq 1$ (line five from top of page 72). These remarks allow it to identify each group in Lie's listing uniquely.

Some applications require the knowledge of all group types with a given number of parameters without duplication. For most groups of the above listing it is easy to identify them. The coefficients of ∂_y for the groups $\mathbf{g}_{17}$ and $\mathbf{g}_{18}$ are obtained by the following observations. The desired coefficients are solutions of a linear homogeneous ode with constant coefficients of the proper order (Lie [112], vol. III, page 54ff). As a consequence they have the general form $x^k e^{\alpha x}$.

Subsequently the infinitesimal generators of various groups of a given size r up to $r \leq 7$ are listed. For $r \leq 4$ this listing is complete. For values higher than four only those groups are listed that occur later on as symmetry groups of differential equations. In addition the Janet bases for the defining equations are given in *grex* term ordering with $\eta > \xi$ and $y > x$. The coefficients ϕ_k for $k = 1, \ldots, r$ occurring in the group types $\mathbf{g}_{15}$ and $\mathbf{g}_{16}$ are undetermined functions of x. The coefficients q_k occurring in the Janet basis for these groups are defined by

$$q_k = -\frac{W_k^{(2)}(\phi_1, \phi_2)}{W^{(2)}(\phi_1, \phi_2)}$$

for $k = 1, \ldots, r$, see eq. (2.4). The algebraic properties of groups with $r \leq 4$ may be obtained from their Lie algebra type which is always given.

One-parameter group

$\quad \mathbf{g}_{27}$: $\{\partial_y\}$. Type $\mathbf{l}_1$. Janet basis $\{\xi, \eta_x, \eta_y\}$.

Two-parameter groups

$\quad \mathbf{g}_4$: $\{\partial_y, y\partial_y\}$. Type $\mathbf{l}_{2,1}$. Janet basis $\{\xi, \eta_x, \eta_{yy}\}$.

$\quad \mathbf{g}_{15}(r = 2)$: $\{\phi_1\partial_y, \phi_2\partial_y\}$. Type $\mathbf{l}_{2,2}$. Janet basis $\{\xi, \eta_y, \eta_{xx} + q_1\eta_x + q_2\eta\}$.

$\quad \mathbf{g}_{25}$: $\{\partial_x, x\partial_x + y\partial_y\}$. Type $\mathbf{l}_{2,1}$. Janet basis $\{\xi_x - \frac{1}{y}\eta, \xi_y, \eta_x, \eta_y - \frac{1}{y}\eta\}$.

$\quad \mathbf{g}_{26}$: $\{\partial_x, \partial_y\}$. Type $\mathbf{l}_{2,2}$. Janet basis $\{\xi_x, \xi_y, \eta_x, \eta_y\}$.

Three-parameter groups

$\quad \mathbf{g}_5$: $\{\partial_x, \partial_y, y\partial_y\}$. Type $\mathbf{l}_{3,4}$. Janet basis $\{\xi_x, \xi_y, \eta_x, \eta_{yy}\}$.

$\quad \mathbf{g}_7$: $\{\partial_x, \partial_y, x\partial_x + cy\partial_y\}$ with $c \neq 0, 1$. Type $\mathbf{l}_{3,2}$.
Janet basis $\{\xi_y, \eta_x, \eta_y - c\xi_x, \xi_{xx}\}$.

$\quad \mathbf{g}_8$: $\{\partial_y, y\partial_y, y^2\partial_y\}$. Type $\mathbf{l}_{3,1}$. Janet basis $\{\xi, \eta_x, \eta_{yyy}\}$.

$\quad \mathbf{g}_{10}$: $\{\partial_x + \partial_y, x\partial_x + y\partial_y, x^2\partial_x + y^2\partial_y\}$. Type $\mathbf{l}_{3,1}$. Janet basis
$\{\xi_y, \eta_x, \eta_y + \xi_x + \frac{2}{x-y}(\eta - \xi), \xi_{xx} - \frac{2}{x-y}\xi_x - \frac{2}{(x-y)^2}(\eta - \xi)\}$.

$\quad \mathbf{g}_{13}$: $\{\partial_x, 2x\partial_x + y\partial_y, x^2\partial_x + xy\partial_y\}$. Type $\mathbf{l}_{3,1}$.
Janet basis $\{\xi_x - \frac{2}{y}\eta, \xi_y, \eta_y - \frac{1}{y}\eta, \eta_{xx}\}$.

$\quad \mathbf{g}_{15}(r = 3)$: $\{\phi_1\partial_y, \phi_2\partial_y, \phi_3\partial_y\}$. Type $\mathbf{l}_{3,6}$.
Janet basis $\{\xi, \eta_y, \eta_{xxx} + q_1\eta_{xx} + q_2\eta_x + q_3\eta\}$.

Lie Transformation Groups

$\mathbf{g}_{16}(r = 2)$: $\{\phi_1\partial_y, \phi_2\partial_y, y\partial_y\}$. Type $\mathbf{l}_{3,2}$ with $c = 1$.
Janet basis $\{\xi, \eta_{xx} + q_1\eta_x + q_2\eta, \eta_{xy}, \eta_{yy}\}$.
$\mathbf{g}_{17}(l = 1, \rho_1 = 1, \alpha_1 = 0)$: $\{\partial_x, \partial_y, x\partial_y\}$. Type $\mathbf{l}_{3,5}$.
Janet basis $\{\xi_x, \xi_y, \eta_y, \eta_{xx}\}$.
$\mathbf{g}_{17}(l = 1, \rho_1 = 1, \alpha_1 = 1)$: $\{\partial_x, e^x\partial_y, xe^x\partial_y\}$. Type $\mathbf{l}_{3,3}$.
Janet basis $\{\xi_x, \xi_y, \eta_y, \eta_{xx} - 2\eta_x + \eta\}$.
$\mathbf{g}_{17}(l = 2, \rho_1 = \rho_2 = 0, \alpha_1 = 0, \alpha_2 = 1)$: $\{\partial_x, \partial_y, e^x\partial_y\}$. Type $\mathbf{l}_{3,4}$.
Janet basis $\{\xi_x, \xi_y, \eta_y, \eta_{xx} - \eta_x\}$.
$\mathbf{g}_{17}(l = 2, \rho_1 = \rho_2 = 0, \alpha_1 = 1, \alpha_2 = c)$: $\{\partial_x, e^x\partial_y, e^{cx}\partial_y\}$, $c \neq 0, 1$.
Type $\mathbf{l}_{3,2}$. Janet basis $\{\xi_x, \xi_y, \eta_y, \eta_{xx} - (c + 1)\eta_x + c\eta\}$.
$\mathbf{g}_{20}(r = 1)$: $\{\partial_x, \partial_y, x\partial_x + (x + y)\partial_y\}$. Type $\mathbf{l}_{3,3}$.
Janet basis $\{\xi_y, \eta_x - \xi_x, \eta_y - \xi_x, \xi_{xx}\}$.
$\mathbf{g}_{24}$: $\{\partial_x, \partial_y, x\partial_x + y\partial_y\}$. Type $\mathbf{l}_{3,2}$ with $c = 1$.
Janet basis $\{\xi_y, \eta_x, \eta_y - \xi_x, \xi_{xx}\}$.

Four-parameter groups

$\mathbf{g}_6$: $\{\partial_x, \partial_y, x\partial_x, y\partial_y\}$. Type $\mathbf{l}_{4,10}$. Janet basis $\{\xi_y, \eta_x, \xi_{xx}, \eta_{yy}\}$.
$\mathbf{g}_9$: $\{\partial_x, \partial_y, y\partial_y, y^2\partial_y\}$. Type $\mathbf{l}_{4,1}$. Janet basis $\{\xi_x, \xi_y, \eta_x, \eta_{yyy}\}$.
$\mathbf{g}_{14}$: $\{\partial_x, x\partial_x, x^2\partial_x + xy\partial_y, y\partial_y\}$. Type $\mathbf{l}_{4,1}$.
Janet basis $\{\xi_y, \eta_y - \frac{1}{y}\eta, \xi_{xx} - \frac{2}{y}\eta_x, \eta_{xx}\}$.
$\mathbf{g}_{15}$ with $r = 4$: $\{\phi_1\partial_y, \phi_2\partial_y, \phi_3\partial_y, \phi_4\partial_y\}$. Type $\mathbf{l}_{4,17}$.
Janet basis $\{\xi, \eta_y, \eta_{4x} + q_1\eta_{xxx} + q_2\eta_{xx} + q_3\eta_x + q_4\eta\}$.
$\mathbf{g}_{16}$ with $r = 3$: $\{\phi_1\partial_y, \phi_2\partial_y, \phi_3\partial_y, y\partial_y\}$. Type $\mathbf{l}_{4,6}$ with $a = b = 1$.
Janet basis $\{\xi, \eta_{xy}, \eta_{yy}, \eta_{xxx} + q_1\eta_{xx} + q_2\eta_x + q_3\eta\}$.
$\mathbf{g}_{17}$ with $l = 1$, $\rho_1 = 2$, $\alpha_1 = 0$: $\{\partial_x, \partial_y, x\partial_y, x^2\partial_y\}$. Type $\mathbf{l}_{4,13}$.
Janet basis $\{\xi_x, \xi_y, \eta_y, \eta_{xxx}\}$.
$\mathbf{g}_{17}$ with $l = 1$, $\rho_1 = 2$, $\alpha_1 = 1$: $\{\partial_x, e^x\partial_y, xe^x\partial_y, x^2e^x\partial_y\}$. Type $\mathbf{l}_{4,2}$.
Janet basis $\{\xi_x, \xi_y, \eta_y, \eta_{xxx} - 3\eta_{xx} + 3\eta_x - \eta\}$.
$\mathbf{g}_{17}$ with $l = 2$, $\rho_1 = 0, \rho_2 = 1$, $\alpha_1 = 0, \alpha_2 = 1$: $\{\partial_x, \partial_y, e^x\partial_y, xe^x\partial_y\}$.
Type $\mathbf{l}_{4,8}$. Janet basis $\{\xi_x, \xi_y, \eta_y, \eta_{xxx} - 2\eta_{xx} + \eta_x\}$.
$\mathbf{g}_{17}$ with $l = 2$, $\rho_1 = 1, \rho_2 = 0$, $\alpha_1 = 0, \alpha_2 = 1$: $\{\partial_x, \partial_y, x\partial_y, e^x\partial_y\}$.
Type $\mathbf{l}_{4,11}$. Janet basis $\{\xi_x, \xi_y, \eta_y, \eta_{xxx} - \eta_{xx}\}$.
$\mathbf{g}_{17}$ with $l = 2$, $\rho_1 = 1, \rho_2 = 0$, $\alpha_1 = 1, \alpha_2 = \alpha$: $\{\partial_x, e^x\partial_y, xe^x\partial_y, e^{\alpha x}\partial_y\}$.
Type $\mathbf{l}_{4,3}$. Janet basis $\{\xi_x, \xi_y, \eta_y, \eta_{xxx} - (\alpha + 1)\eta_{xx} + (2\alpha + 1)\eta_x - \alpha\eta\}$.
$\mathbf{g}_{17}$ with $l = 3$, $\rho_1 = \rho_2 = \rho_3 = 0, \alpha_1 = 0, \alpha_2 = 1, \alpha_3 = \alpha$:
$\{\partial_x, \partial_y, e^x\partial_y, e^{\alpha x}\partial_y\}$. Type $\mathbf{l}_{4,9}$.
Janet basis $\{\xi_x, \xi_y, \eta_y, \eta_{xxx} - (\alpha + 1)\eta_{xx} + \alpha\eta_x\}$.
$\mathbf{g}_{17}$ with $l = 3, \rho_1 = \rho_2 = \rho_3 = 0, \alpha_1 = 1, \alpha_2 = \alpha, \alpha_3 = \beta$: $\{\partial_x, \partial_y, e^{\alpha x}\partial_y, e^{\beta x}\partial_y\}$.
Type $\mathbf{l}_{4,6}$. Janet basis $\{\xi_x, \xi_y, \eta_y, \eta_{xxx} - (\alpha + \beta)\eta_{xx} + \alpha\beta\eta_x\}$.
$\mathbf{g}_{18}$ with $l = 1$, $\rho_1 = 1$, $\alpha_1 = 0$: $\{\partial_y, x\partial_y, y\partial_y, \partial_x\}$. Type $\mathbf{l}_{4,12}$.
Janet basis $\{\xi_x, \eta_y, \xi_{yy}, \eta_{xx}\}$.
$\mathbf{g}_{18}$ with $l = 2$, $\rho_1 = \rho_2 = 0$, $\alpha_1 = 0$, $\alpha_2 = 1$: $\{\partial_y, e^x\partial_y, y\partial_y, \partial_x\}$.
Type $\mathbf{l}_{4,10}$, Janet basis $\{\xi_x, \xi_y, \eta_{xx} - \eta_x, \eta_{xy}, \eta_{yy}\}$.
$\mathbf{g}_{19}$ with $r = 2$: $\{\partial_y, x\partial_y, \partial_x, x\partial_x + cy\partial_y\}$. Type $\mathbf{l}_{4,5}(c)$.
Janet basis $\{\xi_y, \eta_y - c\xi_x, \xi_{xx}, \eta_{xx}\}$.

$\mathbf{g}_{20}$ with $r = 2$: $\{\partial_x, \partial_y, x\partial_y, x\partial_x + (x^2 + 2y)\partial_y\}$. Type $\mathbf{l}_{4,4}$.
Janet basis $\{\xi_y, \eta_y - 2\xi_x, \xi_{xx}, \eta_{xx} - 2\xi_x\}$.

Selected Five-parameter groups

$\mathbf{g}_1$: $\{\partial_x, \partial_y, x\partial_y, x\partial_x - y\partial_y, y\partial_x\}$.
Janet basis $\{\eta_y + \xi_x, \xi_{xx}, \xi_{xy}, \xi_{yy}, \eta_{xx}\}$.
$\mathbf{g}_{11}$: $\{\partial_x, \partial_y, y\partial_y, x\partial_x, y^2\partial_x\}$.
Janet basis $\{\eta_y, \xi_{xx}, \xi_{xy}, \xi_{yy} - \frac{1}{y}\xi_y, \eta_{xx}\}$.
$\mathbf{g}_{15}(r = 5)$: $\{\phi_1\partial_y, \phi_2\partial_y, \phi_3\partial_y, \phi_4\partial_y, \phi_5\partial_y\}$.
Janet basis $\{\xi, \eta_y, \eta_{5x} + q_1\eta_{4x} + q_2\eta_{xxx} + q_3\eta_{xx} + q_4\eta_x + q_5\eta\}$.
$\mathbf{g}_{16}(r = 4)$: $\{\phi_1\partial_y, \phi_2\partial_y, \phi_3\partial_y, \phi_4\partial_y, y\partial_y\}$.
Janet basis $\{\xi, \eta_{xy}, \eta_{yy}, \eta_{4x} + q_1\eta_{xxx} + q_2\eta_{xx} + q_3\eta_x + q_4\eta\}$.
$\mathbf{g}_{18}(l = 1, \rho_1 = 3, \alpha_1 = 0)$: $\{\partial_y, x\partial_y, x^2\partial_y, y\partial_y, \partial_x\}$. Janet basis $\{\xi_x, \xi_y, \eta_{xy}, \eta_{yy}, \eta_{xx}$
$\mathbf{g}_{18}(l = 2, \rho_1 = 1, \rho_2 = 0, \alpha_1 = 0, \alpha_2 = 1)$: $\{\partial_y, x\partial_y, e^x\partial_y, y\partial_y, \partial_x\}$.
Janet basis $\{\xi_x, \xi_y, \eta_{xy}, \eta_{yy}, \eta_{xxx} - \eta_{xx}\}$.
$\mathbf{g}_{18}(l = 3, \rho_1 = \rho_2 = \rho_3 = 0, \alpha_1 = 0, \alpha_2 = 1, \alpha_3 = c)$, $c \neq 0, 1$:
$\{\partial_y, x\partial_y, e^{cx}\partial_y, y\partial_y, \partial_x\}$.
Janet basis $\{\xi_x, \xi_y, \eta_{xy}, \eta_{yy}, \eta_{xxx} - (c + 1)\eta_{xx} + c\eta_x\}$.
$\mathbf{g}_{19}(r = 3)$: $\{\partial_y, x\partial_y, x^2\partial_y, \partial_x, x\partial_x + cy\partial_y\}$. Janet basis $\{\xi_y, \eta_y - c\xi_x, \xi_{xx}, \eta_{xxx}\}$.
$\mathbf{g}_{20}(r = 3)$: $\{\partial_y, x\partial_y, x^2\partial_y, \partial_x, x\partial_x + (3y + x^3)\partial_y\}$.
Janet basis $\{\xi_y, \eta_y - 3\xi_x, \xi_{xx}, \eta_{xxx} - 6\xi_x\}$.
$\mathbf{g}_{21}(r = 2)$: $\{\partial_x, \partial_y, x\partial_y, y\partial_y, x\partial_x\}$. Janet basis $\{\xi_y, \xi_{xx}, \eta_{xx}, \eta_{xy}, \eta_{yy}\}$.
$\mathbf{g}_{22}(r = 2)$: $\{\partial_x, \partial_y, x\partial_y, 2x\partial_x + y\partial_y, x^2\partial_x + xy\partial_y\}$.
Janet basis $\{\xi_y, \eta_y - \frac{1}{2}\xi_x, \eta_{xx}, \xi_{xxx}\}$.

Selected Six-parameter groups

$\mathbf{g}_2$: $\{\partial_x, \partial_y, x\partial_y, y\partial_y, x\partial_x, y\partial_x\}$.
Janet basis $\{\xi_{xx}, \xi_{xy}, \xi_{yy}, \eta_{xx}, \eta_{xy}, \eta_{yy}\}$.
$\mathbf{g}_{12}$: $\{\partial_x, \partial_y, y\partial_y, x\partial_x, y^2\partial_y, x^2\partial_x\}$.
Janet basis $\{\xi_y, \eta_x, \xi_{xxx}, \eta_{yyy}\}$.
$\mathbf{g}_{15}(r = 6)$: $\{\phi_1\partial_y, \phi_2\partial_y, \phi_3\partial_y, \phi_4\partial_y, \phi_5\partial_y, \phi_6\partial_y\}$.
Janet basis $\{\xi, \eta_y, \eta_{6x} + q_1\eta_{5x} + q_2\eta_{4x} + q_3\eta_{xxx} + q_4\eta_{xx} + q_5\eta_x + q_6\eta\}$.
$\mathbf{g}_{16}(r = 5)$: $\{\phi_1\partial_y, \phi_2\partial_y, \phi_3\partial_y, \phi_4\partial_y, \phi_5\partial_y, y\partial_y\}$.
Janet basis $\{\xi, \eta_{xy}, \eta_{yy}, \eta_{4x} + q_1\eta_{xxx} + q_2\eta_{xx} + q_3\eta_x + q_4\eta\}$.
$\mathbf{g}_{19}(r = 4)$: $\{\partial_y, x\partial_y, x^2\partial_y, x^3\partial_y, \partial_x, x\partial_x + cy\partial_y\}$.
Janet basis $\{\xi_y, \xi_{xx}, \eta_{xy}, \eta_{yy}, \eta_{xxxx}\}$.

Selected Seven- and Eight-parameter groups

$\mathbf{g}_{23}(r = 3) = \{\partial_y, x\partial_y, x^2\partial_y, y\partial_y, \partial_x, x\partial_x, x^2\partial_x + 2xy\partial_y\}$.
Janet basis $\{\xi_y, \eta_{xy} - \xi_{xx}, \eta_{yy}, \xi_{xxx}, \eta_{xxx}\}$.
$\mathbf{g}_3$: $\{\partial_x, \partial_y, x\partial_y, y\partial_y, x\partial_x, y\partial_x, x^2\partial_x + xy\partial_y, xy\partial_x + y^2\partial_y\}$.
Janet basis $\{\xi_{yy}, \eta_{xx}, \eta_{xy}, \eta_{yy}, \xi_{xxx}, \eta_{xxy}\}$.

In addition to this classification of the finite groups of a two-dimensional manifold, Lie [110] derived a classification of its infinite groups. This listing is not given here because it is not required for the applications in this book. The only exception is group $\mathbf{G}$, no. XXXV in Lie's notation, which occurs in the symmetry analysis of first order equations.

G: $\{\xi(x,y)\partial_x, \eta(y)\partial_y\}$. Janet basis $\{\eta_x\}$.

3.5 Lie Systems

In Chapter 2 general systems of linear homogeneous pde's have been discussed without any assumptions on their origin or special properties of their solutions. The uniqueness of the Janet basis representation makes it possible to express additional knowledge in terms of the Janet basis coefficients. For the applications in this book, determining the symmetries of ordinary differential equations, this additional knowledge is of threefold origin.

> ▷ *Lie's relations* must always be true because they express the general properties of commutators of vector fields.

> ▷ *Geometric relations* express geometric properties of particular group types, e. g. the number of systems of imprimitivity.

> ▷ *Algebraic relations* express algebraic properties of particular Lie algebra types, e. g. in terms of the structure of its characteristic polynomial.

The importance of knowing the explicit form of the coherence conditions given in Theorems 2.15, 2.16 and 2.17 becomes obvious at this point because only this knowledge, combined with the additional constraints, allows it to identify the symmetry classes of ode's.

Constraints from Lie's Relations. Let a Janet basis with $k \geq 2$ parametric derivatives be given, whose solutions are the coefficients $\xi(x,y)$ and $\eta(x,y)$ of k vector fields generating a finite group of the plane with coordinates x and y. Let (ξ_1, η_1) and (ξ_2, η_2) be any pair of solutions. Then according to (3.11) a third solution (ξ_3, η_3) is given by

$$\xi_3 = \xi_1\xi_{2,x} - \xi_{1,x}\xi_2 + \eta_1\xi_{2,y} - \xi_{1,y}\eta_2,$$
$$\eta_3 = \xi_1\eta_{2,x} - \eta_{1,x}\xi_2 + \eta_1\eta_{2,y} - \eta_{1,y}\eta_2. \tag{3.28}$$

This property leads to severe additional restrictions for the coefficients of the corresponding Janet basis. In order to obtain them explicitly, the condition that ξ_3 and η_3 as defined by (3.28) be a solution would have to be expressed in terms of these coefficients. To this end at first the right hand sides of (3.28) are reduced w.r.t. the Janet basis at hand. From the resulting expressions those derivatives of ξ_3 and η_3 that occur in the Janet basis are calculated and are again reduced w.r.t. to the Janet basis. The result is substituted into the Janet basis. Its elements become linear and homogeneous in $\frac{1}{2}k(k-1)$ bilinear expressions of the form $p_iq_j - p_jq_i$ where the p_j, q_j traverse the k parametric derivatives involved. Their coefficients yield the desired constraints. Due to

144

its importance in later parts of this book, a special term for systems of pde's obeying Lie's relations is introduced.

DEFINITION 3.8 *(Lie system). A system of linear homogeneous pde's with the property that (ξ_3, η_3) defined by (3.28) is a solution if this is true for (ξ_1, η_1) and (ξ_2, η_2), is called a Lie system.*

In the remaining part of this subsection the complete answer for several Janet basis types is given explicitly. Obviously this concept makes sense only for Janet bases of type $\mathcal{J}^{(2,2)}$ of order not less than two. For Janet bases of order two the complete answer is given next.

THEOREM 3.17 *For type $\mathcal{J}_{2,1}^{(2,2)}, \ldots, \mathcal{J}_{2,5}^{(2,2)}$ Janet bases Lie's relations are as follows. The Janet basis*

$$\mathcal{J}_{2,1}^{(2,2)}: \quad \xi = 0, \quad \eta_y + a_1\eta_x + a_2\eta = 0, \quad \eta_{xx} + b_1\eta_x + b_2\eta = 0 \qquad (3.29)$$

is a Lie system if its coefficients satisfy the system of pde's

$$a_{2,x} - \tfrac{1}{2}b_{1,y} - \tfrac{1}{2}a_1 b_2 = 0, \quad a_{1,y} - a_1 a_2 = 0,$$
$$b_{1,xy} - 2a_{1,x}b_2 + b_{1,y}b_1 - 2b_{2,y} - b_{2,x}a_1 + a_1 b_1 b_2 = 0,$$
$$a_{1,xx} - a_{1,x}b_1 - b_{1,x}a_1 + a_1 b_2 = 0.$$

The Janet basis

$$\mathcal{J}_{2,2}^{(2,2)}: \quad \xi = 0, \quad \eta_x + a_1\eta = 0, \quad \eta_{yy} + b_1\eta_y + b_2\eta = 0 \qquad (3.30)$$

is a Lie system if its coefficients satisfy the system of pde's

$$a_1 = 0, \quad b_{2,x} = 0, \quad b_{1,x} = 0, \quad b_{1,y} - b_2 = 0.$$

The Janet basis

$$\mathcal{J}_{2,3}^{(2,2)}: \quad \begin{matrix} \xi_x + a_1\eta + a_2\xi = 0, & \xi_y + b_1\eta + b_2\xi = 0, \\ \eta_x + c_1\eta + c_2\xi = 0, & \eta_y + d_1\eta + d_2\xi = 0 \end{matrix} \qquad (3.31)$$

is a Lie system if its coefficients satisfy the system of pde's

$$d_{1,x} - d_{2,y} + b_1 c_2 - b_2 c_1 = 0, \quad c_{1,y} - d_{2,y} + a_1 d_2 - b_2 c_1 = 0,$$
$$c_{2,y} - d_{2,x} + a_2 d_2 - b_2 c_2 - c_1 d_2 + c_2 d_1 = 0,$$
$$c_{1,x} - d_{2,x} + a_1 c_2 - a_2 c_1 + a_2 d_2 - b_2 c_2 = 0,$$
$$b_{1,x} - b_{2,y} - a_1 b_2 + a_2 b_1 - b_1 d_2 + b_2 d_1 = 0,$$
$$a_{2,y} - b_{2,x} - a_1 d_2 + b_1 c_2 = 0, \quad a_{1,x} - b_{2,x} - a_1 d_2 + b_2 c_1 = 0,$$
$$a_{1,y} - b_{2,y} - a_1 d_1 + b_1 c_1 - b_1 d_2 + b_2 d_1 = 0.$$

$$(3.32)$$

The Janet basis

$$\mathcal{J}_{2,4}^{(2,2)}: \quad \eta + a_2\xi = 0, \quad \xi_y + b_1\xi_x + b_2\xi = 0, \quad \xi_{xx} + c_1\xi_x + c_2\xi = 0 \qquad (3.33)$$

is a Lie system if its coefficients satisfy the system of pde's

$$a_{2,y}b_1 + a_{2,x}b_1^2 + b_{1,y}a_2 - b_{1,x} - a_2b_1b_2 - b_2,$$
$$b_{2,xx} - 2b_{1,x}c_2 + b_{2,x}c_1 - c_{2,y} - c_{2,x}b_1 = 0,$$
$$b_{1,xx} - b_{1,x}c_1 + 2b_{2,x} - c_{1,y} - c_{1,x}b_1 = 0,$$
$$a_{2,xx} + a_{2,x}(2b_{1,x} - b_1c_1) - (2b_{2,x} - c_{1,y})a_2 - c_{1,x} + (a_2b_1 + 1)c_2 = 0.$$

The Janet basis

$$\mathcal{J}_{2,5}^{(2,2)}: \quad \eta + a_1\xi = 0, \ \xi_x + b_1\xi = 0, \ \xi_{yy} + c_1\xi_y + c_2\xi = 0 \qquad (3.34)$$

is a Lie system if its coefficients satisfy the system of pde's

$$b_{1,y} - \tfrac{1}{2}c_{1,x} = 0, \ a_{1,x} - a_1b_1 = 0,$$
$$c_{1,xy} + c_{1,x}c_1 - 2c_{2,x} = 0, \ a_{1,yy} - a_{1,y}c_1 - c_{1,y}a_1 + a_1c_2 = 0.$$

PROOF   The proof is similar for all five cases. Therefore it is given only for type $\mathcal{J}_{2,3}^{(2,2)}$ Janet bases. Reduction of the right hand sides of (3.28) w.r.t. (3.31) where ξ and η are replaced by (ξ_1, η_1) or (ξ_2, η_2), yields

$$\xi_3 = (b_2 - a_1)B, \quad \eta_3 = (d_2 - c_1)B \ \text{ where } \ B \equiv \xi_1\eta_2 - \eta_1\xi_2 \qquad (3.35)$$

and the useful relations $B_x = -(a_2 + c_1)B$, $B_y = -(b_2 + d_1)B$. The reduced first order partial derivatives of ξ_3 and η_3 are

$$\begin{aligned}
\xi_{3,x} &= \big[b_{2,x} - a_{1,x} - (b_2 - a_1)(a_2 + c_1)\big]B, \\
\xi_{3,y} &= \big[b_{2,y} - a_{1,y} - (b_2 - a_1)(b_2 + d_1)\big]B, \\
\eta_{3,x} &= \big[d_{2,x} - c_{1,x} - (d_2 - c_1)(a_2 + c_1)\big]B, \\
\eta_{3,y} &= \big[d_{2,y} - c_{1,y} - (d_2 - c_1)(b_2 + d_1)\big]B
\end{aligned} \qquad (3.36)$$

with B as above. Substituting (3.35) and (3.36) into the Janet basis (3.31) and equating the coefficients of the parametric derivatives to zero yields the following relations for the coefficients $a_1, a_2, \ldots, d_2$ and its first derivatives.

$$\begin{aligned}
b_{2,x} - a_{1,x} - (b_2 - a_1)c_1 + (d_2 - c_1)a_1 &= 0, \\
b_{2,y} - a_{1,y} - (b_2 - a_1)d_1 + (d_2 - c_1)b_1 &= 0, \\
d_{2,x} - c_{1,x} + (b_2 - a_1)c_2 - (d_2 - c_1)c_1 &= 0, \\
d_{1,y} - c_{1,y} + (b_2 - a_1)d_2 - (d_2 - c_1)b_2 &= 0.
\end{aligned}$$

These conditions are necessary and sufficient for (3.31) to be a Lie system. They have to be supplemented by the integrability conditions of Theorem 2.17. After some final reductions have been performed, the above system (3.32) of pde's for the coefficients is obtained. $\square$

146

EXAMPLE 3.32   Consider the type $\mathcal{J}_{2,3}^{(2,2)}$ system

$$z_{1,x} + \frac{1}{y(y^2-2)}z_2 - \frac{2(y^2-1)}{x(y^2-2)}z_1 = 0,\ z_{1,y} + \frac{x}{y^2(y^2-2)}z_2 - \frac{2}{y(y^2-2)}z_1 = 0,$$

$$z_{2,x} + \frac{2}{x(y^2-2)}z_2 - \frac{2y^3}{x^2(y^2-2)}z_1 = 0,\ z_{2,y} - \frac{2(y^2-1)}{y(y^2-2)}z_2 + \frac{2y^2}{x(y^2-2)}z_1 = 0.$$

Because its coefficients satisfy the coherence conditions of Theorem 2.16 it is
a Janet basis and has a two-dimensional solution space, a basis is $(x^2, 2xy)$
and $(\frac{x}{y}, y^2)$. Because the constraints (3.32) are violated it is *not* a Lie system.
Consequently, the commutator of $U_1 = x^2\partial_x + 2xy\partial_y$ and $U_2 = \frac{x}{y}\partial_x + y^2\partial_y$
which is $[U_1, U_2] = -\frac{3x^2}{y}\partial_x + 2x(y^2-1)\partial_y$ is not a linear combination of U_1
and U_2 over the constants, i.e., U_1 and U_2 do not generate a Lie algebra.   □

EXAMPLE 3.33   Consider the type $\mathcal{J}_{2,3}^{(2,2)}$ system

$$\xi_x - \frac{1}{x}\xi = 0,\ \xi_y = 0,\ \eta_x - \frac{1}{x+y}\eta + \frac{y}{x(x+y)}\xi = 0,\ \eta_y - \frac{1}{x+y}\eta - \frac{1}{x+y}\xi = 0.$$

It satisfies the constraints (3.32) that include the coherence conditions of
Theorem 2.16, consequently it is a Lie system. A solution basis is (x, y) and
$(x, -x)$ from which the generators $U_1 = x\partial_x + y\partial_y$ and $U_2 = x(\partial_x - \partial_y)$ follow
with $[U_1, U_2] = 0$.   □

EXAMPLE 3.34   Consider the type $\mathcal{J}_{2,3}^{(2,2)}$ system

$$\xi_x - \frac{3x^2+1}{x^2+x}\xi = 0,\ \xi_y = 0,\ \eta_x + \frac{2y}{x^2+1}\xi = 0,\ \eta_y - \frac{1}{y}\eta = 0.$$

It satisfies the constraints (3.32) that include the coherence conditions of
Theorem 2.16, consequently it is a Lie system. A solution basis is $(0, y)$ and
$(x^3 + x, -x^2 y)$. It is obtained from the preceding example by the variable
transformation $x = \bar{x}\bar{y}$ and $y = \frac{\bar{y}}{\bar{x}}$ and replacing the barred variables by x
and y.   □

As these examples show the main advantage of applying Theorem 3.17 is
that it does not require one to know the solutions; it uses only the coefficients
of a given system and arithmetic operations and differentiations.

For higher order systems Lie's relations become increasingly more complicated due to the larger number of coefficients involved, therefore they cannot
be given explicitly in this book. Only a single third order system will be
treated in detail because it occurs frequently in later applications.

THEOREM 3.18   *The type* $\mathcal{J}_{3,4}^{(2,2)}$ *Janet basis*

$$\xi_x + a_1\eta + a_2\xi = 0,\ \xi_y + b_1\eta + b_2\xi = 0,$$
$$\eta_y + c_1\eta_x + c_2\eta + c_3\xi = 0,\ \eta_{xx} + d_1\eta_x + d_2\eta + d_3\xi = 0$$

(3.37)

Lie Transformation Groups

147

is a Lie system if its coefficients satisfy the following system of pde's.

$$b_2 = 0, \quad b_3 = 0, \quad a_2 c_1 = 0, \quad d_{2,x} - d_{3,y} + (2a_2 - c_2)d_3 - (2a_3 - c_3)d_2 = 0,$$
$$c_{3,x} c_1 + a_2(a_3 - 2c_3 + d_1) - c_1(c_3^2 + c_1 d_3 - c_3 d_1 + d_2) = 0,$$
$$c_{3,y} - \tfrac{1}{2}d_{1,y} + \tfrac{1}{2}a_2(a_3 - 3c_3 + d_1) - \tfrac{1}{2}c_1 d_2 = 0,$$

$$c_{2,x} - \tfrac{1}{2}d_{1,y} + \tfrac{1}{2}a_2(a_3 - 3c_3 + d_1) - \tfrac{1}{2}c_1 d_2 = 0,$$
$$a_{3,x} - 2c_{3,x} + d_{1,x} + a_3(2c_3 - a_3 - d_1) + c_1 d_3 = 0, \quad a_{3,y} - a_2 c_3 = 0,$$
$$c_{1,x} + c_1(a_3 - c_3) = 0, \quad c_{1,y} - c_1 c_2 = 0, \quad a_{2,y} - a_2 c_2 = 0,$$
$$d_{1,xy} - 3c_{3,x}a_2 + d_{1,y}d_1 - 2d_{2,y} - d_{3,y}c_1 - a_2 a_3(a_3 - 3c_3 + 2d_1)$$
$$+ c_1 d_2(2c_3 + a_3 + d_1) + a_2 c_3^2 - c_1 c_2 d_3 - a_2 d_1^2 = 0,$$
$$c_{3,xx} + c_{3,x}(d_1 - 2c_3 - 2a_3) + d_{1,x}c_3 - d_{3,y} - d_{3,x}c_1$$
$$+ 2a_3 c_3(c_3 - d_1) + c_1 d_3(3a_3 - c_3) - c_2 d_3 + c_3 d_2 = 0.$$
$$\tag{3.38}$$

PROOF   Proceeding similarly to the preceding theorem, at first ξ_3 and η_3 are obtained in the form

$$\xi_3 = (a_2 - b_3)B_3, \quad \eta_3 = c_1 B_1 - B_2 - c_3 B_3 \tag{3.39}$$

where the three bilinear expressions B_1, B_2 and B_3 are defined in terms of the parametric derivatives η_x, η, ξ with the respective index, and

$$B_1 = \eta_{1,x}\eta_2 - \eta_1\eta_{2,x}, \quad B_2 = \eta_{1,x}\xi_2 - \xi_1\eta_{2,x}, \quad B_3 = \eta_1\xi_2 - \xi_1\eta_2. \tag{3.40}$$

The following relations are useful for the subsequent calculations.

$$B_{1,x} = -d_1 B_1 + d_3 B_3,$$
$$B_{1,y} = -(c_{1,x} - 2c_2 - c_1 d_1)B_1 - c_3 B_2 + (c_{3,x} - a_3 c_3 - c_1 d_3)B_3,$$
$$B_{2,x} = -a_2 B_1 - (a_3 + d_1)B_2 - d_2 B_3 \tag{3.41}$$
$$B_{2,y} = -b_2 B_1 - (c_{1,x} + b_3 + c_2 - c_1 d_1)B_2 - (c_{2,x} - a_1 c_3 - c_1 d_2)B_3,$$
$$B_{3,x} = B_2 - a_3 B_3, \quad B_{3,y} = -c_1 B_2 - (b_3 + c_2)B_3.$$

Applying these expressions, the necessary derivatives of ξ_3 and η_3 may be computed from (3.39). If they are substituted into (3.37), the coefficients of the $B's$ yield a system of equations for the coefficients $a_1, a_2, \ldots, d_3$. Combined with the integrability conditions of Theorem 2.17 they represent the necessary and sufficient conditions (3.38) such that the general solution of (3.37) generates a three-dimensional Lie algebra of vector fields.   ▯

The system of pde's (3.38) suggests that it may be split into several alternatives due to algebraic factorization of some of its members. It turns out that altogether three systems are generated in this way as is shown next.

148

COROLLARY 3.1   *The system (3.38) splits into the following alternatives.*

i)
$$c_1 = 0, \ b_3 = 0, \ b_2 = 0, \ a_3 - 2c_3 + d_1 = 0,$$
$$d_{2,x} - d_{3,y} + d_3(2a_2 - c_2) + d_2(2d_1 - 3c_3) = 0,$$
$$c_{3,y} - \tfrac{1}{2}d_{1,y} - \tfrac{1}{2}a_2c_3 = 0, \ c_{2,x} - \tfrac{1}{2}d_{1,y} - \tfrac{1}{2}a_2c_3 = 0,$$
$$a_{2,x} - a_2c_3 = 0, \ a_{2,y} - a_2c_2 = 0,$$
$$d_{1,xy} - 3a_2c_{3,x} + d_{1,y}d_1 - 2d_{2,y} + 3a_2c_3(c_1 - d_1) = 0,$$
$$c_{3,xx} - 3c_{3,x}(2c_3 - d_1) + d_{1,x}c_3 - d_{3,y}$$
$$+2c_3^2(2c_3 - 3d_1) + 2c_3d_1^2 - c_2d_3 + c_3d_2 = 0.$$

ii)
$$c_1 = 0, \ b_3 = 0, \ b_2 = 0, \ a_2 = 0,$$
$$d_{2,x} - d_{3,y} - 2a_3d_2 - c_2d_3 + c_3d_2 = 0,$$
$$c_{3,y} - \tfrac{1}{2}d_{1,y} = 0, \ c_{2,x} - \tfrac{1}{2}d_{1,y} = 0,$$
$$a_{3,x} - 2c_{3,x} + d_{1,x} - a_3^2 + 2a_3c_3 - a_3d_1 = 0, \ a_{3,y} = 0,$$
$$d_{1,xy} + d_1d_{1,y} - 2d_{2,y} = 0,$$
$$c_{3,xx} - c_{3,x}(2a_3 + 2c_3 - d_1) + d_{1,x}c_3 - d_{3,y}$$
$$+2a_3c_3(c_3 - d_1) - c_2d_3 + c_3d_2 = 0.$$

iii)
$$b_3 = 0, \ b_2 = 0, \ a_2 = 0,$$
$$d_{2,x} - d_{3,y} - 2a_3d_2 - c_2d_3 + c_3d_2 = 0,$$
$$c_{3,x} - c_1d_3 - c_3^2 + c_3d_1 - d_2 = 0, \ c_{3,y} - \tfrac{1}{2}d_{1,y} - \tfrac{1}{2}c_1d_2 = 0,$$
$$c_{2,x} - \tfrac{1}{2}d_{1,y} - \tfrac{1}{2}c_1d_2 = 0,$$
$$c_{1,x} + c_1(a_3 - c_3) = 0, \ c_{1,y} - c_1c_2 = 0,$$
$$a_{3,x} + d_{1,x} + (a_3 + d_1)(2c_3 - a_3) = 0, \ a_{3,y} = 0,$$
$$d_{1,xy} + d_1d_{1,y} - 2d_{2,y} - d_{3,y}c_1 + c_1d_2(a_3 + d_1 - 2c_3) - c_1c_2d_3 = 0.$$

The next example shows how additional information beyond Lie's conditions is necessary in order to distinguish certain group types.

EXAMPLE 3.35   Three type $\mathcal{J}_{3,4}^{(2,2)}$ Lie systems are defined by nonvanishing coefficients of (3.37) as follows.
$$a) \ a_3 = -\tfrac{2}{x}, \ d_1 = \tfrac{2}{x}, \quad b) \ a_3 = -\tfrac{2}{x}, \ d_1 = \tfrac{2}{x} + \tfrac{2}{x^2}, \ d_2 = \tfrac{1}{x^4},$$
$$c) \ a_3 = -\tfrac{2}{x}, \ d_1 = \tfrac{2}{x} + \tfrac{1}{x^2}.$$
All three cases satisfy system (3.38). Case $a)$ satisfies only alternative $i)$ of Corollary 3.1, case $b)$ satisfies only alternative $ii)$ whereas case $c)$ satisfies alternatives $ii)$ and $iii)$.   ∎

Lie Transformation Groups

Geometric Constraints. In the symmetry analysis of ode's the most important distinguishing feature for the type of a symmetry group which is of geometric origin is its system of imprimitivity. Therefore for those Janet basis types that occur in the symmetry analysis the constraints determining their systems of imprimitivity are described in this subsection. They are based on Theorem 3.16. For the two first order bases the answer is rather simple.

THEOREM 3.19   *The Lie systems of types*

$$\mathcal{J}_{1,1}^{(2,2)} : \quad \xi = 0, \ \eta_x + a\eta = 0, \ \eta_y + b\eta = 0,$$
$$\mathcal{J}_{1,2}^{(2,2)} : \quad \eta + a\xi = 0, \ \xi_x + b\xi = 0, \ \xi_y + c\xi = 0$$

define one-parameter groups allowing a system of imprimitivity depending on an undetermined function.

PROOF   The systems of imprimitivity are determined by

$$\omega_y + b\omega + a = 0 \ \text{ or } \ \omega_x + a\omega_y + c\omega^2 + (a_y - b - c)\omega + a_x - ab = 0$$

respectively. The general solution of either of these pde's for $\omega(x, y)$ contains an undetermined function of a single argument. Consequently, they determine systems of imprimitivity depending on a function. ☐

This result was to be expected because one-parameter groups according to the tabulation on page 139 always allow a system of imprimitivity of this kind.

THEOREM 3.20   *The Lie systems of types $\mathcal{J}_{2,1}^{(2,2)}, \ldots, \mathcal{J}_{2,5}^{(2,2)}$ define vector fields for two-parameter groups with the following systems of imprimitivity.*

$\mathcal{J}_{2,1}^{(2,2)}$: *System (3.29) does not allow a one-parameter system of imprimitivity. It allows two systems of imprimitivity if $a_1 \neq 0$, and a single one if $a_1 = 0$.*

$\mathcal{J}_{2,2}^{(2,2)}$: *System (3.30) always allows two systems of imprimitivity.*

$\mathcal{J}_{2,3}^{(2,2)}$: *System (3.31) always defines a one-parameter system of imprimitivity.*

$\mathcal{J}_{2,4}^{(2,2)}$: *System (3.33) always allows two systems of imprimitivity.*

$\mathcal{J}_{2,5}^{(2,2)}$: *System (3.34) does not allow a one-parameter system of imprimitivity. It allows two systems of imprimitivity if $a_1 \neq 0$ and a single one if $a_1 = 0$.*

PROOF   Type $\mathcal{J}_{2,1}^{(2,2)}$: There is always the system of imprimitivity corresponding to $\xi_y = 0$. Any other system of imprimitivity is determined by the equations $a_1 \omega - 1 = 0$, $\omega_y + a_2 \omega = 0$. If $a_1 \neq 0$ there is one solution, if $a_1 = 0$ the first equation becomes inconsistent, consequently a solution does not exist.

150

Type $\mathcal{J}_{2,2}^{(2,2)}$: There are always the systems of imprimitivity corresponding to $\xi_y = 0$, and a second one that follows from the equations $\omega = 0$, $\omega_y + a_1 = 0$ with the constraint $a_1 = 0$ according to Theorem 3.17.

Type $\mathcal{J}_{2,3}^{(2,2)}$: Any system of imprimitivity is determined by the equations

$$\omega_x - b_2\omega^2 - (a_2 - d_2)\omega + c_2 = 0, \quad \omega_y - b_1\omega^2 - (a_1 - d_1)\omega + c_1 = 0.$$

Due to Theorem 3.17 it satisfies the coherence conditions (2.51). Consequently, it determines always a system of imprimitivity depending on a parameter.

Type $\mathcal{J}_{2,4}^{(2,2)}$: Any system of imprimitivity is determined by the equations

$$b_1\omega^2 + (a_2b_1 - 1)\omega = 0, \quad a_2\omega_y - \omega_x + b_2\omega^2 - (a_{2,y} - a_2b_2)\omega - a_{2,x} = 0.$$

If $b_1 \neq 0$, the first equation determines two systems of imprimitivity because $a_1b_1 + 1 \neq 0$. If $b_1 = 0$, by Theorem 3.17 it follows $b_2 = 0$, i.e., there is one system of imprimitivity corresponding to $\xi_y = 0$, and a second one follows from $\omega = 0$.

Type $\mathcal{J}_{2,5}^{(2,2)}$: Any system of imprimitivity is determined by the equations

$$\omega(\omega + a_1) = 0, \quad a_1\omega_y - \omega_x - (a_{1,y} + b_1)\omega - a_{1,x} + a_1b_1 = 0.$$

There is always the solution $\omega = 0$. If $a_1 \neq 0$ there is a second system of imprimitivity corresponding to $\omega = -a_1$. $\quad\square$

There are three Janet basis types that occur in the symmetry analysis of second and third order ode's in Chapter 5, their systems of imprimitivity are considered next.

THEOREM 3.21 *The Lie systems of types $\mathcal{J}_{3,4}^{(2,2)}$, $\mathcal{J}_{3,6}^{(2,2)}$ and $\mathcal{J}_{3,7}^{(2,2)}$ define vector fields for three-parameter groups with the following systems of imprimitivity.*

$\mathcal{J}_{3,4}^{(2,2)}$: *The Lie system*

$$\xi_x + a_2\eta + a_3\xi = 0, \quad \xi_y + b_2\eta + b_3\xi = 0,$$
$$\eta_y + c_1\eta_x + c_2\eta + c_3\xi = 0, \quad \eta_{xx} + d_1\eta_x + d_2\eta + d_3\xi = 0 \tag{3.42}$$

defines a three-parameter group allowing two systems of imprimitivity if $c_1 \neq 0$ and a single one if $c_1 = 0$. There cannot be a system of imprimitivity depending on a parameter.

$\mathcal{J}_{3,6}^{(2,2)}$: *The Lie system*

$$\xi_y + a_1\xi_x + a_2\eta + a_3\xi = 0, \quad \eta_x + b_1\xi_x + b_2\eta + b_3\xi = 0,$$
$$\eta_y + c_1\xi_x + c_2\eta + c_3\xi = 0, \quad \xi_{xx} + d_1\xi_x + d_2\eta + d_3\xi = 0 \tag{3.43}$$

Lie Transformation Groups

defines a three-parameter group allowing a system of imprimitivity depending on a parameter if $a_1 = b_1 = 0$ and $c_1 = -1$, two systems of imprimitivity if either $a_i = b_i = 0$, $i = 1, 2, 3$ or $D \equiv 4a_1b_1 + (c_1 + 1)^2 \neq 0$, and a single one if $D = 0$ and $a_1 \neq 0$ or $b_1 \neq 0$ or $c_1 \neq -1$.

$\mathcal{J}_{3,7}^{(2,2)}$: *The Lie system*

$$\xi_x + a_2\eta + a_3\xi = 0, \quad \eta_x + b_1\xi_y + b_2\eta + b_3\xi = 0,$$

$$\eta_y + c_1\xi_y + c_2\eta + c_3\xi = 0, \quad \xi_{yy} + d_1\xi_y + d_2\eta + d_3\xi = 0 \tag{3.44}$$

defines a two-parameter group which does not allow a system of imprimitivity depending on a parameter. It allows two systems of imprimitivity if $D \equiv b_1 - \frac{1}{4}c_1^2 \neq 0$ and a single one if $D = 0$.

PROOF   Type $\mathcal{J}_{3,4}^{(2,2)}$: According to Theorem 3.18, for any Janet basis of type $\mathcal{J}_{3,4}^{(2,2)}$ there holds $b_2 = b_3 = 0$, i.e., there is always a system of imprimitivity corresponding to $\xi_y = 0$. A possible second one is determined by the equations

$$c_1\omega - 1 = 0, \quad \omega_x + (c_3 - a_3)\omega = 0, \quad \omega_y + (c_2 - a_2)\omega = 0.$$

The coherence conditions for the system of first order pde's are satisfied due to Theorem 3.18. Consequently, it allows an additional system of imprimitivity if and only if $c_1 \neq 0$.

$\mathcal{J}_{3,6}^{(2,2)}$: Any system of imprimitivity is determined by the equations

$$a_1\omega^2 - (c_1 + 1)\omega - b_1 = 0, \quad \omega_x - a_3\omega^2 + c_3\omega + b_3 = 0, \quad \omega_y - a_2\omega^2 + c_2\omega + b_2 = 0.$$

If the first equation vanishes identically the remaining system of pde's which is coherent defines a system of imprimitivity depending on a parameter. If not, the discriminant of the first equation determines the number of different solutions and thereby the systems of imprimitivity. If the first equation reduces to $\xi_y = 0$ or the second one to $\eta_x = 0$, an additional system of imprimitivity is obtained.

$\mathcal{J}_{3,7}^{(2,2)}$: Any system of imprimitivity is determined by the equations

$$\omega^2 + c_1\omega + b_1 = 0,$$

$$\omega_x + (c_3 - a_3)\omega + b_3 = 0, \quad \omega_y + (c_2 - a_2)\omega + b_2 = 0.$$

Because the quadratic equation for ω cannot vanish identically and the system of pde's for ω is coherent, the discriminant D of the quadratic equation determines the number of imprimitivity systems. $\quad\square$

EXAMPLE 3.36   For all three alternatives of Corollary 3.1 there holds $b_2 = b_3 = 0$, i.e., by the above lemma there is always one system of imprimitivity corresponding to $\xi_y = 0$. Reducing condition (3.26) by (3.42) and

152

separating the result w.r.t. the parametric derivatives leads to the system

$$c_1\omega - 1 = 0, \quad \omega_y - b_2\omega^2 + (c_2 - a_2)\omega = 0, \quad \omega_x - b_3\omega^2 + (c_3 - a_3)\omega = 0.$$

If $c_1 = 0$ the first equation becomes inconsistent and a second system of imprimitivity does not exist. If $c_1 \neq 0$, $\omega = \frac{1}{c_1}$ is the unique solution of the full system and generates a second system of imprimitivity. Summing up this means, alternatives $i)$ and $ii)$ allow a single system of imprimitivity, alternative $iii)$ allows two of them. ⬚

There occur two Lie systems of order four which have to be analyzed w.r.t. to their group properties. Their systems of imprimitivity are considered next.

THEOREM 3.22   *The Lie systems of types $\mathcal{J}_{4,9}^{(2,2)}$ and $\mathcal{J}_{4,14}^{(2,2)}$ define vector fields for four-parameter groups with the following systems of imprimitivity.*

$\mathcal{J}_{4,9}^{(2,2)}$: *The Lie system*

$$\xi_y + a_2\xi_x + a_3\eta + a_4\xi = 0, \quad \eta_y + b_1\eta_x + b_2\xi_x + b_3\eta + b_4\xi = 0,$$
$$\xi_{xx} + c_1\eta_x + c_2\xi_x + c_3\eta + c_4\xi = 0, \quad \eta_{xx} + d_1\eta_x + d_2\xi_x + d_3\eta + d_4\xi = 0$$
$$(3.45)$$

defines a four-parameter group which does not allow a system of imprimitivity depending on a parameter. It allows two systems of imprimitivity if $a_1 = 0$, $a_2 = a_3 = a_4 = 0$, $b_2 + 1 = 0$ and $b_1 \neq 0$, and a single one if $a_2 = a_3 = a_4 = b_1 = 0$.

$\mathcal{J}_{4,14}^{(2,2)}$: *The Lie system*

$$\eta_x + a_1\xi_y + a_2\xi_x + a_3\eta + a_4\xi = 0, \quad \eta_y + b_1\xi_y + b_2\xi_x + b_3\eta + b_4\xi = 0,$$
$$\xi_{xx} + c_1\xi_y + c_2\xi_x + c_3\eta + c_4\xi = 0, \quad \xi_{xy} + d_1\xi_y + d_2\xi_x + d_3\eta + d_4\xi = 0,$$
$$\xi_{yy} + e_1\xi_y + e_2\xi_x + e_3\eta + e_4\xi = 0$$
$$(3.46)$$

defines a four-parameter group which does not allow a system of imprimitivity depending on a parameter. It allows two systems of imprimitivity if $a_1 = 0$, $a_2 = a_3 = a_4 = 0$, $b_2 + 1 = 0$ and $b_1 \neq 0$ or $a_2 = 0$, $b_2 + 1 = 0$, $b_1^2 - 4a_1 \neq 0$ and $a_1 \neq 0$ or $a_3 \neq 0$ or $a_4 \neq 0$.

The proof is similar as for the preceding theorems and is therefore skipped.

Algebraic Relations. The third set of constraints for the coefficients of Janet bases occurring in the symmetry analysis of ode's originates from the algebraic properties of their Lie algebras. They are based on Theorem 3.7, in particular on part $ii)$ describing the properties of characteristic polynomials. The Lie algebra types $\mathbf{l}_{i,j}$ have been defined in Section 3.4 on page 135 ff.

THEOREM 3.23   *The Lie system (3.31) of type $\mathcal{J}_{2,3}^{(2,2)}$ may define a Lie algebra of vector fields of the following types:*

$$\mathbf{l}_{2,1} \text{ if } a_1 \neq b_2 \text{ or } c_1 \neq d_2. \quad \mathbf{l}_{2,2} \text{ if } a_1 = b_2 \text{ and } c_1 = d_2.$$

Lie Transformation Groups

PROOF For the general Lie system of type $\mathcal{J}_{2,3}^{(2,2)}$ the characteristic polynomial has the form $\omega^2 - [(c_1 - d_2)E_2 - (a_1 - b_2)E_1]\omega$. The above coefficient constraints generate the structure of the two-dimensional Lie algebras $l_{2,1}$ or $l_{2,2}$ respectively listed on page 135. □

THEOREM 3.24   *The Lie system (3.42) of type $\mathcal{J}_{3,4}^{(2,2)}$ may define a Lie algebra of vector fields of the following types:*

$l_{3,1}$ if $c_1 = 0$, $a_3 - 2c_3 + d_1 = 0$ and either
$$a_2 \neq 0 \ \text{ or } \ c_{3,x} - c_3^2 + c_3 d_1 - d_2 \neq 0.$$

$l_{3,2}(c)$ if $c_{3,x} + \frac{1}{4}a_3(a_3 - 4c_3 + 2d_1) + \frac{1}{4}d_1^2 - d_2 \neq 0$, $c \neq 0$,
$$c^2 + \frac{2c_{3,x} + 2c_3^2 - 2c_3 d_1 - 4a_3 c_3 + a_1^2 + d_1^2 - 2d_2}{c_{3,x} - c_3^2 + c_3 d_1 - d_2}c + 1 = 0.$$

$l_{3,3}$ if $a_1 = 0$, $a_2 = 0$, $c_{3,x} + \frac{1}{4}a_3(a_3 - 4c_3 + 2d_1 + \frac{1}{4}d_1^2 - d_2 = 0$,
$$a_3 - 2c_3 + d_1 \neq 0.$$

$l_{3,4}$ if $a_2 = 0$, $c_{3,x} - c_3^2 + c_3 d_1 - c_1 d_3 - d_2 = 0$ and either
$$c_1 \neq 0 \ \text{ or } \ a_3 - 2c_3 + d_1 \neq 0.$$

$l_{3,5}$ if $c_1 = 0$, $a_2 = 0$, $a_3 - 2c_3 + d_1 = 0$, $c_{3,x} - c_3^2 + c_3 d_1 - c_1 d_3 - d_2 = 0$.

$l_{3,6}$ not possible.

PROOF According to the definition (3.17) the characteristic polynomial for any three-dimensional Lie algebra has the form $\omega^3 + C_2\omega^2 + C_1\omega$. In terms of a Lie system of type $\mathcal{J}_{3,4}^{(2,2)}$ the coefficients may be expressed as follows.

$$C_2 = c_1 E_1 + (a_3 - c_3 + d_1)c_1 E_2 - (a_3 - 2c_3 + d_1)E_3,$$

$$C_1 = a_2 c_1 E_1 E_2 - a_2(a_2 - a_3 c_1 + c_1 c_3 - c_1 d_1)E_2^2 - 2a_2 E_1 E_3$$
$$+ \left[c_{3,x} - c_1 d_3 - c_3^2 + c_3 d_1 - d_2)c_1 - a_2(a_3 + d_1)\right]E_2 E_3$$
$$-(c_{3,x} - c_1 d_3 - c_3^2 + c_3 d_1 - d_2)E_3^2.$$

Imposing the constraints on C_1 and C_2 that follow from the structure given in the listing on page 135, the above conditions are obtained after some simplifications. The condition for $l_{3,6}$ follows from the fact that some structure constants are numbers different from zero. □

154

EXAMPLE 3.37   The three alternatives obtained in Corollary 3.1 yield the characteristic equations

$$\omega^3 - \big[(c_{3,x} - c_3^2 + c_3 d_1 - d_2)E_3^2 + 2a_2 c_3 E_3 E_2 + 2a_2 E_3 E_1 + a_2^2 E_2^2\big]\omega = 0,$$
$$\omega^3 - (a_3 - 2c_3 + d_1)E_3\omega^2 - (c_{3,x} - c_3^2 + c_3 d_1 - d_2)E_3^2\omega = 0,$$
$$\omega^3 - \big[(a_3 - 2c_3 + d_1)E_3 - (a_3 c_1 - c_1 c_3 + c_1 d_1)E_2 - c_1 E_1\big]\omega^2 = 0$$

respectively. The first alternative does not contain a quadratic term but a linear one, therefore generically it corresponds to a type $l_{3,1}$ Lie algebra. A type $l_{3,2}(c = -1)$ Lie algebra is excluded because the coefficient of ω is not a complete square of a linear form in the $E's$. If $c_{3,x} - c_3^2 + c_2 d_1 - d_2 = 0$ and $a_2 = 0$, a type $l_{3,5}$ or type $l_{3,6}$ Lie algebra is possible.

For the second alternative, define $C_1 \equiv c_{3,x} - c_3^2 + c_3 d_1 - d_2$, $C_2 \equiv a_3 - 2c_3 + d_1$ and the discriminant $D \equiv C_2^2 - 4C_1$. The coefficient of the first order term is proportional to a complete square of a linear form in the $E's$. As a consequence type $l_{3,1}$ is excluded. If $C_1 \neq 0$ and $C_2 \neq 0$, the condition $D \neq 0$ corresponds to type $l_{3,2}(c \neq -1)$ and $D = 0$ to type $l_{3,3}$. Finally $C_1 = 0$, $C_2 \neq 0$ corresponds to the type $l_{3,4}$, and $C_1 = C_2 = 0$ to types $l_{3,5}$ or $l_{3,6}$.

Due to the constraint $c_1 \neq 0$ for the last alternative, there is always a quadratic term and no linear term. Consequently it corresponds uniquely to a Lie algebra of type $l_{3,4}$. $\quad\square$

There are four additional Janet basis types the geometric properties of which have been determined in Theorems 3.21, 3.22 and 3.22. Their Lie algebra types are described in the following two theorems without proof.

EXAMPLE 3.38   The Lie algebra types corresponding to the three Lie systems considered in Example 3.35 are easiliy identified by the above theorem as follows: $a)$ type $l_{3,5}$, $b)$ type $l_{3,3}$ and $c)$ type $l_{3,4}$. $\quad\square$

THEOREM 3.25   A Lie system (3.43) of type $\mathcal{J}_{3,6}^{(2,2)}$ may define a Lie algebra of vector fields of the following types:

$l_{3,1}$ if $c_1 = 1$, $a_3 b_1 - a_1 b_3 + b_2 - c_3 + d_1 = 0$ and either

$$a_1 b_1 + 1 \neq 0 \text{ or } a_1(b_1 - c_3) + a_3 \neq 0 \text{ or } a_3 b_1 - b_2 + c_3 \neq 0$$
$$\text{or } a_{3,x} - a_3(a_1 b_3 + c_3 - d_1) - a_1 d_3 - d_2 \neq 0$$
$$\text{or } a_{3,x} b_1 + (a_1 b_3 + c_3 - d_1)(c_3 - b_2) - b_1 d_2 + d_3 \neq 0$$
$$\text{or } a_{3,x} + \tfrac{1}{2}(a_1 b_2 - a_1 c_3 - a_3)(a_1 b_3 + c_3 - d_1) \neq 0.$$

$l_{3,2}(c)$ if $c^2 + \dfrac{2a_1 b_1 + c_1^2 + 1}{a_2 b_1 + c_1}c + 1 = 0$, $c_1 \neq 1$.

Lie Transformation Groups

$l_{3,3}$ if any of the following set of conditions is satisfied.

$$c_1 + 1 = 0, \ b_1 = 0, \ a_1b_3 + b_2 - c_3 - d_1 = 0,$$
$$a_{3,x} + a_1(b_2 - c_3)^2 - a_3b_2 - a_1d_3 - d_2 = 0$$

or $c_1 + 1 = 0, \ b_1 = 0,$

$$a_{3,x}b_1 - \tfrac{1}{4}a_3^2b_1^2 - \tfrac{1}{4}(b_2 - c_1 - d_1)^2 - \tfrac{1}{2}a_3b_1(b_2 + c_3 - d_1) - b_1d_1 = 0$$

or $4a_1b_1 + (c_1 + 1)^2 = 0, \ b_1 \neq 1,$

$$(c_1 + 3)(b_2 - c_3) + (c_1 - 1)(d_1 - a_1b_3) + (c_1 - 5)a_1b_1 = 0,$$
$$a_3(c_1 + 1)^2(c_1 - 5) + 4a_1(a_1b_3 - d_1)(c_1 - 1) - 4a_1(c_1 + 3)(b_2 - c_3) = 0,$$

$$a_{3,x}b_1 - \tfrac{1}{8}(a_3b_3 - 2d_3)(c_1 + 1)^2 - \tfrac{1}{4}(b_2 - d_1)^2 - \tfrac{1}{4}(a_1^2b_3^2 + a_3^2b_1^2) - b_1d_2$$
$$- \tfrac{1}{2}a_3b_1(b_2 + c_3 + d_1) - \tfrac{1}{2}a_1b_3(b_2 - c_3 - d_1) + \tfrac{1}{4}c_3(2b_2 - c_3 - 2d_1) = 0.$$

$l_{3,4}$ if any of the following set of conditions is satisfied.

$$a_1 = 0, \ c_1 = 0, \ a_3b_1 = b_2 - c_3, \ a_{3,x} = a_3(c_3 - d_1) + d_2, \ c_3 - b_2 \neq \tfrac{1}{2}d_1$$

or $a_1b_1 + 1 = 0, \ a_3c_1 + a_1(b_2 - c_3) = 0, \ a_3b_1 = b_2 - c_3,$

$$a_{3,x} - a_3(a_1b_3 + c_3 - d_1) - a_1d_3 - d_2 = 0,$$
$$a_1b_3 - 2(b_2 - c_3) \neq d_1, \ a_3 + a_1(a_1b_3 + c_3 - b_2 - d_1) \neq 0.$$

$l_{3,5}$ if $c_1 = 1, \ a_1b_1 + 1 = 0, \ a_1b_3 - 2(b_2 - c_3) - d_1 = 0, \ a_3 + a_1(b_2 - c_3) = 0,$

$$(b_{2,x} - c_{3,x})a_1 + (b_2 - c_3)(a_1d_1 - a_3c_1) + a_1d_3 + d_2 = 0.$$

$l_{3,6}$ not possible.

THEOREM 3.26 *A Lie system (3.44) of type $\mathcal{J}_{3,7}^{(2,2)}$ may define a Lie algebra of vector fields of the following types:*

$l_{3,1}$ if $c_1 = 0, \ a_2 - \tfrac{1}{2}(c_2 + d_1) = 0$ and either

$$b_1 \neq 0 \ \text{or} \ c_3 \neq 0 \ \text{or} \ c_{2,y} + d_{1,y} - \tfrac{1}{2}c_2^2 + \tfrac{1}{2}d_1^2 - 2d_3 \neq 0.$$

$l_{3,2}(c)$ if $c^2 + \dfrac{2b_1 - c_1^2}{b_1}c + 1 = 0, \ c_1 \neq 1.$

$l_{3,3}$ if $b_1 - \tfrac{1}{4}c_1^2 = 0, \ c_3 - \tfrac{1}{2}c_1(a_2 + 2c_2 + 2d_1) = 0,$

$$c_{2,y} + d_{1,y} + c_1d_2 - \tfrac{1}{2}c_2^2 + \tfrac{1}{2}d_1^2 - 2d_3 = 0.$$

$l_{3,4}$ if $c_{2,y} + d_{1,y} - 2a_2(a_2 + c_2 + d_1) + c_1d_2 - c_2d_1 - c_2^2 - 2d_3 = 0,$

$$b_1 = 0, \ c_3 - a_2c_1 = 0 \ \text{and either} \ c_1 \neq 0 \ \text{or} \ a_2 - \tfrac{1}{2}(c_2 + d_1) \neq 0.$$

$l_{3,5}$ if $b_1 = 0, \ c_1 = 0, \ c_3 = 0, \ a_2 - \tfrac{1}{2}(c_2 + d_1) = 0,$

$$c_{2,y} + d_{1,y} - \tfrac{1}{2}c_2^2 + \tfrac{1}{2}d_1^2 - 2d_3 = 0.$$

$l_{3,6}$ not possible.

THEOREM 3.27 *A Lie system (3.45) of type $\mathcal{J}_{4,9}^{(2,2)}$ may define a Lie algebra of vector fields of the following types:*

$l_{4,1}$ if $a_2 = b_1$, $b_2 = 0$, $b_1 \neq 0$.

$l_{4,4}$ if $b_2 = -2$, $b_1 = a_2 = 0$, $a_4 = \frac{1}{2}c_1$.

$l_{4,5}(c)$ if $b_2 \neq 0$, $b_2 + 1 \neq 0$. Then $c = -b_2$.

THEOREM 3.28 *A Lie system (3.46) of type $\mathcal{J}_{4,14}^{(2,2)}$ may define a Lie algebra of vector fields of the following types:*

$$l_{4,4} \text{ if } b_2 = -\tfrac{1}{2}, \quad l_{4,5}(c) \text{ if } b_2 = -\tfrac{1}{c}, \ c \neq 0, \ c \neq -2.$$

The proofs of the preceding three theorems are similar to those of Theorems 3.23 and 3.24 and are therefore omitted. The theorems of this subsection provide the interface between geometric and algebraic properties of Lie transformation groups of the plane on the one hand, and the symmetries of ode's expressed in terms of the Janet bases of its determining systems on the other. They will be applied frequently later in this book.

It turns out that not only the Lie algebra type of the vector fields defined by a given Lie system may be obtained from its coefficients, but even its structure constants c_{ij}^k w.r.t. some basis. The subsequent discussion is based on articles by Neumer [135] and Reid et al. [152].

Assume a Lie system for two functions $\xi(x, y)$ and $\eta(x, y)$ is given as a Janet basis. Let this system be of finite rank r, i.e., its general solution determines the infinitesimal generators of a group with r parameters. A special solution of this system is uniquely determined by assigning fixed values to the parametric derivatives $p_1, \ldots, p_r$. Let two special solutions ξ_1, η_1 and ξ_2, η_2 be determined by $p_{1,1}, \ldots, p_{1,r}$ and $p_{2,1}, \ldots, p_{2,r}$ respectively. The Lie system property assures that there exists a third solution ξ_3, η_3 determined by $p_{3,1}, \ldots, p_{3,r}$ corresponding to the commutator of the vector fields $\xi_1 \partial_x + \eta_1 \partial_y$ and $\xi_2 \partial_x + \eta_2 \partial_y$ and is explicitly given by (3.28). Its parametric derivatives may be computed by deriving the right hand sides of (3.28) and reducing them w.r.t. to the given Lie system. Due to the linearity of all operations involved, they are alternating bilinear forms in the parametric derivatives $p_{1,1}, \ldots, p_{1,r}$ and $p_{2,1}, \ldots, p_{2,r}$ and may be expressed in terms of $B_k = p_{1,i}p_{2,j} - p_{1,j}p_{2,i}$ for $k = 1, \ldots, \frac{1}{2}r(r-1)$, $i, j = 1, \ldots, r$, $i < j$. This proceeding is illustrated by a three-parameter group.

EXAMPLE 3.39 Consider again the Janet basis $\{\xi_x, \eta_x + \xi_y, \eta_y, \xi_{yy}\}$ of the preceding example. The parametric derivatives ξ_y, η and ξ lead to

$$B_1 = \xi_{1,y}\eta_2 - \eta_1\xi_{2,y}, \quad B_2 = \xi_{1,y}\xi_2 - \xi_{2,y}\xi_1, \quad B_3 = \eta_1\xi_2 - \xi_1\eta_2.$$

Differentiating and reducing the right hand sides yields $\xi_{3,y} = 0$, $\eta_3 = -B_2$, $\xi_3 = B_1$, i.e., the only nonvanishing structure constants are $c_{32}^1 = -1$ and $c_{33}^2 = 1$. The same answer is of course obtained if they are computed from the generators ∂_x, ∂_y and $y\partial_x - x\partial_y$. □

Exercises

EXERCISE 3.1   Show that the transformations $\bar{x} = x+y+a$, $\bar{y} = x-y+b$ do not define a group. Which part of the *First Fundamental Theorem* is violated? (Lie [112], vol. I, page 72).

EXERCISE 3.2   Determine the finite transformations of the four-parameter group generated by $U_1 = \partial_x$, $U_2 = x\partial_x$, $U_3 = y\partial_y$, $U_4 = x^2\partial_x + xy\partial_y$.

EXERCISE 3.3   Determine the order of transitivity of the group $\mathbf{g}_2$.

EXERCISE 3.4   Show that any group of $F_1(x)\partial_y, \ldots, F_r(x)\partial_y$ with $r \geq 3$ of the type $\mathbf{g}_{15}$ is similar to a group of the form $\partial_v, u\partial_v, G_3(u)\partial_v, \ldots, G_r(u)\partial_v$.

EXERCISE 3.5   If the Lie algebra type of the vector fields in Example 3.1 is not a priori known, one may start with some other algebra of the listing on page 135, e. g. with type $l_{3,1}$. How do the calculations proceed in this case?

EXERCISE 3.6   Applying (3.17) determine the characteristic polynomial for $r = 2$ and $r = 3$ explicitly.

EXERCISE 3.7   Describe the groups $\mathbf{g}_{17}$ for $l = 1, 2$ or 3 in detail, i.e., list all possible groups, its generators and put up its commutator table.

EXERCISE 3.8   For the three-parameter groups listed above determine the assignment of vector fields to the generators U_1, U_2 and U_3 such that the commutation relations of the respective Lie algebra have the canonical form as given in the tabulation on page 135.

EXERCISE 3.9   The same problem for the four-parameter groups.

EXERCISE 3.10   The special structure of the characteristic equation entails algebraic relations for its coefficients in expanded form that are invariant w.r.t. basis transformations of the Lie algebra. Determine these relations for the algebras $l_{4,4}$ and $l_{4,5}$.

EXERCISE 3.11   Determine the assignment of vector fields of gl_2 and sl_2 to the generators U_k such that the commutation relations of the respective Lie algebra $l_{4,1}$ and $l_{3,1}$ in canonical form are obeyed.

Chapter 4

Equivalence and Invariants of Differential Equations

The two main topics of this chapter, equivalence and invariance of differential equations w.r.t. certain transformation groups, have their origin in the theory of algebraic forms as developed in the 19th century. Comprehensive introductions to the latter may be found in one of the classical text books like Clebsch [30] or Gordan [55], or in the expository article by Kung and Rota [97]. There was a general belief around the middle of the 19th century that many concepts in algebra must have an important meaning for differential equations if they were appropriately generalized. Following these ideas Cockle [31] introduced a notion that came fairly close to an invariant of a differential equation in terms of his so-called *criticoids*. He obtained them for linear second order ode's by elementary methods without applying the notion of a group of transformations. Laguerre [98] was the first to realize the close connection between the type of transformations admitted by a linear second order equation and the invariants belonging to it. Brioschi [19] and Halphen [65] generalized these results to equations of third and fourth order. For quasilinear equations of first and second order Liouville [117, 120] determined various classes of invariants and distinguished *absolute* and *relative* invariants. Later, Tresse [181] in his *Preisschrift* presented an extensive discussion of the invariants of the equation $y'' = F(x, y, y')$ and solved the corresponding equivalence problem. Some of his results have been generalized to third order equations of the form $y''' = F(x, y, y', y'')$ by Leja [103]. A survey on the history of the subject, including some comments and reprints of relevant articles may be found in the booklet by Czichowski and Fritzsche [36]. A more modern treatement of the subject is given in the book by Olver [141].

In the context of differential equations the diffeomorphisms between coordinates are usually called *point transformations* in order to distinguish them from so-called *contact transformations*. The latter involve certain transformations of the first derivatives in addition to the variables themselves. This terminology will frequently be applied in the remaining part of this book.

Whenever dealing with differential equations, it is essential to know whether two given equations may be transformed into each other by a suitable variable change. A special name for this fundamental concept is introduced next.

160

DEFINITION 4.1 *(Equivalence problem). Given two differential equations, decide whether there exists a transformation which takes one of them into the other. If the answer is affirmative the two equations are said to be equivalent under the type of transformations admitted.*

This definition creates a partition of all ode's into equivalence classes. Solving an individual equation may be generalized to solving equations of a full equivalence class. To this end, in the first place it must be possible to decide membership in any given equivalence class algorithmically. As shown below, the invariants of a differential equation are a useful device for answering this question. Secondly, within a given equivalence class, a canonical representative is defined which may be applied to characterize this class. An important subproblem of this latter step is to identify the smallest function field in which the transformation functions are contained. In Chapter 5 it will be seen how the symmetry type of an equation narrows down the possible equivalence classes to which it may belong.

In order to make the invariants of a differential equation into a meaningful concept, it is necessary to identify those transformations leading to a well-defined behavior of its coefficients. For linear equations this has been noticed first by Stäckel [173] who obtained the complete answer in this case. For a general equation Goursat [56] gave a definition of the invariants based on the behavior of its coefficients under certain groups of transformations. The subsequent discussion follows closely this latter reference.

Let an n-th order ode $\omega(x, y, y', \dots, y^{(n)}) = 0$ be given with independent variable x and dependent variable y. Assume that it is polynomial in the derivatives, with coefficients $a_1, a_2, \dots, a_N$ depending on x and y. Let a transformation $x = \phi(\bar{x}, \bar{y}; \theta)$ and $y = \psi(\bar{x}, \bar{y}; \theta)$ of a certain group be applied to it where θ denotes collectively the parameters or functions specifying the group element. They are limited by the requirement that the *structure* of the differential equation remains unchanged under its action. The structure of a differential equation may be expressed in terms of certain restrictions on its order and the power to which the various derivatives may occur. A linear equation, for example, is determined by the requirement that the dependent variable and its derivatives occur only to the first power in each term. For the largest group leaving the structure unchanged a proper name is introduced by the following definition.

DEFINITION 4.2 *(Structure invariance group). Let a class of differential equations be defined in terms of certain restrictions on its structure. The largest group of transformations leaving this structure unchanged is called the structure invariance group of this class. In general it will be an infinite dimensional Lie group.*

Originally the concept of a structure invariance group goes back to Ovsiannikov [144], Chapter II, Section 4; he called them *equivalence transformations.* The name structure invariance group, however, appears to be more suggestive of the basic idea behind it. By definition the elements of the structure invariance group operate only on the unspecified elements of an equation, i.e., the parameters and undetermined functions that it may contain. If an equation does not contain any of them, its structure invariance group specializes to its symmetry group to be explained in the next chapter.

If the structure remains unchanged under a certain transformation, it generates a new differential equation of the form $\omega(\bar{x}, \bar{y}, \bar{y}', \ldots, \bar{y}^{(n)}) = 0$ with coefficients

$$\bar{a}_k = f_k(a_1, \ldots, a_N, a_{1,x}, a_{1,y}, \ldots, a_{N,x}, a_{N,y}, \ldots, \theta)$$

for $k = 1, \ldots, N$ where the functions f_k are uniquely determined by the transformation corresponding to θ. Consequently, it generates a group of transformations among the coefficients $a_1, \ldots, a_N$ that may be prolonged to its derivatives. According to Lie [112], vol. I, pages 212 and 523, any such group allows an infinite series of differential invariants.

DEFINITION 4.3 *(Invariants of an ode) Let $\omega(x, y, y', \ldots, y^{(n)}) = 0$ be an ode which is polynomial in $y, y', \ldots, y^{(n)}$ with coefficients $a_1, \ldots, a_N$ depending on x and y. If a transformation of the structure invariance group of ω with group parameters θ is applied, a new equation with transformed coefficients $\bar{a}_1, \ldots, \bar{a}_N$ is obtained. An expression Φ that is transformed like*

$$\Phi(\bar{a}_1, \bar{a}_2, \ldots, \bar{a}_N) = w(\theta) \cdot \Phi(a_1, a_2, \ldots, a_N).$$

is called an invariant of the ode ω with weight w. More precisely, if w is equal to 1, Φ is an absolute invariant or simply invariant, otherwise it is a relative invariant with weight w.

The meaning of this relation is as follows. If the left hand side Φ is evaluated with arguments a_k, an expression depending on x and y is obtained. If at the right hand side the same function Φ is evaluated with $\bar{a}_k = f_k$, the same expession is obtained, possibly multiplied with a factor depending only on the group parameters θ. Sometimes also a dependence on x and y is allowed.

Subsequently the term *invariant of a differential equation* without further specification means always an invariant w.r.t. the structure invariance group of the equation under consideration. These invariants will usually be denoted by Φ. Any other invariants are denoted by small Greek letters ϕ, ψ, etc. The subgroup to which they correspond will be specified explicitly, or it will be clear from the context. Phrases like *semi-invariants* will not be applied.

The problem is now to develop a set of criteria for deciding equivalence for given pairs of differential equations. It turns out that an answer may only be obtained for special classes of equations, e. g. linear equations or special types of first or second order equations. Moreover, if two equations have been shown to be equivalent, it is still an additional task to determine a transformation between them. It turns out that the necessary computations quickly lead to unmanageable equations. Therefore the following observation is vital in many

162

cases in order to replace a given problem by an equivalent one which is much easier to handle.

The complexity of dealing with any class of ode's is determined above all by the number of undetermined elements, and by the function field in which they are contained. This suggests using the undetermined elements in the admitted transformations to eliminate as many coefficients in the equation as possible, *without* enlarging the function field determined by the initially given equation. For the result of this transformation a special term is introduced in the following definition.

DEFINITION 4.4 *(Rational normal form) A rational normal form of an ordinary differential equation that is polynomial in the dependent variable and its derivatives is an equation with the minimal number of nonconstant coefficients that may be obtained from it by a point transformation in the dependent and the independent variable.*

In general, a rational normal form is not unique, but defines a new structure with a structure invariance group of its own. Applying these concepts, a fairly complete discussion of linear equations is given in Section 4.1. Nonlinear equations of first order are the subject of Section 4.2. In the last Section 4.3 quasilinear equations of second order are considered.

4.1 Linear Equations

As mentioned in the introduction to this chapter, linear ode's were historically the first class of differential equations for which a systematic study of the invariants corresponding to a group of transformations had been made in analogy to algebraic invariants. As a consequence the results are more complete for linear equations than for nonlinear ones. As usual, the general linear homogeneous equation will be written in the form

$$y^{(n)} + q_1 y^{(n-1)} + \ldots + q_{n-1} y' + q_n y = 0. \tag{4.1}$$

Structure Invariance, Normal Forms and Invariants. The subsequent theorem is originally due to Stäckel [173]. The proof follows essentially an article of Neumer [137].

THEOREM 4.1 *(Stäckel 1893, Neumer 1937) The structure invariance group of a linear homogenous equation (4.1) consists of transformations of the form $x = F(u)$, $y = G(u)v + \mathcal{L}(x)$ with F and G undetermined functions and $\mathcal{L}$ the general solution.*

PROOF Let $x = \phi(u, v)$ and $y = \psi(u, v)$ be a general point transformation. An undetermined dependence on v in the coefficients of the transformed

Equivalence and Invariants

equation requires $\phi_v = 0$, i.e., $x = F(u)$. Under this constraint the expression (B.9) for $y^{(k)}$ shows that an unmatched term proportional to $v'v^{(k-1)}$ can only be avoided if $\psi_{vv} = 0$, i.e., $\psi(u, v) = G(u)v + H(u)$. Furthermore, the expression (B.9) shows that the homogeneity requires $H = 0$ in order to avoid any term not proportional to v or any derivative of it. $\mathcal{L}(x)$ is a linear combination of a fundamental system, therefore its contribution vanishes. $\Box$

Although the term $\mathcal{L}(x)$ leaves the equation invariant and therefore may be considered as trivial, it is important to recognize its existence because in some applications the maximal number of parameters involved in the structure invariance group is an important invariant. For later reference the following special case due to Neumer is formulated as a corollary.

COROLLARY 4.1   *(Neumer 1937) The structure invariance group of a linear homogeneous equation with constant coefficients is $x = \alpha u$, $y = e^{\beta u}v + \mathcal{L}(x)$ with α and β constants.*

THEOREM 4.2   *Equation (4.1) has the rational normal form*

$$\bar{y}^{(n)} + q_2\bar{y}^{(n-2)} + \ldots + q_{n-1}\bar{y}' + q_n\bar{y} = 0. \tag{4.2}$$

It is obtained from (4.1) by the transformations

$$x = F(\bar{x}), \quad y = CF'^{(n-1)/2}\exp\left(-\frac{1}{n}\int q_1(F)F'd\bar{x}\right)\bar{y} \tag{4.3}$$

where C is a constant and $F(\bar{x})$ an undetermined function. The structure invariance group of (4.2) is

$$\bar{x} = F(u), \quad \bar{y} = CF'(u)^{(n-1)/2}v + \mathcal{L}(u) \tag{4.4}$$

where C is a constant and $\mathcal{L}$ is a linear combination of a fundamental system.

PROOF   Transforming equation (4.1) according to its structure invariance group by $x = F(\bar{x})$, $y = G(\bar{x})\bar{y}$ leads to a new equation (see equation (B.16) of Appendix B) with the coefficient of $y^{(n-1)}$ equal to

$$n\frac{G}{F'^n}\left(\frac{G'}{G} - \frac{n-1}{n}\frac{F''}{F'}\right) + q_1\frac{G}{F'^{n-1}}.$$

It vanishes if

$$G = CF'^{(n-1)/2}\exp\left(-\frac{1}{n}\int q_1F'd\bar{x}\right)$$

with C a constant. This yields (4.3).

Any transformation of the structure invariance group of (4.2) must have the form $\bar{x} = F(u)$, $\bar{y} = G(u)v$. Applying it to (4.2) yields an equation for v containing the term

$$n\frac{G}{F'^n}\left(\frac{G'}{G} - \frac{n-1}{2}\frac{F''}{F'}\right)v^{(n-1)}. \tag{4.5}$$

164

It vanishes if G is proportional to $F'^{(n-1)/2}$. Any contribution originating from $\mathcal{L}$ obviously drops in the final result. This yields (4.4). □

The transformation (4.3) is the most general one leading to rational normal form. Different choices of $F(\bar{x})$ lead to rational normal forms that are related by a transformation of its structure invariance group (4.4). Often the special choice $x = \bar{x}$ is made with the corresponding transformation $y = \exp\left(-\frac{1}{n}\int q_1(x)dx\right)\bar{y}$ as in Section 2.1.

If the coefficients of the transformed equations are allowed to be in an extension of the base field, a more special canonical form, the so-called *Laguerre-Forsyth canonical form* (Laguerre [99], Forsyth [47]), of equation (4.1) may be obtained.

THEOREM 4.3   *(Laguerre 1879, Forsyth 1888) Any linear homogeneous equation in rational normal form*

$$y^{(n)} + q_2 y^{(n-2)} + \ldots + q_{n-1}y' + q_n y = 0 \tag{4.6}$$

is equivalent to an equation in Laguerre-Forsyth canonical form

$$v^{(n)} + p_3 v^{(n-3)} + \ldots + p_{n-1}v' + p_n v = 0 \tag{4.7}$$

with $v \equiv v(u)$ and $p_k \equiv p_k(u)$. It is obtained by the transformation $x = F(u)$, $y = F'(u)^{(n-1)/2}v$ where the inverse function $u = \bar{F}(x)$ is

$$\bar{F}(x) = C_1 \exp\int z(x)dx + C_2 \ \ with \ \ z' - \frac{1}{2}z^2 - \frac{12q_2(x)}{n(n^2-1)} = 0. \tag{4.8}$$

The structure invariance group of (4.7) is

$$u = \frac{a_1\bar{u} + a_2}{a_3\bar{u} + a_4}, \quad v = \frac{\bar{v}}{(a_3\bar{u} + a_4)^{n-1}} + \mathcal{L}(u)$$

where $a_1, \ldots, a_4$ are constants and $\mathcal{L}(u)$ is a linear combination of a fundamental system.

PROOF   Due to Theorem 4.2, the desired transformation must be of the form $x = F(u)$, $y = F'(u)^{(n-1)/2}v$. Substituting it into (4.6) (for the required derivatives see (B.16) of Appendix B) and imposing the constraint $p_2(u) = 0$ yields the equation

$$\left(\frac{F''}{F'}\right)' - \frac{1}{2}\left(\frac{F''}{F'}\right)^2 + \frac{12q_2(F)F'^2}{n(n^2-1)} = 0$$

for $F(u)$ with $F' = \frac{dF}{du}$. For the inverse function $u = \bar{F}(x)$ there follows

$$\left(\frac{\bar{F}''}{\bar{F}'}\right)' - \frac{1}{2}\left(\frac{\bar{F}''}{\bar{F}'}\right)^2 - \frac{12q_2(x)}{n(n^2-1)} = 0 \tag{4.9}$$

Equivalence and Invariants 165

where $\bar{F}' = \frac{d\bar{F}}{dx}$. This first order Riccati equation for $z \equiv \frac{\bar{F}''}{\bar{F}'}$ has always a solution.

Any transformation of the structure invariance group of (4.7) is obtained by further specialization of (4.4). Substituting the corresponding relation $\frac{G'}{G} = \frac{n-1}{n}\frac{F''}{F'}$ into the coefficient of $\bar{y}^{(n-2)}$ (see (B.16) of Appendix B) leads after some simplifications to

$$\left(\frac{F''}{F'}\right)' - \frac{1}{2}\left(\frac{F''}{F'}\right)^2 = 0.$$

From its general solution and after a further integration for obtaining G the final answer is

$$F = \frac{C_1 u + C_2}{u + C_3}, \qquad G = \frac{C_4}{(u + C_3)^{n-1}}.$$

Introducing new constants by

$$C_1 = \frac{a_1}{a_3}, \quad C_2 = \frac{a_1}{a_3}, \quad C_3 = \frac{a_4}{a_3}, \quad C_4 = \frac{1}{a_3^n}$$

the above form of the structure invariance group corresponding to parameters $a_1, \ldots, a_4$ is obtained. ▯

The structure invariance group of the Laguerre-Forsyth canonical form has been given by Bouton [18], Chapter 5, equation (46). This is the first example in this chapter where the structure invariance group is a finite dimensional group. In order to determine the transformation to Laguerre-Forsyth canonical form explicitly, the Riccati equation (4.8) for z has to be solved. Substituting $\bar{F}' = \frac{1}{\theta(x)^2}$ the linear second order equation

$$\theta'' + \frac{6q_2}{n(n^2 - 1)}\theta = 0 \tag{4.10}$$

for θ is obtained. Then the transformation function $F(u)$ is the inverse of

$$u = \bar{F}(x) = \int \frac{dx}{\theta(x)^2}.$$

The above equation (4.10) for θ shows that determining the Laguerre-Forsyth normal form is equivalent to solving a general second order linear equation as it has already been shown by Schlesinger [160], vol. II, page 198. For $n = 2$ this is equivalent to solving the original equation and no simplification of the solution procedure is obtained. This was to be expected because otherwise the general second order equation would be solvable which is known to be impossible. For equations of order higher than the second, however, the transformation to canonical form is a problem of more minor complexity than solving the given equation itself.

166

EXAMPLE 4.1   Consider $y''' + \frac{1}{x}y = 0$. Applying a general transformation of its structure invariance group yields

$$y''' + \frac{(a_1a_4 - a_2a_3)^3}{(a_1x + a_2)(a_3x + a_4)^5}y = 0.$$

In addition there are the invariance transformations $x = \bar{x}$, $y = a_5\bar{y}$, and those corresponding to $\mathcal{L}(x)$.   ⬜

The third subject to be treated in this subsection are invariants of linear ode's. A first result due to Laguerre is given next.

THEOREM 4.4   (Laguerre 1879) Any third order linear homogeneous equation

$$y''' + p_1y'' + p_2y' + p_3y = 0, \qquad p_k \equiv p_k(x) \tag{4.11}$$

allows the relative second order invariant

$$\Theta_3 = p_1'' + 2p_1'p_1 - 3p_2' + \tfrac{4}{9}p_1^3 - 2p_1p_2 + 6p_3 \tag{4.12}$$

w.r.t. to the structure invariance group $x = F(u)$, $y = G(u)v$. It is transformed according to $\Theta_3(x)|_{x=F(u)} \cdot F'(u)^3 = \Theta_3(u)$.

PROOF   The infinitesimal variations corresponding to the group $F = u + f \cdot \delta t$, $G = 1$ are $\delta x = f \cdot \delta t$, $\delta y = 0$, $\delta y' = -f'y' \cdot \delta t$, $\delta y'' = -f''y' - 2f'y''$ and $\delta y''' = -f'''y' - 3f''y'' - 3f'y'''$. These values are substituted into the equation

$$\delta y''' + p_1\delta y'' + p_2\delta y' + p_3\delta y + \delta p_1 y'' + \delta p_2 y' + \delta p_3 y = 0.$$

The third derivative y''' is substituted by those of lower order applying the given ode. The resulting expression can only vanish if the coefficient of each derivative including y itself vanishes. This leads to a system of linear equations from which the variations of the coeffcients may be determined. The result is

$$\delta p_1 = 3f'' - p_1f', \quad \delta p_2 = f''' + p_1f'' - 2p_2f', \quad \delta p_3 = -3p_3f'.$$

The variations of p_k' and p_k'' are obtained from δp_k by repeated application of Lie's relation (Lie [114], page 670)

$$\delta\phi' = \frac{d}{dx}\delta\phi - \phi'\frac{d}{dx}\delta x \tag{4.13}$$

which is valid for any ϕ. If invariants Φ of second order are to be determined, these variations have to be substituted into the equation

$$\sum_{k=1,2,3} \left(\frac{\partial\Phi}{\partial p_k}\delta p_k + \frac{\partial\Phi}{\partial p_k'}\delta p_k' + \frac{\partial\Phi}{\partial p_k''}\delta p_k'' \right) = \alpha\Phi.$$

The coefficients of the various derivatives of f yield a linear first order system for Φ.

$$\text{Equivalence and Invariants} \qquad\qquad 167$$

For the group generated by $G = (1 + g \cdot \delta t)v$, $F = u$, the infinitesimal variations are $\delta x = 0$, $\delta y = gy \cdot \delta t$, $\delta y' = (g'y + gy') \cdot \delta t$, $\delta y'' = g''y + 2g'y' + gy''$ and $\delta y''' = g'''y + 3g''y' + 3g'y'' + gy'''$. By a similar analysis as above the variations

$$\delta p_1 = -3g', \quad \delta p_2 = -4g'' - 2p_1 g', \quad \delta p_3 = -g''' - p_1 g'' - p_2 g'$$

are obtained. Together with its first and second derivatives they yield another linear first order system for Φ. The two systems have to be combined. After a lengthy calculation the following Janet basis is obtained.

$$\Phi_{p_3''} = 0, \quad \Phi_{p_2''} = 0, \quad \Theta_3 \Phi_{p_1''} = \Phi,$$

$$\Phi_{p_3'} = 0, \quad \Theta_3 \Phi_{p_2'} = -3\Phi, \quad \Phi_{p_1'} = 2p_1 \Phi,$$

$$\Theta_3 \Phi_{p_3} = 6\Phi, \quad \Theta_3 \Phi_{p_2} = 2p_1 \Phi, \quad \Theta_3 \Phi_{p_1} = \left(2P_1' + \tfrac{4}{3}p_1^2 - 2p_2\right)\Phi$$

with Θ_3 given by (4.12). This system has $\Phi = \Theta_3$ as solution which is obtained by integration. $\qquad\qquad$ □

EXAMPLE 4.2 $\quad$ Consider $y''' + \left(\frac{1}{x} + \frac{1}{x^2}\right)y'' - \frac{1}{x^2}y = 0$ which is equation 3.41 from the collection by Kamke with $\Theta_3(x) = -\frac{1}{x^6}\left(6x^4 - \tfrac{4}{9}x^3 - \tfrac{4}{3}x^2 + \tfrac{8}{3}x - \tfrac{4}{9}\right)$. In new variables $x = \frac{1}{u}$, $y = uv$ with $v \equiv v(u)$ the equation becomes

$$v''' - \left(1 - \frac{8}{u}\right)v'' - \left(\frac{4}{u} - \frac{14}{u^2}\right)v' - \left(\frac{2}{u^2} - \frac{4}{u^3} - \frac{1}{u^4}\right)v = 0$$

with $\Theta_3(u) = -\frac{1}{u^4}\left(\tfrac{4}{9}u^4 - \tfrac{8}{3}u^3 + \tfrac{4}{9}u^2 + \tfrac{4}{9}u - 6\right)$. A simple calculation shows that the two invariants obey the relation (4.12). $\qquad\qquad$ □

Laguerre's Theorem 4.4 does not require the equation to be in rational normal form. As explained in the introduction to this chapter however, the complexity of any calculation involving differential equations increases significantly if the number of nonvanishing coefficients increases. This is particularly true for any kind of invariants. Therefore from now on in this section any linear ode will be assumed in rational normal form.

THEOREM 4.5 $\quad$ (Schlesinger 1897) Any linear homogeneous equation in rational normal form

$$y^{(n)} + q_2 y^{(n-2)} + \ldots + q_{n-1} y' + q_n y = 0$$

allows a series of $n-2$ relative invariants Θ_k, $k = 3, \ldots, n$ w.r.t. to the group $x = F(u)$, $y = F'(u)^{(n-1)/2}v$ transforming like $\Theta_k(u) = F'(u)^k \Theta_k(x)|_{x=F}$. Explicitly the three lowest invariants are

$$\Theta_3 = q_3 - \frac{n-2}{2}q_2',$$

$$\Theta_4 = q_4 - \frac{n-3}{2}q_3' + \frac{(n-2)(n-3)}{10}q_2'' - \frac{1}{10}\frac{5n+7}{n+1}\frac{(n-2)(n-3)}{n(n-1)}q_2^2, \quad (4.14)$$

$$\Theta_5 = q_5 - \frac{n-4}{2}q_4' + \frac{3(n-3)(n-4)}{28}q_3'' - \frac{(n-2)(n-3)(n-4)}{84}q_2''$$

$$-\frac{1}{7}\frac{7n+13}{n+1}\frac{(n-3)(n-4)}{n(n-1)}q_2q_3 + \frac{1}{14}\frac{7n+13}{n+1}\frac{(n-2)(n-3)(n-4)}{n(n-1)}q_2q_2'.$$

The proof may be found in Schlesinger [160], vol. II, page 147. If $p_1 = 0$, $p_2 = q_2$ and $p_3 = q_3$ are substituted into (4.12), it is identical to Θ_3 defined in (4.12), up to a constant factor. This example shows the considerable simplification due to rational normal form. For higher order invariants the difference in size is even more prominent.

THEOREM 4.6   *(Forsyth 1888) Any linear homogeneous equation in rational normal form*

$$y^{(n)} + q_2 y^{(n-2)} + \ldots + q_{n-1}y' + q_n y = 0$$

allows a series of $n-2$ relative invariants Φ_k, $k = 3, \ldots, n$ w.r.t. to the group $x = F(u)$, $y = F'(u)^{(n-1)/2}v$ transforming like $\Phi_k(u) = F'(u)^2 \Phi_k(x)|_{x=F}$. In terms of the invariants Θ_k they are

$$\Phi_k(x) = 2k\frac{\Theta_k''}{\Theta_k} - (2k+1)\left(\frac{\Theta_k'}{\Theta_k}\right)^2 - \frac{24k^2 q_2}{n(n^2-1)}.$$

Combining the Θ_k and the Φ_k, the following series of absolute invariants

$$\Theta_{j,k} \equiv \frac{\Theta_k^j}{\Theta_j^k} \quad and \quad \Psi_k \equiv \frac{\Phi_k^k}{\Theta_k^2}$$

is obtained.

The invariants Φ_k are obtained by differentiation and substitution from the invariants Θ_k. The details may be found in the original article by Forsyth [47], a good account is also given in the book by Schlesinger [160], part II, Chapter 6, page 200.

Equivalence of Second Order Equations. In order to proceed further with identifying equivalence classes of linear ode's, equations of fixed order are considered. The complete answer for second order linear ode's is given next.

THEOREM 4.7   *Any linear second order ode $y'' + p_1 y + p_2 y = 0$ is equivalent to $v'' = 0$. As a consequence all second order linear ode's form a single equivalence class with canonical representative $v'' = 0$.*

PROOF   At first the given equation is transformed into its rational normal form $y'' + py = 0$ with $p = p_2 - \frac{1}{2}p_1' - \frac{1}{4}p_1^2$. It remains to be shown that

Equivalence and Invariants 169

$v'' = 0$ may be transformed into it by a transformation $u = f(x)$, $v = g(x)y$. According to (B.15) the second derivative is

$$v'' = \frac{g}{f'^2}\left[y'' + \left(2\frac{g'}{g} - \frac{f''}{f'}\right)y' + \left(\frac{g''}{g} - \frac{f''}{f'}\frac{g'}{g}\right)y\right].$$

It yields the equation $y'' + py = 0$ if

$$2\frac{g'}{g} - \frac{f''}{f'} = 0, \quad \frac{g''}{g} - \frac{f''}{f'}\frac{g'}{g} = p.$$

Substituting $\frac{g'}{g} = 2\frac{f''}{f'}$ and $\frac{g''}{g} = \left(\frac{g'}{g}\right)' + \left(\frac{g'}{g}\right)^2$ into the latter equation, the Riccati equation $z' - z^2 - p_2 + \frac{1}{2}p_1' + \frac{1}{4}p_1^2 = 0$ for $z \equiv \frac{g'}{g}$ is obtained. Solving for f and g the expressions

$$f = C_1 \int \exp\left(2\int z\,dx\right)dx + C_2, \quad g = C_3 \exp\left(\int z\,dx\right)$$

follow; z is a solution of the above Riccati equation. ☐

EXAMPLE 4.3   Let $\{1, u\}$ and $\{y_1, y_2\}$ be fundamental systems for $v'' = 0$ and $y'' + p_1 y' + p_2 y = 0$ respectively. The transformation $u = \sigma \equiv \frac{y_1}{y_2}$ and $v = \rho \equiv \frac{y}{y_2}$ transforms the latter into the former. Correspondingly $v'' = 0$ is transformed into the latter equation in the form (2.3). ☐

This result has been known for a long time; see for example Schlesinger [160], vol. II, page 184, equation (8). It is mainly of theoretical interest because it does not help in solving any second order equation. Solving the Riccati equation that occurs at the end of the proof of Theorem 4.7 comes down to solving the original linear second order ode.

Numerous higher transcendents or *special functions* are defined in terms of linear second order ode's, e. g. Bessel functions, hypergeometric functions, etc. In applications there often arises the problem of recognizing whether the solution of a given differential equation may be expressed in terms of particular special functions. The answer may be obtained by the following result which generalizes the above theorem.

THEOREM 4.8   *A linear second order ode $y'' + p(x)y = 0$ is transformed into $v'' + q(u)v = 0$ by $x = f(u)$, $y = f(u)'^{1/2}v$ if a function $f(u)$ may be found that obeys*

$$\left(\frac{f''}{f'}\right)' - \frac{1}{2}\left(\frac{f''}{f'}\right)^2 + 2f'^2 p(f) = 2q(u). \tag{4.15}$$

This third order ode has always a six-parameter group of symmetries.

PROOF   By Theorem 4.2 the desired equivalence transformation must have the form (4.4) for $n = 2$. Substituting it into $y'' + p(x)y = 0$, applying (B.15)

170

and equating the result to $v'' + q(u)v = 0$ leads immediately to the above equation for f. A simple calculation shows that the conditions for case ii) of Theorem 5.15 are satisfied, i.e., (4.15) belongs to symmetry class $\mathcal{S}_6^3$. $\square$

In order to determine an equivalence transformation to any equation defining a special function by Theorem 4.8, a rational normal form must be known for it. The most important ones are given next. The defining equation is written in terms of the independent variable z and dependent variable w. For the rational normal forms the dependent variable is $\bar{w}$. Further details may be found in Abramowitz [2]. It should be emphasized again that the given rational normal forms are not unique. The Bessel equation

$$w'' + \frac{1}{z}w' + \left(1 - \frac{\nu^2}{z^2}\right)w = 0 \tag{4.16}$$

with fundamental system $J_{\pm\nu}(z)$. A rational normal form is

$$\bar{w}'' + \left(1 - \frac{\nu^2 - \frac{1}{4}}{z^2}\right)\bar{w} = 0 \quad \text{with} \quad \bar{w} = \sqrt{z}w. \tag{4.17}$$

The hypergeometrical equation

$$w'' + \frac{(\alpha + \beta + 1)z - \gamma}{z(z-1)}w' + \frac{\alpha\beta}{z(z-1)}w = 0. \tag{4.18}$$

A rational normal form is

$$\bar{w}'' - \frac{[(\alpha - \beta)^2 - 1]z^2 + 2[(\alpha + \beta - 1)\gamma - 2\alpha\beta]z - \gamma(\gamma - 2)}{4z^2(z-1)^2}\bar{w} = 0 \tag{4.19}$$

with $w = z^{\gamma/2}(z-1)^{(\alpha+\beta+1-\gamma)/2}\bar{w}$. The Legendre equation

$$w'' + \frac{2z}{z^2 - 1}w' - \frac{\nu(\nu + 1)}{z^2 - 1}w = 0. \tag{4.20}$$

A rational normal form is

$$\bar{w}'' - \frac{\nu(\nu - 1)(z^2 - 1) - 1}{(z^2 - 1)^2}\bar{w} = 0 \quad \text{with} \quad \bar{w} = \sqrt{z^2 - 1}w. \tag{4.21}$$

Weber's equation

$$w'' - zw' - aw = 0. \tag{4.22}$$

A rational normal form is

$$\bar{w}'' - \tfrac{1}{4}[z^2 + 2(a-1)]\bar{w} = 0 \quad \text{with} \quad \bar{w} = \exp\left(-\tfrac{1}{4}z^2\right)w. \tag{4.23}$$

The confluent hypergeometric equation or Whittaker's equation is already in rational normal form.

$$y'' - \left(\frac{1}{4} - \frac{k}{x} + \frac{4m^2 - 1}{4x^2}\right)y = 0. \tag{4.24}$$

Equivalence and Invariants

171

Despite the large symmetry group of equation (4.15), the solution algorithm for this symmetry type described in Chapter 7 on page 344 is of no help here. The reason is that it requires solving the second order linear ode's for which the equivalence transformation is desired. Details are discussed in Exercise 4.8.

In order to proceed, the problem has to be further specialized. It is based on the observation that equivalence to an equation defining a special function is only of interest if the transformation is *simple* in some sense, e. g. it should be rational or algebraic. For equivalence to Bessel's equation the following result is useful.

LEMMA 4.1   *The transformations $z = ax^\alpha$, $\bar{w} = (\alpha a x^{\alpha-1})^{1/2}\bar{y}$ generate the equations*

$$\bar{y}'' + \left[\alpha^2 a^2 x^{2\alpha-2} + \left(\tfrac{1}{4} - \alpha^2 \nu^2\right)\tfrac{1}{x^2} \right] \bar{y} = 0 \qquad (4.25)$$

from the rational normal form (4.17) of Bessel's equation.

It is easily proved by applying the formulas (B.15) given in Appendix B to the suggested transformations. Let $y'' + q_1 y' + q_2 y = 0$ be a given equation, $\bar{y}'' + q(x)\bar{y} = 0$ its rational normal form obtained by $y = \exp\left(-\tfrac{1}{2}\int q_1 dx\right)\bar{y}$. If $q(x)$ has the form of the square bracket in (4.25), a fundamental system of the original equation in $y(x)$ may be written as $\sqrt{x}\exp\left(-\tfrac{1}{2}\int q_1 dx\right)w_{1,2}(ax^\alpha)$ if $w_{1,2}(z)$ is a fundamental system of (4.16).

It turns out that the class of transformations admitted in Lemma 4.1 covers virtually all equivalence problems for Bessel's equation listed in Kamke's collection. A few examples are given next.

EXAMPLE 4.4   Equation 2.347 from Kamke's collection $y'' + \tfrac{1}{x}y' + \tfrac{1}{x^4}y = 0$ has the rational normal form $\bar{y}'' + \left(\tfrac{1}{x^4} + \tfrac{1}{4x^2}\right)\bar{y} = 0$ if $y = \tfrac{1}{\sqrt{x}}\bar{y}$ is chosen. The above lemma yields $\alpha = -1$, $a = \pm 1$ and $\nu = 0$. With $a = -1$ there follows $z = \tfrac{1}{x}$, $\bar{w} = \tfrac{\bar{y}}{x}$. Finally the full transformation from the originally given equation to (4.16) with $\nu = 0$ is $x = \tfrac{1}{z}$, $y = w$, consequently, it has a fundamental system $J_0\left(\tfrac{1}{x}\right)$ and $Y_0\left(\tfrac{1}{x}\right)$. ▯

EXAMPLE 4.5   Let the irreducible Airy equation $y'' + xy = 0$ be given. It is already in rational normal form. By Lemma 4.1 $\alpha = \tfrac{3}{2}$, $a = \tfrac{2}{3}$ and $\nu = \tfrac{1}{3}$ is obtained. Consequently, a fundamental system for the original equation is $\sqrt{x}J_{\pm 1/3}\left(\tfrac{2}{3}x^{3/2}\right)$. ▯

For Weber's equation the equivalent of Lemma 4.1 is given next. Again this class of transformations covers all cases listed in Kamke's collection.

LEMMA 4.2   *The transformation $z = ax^\alpha$, $\bar{w} = (\alpha a x^{\alpha-1})^{1/2}\bar{y}$ generates the equation*

$$\bar{y}'' + \left[\tfrac{1}{4}\alpha^2 a^4 x^{4\alpha-2} + \tfrac{1}{2}\alpha^2 a^2(k-1)x^{2\alpha-2} + \tfrac{a^2-1}{4x^2} \right] \bar{y} = 0 \qquad (4.26)$$

from the rational normal form (4.17) of Bessel's equation.

172

Equivalence of Third Order Equations. For linear equations of third order it occurs for the first time that there is more than a single equivalence class. This is because a linear homogeneous equation of order n contains n coefficient functions. Only two of them may be transformed into each other by the two functions $F(u)$ and $G(u)$ contained in the structure invariance group given in Theorem 4.1. Obviously this is only possible for $n = 2$. The next theorem applies the invariants defined in the preceding subsection in order to identify all equivalence classes of linear third order equations.

THEOREM 4.9   *Let two third order linear homogeneous ode's*

$$y''' + A(x)y' + B(x)y = 0 \ \text{and} \ v''' + P(u)v' + Q(u)v = 0 \qquad (4.27)$$

in rational normal form be given. The four alternatives below define equivalence classes, canonical forms and give the equations that are obeyed by the transformation function $x = F(u)$, $y = F'(u)v$ to canonical form, or its inverse $u = \bar{F}(x)$, $v = \bar{F}'(x)y$. The required invariants

$$\Theta_3 = B - \tfrac{1}{2}A', \quad \Phi_3 = 6\frac{\Theta_3''}{\Theta_3} - 7\left(\frac{\Theta_3'}{\Theta_3}\right)^2 - 9A \ \text{and} \ \Psi_3 = \frac{\Phi_3^3}{\Theta_3^2}$$

are obtained from the preceding Theorems 4.14 and 4.6 for $n = 3$.

$\mathcal{E}_1^3 : \Theta_3 = 0$, *canonical form* $v''' = 0$, $\bar{F}(x)$ *obeys*

$$\left(\frac{\bar{F}''}{\bar{F}'}\right)' - \frac{3}{2}\left(\frac{\bar{F}''}{\bar{F}'}\right)^2 - \frac{1}{2}A = 0. \qquad (4.28)$$

$\mathcal{E}_2^3 : \Theta_3 \neq 0$, $\Phi_3 = 0$, *canonical form* $v''' + v = 0$, $\bar{F}$ *obeys* $\bar{F}' = \Theta_3(x)^{\frac{1}{3}}$.

$\mathcal{E}_3^3(c) : \Theta_3 \neq 0$, $\Phi_3 \neq 0$, $\Psi_3 = const$, *canonical form* $v''' + cv' + v = 0$, $c \neq 0$. *The constant c parametrizes equivalence classes, while $F(u)$ is determined by integration from* $F'^2 \cdot \Phi_3(x)|_{x=F} = -9c$.

$\mathcal{E}_4^3(Q) :$ *If none of the preceding three cases applies, the two equations (4.27) are equivalent to each other if both*

$$\Theta_3(x)|_{x=F}F'(u)^3 = \Theta_3(u) \ \text{and} \ \Psi_3(x)|_{x=F} = \Psi_3(u)$$

are valid. The equation for y is equivalent to $v''' + Q(u)v = 0$ *if $\bar{F}(x)$ is determined from (4.28), and its inverse $F(u)$ is substituted into the second equation of (4.29) from which Q is obtained.*

PROOF   Applying a general transformation of the structure invariance group $x = F(u)$ and $y = F'(u)v$ to the above equation for $y(x)$ yields

$$v''' + \left[2\frac{F'''}{F'} - 3\left(\frac{F''}{F'}\right)^2 + A(F)F'^2 \right] v'$$

$$+ \left[\frac{F^{(4)}}{F'} + 3\left(\frac{F''}{F'}\right)^3 - 4\frac{F'''F''}{F'^2} + A(F)F'F'' + B(F)F'^3 \right] v = 0.$$

Equivalence and Invariants

In order to generate the equation for $v(u)$, $F(u)$ has to obey the constraints

$$\frac{F'''}{F'} - \frac{3}{2}\left(\frac{F''}{F'}\right)^2 + \frac{1}{2}A(F)F'^2 = \frac{1}{2}P,$$

$$\frac{F^{(4)}}{F'} + 3\left(\frac{F''}{F'}\right)^3 - 4\frac{F'''F''}{F'^2} + A(F)F'F'' + B(F)F'^3 = Q. \tag{4.29}$$

Reduction of the second equation w.r.t. the first one yields

$$\left[B(F) - \frac{1}{2}A'(F)\right]F'^3 = Q - \frac{1}{2}P' \quad \text{or} \quad \Theta_3(x)\Big|_{x=F} F'(u)^3 = \Theta_3(u) \tag{4.30}$$

where $A'(F) = \dfrac{dA(F)}{dF}$. If $\Theta_3(x) = \Theta_3(u) = 0$, the second equation of (4.29) is the derivative of the first and the system reduces to a single third order ode for F which has always a solution. If $P = Q = 0$ is chosen for the canonical form equation, $F(u)$ satisfies

$$\left(\frac{F''}{F'}\right)' - \frac{1}{2}\left(\frac{F''}{F'}\right)^2 + \frac{1}{2}A(F)F'^2 = 0$$

from which the Riccati equation (4.28) for the inverse function $\bar{F}(x)$ is obtained. If $\Theta_3(x) = 0$ and $\Theta_3(u) \neq 0$ or $\Theta_3(x) \neq 0$ and $\Theta_3(u) = 0$, the above system (4.29) for F is inconsistent and the two equations for y and v cannot be equivalent to each other. Consequently, the condition $\Theta_3 = 0$ characterizes the single equivalence class $\mathcal{E}_1^3$.

If both $\Theta_3(x) \neq 0$ and $\Theta_3(u) \neq 0$, reducing the first equation of (4.29) w.r.t. (4.30) yields

$$6\frac{\Theta_3''(u)}{\Theta(u)} - 7\left(\frac{\Theta_3'(u)}{\Theta_3(u)}\right)^2 - 9P = \left[6\frac{\Theta_3''(F)}{\Theta(F)} - 7\left(\frac{\Theta_3'(F)}{\Theta_3(F)}\right)^2 - 9A(F)\right]F'^2$$

where $\Theta_3'(F) = \dfrac{d\Theta_3(F)}{dF}$, or

$$\Phi_3(u) = \Phi_3(x)|_{x=F} \cdot F'(u)^2. \tag{4.31}$$

If $\Phi_3(x) = \Phi_3(u) = 0$, F is determined by the first order ode (4.30) which has always a solution. If $P = 0$, $Q = 1$ is chosen for the canonical form equation, $\theta_3(u) = 1$ and the equation $\bar{F}(x)' = \Theta_3(x)^{\frac{1}{3}}$ is obtained. If $\Phi_3(x) = 0$ and $\Phi_3(u) \neq 0$ or $\Phi_3(x) \neq 0$ and $\Phi_3(u) = 0$, the system for F is again inconsistent and the two equations are not equivalent. Consequently, $\Theta_3 = 0$, $\Phi_3 = 0$ characterize another equivalence class $\mathcal{E}_2^3$.

If both $\Phi_3(x) \neq 0$ and $\Phi_3(u) \neq 0$, consistency of (4.30) and (4.31) requires $\Psi_3(u) = \Psi_3(x)|_{x=F}$. If both $\Psi_3(x)$ and $\Psi_3(u)$ are constant, a possible choice for the canonical form is $Q = 1$ and $P = c = const$, the value of c is determined by $\Psi_3(u) = -729c^3 = \Psi_3(x)$. In general, two canonical equations with $c \neq \bar{c}$ are not equivalent, i.e., any given value of c determines an equivalence class

174

$\mathcal{E}_3^3(c)$. Substituting the corresponding value for $\Phi_3(u) = -9c$ into (4.31) leads to the equation given for F given above.

If both invariants $\Psi_3(x)$ and $\Psi_3(u)$ are not constant, the transformation function $F(u)$ has to be determined as a solution of the system (4.29) or equivalently (4.30) and (4.31). Its consistency condition $\Psi_3(u) = \Psi_3(x)|_{x=F}$ is necessary for equivalence. If it is satisfied, the validity of (4.30) is sufficient for equivalence because in this case (4.31) is a consequence of it. Choosing $P = 0$ for the canonical form equation, the first equation determines $F(u)$; as a consequence $\bar{F}(x)$ satisfies the same equation as in case i). Q is obtained by substituting the solution for $F(u)$ into the second equation of (4.29). $\square$

The canonical form for $\mathcal{E}_4^3(Q)$ is actually the Laguerre-Forsyth canonical form of Theorem 4.3, the Riccati equation of the above theorem corresponds to the equation (4.8) for z given there. In order to decide equivalence of two given equations in $\mathcal{E}_3^3(Q)$ and $\mathcal{E}_3^3(\bar{Q})$, it is not necessary to solve a Riccati equation. Rather a possible transformation function is determined from the condition $\Psi_3(u) = \Psi_3(x)|_{x=F}$ for the invariants which is a purely algebraic problem. If the resulting function $F(u)$ satisfies (4.30) and (4.31), equivalence is established, otherwise the two equations are not equivalent to each other. The following examples illustrate this case.

EXAMPLE 4.6   Let the two equations $y''' + xy' + y = 0$ with $\Theta_3(x) = \frac{1}{2}$, $\Phi_3(x) = -9x$, $\Psi_3(x) = -2916x^3$ and $v''' + \frac{1}{u^5}v' - \frac{3}{u^6}v = 0$ with $\Theta_3(u) = -\frac{1}{2u^6}$, $\Phi_3(u) = -\frac{9}{u^5}$ and $\Psi_3(u) = -\frac{2916}{u^3}$ be given, i.e., they belong to $\mathcal{E}_4^3(Q)$. By Theorem 4.9 the differential equation $F' = -\frac{F}{u}$ with the solution $F = \frac{C}{u}$ is obtained. The condition for the invariants Ψ_3 leads to $F = \frac{1}{u}$, i.e., the two equations are in the same equivalence class; $x = \frac{1}{u}$ and $y = -\frac{v}{u^2}$ establish the transformation. $\square$

EXAMPLE 4.7   Let the equations $y''' + xy' + y = 0$ like in the preceding example and $v''' + v' + uv = 0$ with $\Theta_3(u) = u$, $\Phi_3(u) = -\frac{9u^2 + 7}{u^2}$ and $\Psi_3(u) = -\frac{(9u^2 + 7)^3}{u^8}$ be given. By Theorem 4.9 the constraints

$$F^3 = \frac{(9u^2 + 7)^3}{2916u^8}, \quad F'(u) = \frac{18u^3 F}{9u^2 + 7} \longrightarrow F^9 = \frac{Ce^{9u^2}}{(9u^2 + 7)^7}$$

are obtained which are inconsistent for any value of the integration constant C. As a consequence the two equations are not equivalent to each other. $\square$

EXAMPLE 4.8   The equation $y''' - y'' = 0$ with rational normal form $y''' - \frac{1}{3}y' - \frac{2}{27}y = 0$ has invariants $\Theta_3 = -\frac{2}{27}$, $\Phi_3 = 3$ and $\Psi_3 = \frac{19683}{4}$. Transformation to canonical form $v''' - \frac{3}{2}\sqrt[3]{2}v' + v = 0$ is achieved by $F(u) = -\frac{3}{2}\sqrt[3]{4}u$, i.e., it belongs to $\mathcal{E}_3^3(-\frac{3}{2}\sqrt[3]{2})$. Similarly, $y''' - \frac{1}{x}y'' = 0$ with rational normal

Equivalence and Invariants 175

form $y''' - \frac{4}{3x^2}y' + \frac{16}{27x^3}y = 0$ and invariants $\Theta_3 = -\frac{20}{27x^3}$, $\Phi_3 = \frac{21}{x^2}$ and $\Psi_3 = 137781400$ belongs to equivalence class $\mathcal{E}_3^3(-\frac{21}{2\sqrt[3]{50}})$, because it is transformed to canonical form $v''' - \frac{21}{2\sqrt[3]{50}}v' + v = 0$ by $F(u) = \exp\left(-\frac{3}{10}\sqrt[3]{50}u\right)$. □

In the next chapter it will be shown how the various equivalence classes described in Theorem 4.9 combine into symmetry classes. In this context equations with constant coefficients play an important role. They are the subject of the next example.

EXAMPLE 4.9   For linear ode's with constant coefficients the results of Theorem 4.9 may be described more explicitly. The generic equation of the equivalence class $\mathcal{E}_1^3$ is $y''' + ay' = 0$ with a nonzero constant a. By $x = \frac{u}{i\sqrt{a}}\log i\sqrt{a}$, $y = \frac{v}{u}$ it is transformed into the canonical form $v''' = 0$. The generic equation for the equivalence class $\mathcal{E}_2^3$ is $y''' + by = 0$ where $b \neq 0$ is again a constant. By $x = \left(\frac{q}{b}\right)^{1/3}u$, $y = \left(\frac{q}{b}\right)^{1/3}v$ it is transformed into $v''' + qv = 0$ where q is an arbitrary constant. The canonical form $v''' + v = 0$ is obtained by $x = \left(\frac{1}{a}\right)^{1/3}u$, $y = \left(\frac{1}{a}\right)^{1/3}v$. The generic equation for case $\mathcal{E}_3^3(c)$ has the form $y''' + ay' + by = 0$ with both a, b nonvanishing constants. It cannot be transformed into an arbitrary equation of the same kind, only one of its coefficients may be chosen at will, e. g. the coefficient of v may be made to unity as in the preceding case with the result $v''' + \frac{a}{b^{2/3}}v' + v = 0$. □

In later chapters different canonical forms will be applied for equivalence classes $\mathcal{E}_2^3$ and $\mathcal{E}_3^3(c)$. They are described in the subsequent lemma.

LEMMA 4.3   *A third order equation* $y''' + cy' + y = 0$ *with* $c \neq 0$ *and* $c \neq -\frac{3}{2}\sqrt[3]{2}$ *is equivalent to* $v''' - (a+1)v'' + av' = 0$, *if the constants* a *and* c *are related to each other by*

$$(a - \tfrac{1}{2})^2(a+1)^2(a-2)^2c^3 + \tfrac{27}{4}(a^2 - a + 1)^3 = 0. \qquad (4.32)$$

The transformation between the two equations is achieved by $x = \alpha u$ *and* $y = e^{\beta u}v$ *where*

$$\alpha = \tfrac{1}{9}\frac{(a+1)(a-2)(2a-1)c}{a^2 - a + 1}, \qquad \beta = -\tfrac{1}{3}(a+1).$$

If $c = 0$ *it is equivalent to* $v''' - (a+1)v'' + av' = 0$ *with* $a = \frac{1}{2}(1 \pm i\sqrt{3})$ *whereby* $\alpha = \frac{1}{3}(a+1)$, $\beta = -\frac{1}{3}(a+1)$. *If* $c = -\frac{3}{2}\sqrt[3]{2}$ *it is equivalent to* $v''' - v'' = 0$ *whereby* $\alpha = -\frac{1}{3}\sqrt[3]{2}$, $\beta = -\frac{1}{3}$.

PROOF   By Corollary 4.1 the transformation between the two linear ode's must have the form $x = \alpha u$, $y = e^{\beta u}v$. Substitution into the first equation yields the algebraic system

$$3\beta + a + 1 = 0, \quad c\alpha^2 + 3\beta^2 - a = 0, \quad \alpha^3 + c\alpha^2\beta + \beta^3 = 0$$

176

from which α and β are obtained by elimination if the constraint (4.32) is satisfied. The calculations in the remaining cases are similar. ⬚

The relation between equivalence classes and the value of c in the canonical form representative $y''' + cy' + y = 0$ is not one-to-one. The same is true for the canonical form $y''' - (a + 1)y'' + ay' = 0$ in the above lemma. These relations are discussed in more detail in the Exercises 4.6 and 4.7.

4.2 Nonlinear First Order Equations

The first result in this section is mainly of theoretical interest. It shows that point symmetries will be of limited help in solving first order ode's because they are members of a single equivalence class. Consequently, they all share the same symmetry type.

THEOREM 4.10 *Any quasilinear first order equation $y' + r(x, y) = 0$ is equivalent to $v' = 0$, i.e., all first order equations of this kind form a single equivalence class under point transformations for which $v' = 0$ may be chosen as a canonical form.*

PROOF The point transformation $x = \phi(u, v)$, $y = \psi(u, v)$ transforms the given equation into $v' + \dfrac{\psi_u + r(\phi, \psi)\phi_u}{\psi_v + r(\phi, \psi)\phi_v} = 0$. The desired canonical form is obtained if $\psi_u + r(\phi, \psi)\phi_u = 0$. This equation always has a solution. ⬚

Although this result guarantees that an equivalence transformation to canonical form $v' = 0$ does exist for any given first order ode, it does not provide a method for finding such a transformation. In general it comes down to solving the given equation as is obvious from rewriting the condition in the proof as $\dfrac{d\psi}{d\phi} + r(\phi, \psi) = 0$. In Exercise 4.2 the reverse transformation from the canonical form to a general first order equation is considered.

EXAMPLE 4.10 The Riccati equation $y' + y^2 + \frac{4}{x}y + \frac{2}{x^2} = 0$ has the general solution $y = \dfrac{1}{x + C} - \dfrac{2}{x}$. It yields the first integral $\dfrac{x(xy + 1)}{xy + 2} = C$. According to Exercise 4.2 this yields the canonical variables $v \equiv \rho(x, y) = \dfrac{x(xy + 1)}{xy + 2}$ and $u \equiv \sigma(x, y)$ where $\sigma(x, y)$ may be chosen arbitrarily. For $u = \frac{1}{x}$ one obtains the transformation $x = \frac{1}{u}$, $y = \dfrac{u(1 - 2uv)}{uv - 1}$. For $u = x$ the transformation is $x = u$, $y = \dfrac{2v - u}{u(u - v)}$. ⬚

From this result follows that in order to proceed, more special classes of

Equivalence and Invariants 177

equations have to be considered, e. g. by limiting the admitted transformations to the corresponding structure invariance groups. The two major classes are Riccati's equation containing a quadratic term in the dependent variable, and Abel's equation with a cubic term.

Riccati's Equation. Algorithms for finding special solutions of these equations have been described already in Chapter 2. Here they will be reconsidered in the context which is the subject of this chapter. Its structure invariance is determined first.

THEOREM 4.11   *The structure invariance group of Riccati's equation*

$$y' + a_2 y^2 + a_1 y + a_0 = 0, \quad a_k \equiv a_k(x) \ for \ k = 0, 1, 2 \tag{4.33}$$

is $x = F(u)$, $y = G(u)v + H(u)$ *where* $v \equiv v(u)$ *and* F, G *and* H *are undetermined functions of its argument.*

PROOF   A general point transformation from x and y to new variables u and v with $v \equiv v(u)$

$$x = \phi(u, v), \quad y = \psi(u, v), \quad y' = \frac{\psi_u + \psi_v v'}{\phi_u + \phi_v v'}$$

generates the equation

$$v' + \frac{(b_2 \psi^2 + b_1 \psi + b_0)\phi_u + \psi_u}{(b_2 \psi^2 + b_1 \psi + b_0)\phi_v + \psi_v} = 0 \tag{4.34}$$

where $b_k \equiv a_k(\phi)$. In order to avoid the occurrence of an unspecified dependence on v via the coefficients a_k, $\phi_v = 0$ is required, i.e., $\phi = F(u)$. The denominator must be independent of v, this requires ψ_v to be a function of u alone, i.e., it must be linear in v and has the form $\psi = G(u)v + H(u)$. Under these constraints (4.34) simplifies to

$$v' + b_2 F'Gv^2 + \left[\frac{G'}{G} + (2b_2 H + b_1)F' \right] v + (b_2 H^2 + b_1 H + b_0)\frac{F'}{G} + \frac{H'}{G} = 0, \tag{4.35}$$

i.e., it has the desired structure. □

THEOREM 4.12   *The rational normal form of Riccati's equation (4.33) is* $y' + y^2 + A(x) = 0$. *It has the structure invariance group*

$$x = F(u), \quad y = \frac{1}{F'} \left(v + \frac{1}{2}\frac{F''}{F'} \right). \tag{4.36}$$

PROOF   By Theorem 4.11 the desired transformation must have the general form $x = F(u)$, $y = G(u)v + H(u)$. From (4.35) the constraints

$$b_2 F'G = 1, \quad \frac{G'}{G} + (2b_2 H + b_1)F' = 0$$

178

are obtained with $b_k \equiv a_k(F)$. Eliminating G and H yields for the transformation to rational normal form

$$x = F, \quad y = \frac{v}{b_2 F'} + \frac{b_2 F'' + b_2' F' - b_1 b_2 F'^2}{2b_2^2 F'^2}. \tag{4.37}$$

If $F \equiv F(u)$ is in the base field of (4.33), the same is true for y and y' obtained from it. On the other hand, any additional constraint on the term which is free of the dependent variable in order to obtain a particular form $A(x)$ requires solving a Riccati equation for H which in general is not possible within the base field. This proves the first part.

Any transformation of the structure invariance group of $y' + y^2 + A = 0$ must have the general form (4.35) with the additional constraints $F'G = 1$, $\frac{G'}{G} + 2HF' = 0$. This yields (4.36). $\quad\square$

Sometimes it occurs that the rational normal form allows one to obtain the solution immediately as the following examples show.

EXAMPLE 4.11   The two equations from Kamke's collection

$$\text{no.1.140}: \quad y' + y^2 + \tfrac{4}{x}y + \tfrac{2}{x^2} = 0 \text{ and}$$
$$\text{no.1.165}: \quad y' + \frac{1}{2x^2 - x}y^2 - \frac{4x+1}{2x^2 - x}y + \frac{4}{2x-1} = 0$$

have rational normal form $v' + v^2 = 0$ with general solution $v = \frac{1}{u+C}$. $\quad\square$

If the rational normal form does not contain the independent variable explicitly, the solution may be obtained as explained on page 18. Two examples are given next.

EXAMPLE 4.12   Equation 1.15 $y' + y^2 - 2x^2 y + x^4 - 2x - 1 = 0$ from Kamke's collection has rational normal form $v' + v^2 - 1 = 0$ with a special solution $v = 1$. Equation 1.140 $y' + y^2 + \tfrac{4}{x}y + \tfrac{2}{x^2} = 0$ with rational normal form $v' + v^2 = 0$ has the special solution $v = 0$. In either case, the general solution may be obtained by Corollary 2.1 or by direct integration. $\quad\square$

COROLLARY 4.2   *Any Riccati equations $y' + y^2 + A(x) = 0$ and $v' + v^2 + P(u) = 0$ are equivalent w.r.t. to the structure invariance group (4.36).*

PROOF   Applying a transformation (4.36) to the first equation yields

$$v' + v^2 + \frac{1}{2}\left(\frac{F''}{F'}\right)' - \frac{1}{4}\left(\frac{F''}{F'}\right)^2 + A(F)F'^2 = 0.$$

This represents the second equation if $\frac{1}{2}\left(\frac{F''}{F'}\right)' - \frac{1}{4}\left(\frac{F''}{F'}\right)^2 + A(F)F'^2 = P(u)$. Because this third order ode always has a solution, equivalence is always assured. $\quad\square$

Equivalence and Invariants

This result is an immediate consequence of Theorem 4.10. It shows that also for Riccati equations an equivalence transformation in general cannot be found.

Abel's Equation. The simplest first order equation that cannot be transformed easily into a linear one as is true for Riccati's equation is the equation introduced about 200 years ago by the famous Norwegian mathematician Abel. It has the form

$$y' + a_3 y^3 + a_2 y^2 + a_1 y + a_0 = 0 \tag{4.38}$$

with $a_k \equiv a_k(x)$ for $k = 0, \ldots, 3$, $a_3 \neq 0$. The more general equation

$$y' + \frac{f_3 y^3 + f_2 y^2 + f_1 y + f_0}{y + g} = 0 \tag{4.39}$$

which is usually called Abel's equation of second kind may be reduced to (4.38) by changing the dependent variable $y = \dfrac{1}{v(x)} - g$ with the result

$$a_3 = f_3 g^3 - f_2 g^2 + f_1 g - f_0, \quad a_2 = g' - 3f_3 g^2 + 2f_2 g - f_1,$$
$$a_1 = 3f_3 g - f_2, \quad a_0 = -f_3. \tag{4.40}$$

If in (4.39) the numerator is of degree two in y, i.e., if $f_3 = 0$, it follows that $a_0 = 0$.

As usual by now, the structure invariance group of (4.38) will be identified first. Although the result has been known for a long time, for example to Appell and Painlevé, it has never really been proved. Therefore it is derived next.

THEOREM 4.13   *The structure invariance group of Abel's equation (4.38) is*

$$x = F(u), \quad y = G(u)v + H(u) \tag{4.41}$$

where $v \equiv v(u)$ and F, G and H are undetermined functions of its argument.

PROOF   The first part of the proof is identical to the proof of Theorem 4.11 with the result $\phi = F(u)$ and $\psi = G(u)v + H(u)$. Applying this transformation to (4.38) yields

$$v' + b_3 F' G^2 v^3 + (3b_3 H + b_2) F' G v^2 + \left(3b_3 H^2 + 2b_2 H + b_1 + \frac{G'}{F'G} \right) F'v$$
$$+ (b_3 H^3 + b_2 H^2 + b_1 H + b_0) \frac{F'}{G} + \frac{H'}{G} = 0 \tag{4.42}$$

with $b_k(u) \equiv a_k(F)$, i.e., it has the desired structure. $\quad\square$

Abel's equation (4.38) contains four coefficient functions, whereas there are only three in the structure invariance group (4.41). This suggests that Abel's

180

equation may be transformed into an equation with only a single nonconstant coefficient involved. This transformation was performed by Appell [6] with the result $y' + y^3 + J(x) = 0$. However, in general $J(x)$ is not contained in the base field; it is left as an exercise to determine it explicitly. The optimal result that may be achieved *without* leaving it is obtained next.

THEOREM 4.14   *There are two different rational normal forms of Abel's equation (4.38). Let $A \equiv A(x)$ and $B \equiv B(x)$ be rational functions of $a_0, \ldots, a_3$ and its derivatives, $A \neq 0$.*

i) $y' + Ay^3 + By = 0$ *with structure invariance group* $x = F(u)$, $y = G(u)v$.

ii) $y' + Ay^3 + By + 1 = 0$ *with structure invariance group* $x = F(u)$, $y = F'(u)v$.

PROOF   If any transformation (4.41) is applied to (4.38), it assumes the form (4.42). The choice $H = -\dfrac{b_2}{3b_3}$ entails the vanishing of the coefficient of v^2. As a consequence applying the transformation

$$y = v - \frac{a_2}{3a_3}, \quad y' = v' - \frac{1}{3} \left(\frac{a_2}{a_3} \right)'$$

which leaves the independent variable unchanged, Abel's equation assumes the form $v' + b_3 v^3 + b_1 v + b_0 = 0$ with

$$b_3 = a_3, \ b_1 = a_1 - \frac{a_2^2}{3a_3}, \ b_0 = a_0 - \frac{a_1 a_2}{3a_3} + \frac{2a_2^3}{27a_3^2} - \frac{1}{3} \left(\frac{a_2}{a_3} \right)'. \quad (4.43)$$

If $b_0 = 0$, the first alternative is obtained. Otherwise introducing again a new function w by $v = b_0 w$, $v' = b_0 w' + b_0' w$ leads to

$$w' + Aw^3 + Bw + 1 = 0 \ \text{with} \ A = b_0^2 b_3, \ B = b_1 + \frac{b_0'}{b_0}.$$

This is the canonical form for case *ii*). For the structure invariance group of either case the terms independent of v in (4.42) lead immediately to the constraint $H = 0$. If $b_0 = 1$ the additional constraint $G = F'$ is necessary in order not to change this value. ⬜

The rational normal form of case *i*) is actually a Bernoulli equation. It is a pleasant feature that this special case is identified without any additional effort. As is well known, if the new function v is replaced by $\dfrac{1}{v^2}$ the linear equation

$$v' - 2Bv - 2A = 0$$

is obtained that is easily solved in terms of quadratures with the result

$$v = \exp \left(2 \int B dx \right) \left[2 \int A \exp \left(-2 \int B dx \right) dx + C \right]. \quad (4.44)$$

Equivalence and Invariants 181

As will be seen later on, this is a consequence of a two-parameter symmetry group that may always be determined explicitly for a Bernoulli equation.

EXAMPLE 4.13   Equation 1.146 $y' - x^2 y^3 + 3y^2 - \frac{1}{x^2}y + \frac{2x - 1}{x^4} = 0$ from Kamke's collection has rational normal form $v' - u^2 v^3 + \frac{2}{u^2}v = 0$, i.e., it is a Bernoulli equation the solution of which may be obtained by integration as explained above. ⬚

The invariants of Abel's equation with respect to the group (4.41) were given for the first time by Liouville [119]. He did not mention, however, how he obtained them. For the applications in this book only the two lowest invariants derived in the next theorem are needed.

THEOREM 4.15   *(Liouville 1887) For Abel's equation*

$$y' + a_3 y^3 + a_2 y^2 + a_1 y + a_0 = 0 \tag{4.45}$$

where $a_k \equiv a_k(x)$ a relative invariant w.r.t. its structure invariance group $x = F(u)$, $y = G(u)v + H(u)$ with $v \equiv v(u)$ is

$$\Phi_3 = a_2 a_3' - a_2' a_3 + 3 a_0 a_3^2 - a_1 a_2 a_3 + \tfrac{2}{9} a_2^3. \tag{4.46}$$

By differentiation the series of higher order relative invariants

$$\Phi_{2m+1} = a_3 \Phi_{2m-1}' - (2m - 1)[a_3' + \tfrac{1}{3}a_2^2 - a_1 a_3]\Phi_{2m-1} \tag{4.47}$$

for $m = 2, 3, \ldots$ is obtained. To the invariant Φ_k the weight k is assigned. Denoting the coefficients of the original and the transformed equation collectively by a and b, the transformation law between the invariants is

$$\Phi_k(u) = F(u)'^k \cdot G(u)^k \cdot \Phi_k(x)|_{x=F(u)}.$$

Rational expressions of these relative invariants with the property that the sum of the products of their order and the corresponding exponent are the same in the numerator and the denominator are absolute invariants.

PROOF   The expression (4.46) for the first order invariant Φ_3 will be derived in detail. The infinitesimal variations corresponding to the group $F = u + f \cdot \delta t$, $G = 1$, $H = 0$ are $\delta x = f \cdot \delta t$, $\delta y = 0$, $\delta y' = -f'y' \cdot \delta t$. Substituting these values into the variation of Abel's equation yields

$$\delta y' + (3a_3 y^2 + 2a_2 y + a_1)\delta y + \delta a_3 y^3 + \delta a_2 y^2 + \delta a_1 y + \delta a_0 =$$
$$-f'y' + \delta a_3 y^3 + \delta a_2 y^2 + \delta a_1 y + \delta a_0 = 0.$$

If in the latter expression y' is eliminated by means of Abel's equation and the result is separated w.r.t. to y, a system of equations is obtained from which the variations of the coefficients may be determined as $\delta a_k = -f'a_k \cdot \delta t$. The variations of the derivatives a_k' follow from these expressions by means of the

182

identity (4.13) with the result $\delta a'_k = -(a_k f'' + 2a_k f') \cdot \delta t$. Accordingly a relative invariant Φ for this group has to obey the system of equations

$$a_3\Phi_{a'_3} + a_2\Phi_{a'_2} + a_1\Phi_{a'_1} + a_0\Phi_{a'_0} = 0,$$

$$2a'_3\Phi_{a'_3} + 2a'_2\Phi_{a'_2} + 2a'_1\Phi_{a'_1} + 2a'_0\Phi_{a'_0} + a_3\Phi_{a_3} + a_2\Phi_{a_2} + a_1\Phi_{a_1} + a_0\Phi_{a_0} = 3\Phi.$$
(4.48)

The term 3Φ at the right hand side originates from the variation of the factor F'^m that yields the contribution $mf' \cdot \delta t$. It turns out that only $m = 3$ allows a nontrivial solution for Φ.

For the group generated by $G = (1 + g \cdot \delta t)v$, $F = u$ and $H = 0$, the infinitesimal variations are $\delta x = 0$, $\delta y = gy \cdot \delta t$, $\delta y' = (g'y + gy') \cdot \delta t$. By a similar analysis to the above the variations

$$\delta a_0 = ga_0, \quad \delta a_1 = -g', \quad \delta a_2 = -ga_2, \quad \delta a_3 = -2ga_3,$$

$$\delta a'_0 = g'a_0 + ga'_0, \quad \delta a'_1 = -g'', \quad \delta a'_2 = -ga'_2 - g'a_2, \quad \delta a'_3 = -2g'a_3 - 2ga'_3$$

of the coefficients and its derivatives are obtained. They yield the system

$$\Phi_{a'_1} = 0, \quad 2a_3\Phi_{a'_3} + a_2\Phi_{a'_2} - a_0\Phi_{a'_0} + \Phi_{a_1} = 0,$$
(4.49)

$$2a'_3\Phi_{a'_3} + a'_2\Phi_{a'_2} - a'_0\Phi_{a'_0} + 2a_3\Phi_{a_3} + a_2\Phi_{a_2} - a_0\Phi_{a_0} = 3\Phi$$

for the invariants. Finally the group $F = u$, $G = 1$ and $H = h \cdot \delta t$ generates the variations $\delta x = 0$, $\delta y = h \cdot \delta t$ and $\delta y' = h' \cdot \delta t$ of the variable and the first derivative. The variations of the coefficients and its derivatives are

$$\delta a_0 = ha_0, \quad \delta a_1 = -h', \quad \delta a_2 = -a_2h, \quad \delta a_3 = -2ha_3,$$

$$\delta a'_0 = h'a_0 + ha'_0, \quad \delta a'_1 = -h'', \quad \delta a'_2 = -a'_2h - a_2h', \quad \delta a'_3 = -2h'a_3 - 2ha'_3.$$

They yield the system

$$\Phi_{a'_0} = 0, \quad 3a_3\Phi_{a'_2} + a_2\Phi_{a'_1} + \Phi_{a_0} = 0,$$
(4.50)

$$3a'_3\Phi_{a'_2} + 2a'_2\Phi_{a'_1} + 3a_3\Phi_{a_2} + 2a_2\Phi_{a_1} + a_1\Phi_{a_0} = 0.$$

The full system comprising (4.48), (4.49) and (4.50) determines the desired invariants. It is arranged in *grlex* term order with $a_k > a_j$ for $k > j$. Due to the group properties of the transformations of x and y it is a complete system. As a consequence, a Janet basis is obtained by purely algebraic autoreduction steps with the result

$$\Phi_{a_0} - \frac{3a_3^2}{\Phi_3}\Phi = 0, \quad \Phi_{a_1} + \frac{a_2a_3}{\Phi_3}\Phi = 0,$$

$$\Phi_{a_2} - \frac{3a_1a_3 - 2a_2^2 - 3a'_3}{3\Phi_3}\Phi = 0, \quad \Phi_{a_3} - \frac{6a_0a_3 - a_1a_2 - a'_2}{\Phi_3}\Phi = 0,$$

$$\Phi_{a'_0} = 0, \quad \Phi_{a'_1} = 0, \quad \Phi_{a'_2} + \frac{a_3}{\Phi_3}\Phi = 0, \quad \Phi_{a'_3} - \frac{a_2}{\Phi_3}\Phi = 0$$

Equivalence and Invariants 183

with Φ_3 given by (4.46). It is easily seen that $\Phi \equiv \Phi_3$ solves this system.

The higher order invariants (4.47) may be verified by differentiation and elimination.

If in a rational expression involving the invariants the powers of $F(u)'$ in the numerators and the denominator are the same, $F(u)'$ drops from this expression. The same is true for $G(u)$, i.e., a rational expression with this property is an absolute invariant. ☐

Due to the fact that the structure invariance group of Abel's equations in rational normal form is a subgroup of (4.41), their invariants may be obtained by specialization of the invariants determined in the above theorem. Of particular interest for solving equivalence problems are the lowest absolute invariants.

COROLLARY 4.3   *Two absolute invariants for Abel's equation in rational normal form case ii) are*

$$K(A, B) = \frac{1}{A} \left(3B - \frac{A'}{A} \right)^3 \quad \text{and} \tag{4.51}$$

$$J(A, B) = \frac{1}{A} \left(3B - \frac{A'}{A} \right) \left[\left(3B - \frac{A'}{A} \right)' + \left(5B - 2\frac{A'}{A} \right) \left(3B - \frac{A'}{A} \right) \right]. \tag{4.52}$$

For the rational normal form in case i) J and K are not defined.

PROOF   For case *ii*), $a_3 = A$, $a_2 = 0$, $a_1 = B$ and $a_0 = 1$ have to be substituted into (4.46) with the result $\Phi_3 = 3A^2$, and into (4.47) with the result

$$\Phi_5 = 3A^3 \left(3B - \frac{A'}{A} \right) \quad \text{and}$$

$$\Phi_7 = 3A^4 \left[\left(3B - \frac{A'}{A} \right)' + \left(5B - 2\frac{A'}{A} \right) \left(3B - \frac{A'}{A} \right) \right].$$

The ratio $\dfrac{\Phi_5^3}{\Phi_3^5}$ is the above invariant $K(A, B)$ up to an irrelevant factor 9 which is due to a different normalization. The quotient $\dfrac{\Phi_5\Phi_7}{\Phi_3^4}$ is $J(A, B)$. For the rational normal form of case *i*), $a_2 = a_0 = 0$ leads to $\Phi_3 = \Phi_5 = 0$, i.e., J and K are not defined. ☐

The absolute invariants (4.51) and (4.52) are the decisive tool for solving equivalence problems of case *ii*).

THEOREM 4.16   *The two rational normal forms have to be considered separately.*

i) *Any two Abel's equations in rational normal form of first kind*

$$y' + A(x)y^3 + B(x)y = 0 \quad \text{and} \quad v' + P(u)v^3 + Q(u)v = 0$$

are equivalent to each other w.r.t. the group $x = F(u)$, $y = G(u)v$.

184

ii) Any two equations in rational normal form of second kind

$$y' + A(x)y^3 + B(x)y + 1 = 0 \ \ and \ \ v' + P(u)v^3 + Q(u)v + 1 = 0$$

are equivalent to each other w.r.t. the group $x = F(u)$, $y = F'(u)v$ if and only if for the absolut invariants (4.51) and (4.52) there holds

$$K(A, B)|_{x=F(u)} = K(P, Q) \tag{4.53}$$

and

$$\frac{[3J(A, B) - 5K(A, B)]^3}{K(A, B)} \bigg|_{x=F(u)} = \frac{[3J(P, Q) - 5K(P, Q)]^3}{K(P, Q)}. \tag{4.54}$$

PROOF In case *i)* the first equation is transformed by $x = F(u)$ and $y = G(u)v$ with the result

$$v' + A(F)F'G^2v^3 + \left[B(F)F' + \frac{G'}{G} \right] v = 0$$

with $' = \frac{d}{du}$. The conditions for equivalence are therefore

$$A(F)F'G^2 = P, \quad B(F)F' + \frac{G'}{G} = Q$$

or

$$F' = \frac{P}{A(F)G^2}, \quad G' = QG - \frac{B(F)}{A(F)} \frac{P}{G}.$$

This first order system for the two undetermined functions F and G always has a solution depending on two constants.

In case *ii)* the first equation is transformed by $x = F(u)$ and $y = F'(u)v$. This leads to

$$v' + A(F)F'^3v^3 + \left[B(F)F' + \frac{F''}{F'} \right] v + 1 = 0$$

from which the conditions for equivalence

$$A(F)F'^3 = P, \quad B(F)F' + \frac{F''}{F'} = Q \tag{4.55}$$

follow. This is a system of ode's for the unknown function F. It may be transformed into a Janet basis by two reductions with the result

$$\frac{1}{A} \left(3B - \frac{A'}{A} \right)^3 \bigg|_{x=F(u)} = \frac{1}{P} \left(3Q - \frac{P'}{P} \right)^3,$$

$$\frac{1}{A^{2/3}}\left[\left(3B - \frac{A'}{A}\right)' - \frac{1}{3}\frac{A'}{A}\left(3B - \frac{A'}{A}\right)\right]\Big|_{x=F(u)} =$$
$$\frac{1}{P^{2/3}}\left[\left(3Q - \frac{P'}{P}\right)' - \frac{1}{3}\frac{P'}{P}\left(3Q - \frac{Q'}{Q}\right)\right].$$

Expressing these equations in terms of the invariants K and J, the conditions (4.53) and (4.54) are obtained. □

Up to now the following results have been obtained. The totality of Abel's equations (4.45) comprises the following equivalence classes w.r.t. the group (4.41). Bernoulli's equations form a single class, all other classes are parametrized by two rational functions, the absolute invariants (4.51) and (4.52).

EXAMPLE 4.14   Consider the Abel equations $y' + xy^3 + y + 1 = 0$ with invariants

$$K(A, B) = \frac{27}{x^4}\left(x - \tfrac{1}{3}\right)^3 \text{ and } J(A, B) = \frac{45}{x^4}\left(x^2 - \tfrac{11}{15}x + \tfrac{1}{5}\right)\left(x - \tfrac{1}{3}\right),$$

and $v' - \frac{1}{u^7}v^3 - \frac{2u+1}{u^2}v + 1 = 0$ with invariants

$$K(P, Q) = -u(u - 3)^3 \text{ and } J(P, Q) = -3u\left(u^2 - \tfrac{11}{3}u + 5\right)(u - 3)$$

in the notation of the above theorem. The constraints (4.53) and (4.54) lead to

$$u(u - 3)^3 F^4 + 27F^3 - 27F^2 + 9F - 1 = 0,$$
$$u^5(4u - 3)^3 F^8 + 27F^3 - 108F^2 + 144F - 64 = 0.$$

The greatest common divisor of its left hand sides is $uF - 1$; it yields the transformation function $F = \frac{1}{u}$. Consequently, the two equations are equivalent, the transformation is achieved by $x = \frac{1}{u}$ and $y = -\frac{v}{u^2}$. □

EXAMPLE 4.15   In the preceding example the first equation is modified to $y' + xy^3 + 2y + 1 = 0$ with the absolute invariants

$$K(A, B) = \frac{216}{x^4}\left(x - \tfrac{1}{6}\right)^3 \text{ and } J(A, B) = \frac{360}{x^4}\left(x^2 - \tfrac{11}{30}x + \tfrac{1}{20}\right)\left(x - \tfrac{1}{6}\right).$$

The second equation remains unchanged. The constraints (4.53) and (4.54) lead to

$$u(u - 3)^3 F^4 + 216F^3 - 108F^2 + 18F - 1 = 0,$$
$$u^5(4u - 3)^3 F^8 + 216F^3 - 432F^2 + 288F - 64 = 0.$$

The *gcd* of the left hand sides is 1 now. Consequently, the two equations are not equivalent. □

186

4.3 Nonlinear Equations of Second and Higher Order

For nonlinear equations of any order the results are far less complete than for linear or first order equations. They are basically limited to quasilinear equations of order two and three, and equations that are homogeneous in the dependent variable and its derivatives.

Lie's Equation. An important class of second order equations is

$$y'' + Ay'^3 + By'^2 + Cy' + D = 0 \tag{4.56}$$

where $A \equiv A(x, y), \ldots, D \equiv D(x, y)$; almost all equations in Chapter 6 of Kamke's collection are of this type. It is suggested that it is called *Lie's equation* because Lie was the first to investigate it in detail in connection with his symmetry analysis. These aspects will be discussed in the subsequent Chapter 5. Most of the results in this subsection have been given for the first time by Roger Liouville in a series of articles between 1885 and 1900.

THEOREM 4.17   *The structure invariance group of (4.56) is $x = F(u, v)$, $y = G(u, v)$ where $v \equiv v(u)$, i.e., the full group of point-transformations of the plane.*

PROOF   From (B.7) it is obvious that the numerator of y'' is linear in v'' and a third order polynomial in v', (B.6) shows that y' is a fraction with numerator and denominator linear in v'. Consequently, upon substitution into (4.56) an equation of the same structure is generated.   ☐

A special case of this result is considered in Exercise 4.5.

THEOREM 4.18   *(Liouville 1887) Lie's equation (4.56) is equivalent to $v'' = 0$ if and only if $L_1 = L_2 = 0$ where*

$$L_1 \equiv D_{yy} + BD_y - AD_x + (B_y - 2A_x)D + \tfrac{1}{3}B_{xx} - \tfrac{2}{3}C_{xy} + \tfrac{1}{3}C(B_x - 2C_y),$$

$$L_2 \equiv 2AD_y + A_yD + \tfrac{1}{3}C_{yy} - \tfrac{2}{3}B_{xy} + A_{xx} - \tfrac{1}{3}BC_y + \tfrac{2}{3}BB_x - A_xC - AC_x.$$

This important result singles out a single equivalence class of Lie's equations. Later on in Theorem 5.11 it will turn out that the conditions $L_1 = L_2 = 0$ assure the maximal eight-parameter symmetry group for second order equations, i.e., they determine the symmetry class $\mathcal{S}_8^2$. If they are not satisfied another set of invariants may be obtained.

THEOREM 4.19   *(Liouville 1889) Lie's equation allows a series of relative invariants w.r.t. point transformations which are determined by*

$$\nu_5 = L_1(L_2L_{1,y} - L_1L_{2,y}) + L_2(L_1L_{2,x} - L_2L_{1,x})$$

$$-AL_1^3 + BL_1^2L_2 - CL_1L_2^2 + DL_2^3, \tag{4.57}$$

$$\nu_{k+2} = L_1\nu_{k,y} - L_2\nu_{k,x} + k\nu_k(L_{2,x} - L_{1,y}), \quad k = 5, 7, \ldots$$

Equivalence and Invariants

187

If $x = \phi(u, v)$, $y = \psi(u, v)$ and $\Delta \equiv \partial_u \phi \partial_v \psi - \partial_v \phi \partial_u \psi$, these invariants are transformed according to $\nu_k(x, y)|_{x=\phi, y=\psi} = \Delta^k \cdot \nu_k(u, v)$. From these relative invariants the series of absolute invariants $t_k = \dfrac{\nu_k^5}{\nu_5^k}$ is obtained.

The invariants ν_k and t_k are derived in Liouville [125], pages 19-23. Liouville describes a procedure of how to decide equivalence from the absolute invariants t_7 and t_9. To this end, they must be defined and moreover they must be functionally independent. It turns out that this is not true for most equations that occur in actual problems, e.g. for the equations listed in Chapter 6 of Kamke's collection. In these cases another series of invariants may be determined.

THEOREM 4.20   *(Liouville 1889) If $L_1 L_2 \neq 0$ and $\nu_5 = 0$, Lie's equation allows the following series of invariants. At first the invariant*

$$w_1 = \frac{1}{L_1^4}[L_1^3(\alpha_1 L_1 - \alpha_2 L_2) + R_1(L_1^2)_x - L_1^2 R_{1,x} + L_1 R_1(\tfrac{1}{3}CL_1 - DL_2)] \text{ or}$$

$$w_1 = \frac{1}{L_2^4}[L_2^3(\alpha_1 L_2 - \alpha_0 L_1) - R_2(L_2^2)_y + L_2^2 R_{2,y} - L_2 R_2(AL_1 - \tfrac{1}{3}BL_2)]$$

is defined for $L_1 \neq 0$ or $L_2 \neq 0$ respectively. R_1, R_2, α_0, α_1 and α_2 are

$$R_1 = L_1 L_{2,x} - L_2 L_{1,x} + \tfrac{1}{3}BL_1^2 - \tfrac{2}{3}CL_1 L_2 + DL_2^2,$$

$$R_2 = L_1 L_{2,y} - L_2 L_{1,y} + AL_1^2 - \tfrac{2}{3}BL_1 L_2 + \tfrac{1}{3}CL_2^2,$$

$$\alpha_0 = \tfrac{1}{3}B_y - A_x + \tfrac{2}{3}(AC - \tfrac{1}{3}B^2), \quad \alpha_1 = \tfrac{1}{3}(C_y - B_x) + AD - \tfrac{1}{9}BC,$$

$$\alpha_2 = D_y - \tfrac{1}{3}C_x + \tfrac{2}{3}(BD - \tfrac{1}{3}C^2).$$

From w_1 the series of relative invariants

$$w_{k+2} = L_1 w_{k,y} - L_2 w_{k,x} + k(L_{2,x} - L_{1,y})w_k$$

is obtained from which the absolute invariants $u_k = \dfrac{w_k}{w_1^k}$ for $k = 3, 5, \ldots$ follow if $w_1 \neq 0$.

The invariants w_k and u_k are also derived in Liouville [125], pages 38-42. Finally, there is a third series of invariants if in addition there holds $w_1 = 0$.

THEOREM 4.21   *(Liouville 1889) If $L_1 \neq 0$, $\nu_5 = 0$ and $w_1 = 0$, Lie's equation allows the absolute invariant*

$$i_2 = \frac{3R_1}{L_1} + L_{2,x} - L_{1,y}$$

and the series of relative invariants

$$i_{2k+2} = L_1 i_{2k,y} - L_2 i_{2k,x} + 2k i_{2k}(L_{2,x} - L_{1,y}), \quad k \geq 1.$$

188

They yield the series of absolute invariants $j_{2k} = \frac{i_{2k}}{i_2^k}$ if $i_2 \neq 0$.

These invariants may be found in Liouville [125], page 50. The case $\nu_5 = 0$, $w_1 = 0$ deserves special attention because any Lie equation with this property is equivalent to a generalized Emden-Fowler equation.

THEOREM 4.22   *(Babich and Bordag 1997) Any Lie equation (4.56) with $\nu_5 = w_1 = 0$ is equivalent to an equation of the form $v'' + r(u,v) = 0$. If in addition $L_2 = 0$, the transformation to this form is determined by $u = \sigma(x,y)$, $v = \rho(x,y)$ with*

$$\sigma(x,y) = \int \exp\left(2\int B\, dy - \int C\, dx\right) dx \text{ and } \rho(x,y) = \int \exp\left(\int B\, dy\right) dy.$$

The proof may found in the article by Babich [8]. They discuss in detail the application of this result to the six types of Painlevé's equations.

EXAMPLE 4.16   Painlevé's equation III has the form

$$y'' = \frac{y'^2}{y} - \frac{y'}{x} + \frac{\alpha y^2 + \beta}{x} + \gamma y^3 + \frac{\delta}{y}.$$

With $B = -\frac{1}{y}$ and $C = \frac{1}{x}$ one obtains $\sigma = \log x$, $\rho = \log y$ and the canonical form

$$v'' = \alpha e^{u+v} + \beta e^{u-v} + \gamma e^{2(u+v)} + \delta e^{2(u-v)}. \qquad \square$$

EXAMPLE 4.17   Equation 6.159 in Kamke's collection $y'' - \frac{3}{4y}y'^2 - 3y^2 = 0$ fullfills the assumptions of the above theorem. Its coeffients $B = -\frac{3}{4y}$ and $C = 0$ yield $\sigma = x$, $\rho = \frac{1}{256}y^4$ and the canonical form $v'' - \frac{3}{1024}v^5 = 0$. It should be compared with Lie's canonical form obtained in Example 6.9. $\square$

The Emden-Fowler canonical form obtained according to the above theorem is not unique as the next result shows.

THEOREM 4.23   *(Babich and Bordag 1997) The structure invariance group of the Emden-Fowler equation $y'' + r(x,y) = 0$ is*

$$x = c_1 \int a(\bar{x})^2 d\bar{x} + c_2, \quad y = a(\bar{x})\bar{y} + b(\bar{x})$$

where a and b are undetermined functions of its argument, c_1 and c_2 are constants, and $\bar{y}$ depends on $\bar{x}$.

Quasilinear Equations of Second Order.   In his award-winning article for the *Fürstlich Jablonowski'sche Gesellschaft* Tresse [181] (see also [180]) applied Lie's theory of differential invariants to the equivalence problem of second order quasilinear ode's of the form $y'' = \omega(x,y,z)$ with $z \equiv y'$. In order

Equivalence and Invariants

to explain his results some notation has to be introduced first. If $\psi \equiv \psi(x, y, z)$ is any function of x, y and z, define

$$\psi_{10} = \frac{\partial \psi}{\partial x} + z \frac{\partial \psi}{\partial y}, \quad \psi_{01} = \frac{\partial \psi}{\partial y} \quad \text{and} \quad \psi^k = \frac{\partial^k \psi}{\partial z^k}.$$

Due to the commutativity of the first two differential operators, the definition $\psi_{ij}^k \equiv (\psi^k)_{ij}$ is meaningful. To any such symbol the two weights $r = i - k + 2$ and $s = j + k - 1$ are assigned. Furthermore, three *differential parameters* Δ_x, Δ_y and Δ_z are defined in terms of their action on any invariant ψ with weights r and s. They return an invariant the order of which is increased by one. For $\omega \neq 0$ they are defined as follows.

$$\Delta_x \psi = \psi_{10} + \omega \Delta_z \psi + \left[(3r + 2s) \left(\omega^1 + \frac{3\omega\omega^5}{5\omega^4} \right) + (2r + s) \frac{\omega_{10}^4}{\omega^4} \right] \psi$$

returns an invariant with weights $r + 1$ and s,

$$\Delta_y \psi = \psi_{01} + \frac{\omega^5}{5\omega^4} \Delta_x \psi + \frac{\omega_{10}^4 + \omega\omega^5 + 2\omega^1\omega^4}{\omega^4} \Delta_z \psi$$

$$- \left\{ (3r + 2s) \left[\frac{\omega^2}{8} + \frac{3}{20} \frac{\omega^5(\omega_{10}^4 + \omega\omega^5 + 2\omega^1\omega^4)}{\omega^4\omega^4} \right] + \frac{r + 2s}{4} \frac{\omega_{01}^4}{\omega^4} \right\} \psi$$

returns an invariant with weights r and $s + 1$ and

$$\Delta_z \psi = \psi^1 + (r - s) \frac{\omega^5}{5\omega^4} \psi$$

returns an invariant with weights $r - 1$ and $s + 1$. Applying this notation the invariants of $y'' = \omega$ may be described as follows.

THEOREM 4.24 *(Tresse 1896) A quasilinear ode $y'' = \omega(x, y, z)$ with $z \equiv y'$ and $\partial^4 \omega / \partial z^4 \neq 0$ has the following relative invariants w.r.t. point transformations $x = \sigma(\bar{x}, \bar{y})$ and $y = \rho(\bar{x}, \bar{y})$.*

i) *Two invariants of order four:* ω^4 *with* $r = -2$, $s = 3$ *and*

$$H \equiv \omega_{20}^2 - 4\omega_{11}^1 + 6\omega_{02} - \omega(3\omega_{01}^2 - 2\omega_{10}^3) + \omega^1(4\omega_{01}^1 - \omega_{10}^2)$$
$$- 3\omega^2\omega_{01} + \omega^3\omega_{10} + \omega\omega\omega^4 \text{ with } r = 2, \ s = 1.$$

ii) *Three invariants of order five:*

$$H_{10} \equiv \Delta_x H \ \text{ with } \ r = 3, s = 1, \quad H_{01} \equiv \Delta_y H \ \text{ with } \ r = s = 2,$$

$$K \equiv \Delta_z H \ \text{ with } \ r = 1, s = 2.$$

iii) *Eleven invariants of order six:*

$$\Omega^6 \equiv \omega^6 - \frac{6}{5} \frac{\omega^5\omega^5}{\omega^4} \ \text{ with } \ r = -4, s = 5,$$

$$\Omega_{10}^5 \equiv \frac{5\omega^4}{24H}(\Delta_{zz}H - \Delta_{xz}H - \Delta_y H) \quad \text{with} \ \ r = -2, s = 4,$$

$$\Omega_{01}^5 = \tfrac{4}{9}(\Delta_z\Omega_{10}^5 - \Delta_x\Omega^6) \quad \text{with} \ \ r = -3, s = 5,$$

$$\Omega_{20}^4 = \Delta_{zz}H - \frac{\Omega^6}{5\omega^4}H \quad \text{with} \ \ r = 0, s = 3,$$

$$\Omega_{11}^4 = \tfrac{4}{3}(\Delta_z\Omega_{20}^4 - \Delta_x\Omega_{10}^5) \quad \text{with} \ \ r = -1, s = 4,$$

$$\Omega_{02}^4 \equiv \frac{4}{5}\left(\Delta_y\Omega_{10}^5 - \Delta_x\Omega_{01}^5 + \frac{\Omega^6\Omega_{20}^4}{\omega^4} + \frac{\Omega_{10}^5\Omega_{10}^5}{5\omega^4}\right) \quad \text{with} \ \ r = -2, s = 5,$$

$$\Delta_{xz}H \ \ \text{with} \ \ r = s = 2, \ \ \Delta_{yz}H \ \ \text{with} \ \ r = 1, \ s = 3,$$

$$\Delta_{xx}H \ \ \text{with} \ \ r = 4, \ s = 1,$$

$$\Delta_{xy}H \ \ \text{with} \ \ r = 3, s = 2, \ \ \Delta_{yy}H \ \ \text{with} \ \ r = 2, \ s = 3.$$

Any of these invariants I transforms according to

$$\frac{\bar{I}}{I} = \frac{(\sigma_x + y'\sigma_y)^{s-r}}{(\sigma_x\rho_y - \sigma_y\rho_x)^s}.$$

The proof involves rather lengthy calculations. They may be found in the above quoted article by Tresse. By inspection of the invariants it is seen that all invariants may be obtained by application of the differential parameters to ω^4, H, Ω^6 and Ω_{10}^5.

The transformation law for the invariants given in Theorem 4.24 implies that absolute invariants may be formed from suitable powers of three properly chosen relative invariants. If $\omega^4 H \neq 0$ and ψ is any relative invariant with weights r and s, an absolute invariant

$$\hat{\psi} = \frac{\psi}{(\omega^4)^\alpha (H)^\beta}$$

is obtained if $\alpha = \tfrac{1}{8}(2s - r)$ and $\beta = \tfrac{1}{8}(2s + 3r)$ are chosen. In order to obtain rational expressions, they are raised to the power of the least common multiple of the denominators of α and β. For the invariants up to order six the complete answer is given next. Absolute invariants are denoted by a hat on top of the corresponding symbol for the relative invariant.

COROLLARY 4.4 *With the same notation as in Theorem 4.24 and assuming $\omega^4 H \neq 0$, the absolute invariants for $y'' = \omega(x, y, z)$ of order 5 are*

$$\hat{K} = \frac{K^8}{(\omega^4)^3(H)^7}, \quad \hat{H}_{10} = \frac{(H_{10})^8\omega^4}{(H)^{11}}, \quad \hat{H}_{01} = \frac{(H_{01})^4}{\omega^4(H)^5}.$$

The invariants of order 6 are

$$\hat{\Omega}^6 = \frac{(\Omega^6)^4 H}{(\omega^4)^7}, \quad \hat{\Omega}^5_{10} = \frac{(\Omega^5_{10})^4}{(\omega^4)^5 H}, \quad \hat{\Omega}^5_{01} = \frac{(\Omega^5_{01})^8}{(\omega^4)^{13} H},$$

$$\hat{\Omega}^4_{20} = \frac{(\Omega^4_{20})^4}{(\omega^4)^3 (H)^3}, \quad \hat{\Omega}^4_{11} = \frac{(\Omega^4_{11})^8}{(\omega^4)^9 (H)^5}, \quad \hat{\Omega}^4_{02} = \frac{(\Omega^4_{02})^2}{(\omega^4)^3 H},$$

$$\frac{(\Delta_{xz} H)^4}{\omega^4 (H)^5}, \quad \frac{(\Delta_{yz} H)^8}{(\omega^4)^5 (H)^9}, \quad \frac{(\Delta_{xx} H)^4 \omega^4}{(H)^7}, \quad \frac{(\Delta_{xy} H)^8}{\omega^4 (H^{13})}, \quad \frac{(\Delta_{yy} H)^2}{\omega^4 (H^3)}.$$

Some of the results just described will be applied in Chapter 5.3 for determining symmetry classes of second order ode's.

Exercises

EXERCISE 4.1   Let a linear homogeneous ode be given in the form

$$p_0 y^{(n)} + p_1 y^{(n-1)} + \ldots + p_{n-1} y' + p_n y = 0$$

with $p_k \equiv p_k(x)$ polynomials in x for all k. Find a transformation that generates a new equation of the same type from it such that the constraint $p_1 = -p_0'$ is valid. Discuss the result (Hirsch [72]).

EXERCISE 4.2   Determine a transformation from the equation $v' = 0$ to a given equation $y' + r(x, y) = 0$. Compare the result with Theorem 4.10.

EXERCISE 4.3   Derive the rational normal form obtained by choosing $F(u) = u$ in (4.37).

EXERCISE 4.4   Determine the structure invariance group of an Abel equation in Appel's normal form $y' + y^3 + r(x) = 0$.

EXERCISE 4.5   How does Lie's equation (4.56) change if the dependent and the independent variable are exchanged by each other?

EXERCISE 4.6   Show that the three equations $y'' + cy' + y = 0$ and $y'' + \frac{1}{2}(3 \pm i\sqrt{3})cy' + y = 0$ are pairwise equivalent to each other.

EXERCISE 4.7   Show that the six equations $y'' - (a + 1)y'' + ay' = 0$ and $y'' - (\bar{a} + 1)y'' + \bar{a}y' = 0$ with $\bar{a} = \frac{1}{a}$, $\bar{a} = 1 - a$, $\bar{a} = \frac{1}{1-a}$, $\bar{a} = 1 - \frac{1}{a}$ and $\bar{a} = 1 + \frac{1}{a-1}$ are pairwise equivalent to each other.

EXERCISE 4.8   Apply the algorithm *LieSolve3.6* described on page 344 to equation (4.15). Why is it not effective for the particular application of Theorem 4.8 ?

192

EXERCISE 4.9   Design an algorithm that accepts a linear second order ode as input and returns a rational transformation to Weber's equation (4.22) if there exists one or *failed* otherwise.

EXERCISE 4.10   The same problem for the Bessel equation (4.16).

Chapter 5

Symmetries of Differential Equations

This chapter deals with the most important topic of this book, i.e., the non-trivial *symmetries* that a given ordinary differential equation (ode) may admit. Let two coordinate sets of the plane be defined with the additional assumption that in either of them one coordinate variable is dependent on the other. Roughly speaking a symmetry of an ode is a diffeomorphism connecting these coordinates for which this ode is an invariant. In the literature on symmetries these transformations are often called *point transformations* or *variable transformations* in order to distinguish them from more general transformations also involving the first derivative. This notation will frequently be used from now on whenever symmetries of an ode are the main topic.

It is obvious that the entirety of symmetries of any given ode forms a group. The term *symmetry group* of a differential equation is applied to the *largest* group of transformations sharing this property. The Lie algebra of its infinitesimal generators forms the corresponding *symmetry algebra*. If a variable transformation is applied to a given differential equation, the symmetry group of the transformed equation is similar to the original one according to Definition 3.2. The equivalence class to which the symmetry group of a particular ode belongs is called its *symmetry type*. Consequently, all equations contained in an equivalence class have the same symmetry type. The reverse is not true. As a consequence, the entirety of all differential equations allowing the same type of symmetry group is the union of equivalence classes. Krause and Michel [96] called it the *stratum* of ode's corresponding to a symmetry type. In this book it will be called the *symmetry class*.

The symmetries of a differential equation occur in different connections. On the one hand, there is the *classification problem*. Its aim is to determine all possible symmetry types for a family of ode's, e. g. ode's of a fixed order. The starting point for this approach is the listing of groups given in Section 3.4; its differential invariants determine the general form of an ode that may be invariant under the respective group. On the other hand, if any particular ode is given, its symmetry type has to be determined if it is to be applied for finding its solutions.

The symmetry problem for equations of order one is significantly different from that of equations of order two or higher. The main difference is the fact that for first order equations in general there is no algorithm available for determining any symmetry generator; only heuristics or insight into the

193

194

problem from which the equation originates may allow finding them. What makes the problem especially difficult is the fact that any first order ode allows infinitely many symmetries as discussed at the beginning of Section 5.2. Contrary to this, for equations of order two or higher, the symmetry type may always be identified algorithmically, and there is a well-defined procedure for transforming it to a canonical form. The key for this is the Janet basis for the determining system and the theorems derived in Section 3.5.

In the subsequent Section 5.1, general properties of the behavior of differential equations under a change of variables are discussed and various concepts related to its symmetries are defined. Section 5.2 discusses the symmetry structure of first order equations. The most important field of application for Lie himself was equations of second order; they are the subject of Section 5.3. In Section 5.4 a complete discussion of equations of order three is given following a similar approach to the second order equations. Finally in Section 5.5 the symmetries of linear equations of any order are discussed.

5.1 Transformation of Differential Equations

Introducing new variables into a given ode has been a popular method in order to faciliate the solution procedure all the time. Usually this is done in an ad hoc manner without guaranteed success. In particular there is no criterion for deciding whether a certain class of transformations will lead to an integrable equation or not. A critical examination of these methods was the starting point for Lie's symmetry analysis. Therefore the behavior of ode's under various kinds of transformations will be investigated first.

Properties of Point Transformations. Let an ode of order n in the independent variable x and the dependent variable $y \equiv y(x)$ be given as

$$\omega(x, y, y', \ldots, y^{(n)}) = 0. \tag{5.1}$$

If not stated explicitly, it will be assumed that ω is a polynomial in the derivatives with coefficients in some *base field* which is usually the field of rational functions in x and y, i.e., $\omega \in \mathbb{Q}(x, y)[y', \ldots, y^{(n)}]$. Any other field that occurs later on during the solution procedure is an extension of this base field.

A *point-transformation* between two planes with coordinates (x, y) and (u, v), and dependencies $y \equiv y(x)$ and $v \equiv v(u)$ respectively, is considered in the form

$$u = \sigma(x, y), \qquad v = \rho(x, y). \tag{5.2}$$

The term point-transformation expresses the fact that only the coordinates occur as arguments of the transformation functions. If, in addition, the first

Symmetries of Differential Equations

derivative is transformed by an independent function of x, y and y' of a certain form it is called a *contact-transformation*. They are not dicussed in this book. In general equations (5.2) define a diffeomorphism between the two coordinate planes. Depending on the particular situation, the function field in which σ and ρ are contained has to be specified.

Let a curve in the $x - y$-plane described by $y = f(x)$ be transformed under (5.2) into $v = g(u)$. There arises the question of how the derivative $y' = \dfrac{df}{dx}$ corresponds to $v' = \dfrac{dg}{du}$ under this transformation. A simple calculation leads to the *first prolongation*

$$v' = \frac{dv}{du} = \frac{\rho_x + \rho_y y'}{\sigma_x + \sigma_y y'} \equiv \chi_1(x, y, y'). \tag{5.3}$$

It is remarkable that the knowledge of (x, y, y') and the equations of the point transformation (5.2) are sufficient for computing the derivative v', knowing the equation of the curve is not required. This is also expressed by saying that the *line element* (x, y, y') is transformed into the line element (u, v, v') under the action of a point transformation. Similarly the transformation law for derivatives of second order is obtained as

$$v'' = \frac{dv'}{du} = \frac{\chi_{1,x} + \chi_{1,y} y' + \chi_{1,y'} y''}{\sigma_x + \sigma_y y'} \equiv \chi_2(x, y, y', y''). \tag{5.4}$$

Explicitly in terms of σ and ρ it is

$$v'' = \frac{1}{(\sigma_x + \sigma_y y')^3} \{ (\sigma_x \rho_y - \sigma_y \rho_x) y'' + (\sigma_y \rho_{yy} - \sigma_{yy} \rho_y) y'^3$$

$$+ [\sigma_x \rho_{yy} - \sigma_{yy} \rho_x + 2(\sigma_y \rho_{xy} - \sigma_{xy} \rho_y)] y'^2 \tag{5.5}$$

$$+ [\sigma_y \rho_{xx} - \sigma_{xx} \rho_y + 2(\sigma_x \rho_{xy} - \sigma_{xy} \rho_x)] y' + \sigma_x \rho_{xx} - \sigma_{xx} \rho_x \}.$$

In general there holds

$$v^{(n)} = \frac{dv^{(n-1)}}{du} \equiv \chi_n(x, y, y', \ldots, y^{(n)}) \tag{5.6}$$

and

$$v^{(n+1)} = \frac{dv^{(n)}}{du} = \frac{\chi_{n,x} + \chi_{n,y} y' + \ldots + \chi_{n,y^{(n-1)}} y^{(n)}}{\sigma_x + \sigma_y y'}.$$

The form of a differential equation is extremely sensitive to a variable change. The following definition introduces a particular term for the exceptional cases where this is not true.

DEFINITION 5.1   *(Symmetry of an ode) Equation (5.1) is said to be invariant under the transformation*

$$x = \phi(u, v), \quad y = \psi(u, v) \text{ with } v \equiv v(u) \tag{5.7}$$

196

if the functional dependence on $u, v, v', \ldots, v^{(n)}$ of the transformed equation is the same as that of the original equation (5.1) on $x, y, y', \ldots, y^{(n)}$, i.e., if it is a differential invariant for (5.7). Such a transformation is called a symmetry of the differential equation.

The general solution of equation (5.1) is a family of curves

$$\phi(x, y, C_1, \ldots, C_n) = 0 \qquad (5.8)$$

in the $x-y$-plane depending on n constant parameters $C_1, \ldots, C_n$. The transformation (5.7) acts on the curves (5.8) just as on the differential equation. If it is a symmetry, the functional dependence of the transformed curves on u and v is the same as the dependence on x and y in (5.8). This is not necessarily true for the parameters because they do not occur in the differential equation itself. This means, the entirety of curves described by the two equations is the same; to any fixed values of the constants, however, there may correspond a different curve in either set. In other words, the solution curves are *permuted among themselves* by a symmetry transformation.

EXAMPLE 5.1   Consider the ode $2yy'' - y'^2 = 0$ with the general solution $y = (C_1 x + C_2)^2$. Its symmetry group is $x = \dfrac{a_1 u}{1 - a_2 u} + a_3$, $y = \dfrac{a_1 v}{(1 - a_2 u)^2}$ where a_1, a_2 and a_3 are the group parameters. In the transformed variables the solution is $v = (\bar{C}_1 u + \bar{C}_2)^2$ with

$$\bar{C}_1 = \frac{1}{\sqrt{a_1}}[C_1(a_1 - a_2 a_3) - C_2 a_2], \quad \bar{C}_2 = \frac{a_3 + 1}{\sqrt{a_1}} C_2. \qquad \square$$

From the above Definition 5.1 of a symmetry of an ode and the geometric considerations on its solution curves it is obvious that all symmetry transformations of a given ode form a group. This important concept is introduced next.

DEFINITION 5.2   *(Symmetry group, symmetry type) The totality of symmetry transformations of an ode forms a continuous group; it is called the symmetry group of that equation. The type of this group is called its symmetry type.*

At this point the connection to Chapter 3 of this book becomes obvious. All that has been said about groups of transformations of a two-dimensional manifold applies to the symmetry groups of ode's. In particular this applies to the classification of these groups and the resulting constraints for the possible symmetries of a differential equation. In order to establish a systematic theory of symmetries of ode's, it must be known how they behave under variable transformations. The answer is given in the following theorem.

THEOREM 5.1   *(Lie 1891) Equivalent equations have the same symmetry type. The reverse is not true. In general to any symmetry type, there corresponds the union of one or more equivalence classes.*

Symmetries of Differential Equations

PROOF   Let $\omega(x,y,y',\ldots,y^{(n)}) = 0$ be a differential equation in x and $y \equiv y(x)$, and $\bar{\omega}(\bar{x},\bar{y},\bar{y}',\ldots,\bar{y}^{(n)}) = 0$ the transformed equation under the variable change $\Phi : x = \alpha(\bar{x},\bar{y}), y = \beta(\bar{x},\bar{y})$ with $\bar{y} \equiv \bar{y}(\bar{x})$ and inverse Φ^{-1}. Furthermore, let $g : x = \phi(u,v), y = \psi(u,v)$ with $v \equiv v(u)$ be a symmetry transformation of ω with the result $\omega(u,v,v',\ldots,v^{(n)}) = 0$. Then $\bar{g} \equiv \Phi^{-1}g\Phi$ transforms $\bar{\omega}(\bar{x},\bar{y},\bar{y}',\ldots,\bar{y}^{(n)}) = 0$ into $\bar{\omega}(\bar{u},\bar{v},\bar{v}',\ldots,\bar{v}^{(n)}) = 0$, i.e., it defines a symmetry of $\bar{\omega}$ of the same type as g.

On the other hand, by Theorems 6.4 and 6.16, the equations $y''+y'^4+1 = 0$ and $y''' + y''y' + 1$ both belong to symmetry classes $\mathcal{S}_{2,1}^2$ and $\mathcal{S}_{2,1}^3$ respectively with the same symmetry type, yet due to its different order they are obviously not equivalent to each other. $\Box$

In order to utilize the symmetries of an ode for the solution procedure, the geometrical considerations above must be expressed in analytical terms. Of particular importance is the fact that this may be achieved in terms of the infinitesimal generators of a symmetry, the expressions for its finite transformations are not required. Due to its importance for the rest of this book the prolongation of a vector field in x and $y(x)$ is introduced next. It is a specialization of Definition 3.4 in Chapter 3.

DEFINITION 5.3   (Prolongation in the x-y-plane) Let an infinitesimal generator $U = \xi(x,y)\partial_x + \eta(x,y)\partial_y$ be given and $y \equiv y(x)$ depend on x. Its nth prolongation is

$$U^{(n)} = U + \zeta^{(1)}\frac{\partial}{\partial y'} + \zeta^{(2)}\frac{\partial}{\partial y''} + \ldots + \zeta^{(n)}\frac{\partial}{\partial y^{(n)}}. \tag{5.9}$$

The functions $\zeta^{(k)}$ are recursively defined by

$$\zeta^{(1)} = D(\eta) - y'D(\xi), \quad \zeta^{(k)} = D(\zeta^{(k-1)}) - y^{(k)}D(\xi) \tag{5.10}$$

for $k \geq 2$, $D = \partial_x + y'\partial_y + y''\partial_{y'} \ldots$ is the operator of total differentiation with respect to x.

From (5.9) it is obvious that the prolonged generators operate on the space of derivatives up to order n. A few remarkable properties of the functions $\zeta^{(k)}$ are formulated subsequently as a lemma. They follow immediately from its definition and are quite useful for practical calculations.

LEMMA 5.1   The functions $\zeta^{(k)}$ defined by (5.10) have the following properties.

1. They are linear and homogeneous in $\xi(x,y)$, $\eta(x,y)$ and its derivatives up to order k.

2. Its coefficients depend explicitly on $y', y'', \ldots, y^{(k)}$; for $k > 1$, $y^{(k)}$ occurs linearly and y' with power $k+1$. In particular they do not depend explicitly on x or y.

From this lemma it is obvious that the amount of computations grows enormously with the order of the differential equation due to the increase in the number of terms in $\zeta^{(k)}$. A rough estimate for this number is 2^k. For $k = 1, \ldots, 10$ the exact figure is given in the subsequent table.

k	1	2	3	4	5	6	7	8	9	10
Number of terms	4	9	17	29	47	73	110	162	234	332

The three lowest ζ's are

$$\zeta^{(1)} = \eta_x + (\eta_y - \xi_x)y' - \xi_y y'^2,$$

$$\zeta^{(2)} = \eta_{xx} + (2\eta_{xy} - \xi_{xx})y' + (\eta_{yy} - 2\xi_{xy})y'^2 - \xi_{yy}y'^3$$
$$+ (\eta_y - 2\xi_x)y'' - 3\xi_y y'y'',$$

$$\zeta^{(3)} = \eta_{xxx} + (3\eta_{xxy} - \xi_{xxx})y' + 3(\eta_{xyy} - \xi_{xxy})y'^2 + (\eta_{yyy} - 3\xi_{xyy})y'^3$$
$$- \xi_{yyy}y'^4 + 3(\eta_{xy} - \xi_{xx})y'' + 3(\eta_{yy} - 3\xi_{xy})y'y'' - 6\xi_{yy}y'^2y''$$
$$+ (\eta_y - 3\xi_x)y''' - 4\xi_y y'y''' - 3\xi_y y''^2.$$

$$(5.11)$$

It is convenient to have the prolongations for some simple vector fields explicitly available. For linear coefficients $\xi = ax$, $\eta = by$ or $\xi = ay$, $\eta = bx$, a and b constant, the third prolongations are

$$U^{(3)} = ax\partial_x + by\partial_y + (b - a)y'\partial_{y'} + (b - 2a)y''\partial_{y''} + (b - 3a)y'''\partial_{y'''}$$

$$U^{(3)} = ay\partial_x + bx\partial_y + (b - ay'^2)\partial_{y'} - 3ay'y''\partial_{y''} - a(4y'y''' + 3y''^2)\partial_{y'''}.$$

For $\xi = x^2$, $\eta = y^2$ it is

$$U^{(3)} = x^2\partial_x + y^2\partial_y + 2(y - x)y'\partial_{y'} + 2[y'(y' - 1) + (y - 2x)y'']\partial_{y''}$$
$$+ 2[3y''(y' - 1) + (y - 3x)y''']\partial_{y'''}.$$

For later use a relation between applying the prolongation operator and taking the total derivative is given next.

LEMMA 5.2   *The prolongation (5.9) of $U = \xi\partial_x + \eta\partial_y$ and the operator of total differentiation D obey $U^{(n+1)}D(\phi) = D(U^{(n)}\phi) - D(\xi)D(\phi)$; ϕ is an undetermined function of x, $y(x)$ and its derivatives up to any order.*

The proof is based on expanding the left and the right hand side of this relation; it is left as Exercise 5.1.

The next result which is due to Lie [113], Kapitel 16, is fundamental for the further proceeding because it is the basis of a constructive method for determining the symmetries of large classes of ode's. It is an immediate consequence of Definition 5.2.

Symmetries of Differential Equations

THEOREM 5.2 *(Lie 1891) A differential equation $w(x, y, y', \ldots, y^{(n)}) = 0$ with $n \geq 1$ is invariant under a transformation with infinitesimal generator $U = \xi(x, y)\partial_x + \eta(x, y)\partial_y$ if and only if*

$$U^{(n)}w(x, y, y', \ldots, y^{(n)}) = 0 \quad mod \quad (w = 0) \tag{5.12}$$

in the space of variables $x, y, y', \ldots, y^{(n)}$.

There is a remarkable constraint for the number of possible symmetries of an ode if the order is not less than two which is also due to Lie. In addition to the theoretical interest in this result it has the important consequence that the symmetry classification becomes a finite problem.

THEOREM 5.3 *(Lie 1893) The maximal number of symmetry generators of an ode of order $n \geq 2$ is finite. For $n = 2$ it is 8, for $n \geq 3$ it is $n + 4$.*

The proof which is based some geometric considerations may be found in Lie [114], Kapitel 12, Satz 3.

For reasons that will become clear shortly the further discussion is limited to $n \geq 2$. The case $n = 1$ will be considered in Section 5.2. In order to make the symmetry problem for ode's manageable the functional dependence of w on its arguments is restricted as follows. If not stated otherwise it is assumed to be quasilinear, i.e., linear in the highest derivative, and rational in its remaining arguments. Consequently, equations of the form

$$y^{(n)} + r(x, y, y', \ldots, y^{(n-1)}) = 0 \tag{5.13}$$

are considered with $r \in \mathbb{Q}(x, y, y', \ldots, y^{(n-1)})$. This class covers a wide range of interesting equations as it may be seen from the respective chapters of the collection by Kamke [85]. The quasilinearity and the rationality in the derivatives have another important property, they are *invariant under general, i.e., not necessarily rational point transformations* as it may be seen from eqs. (5.3) to (5.4).

Based on Theorem 5.2 a procedure for determining the infinitesimal symmetry generators of an ode (5.13) may be designed as follows. Due to the assumptions on the form of w, the expression at the left hand side of (5.12) is rational in the derivatives with coefficients depending on x and y, the unknown functions $\xi(x, y)$ and $\eta(x, y)$ and derivatives thereof. Because the dependence on the derivatives $y^{(k)}$ is completely explicit, it can only vanish if each coefficient of the monomials in the derivatives occuring in the numerator vanishes. As a consequence of Lemma 5.1, the resulting constraints form a system of linear homogeneous partial differential equations for ξ and η in which derivatives $y^{(k)}$ do not occur. It is called the *determining system*. Its general solution defines the symmetries of the original differential equation. The algorithms described in Chapter 2 are applied in the following algorithm for determining large classes of symmetry generators.

200

ALGORITHM 5.1   *Symmetries(ω)*. Given a quasilinear ode ω of order
$n \geq 2$, its infinitesimal symmetry generators with Liouvillian coefficients are
returned.

S1 : Set up determining system. Reduce the n–th prolongation of ω
w.r.t. ω, collect the coefficients of the monomials in the derivatives and
equate them to zero.

S2 : Janet basis. By means of the algorithm given in Chapter 2 transform
the system obtained in *S1* into a Janet basis.

S3 : Determine coefficients. Decompose the Janet basis obtained in *S2*
applying the algorithms of Chapter 2 and determine the Liouvillian solutions from it. If there is no nontrivial solution, return *failed*.

S4 : Return vector fields. From the solutions obtained in *S3* construct
the infinitesimal generators and return them.

This algorithm will be applied frequently in later parts of this chapter, both
for determining the symmetry generators explicitly, or the symmetry class
to which a given equation belongs. It should be emphasized, however, that
symmetry generators with non Liouvillian coefficients in general cannot be
obtained from it.

EXAMPLE 5.2   Equation 6.159 of Kamke's collection $4y''y - 3y'^2 - 12y^3 = 0$
generates the determining system

$$\xi_{yy} + \frac{3}{4y}\xi_y = 0, \quad \eta_{xx} + 3y^2\eta_y - 6y^2\xi_x - 6y\eta = 0,$$

$$\eta_{xy} - \frac{1}{2}\xi_{xx} - \frac{3}{4y}\eta_x - \frac{9}{2}y^2\xi_y = 0, \quad \eta_{yy} - 2\xi_{xy} - \frac{3}{4y}\eta_y + \frac{3}{4y^2}\eta = 0.$$

This is the result of step *S1* of the above algorithm. The Janet basis

$$\xi_x + \frac{1}{2y}\eta = 0, \quad \xi_y = 0, \quad \eta_x = 0, \quad \eta_y - \frac{1}{y}\eta = 0$$

is obtained in step *S2* from which the solution $\xi = -\frac{1}{2}C_1x + C_2$, $\eta = C_1y$ in
step *S3* follows. Finally in step *S4* the two symmetry generators $U_1 = \partial_x$ and
$U_2 = x\partial_x - 2y\partial_y$ are obtained. Its commutator $[U_1, U_2] = U_1$ shows that they
are canonical generators of the group $\mathbf{g}_{25}$.   □

Classification of Differential Invariants. In order to obtain a classification of possible symmetry types of ode's, the proper differential invariants
of all finite transformation groups of a two-dimensional manifold have to be
determined. The starting point for this classification is the listing of transformation groups as described in Chapter 3. Coordinates are x and y, y depends
on x. For an r-parameter group with infinitesimal generators $\xi_i p + \eta_i q$ for
$i = 1, \ldots, r$, the differential invariants of order m are solutions of the linear

Symmetries of Differential Equations

homogeneous system

$$
\begin{aligned}
\xi_1 \frac{\partial \Phi}{\partial x} + \eta_1 \frac{\partial \Phi}{\partial y} + \zeta_1^{(1)} \frac{\partial \Phi}{\partial y'} + \dots + \zeta_1^{(m)} \frac{\partial \Phi}{\partial y^{(m)}} &= 0, \\
\xi_2 \frac{\partial \Phi}{\partial x} + \eta_2 \frac{\partial \Phi}{\partial y} + \zeta_2^{(1)} \frac{\partial \Phi}{\partial y'} + \dots + \zeta_2^{(m)} \frac{\partial \Phi}{\partial y^{(m)}} &= 0, \\
&\vdots \qquad\qquad \vdots \\
\xi_r \frac{\partial \Phi}{\partial x} + \eta_r \frac{\partial \Phi}{\partial y} + \zeta_r^{(1)} \frac{\partial \Phi}{\partial y'} + \dots + \zeta_r^{(m)} \frac{\partial \Phi}{\partial y^{(m)}} &= 0.
\end{aligned}
\tag{5.14}
$$

In this subsection, all systems of pde's are represented in *grlex* term order with $x > y > y' > y'' > y'''$. The group property guarantees that this is a complete system for Φ with $m + 2 - r$ solutions for $1 \le r \le m + 2$. Due to Lemma 5.1 their dependencies may be chosen such that

$$
\begin{aligned}
\Phi_1 &\equiv \Phi_1(x, y, y', \dots, y^{(r-1)}), \\
\Phi_2 &\equiv \Phi_2(x, y, y', \dots, y^{(r)}), \\
&\vdots \qquad\qquad \vdots \\
\Phi_{m-r+2} &\equiv \Phi_{m-r+2}(x, y, y', \dots, y^{(m)}).
\end{aligned}
$$

The two lowest invariants Φ_1 and Φ_2 are called *fundamental invariants*. They are determined by the methods described in Section 3.2 for solving Jacobian systems. The Jacobian scheme breaks down if the determinant of the coefficient matrix of the first $r - 2$ rows vanishes. Correspondingly the *lower equations* of the group are obtained from the irreducible factors of the *Lie determinant*

$$
\Delta = \begin{vmatrix}
\xi_1 & \eta_1 & \zeta_1^{(1)} & \dots & \zeta_1^{(r-2)} \\
\xi_2 & \eta_2 & \zeta_2^{(1)} & \dots & \zeta_2^{(r-2)} \\
\vdots & \vdots & \vdots & & \vdots \\
\xi_r & \eta_r & \zeta_r^{(1)} & \dots & \zeta_r^{(r-2)}
\end{vmatrix}.
\tag{5.15}
$$

Its irreducible factors may determine additional invariants. According to Lie [109], part I, page 247, the higher invariants may be obtained by differentiation as it is shown next. Lie gave some intuitive geometric arguments in favor of this result; see also the remarks by Krause and Michel [96] on page 254, footnote 4. This important result is proved next.

THEOREM 5.4   *The higher invariants Φ_j for $j \ge 3$ may be obtained by differentiation* $\Phi_{j+1} = \dfrac{d\Phi_j}{d\Phi_1}.$

PROOF   The above relation may be written as $\Phi_{j+1} = \dfrac{D\Phi_j}{D\Phi_1}$ because by definition there holds $\dfrac{d\Phi_j}{dx} = D\Phi_j$ and $\dfrac{d\Phi_1}{dx} = D\Phi_1$. Let Φ_j be an invariant of order n. If Φ_{j+1} is an invariant of order $n+1$, it has to obey $U^{(n+1)}\Phi_{j+1} = 0$.

202

The numerator of the left hand side is

$$D(\Phi_1)U^{(n+1)}D(\Phi_j) - D(\Phi_j)U^{(n+1)}D(\Phi_1) =$$

$$D(\Phi_1)[D(U^{(n)}\Phi_j) - D(\xi)D(\Phi_j)] - D(\Phi_j)[D(U^{(n)}\Phi_1) + D(\xi)D(\phi_1)] = 0.$$

In the last line Lemma 5.2 has been applied. Due to the assumption that Φ_1 and Φ_j are invariants there holds $U^{(n)}\Phi_1 = 0$ and $U^{(n)}\Phi_j = 0$. The remaining terms cancel each other. $\Box$

EXAMPLE 5.3   Extending the three generators of the group $\mathbf{g}_{13}$ defined on page 139 up to third order yields the following system of pde's for any invariant Φ.

$$\Phi_x = 0, \quad 2x\Phi_x + y\Phi_y - y'\Phi_{y'} - 3y''\Phi_{y''} - 5y'''\Phi_{y'''} = 0,$$

$$x^2\Phi_x + xy\Phi_y - (y'x - y)\Phi_{y'} - 3y''x\Phi_{y''} - (5y'''x + 3y'')\Phi_{y'''} = 0$$

with $\Delta = y^2$. Applying the algorithm *JanetBasis* of the previous chapter generates the Jacobian form

$$X_1 \equiv \Phi_x = 0, \quad X_2 \equiv \Phi_{y'} - \frac{3y''}{y}\Phi_{y'''} = 0,$$

$$X_3 \equiv \Phi_y - \frac{3y''}{y}\Phi_{y''} - \frac{5y'''y + 3y''y'}{y^2}\Phi_{y'''} = 0.$$

The first equation implies that the invariants do not depend explicitly on x. In the last equation, the third derivative y''' does not occur in the coefficients of Φ_y and $\Phi_{y''}$. Therefore a solution independent of y''' may be obtained from $3\frac{dy}{y} + \frac{dy''}{y''} = 0$ with the result $\Phi_1 = y''y^3$. Because also $X_1\Phi_1 = 0$ and $X_2\Phi_1 = 0$, Φ_1 is the first invariant searched for.

For the second equation a solution ϕ is obtained from $3\frac{dy'}{y} + \frac{dy'''}{y''} = 0$ in the form $\phi = y'''y + 3y''y'$. Because $X_3\phi = -4\frac{\phi}{y}$, the extension of ϕ to a solution of the full system must have the form $\Phi(\phi, y)$ and satisfies the equation $\frac{\partial\Phi}{\partial\phi}\frac{4\phi}{y} + \frac{\partial\Phi}{\partial y} = 0$ or equivalently $\frac{d\phi}{\phi} + 4\frac{dy}{y} = 0$, i.e., $\phi y^4 = C$. Therefore $\Phi = \phi y^4$ is a solution of the complete system, the second invariant is $\Phi_2 = y'''y^5 + 3y''y'y^4$. $\Box$

EXAMPLE 5.4   Consider the group $\mathbf{g}_{25}$ with generators ∂_y and $x\partial_x + y\partial_y$. The pde's for invariants up to order n are $x\Phi_x - \sum_{k=2}^{n}(k-1)y^{(k)}\Phi_{y^{(k)}} = 0$ and $\Phi_y = 0$, with invariants $\Phi_1 = y'$, $\Phi_k = x^{k-1}y^{(k)}$ for $k = 2, \ldots, n$. Applying Theorem 5.4 yields $\frac{d\Phi_2}{d\Phi_1} = 1 + \frac{\Phi_3}{\Phi_2}$ and $\frac{d\Phi_3}{d\Phi_1} = \frac{2\Phi_3 + \Phi_4}{\Phi_2}$ whereas the above formula leads to $\Phi_2 = xy''$, $\Phi_3 = x^2y'''$ and $\Phi_4 = x^3y^{(4)}$. It shows that the invariants obtained by differentiation may be not in simplest form. $\Box$

Symmetries of Differential Equations

The listing below contains the complete information on the possible groups of point symmetries of ordinary differential equations of arbitrary order. It is due to Lie [109], part I. Some minor corrections have been included that may be found in the *Anmerkungen* by Engel in vol. V of the *Gesammelte Abhandlungen*, pages 676-679. Furthermore, various corrections and improvements described in the dissertation of Heineck [69] have been added. In general the necessary calculations proceed similarly to the above examples.

In the subsequent tabulation the fundamental invariants $\Phi_1^{(k)}$ and $\Phi_2^{(k)}$, and the Lie determinants $\Delta^{(k)}$ for the groups $\mathbf{g}_k$, $k = 1, \ldots, 27$, are explicitly given. Different from Lie's listing, all invariants are *rational*. This faciliates the explicit calculations that will be performed later on. In various cases the Schwarzian derivative of y is denoted by the abbreviation

$$w = \frac{y'''}{y'} - \frac{3}{2}\left(\frac{y''}{y'}\right)^2. \tag{5.16}$$

Group $\mathbf{g}_1$: $\Phi_1^{(1)} = \dfrac{[3y''y^{(4)} - 5y'''^2]^3}{y''^8}$, $\quad \Phi_2^{(1)} = \dfrac{9y''^2y^{(5)} - 45y''y'''y^{(4)} + 40y'''^3}{3y''^4}$,

$\Delta^{(1)} = 9y''^3$.

Group $\mathbf{g}_2$: $\quad \Phi_1^{(2)} = \dfrac{[9y''^2y^{(5)} - 45y''y'''y^{(4)} + 40y'''^3]^2}{3(3y''y^{(4)} - 5y'''^2)^3}$,

$\Phi_2^{(2)} = \dfrac{9y''^3y^{(6)} - 63y''^2y'''y^{(5)} + 105y''y'''^2y^{(4)} - 35y'''^4}{3(3y''y^{(4)} - 5y'''^2)^2}$,

$\Delta^{(2)} = 2y''^2(5y'''^2 - 3y''y^{(4)})$.

Group $\mathbf{g}_3$: $\Phi_1^{(3)} = \dfrac{u^3}{\rho_5^8}$, $\quad \Phi_2^{(3)} = \dfrac{v}{\rho_5^3}$ where

$$\rho_4 = 3y''y^{(4)} - 4y'''^2, \quad \rho_5 = 3y''^2y^{(5)} - 15y''y'''y^{(4)} + \frac{40}{3}y'''^3,$$

$$\rho_6 = 3y''^2y^{(6)} - 24y''^2y'''y^{(5)} + 60y''y'''^2y^{(4)} - 40y'''^4,$$

$$\rho_7 = 9y''^4y^{(7)} - 105y''^3y'''y^{(6)} + 420y''^2y'''^2y^{(5)} - 7000y''y'''^3y^{(4)} + \frac{1120}{3}y'''^5,$$

$$\rho_8 = 27y''^5y^{(8)} - 48y'''\rho_7 - 840y''^2\rho_6 - 2240y'''^3\rho_5 - 2800y'''^4\rho_4 - \frac{2240}{3}y'''^6,$$

$$u = 2\rho_5\rho_7 - 35\rho_4\rho_5^2 - 7(\rho_6 - \tfrac{5}{3}\rho_4^2)^2,$$

$$v = \rho_5(\rho_8 - 84\rho_4\rho_6 + \tfrac{245}{3}\rho_4^3) - 12(\rho_7 - \tfrac{35}{2}\rho_4\rho_5)(\rho_6 - \tfrac{5}{3}\rho_4^2) + \frac{28}{\rho_5}(\rho_6 - \tfrac{5}{3}\rho_4^2)^3,$$

$$\Delta^{(3)} = -2y''(9y''^2y^{(5)} - 45y''y'''y^{(4)} + 40y'''^3)^2.$$

Group $\mathbf{g}_4$: $\quad \Phi_1^{(4)} = x$, $\Phi_2^{(4)} = \dfrac{y''}{y'}$, $\Phi_3^{(4)} = \dfrac{y'''}{y'}$, $\Delta^{(4)} = 0$.

Group $\mathbf{g}_5$: $\quad \Phi_1^{(5)} = \dfrac{y''}{y'}$, $\Phi_2^{(5)} = \dfrac{y'''}{y'}$, $\Delta^{(5)} = y'$.

Group $\mathbf{g}_6$: $\quad \Phi_1^{(6)} = \dfrac{y'y'''}{y''^2}$, $\Phi_2^{(6)} = \dfrac{y'^2y^{(4)}}{y''^3}$, $\Delta^{(6)} = -y'y''$.

Group $\mathbf{g}_7$: $\quad \Phi_1^{(7)} = y''y'^{(2-c)(c-1)}, \quad \Phi_2^{(7)} = y'''y'^{(3-c)(c-1)},$
$$\Delta^{(7)} = (c-1)y', \quad c \neq 1.$$

Group $\mathbf{g}_8$: $\quad \Phi_1^{(8)} = x, \quad \Phi_2^{(8)} = w, \quad \Delta^{(8)} = 0.$

Group $\mathbf{g}_9$: $\Phi_1^{(9)} = w, \quad \Phi_2^{(9)} = w', \quad \Delta^{(9)} = 2y'^3.$

Group $\mathbf{g}_{10}$: $\Phi_1^{(10)} = \dfrac{[(x-y)y'' + 2y'(y'+1)]^2}{y'^3},$

$$\Phi_2^{(10)} = \frac{(x-y)^2 y''' + 6(x-y)(y'+1)y''}{y'^2} + \frac{6(y'^2 + 4y' + 1)}{y'},$$

$$\Delta^{(10)} = 2(y-x)^2 y'.$$

Group $\mathbf{g}_{11}$: $\Phi_1^{(11)} = \dfrac{w'^2}{w^3}, \quad \Phi_2^{(11)} = \dfrac{w''}{w^2}, \quad \Delta^{(11)} = 4y'^2(y'y''' - \frac{3}{2}y''^2).$

Group $\mathbf{g}_{12}$: $\Phi_1^{(12)} = \dfrac{4ww'' - 5w'^2}{w^3}, \quad \Phi_2^{(12)} = \dfrac{(4w^2 w''' - 18ww'w'' + 15w'^3)^2}{w^9},$
$$\Delta^{(12)} = 4y'^2(y'y''' - \tfrac{3}{2}y''^2).$$

Group $\mathbf{g}_{13}$: $\quad \Phi_1^{(13)} = y''y^3, \quad \Phi_2^{(13)} = y'''y^5 + 3y''y'y^4, \quad \Delta^{(13)} = 2y^2.$

Group $\mathbf{g}_{14}$: $\Phi_1^{(14)} = \dfrac{(yy''' + 3y'y'')^2}{yy''^3}, \quad \Phi_2^{(14)} = \dfrac{3yy''y^{(4)} - 4yy'''^2}{y''^3},$
$$\Delta^{(14)} = 2y^2 y''.$$

Group $\mathbf{g}_{15}$: Let the determinants D and D_i, for $i = 1, 2, \ldots$ be defined by

$$D = \begin{vmatrix} \phi_1 & \phi_1' & \cdots & \phi_1^{(r)} \\ \vdots & \vdots & & \vdots \\ \phi_r & \phi_r' & \cdots & \phi_r^{(r)} \\ y & y' & \cdots & y^{(r)} \end{vmatrix}, \quad D_i = \begin{vmatrix} \phi_1 & \phi_1' & \cdots & \phi_1^{(r-1)} & \phi_1^{(r+i)} \\ \vdots & \vdots & \vdots & & \vdots \\ \phi_r & \phi_r' & \cdots & \phi_r^{(r-1)} & \phi_r^{(r+i)} \\ y & y' & \cdots & y^{(r-1)} & y^{(r+i)} \end{vmatrix}.$$

Then $\Phi_1^{(15)} = x, \quad \Phi_2^{(15)} = D, \quad \Phi_3^{(15)} = D_1, \quad \Delta^{(15)} = 0.$

Group $\mathbf{g}_{16}$: $\Phi_1^{(16)} = x, \quad \Phi_2^{(16)} = (\log D)' = \dfrac{D_1}{D}$ with D as above, $\Delta^{(16)} = 0.$

Group $\mathbf{g}_{17}$: $\Phi_1^{(17)} = c_0 y + c_1 y' + \ldots + c_r y^{(r)}, \quad \Phi_2^{(17)} = \Phi_1^{(17)'},$

$$\Delta^{(17)} = \begin{vmatrix} \phi_1 & \phi_1' & \cdots & \phi_1^{(r-1)} \\ \vdots & \vdots & \vdots \\ \phi_r & \phi_r' & \cdots & \phi_r^{(r-1)} \end{vmatrix}.$$

Group $\mathbf{g}_{18}$: In terms of the previously defined determinants D, D_1 and D_2 they may be written as

$$\Phi_1^{(18)} = \frac{D_1}{D}, \quad \Phi_2^{(18)} = \frac{D_2}{D}$$

Symmetries of Differential Equations

205

if the proper values for the ϕ_k are substituted. They are of order $l + \sum \rho_k + 1$ and $l + \sum \rho_k + 2$ respectively, $\Delta^{(18)} = D$.

<u>Group $\mathbf{g}_{19}$:</u> For c arbitrary: $\Delta^{(19)} = (c - r)y^{(r)}$,

for $c \neq r$: $\Phi_1^{(19)} = \dfrac{y^{(r+1)\,c-r}}{y^{(r)\,c-r-1}}$, $\quad \Phi_2^{(19)} = \dfrac{y^{(r+2)\,c-r}}{y^{(r)\,c-r-2}}$,

for $c = r$: $\Phi_1^{(19)} = y^{(r)}$, $\quad \Phi_2^{(19)} = \dfrac{y^{(r+2)}}{y^{(r+1)\,2}}$.

<u>Group $\mathbf{g}_{20}$:</u> $\Phi_1^{(20)} = y^{(r+1)} e^{y^{(r)}/r!}$, $\quad \Phi_2^{(20)} = y^{(r+2)} e^{2y^{(r)}/r!}$, $\quad \Delta^{(20)} = \prod_{k=1}^{r} k!$.

<u>Group $\mathbf{g}_{21}$:</u> $\Phi_1^{(21)} = \dfrac{y^{(r)} y^{(r+2)}}{y^{(r+1)\,2}}$, $\quad \Phi_2^{(21)} = \dfrac{y^{(r)\,2} y^{(r+3)}}{y^{(r+1)\,3}}$, $\quad \Delta^{(21)} = y^{(r)} y^{(r+1)}$.

<u>Group $\mathbf{g}_{22}$:</u> $\Phi_1^{(22)} = \dfrac{v_1^{r+1}}{y^{(r)\,2(r+3)}}$, $\quad \Phi_2^{(22)} = \dfrac{v_2^{r+1}}{y^{(r)\,3(r+3)}}$, $\quad \Delta^{(22)} = y^{(r)\,2}$

where

$$v_1 = (r+1)y^{(r)} y^{(r+2)} - (r+2)y^{(r+1)\,2},$$

$$v_2 = (r+1)^2 y^{(r)\,2} y^{(r+3)} - 3(r+1)(r+3)y^{(r)} y^{(r+1)} y^{(r+2)}$$
$$+ 2(r+2)(r+3)y^{(r+1)\,3}.$$

<u>Group $\mathbf{g}_{23}$:</u> $\Phi_1^{(23)} = \dfrac{v_2^{2}}{v_1^{3}}$, $\quad \Phi_2^{(23)} = \dfrac{v_3}{v_1^{2}}$

where v_1 and v_2 are the same as for the preceding group and

$$v_3 = (r+1)^3 y^{(r)\,3} y^{(r+4)} - 4(r+1)^2(r+4)y^{(r)\,2} y^{(r+1)} y^{(r+3)}$$
$$+ 6(r+1)(r+3)(r+4)y^{(r)} y^{(r+1)\,2} y^{(r+2)} - 3(r+2)(r+3)(r+4)y^{(r+1)\,4},$$

$$\Delta^{(23)} = y^{(r)} \left[(r+2)y^{(r+1)\,2} - (r+1)y^{(r)} y^{(r+2)} \right].$$

<u>Group $\mathbf{g}_{24}$:</u> $\Phi_1^{(24)} = y'$, $\quad \Phi_2^{(24)} = \dfrac{y'''}{y''^{2}}$, $\quad \Delta^{(24)} = 0$.

<u>Group $\mathbf{g}_{25}$:</u> $\Phi_1^{(25)} = y'$, $\quad \Phi_2^{(25)} = xy''$, $\quad \Phi_3^{(25)} = x^2 y'''$, $\quad \Delta^{(25)} = y$.

<u>Group $\mathbf{g}_{26}$:</u> $\Phi_k^{(26)} = y^{(k)}$ for $k \geq 1$, $\Delta^{(26)} = 1$.

<u>Group $\mathbf{g}_{27}$:</u> $\Phi_1^{(27)} = x$, $\Phi_k^{(27)} = y^{(k)}$ for $k \geq 1$, $\Delta^{(27)}$ not defined.

There is an important distinction on whether an invariant equation originates from the invariant of a group or from a lower equation. In the latter case, the left hand side of an invariant equation in normal form is *identical* to the factor of the lower equation as given in the above listing. Invariant equations originating from an invariant may be *any function* of them.

5.2 Symmetries of First Order Equations

Although first order quasilinear equations $\omega \equiv y' + r(x, y) = 0$ are simpler than their higher order counterparts, applying its symmetries for solving them is limited by the fact that there is no algorithmic procedure for obtaining the symmetry generators explicitly. This is obvious from the expression at the left hand side of (5.12) for $n = 1$; it is a single linear homogeneous pde of first order for the two unknown functions $\xi(x, y)$ and $\eta(x, y)$. With ω as given above it is

$$\xi r_x + \eta r_y + \eta_x - (\eta_y - \xi_x)r - \xi_y r^2 = 0. \tag{5.17}$$

There is no solution algorithm available for this pde. What makes things even worse, there are a lot of trivial solutions that cannot be utilized for the solution procedure as it will turn out later on. They yield the *trivial symmetries* $\xi = \phi(x, y)$ and $\eta = -r\phi(x, y)$ with $\phi(x, y)$ an undetermined function of its arguments. The problem is to find the nontrivial solutions that may be applied for the solution procedure.

Classification of Symmetries. According to Theorem 4.10 the quasilinear first oder equations form a single equivalence class for which $v' = 0$ is a canonical form. Consequently, there is only a single symmetry class, i.e. all equations of this kind share the same symmetries up to similarity. They are described in the subsequent theorem.

THEOREM 5.5   *Any symmetry generator of a first order quasilinear ode $v'=0$ in canonical variables u and $v \equiv v(u)$ is an operator $\alpha(u, v)\partial_u + \beta(v)\partial_v$ of the infinite-dimensional Lie algebra corresponding to the transformation group $\mathcal{S}^1 \equiv \mathbf{G}$ with Janet basis $\{\beta_u\}$; α and β undetermined functions of its arguments.*

PROOF   The determining system for the coefficients $\alpha(u, v)$ and $\beta(u, v)$ in canonical variables is immediately obtained from (5.12) in the form $\beta_u = 0$ with the solution $\alpha(u, v)$ and $\beta(v)$. $\qquad \Box$

This theorem describes the symmetries of first order ode's in full generality. Any finite group of invariance transformations of such an equation must be up to similarity contained in the pseudogroup given there. From the listing of groups and invariants in the preceding section it follows that the group $\mathbf{g}_{12}$ with generators $\partial_u, u\partial_u, u^2\partial_u, \partial_v, v\partial_v, v^2\partial_v$ leaves $v' = 0$ invariant.

An arbitrary first order equation $y' + r(x, y) = 0$ not in canonical variables allows the same invariance transformations; its symmetry generators are similar to those just mentioned.

COROLLARY 5.1   *Any symmetry generator of a first order quasilinear ode $y' + r(x, y) = 0$ obtained from $v' = 0$ by the transformation $u = \sigma(x, y)$, $v = \rho(x, y)$ with the inverse $x = \phi(u, v)$, $y = \psi(u, v)$ has the form*

$$[\alpha(u, v)(\phi_u\partial_x + \psi_u\partial_y) + \beta(v)(\phi_v\partial_x + \psi_v\partial_y)]_{u=\sigma(x,y),v=\rho(x,y)}$$

with $\alpha(u, v)$ and $\beta(v)$ undetermined functions of its arguments.

Symmetries of Differential Equations 207

This is a straightforward consequence of Theorem 5.5 and the transformation formulas of Lemma 3.3. The above corollary may be applied to obtain finite groups of invariance transformations for first order ode's as the following example shows.

EXAMPLE 5.5　For the Riccati equation of Example 4.10 the transformation functions to canonical variables were $\sigma = \frac{1}{x}$ and $\rho = \frac{x(xy+1)}{xy+2}$ with the inverse $\phi = \frac{1}{u}$ and $\psi = \frac{u(1-2uv)}{uv-1}$. From the expression in the corollary, any symmetry generator of the Riccati equation has form

$$\alpha\left(\frac{1}{x}, \frac{x(xy+1)}{xy+2}\right)\left[-x^2\partial_x + (x^2y^2 + 4xy + 2)\partial_y\right] + \beta\left(\frac{x(xy+1)}{xy+2}\right)\frac{(xy+2)^2}{x^2}\partial_y.$$

The generators of the group $\mathbf{g}_{12}$ in noncanonical variables are obtained from this expression upon substitution of the proper values for α and β with the following result.

$$\partial_u : \ -x^2\partial_x + (x^2y^2 + 4xy + 2)\partial_y,$$

$$u\partial_u : \ -x\partial_x + (xy^2 + 4y + \tfrac{2}{x})\partial_y, \quad u^2\partial_u : \ -\partial_x + \left(y^2 + \tfrac{4y}{x} + \tfrac{2}{x}\right)\partial_y,$$

$$\partial_v : \ \left(y + \tfrac{2}{x}\right)^2\partial_y, \quad v\partial_v : \ \tfrac{1}{x}(xy+1)(xy+1)\partial_y, \quad v^2\partial_v : \ (xy+1)^2\partial_y. \quad \square$$

Lie gave another characterization of the most general symmetry generator of a first order ode; it is described next.

THEOREM 5.6　*(Lie 1891) Any symmetry generator of $y' + r(x,y) = 0$ may be written in the form $\rho(x,y)A + \Omega(\omega(x,y))U$ where $A = \partial_x - r(x,y)\partial_y$, $A\omega = 0$, $\rho(x,y)$ and $\Omega(\omega)$ are undetermined functions of its arguments. U is a special symmetry generator.*

The proof of this result may be found in Lie [113], Kapitel 7. Both Lie's representation given in this theorem and the expression in Corollary 5.1 contain an undetermined function depending on two arguments, i.e., α or ρ respectively, and one undetermined function depending on a single argument, i.e., β and Ω. Furthermore, in order to obtain a generator explicitly, either the transformation to canonical form must be known, or the general solution of the equation at issue.

EXAMPLE 5.6　Coming back to the Riccati equation of Example 5.5, with $r(x,y) = y^2 + \frac{4}{x}y + \frac{2}{x^2}$, $\omega(x,y) = \frac{x(xy+1)}{xy+2}$ and $U = \xi(x,y)\partial_x + \eta(x,y)\partial_y$ a given symmetry generator for the Riccati equation at issue, the representation for an arbitrary generator is

$$\rho(x,y)\left[\partial_x - \left(y^2 + \frac{4y}{x} + \frac{2}{x^2}\right)\partial_y\right] + \Omega\left(\frac{x(xy+1)}{xy+2}\right)\left[\xi(x,y)\partial_x + \eta(x,y)\partial_y\right].$$

This result should be compared with the representation given in the preceding Example 5.5　$\square$

208

Classification of Invariant Equations. Due to the fact that invariance transformations for first order equations in general cannot be obtained algorithmically, it is of some interest to consider the inverse problem, i.e., to start from any generator and to determine the corresponding invariant equation. Such a tabulation may be applied to raise the intuition in order to guess a generator for a given first order ode. The subsequent discussion follows Lie [113], Kapitel 8; see also Cohen [32], §18 and §19 and the listing given in Table 2 on pages 231-235, and Emanuel [42], Appendix 5.

The type of calculation required to obtain these tabulations is similar to the computation of invariants in the preceding section. However, now the starting points are certain simple symmetry generators for which there is a good chance of occurrence in a significant number of problems as previous experience has shown, in particular those on which the usual heuristic methods of solving first order equations are based. Its first order invariants determine the type of equation taken into account. A few examples of symmetry generators and corresponding invariant equations are given next.

$$U = \partial_x : \quad y' = F(y), \quad U = \partial_y : \quad y' = F(x),$$

$$U = x\partial_x + y\partial_y : \quad y' = F\left(\frac{y}{x}\right), \quad U = x\partial_x - y\partial_y : \quad y' = \frac{y}{x}F(xy),$$

$$U = x\partial_x + ky\partial_y : \quad y' = \frac{y}{x}F\left(\frac{y}{x^k}\right),$$

$$U = \phi(x)(x\partial_x + ky\partial_y) : \quad y' = \frac{y}{x}\left[k + \frac{1}{\phi(x)}F\left(\frac{y}{x^k}\right)\right].$$

The latter result is obtained from the prolongation

$$U^{(1)} = x\phi\partial_x + ky\phi\partial_y + \left[ky\phi' + (k\phi - \phi - x\phi')y'\right]\partial_{y'}$$

with the two invariants $\frac{y}{x^k}$ and $\frac{xy' - ky}{x^k}\phi$. For F linear or quadratic in y it comprises certain classes of Riccati or Abel equations. They will be discussed in more detail in Chapter 7 on solution algorithms.

Abel's Equation. In this subsection the symmetry properties of Abel's equation are considered. It turns out that case i) and case ii) of Theorem 4.14 have a significantly different behavior that is described first.

THEOREM 5.7   *The two rational normal forms of Theorem 4.14 are considered separately.*

i) Equation $y' + Ay^3 + By = 0$ with $A \neq 0$ allows the two symmetry generators

$$U_1 = \frac{\bar{B}}{A}(\partial_x - By\partial_y), \quad U_2 = \frac{\bar{B}}{A}\int \frac{A}{\bar{B}}dx(\partial_x - By\partial_y) - \frac{1}{2}y\partial_y \quad (5.18)$$

with $\bar{B} = \exp\left(2\int Bdx\right)$ and the commutator $[U_1, U_2] = U_1$.

$$\textit{Symmetries of Differential Equations} \qquad 209$$

ii) Equation $y' + Ay^3 + By + 1 = 0$ with $A \neq 0$ allows the symmetry generator

$$U = \frac{1}{A^{1/3}} \left(\partial_x - \frac{1}{3}\frac{A'}{A}y\partial_y \right) \tag{5.19}$$

if and only if its absolute invariant $K = \frac{1}{A}\left(3B - \frac{A'}{A} \right)^3$ is a constant, i.e., if there holds

$$\frac{dK(A,B)}{dx} = 0. \tag{5.20}$$

PROOF   In general a first simplification is obtained by the following observation. Substituting an infinitesimal generator with coefficients

$$\xi = \xi(x), \qquad \eta = \phi_1(x) + \phi_2(x)y$$

into (5.17) and equating the coefficients of y in the resulting polynomial with zero, the coefficient of y^2 yields the equation $\phi_1 A = 0$, i.e., in general $\phi_1 = 0$. Let the simplified form of η be $\eta = \phi(x)y$. For the determining system of these symmetries the lexicographic term ordering with $\phi > \xi$ is always applied, i.e., any term involving ϕ is considered to be higher than any term involving ξ.

For case *i)* substitution of these values for ξ and η into (5.17) leads to

$$\phi + \frac{1}{2}\xi' + \frac{1}{2}\frac{A'}{A}\xi = 0, \quad (\phi + B\xi)' = 0 \tag{5.21}$$

where $' = \frac{d}{dx}$. Reduction with respect to the first equation yields

$$\left[\xi' + \left(\frac{A'}{A} - 2B \right)\xi \right]' = 0, \quad \phi + \frac{1}{2}\xi' + \frac{1}{2}\frac{A'}{A}\xi = 0.$$

The former equation for ξ has the general solution

$$\xi = \frac{1}{A}\exp\left(2\int Bdx \right)\left[C_1 \int A\exp\left(-2\int Bdx \right) + C_2 \right].$$

The coefficients of the generators (5.18) are obtained by choosing $C_1 = 0$, $C_2 = 1$ or $C_1 = 1$, $C_2 = 0$ respectively, applying the latter equation for determining ϕ and defining $\bar{B} = \exp\left(2\int Bdx \right)$.

For case *ii)* there is the additional equation $\phi - \xi' = 0$. It is used for eliminating ϕ from system (5.21) with the result

$$\xi' + \frac{1}{3}\frac{A'}{A}\xi = 0, \quad (\xi' + B\xi)' = 0, \quad \phi - \xi' = 0. \tag{5.22}$$

Two further reductions with respect to the first equation yield

$$\left[\left(B - \frac{1}{3}\frac{A'}{A} \right)\xi \right]' = 0, \quad \xi' + \frac{1}{3}\frac{A'}{A}\xi = 0, \quad \phi + \frac{1}{3}\frac{A'}{A}\xi = 0. \tag{5.23}$$

210

The latter two equations yield $\xi = \dfrac{C}{A^{1/3}}$ and $\phi = \dfrac{1}{3}\dfrac{C}{A^{1/3}}\dfrac{A'}{A}$, i.e., the generator (5.19), if and only if (5.20) is satisfied. This condition is obtained from the first equation (5.22) upon substituting ξ. □

The basic achievement of this theorem is the fact that certain symmetry generators with Liouvillian coefficients may be determined algorithmically. Condition (5.20) of this theorem is convenient in applications for deciding the existence of such a symmetry. It is interesting to note that it allows an explicit parametrization of all Abel equations with rational normal form of case ii) with a symmetry (5.19) in terms of an unconstrained coefficent A. To this end it may be explicitly solved for B with the result

$$B = \frac{1}{3}\frac{A'}{A} + const. \cdot A^{1/3}.$$

If condition (5.20) is not satisfied the above theorem does not apply. It is possible that different types of symmetries may exist.

EXAMPLE 5.7   Equation 1.46 of Kamke's collection

$$y' - x^2 y^3 + 3y^2 - \frac{1}{x^2}y + \frac{2x-1}{x^4} = 0 \tag{5.24}$$

assumes the rational normal form $z' - x^2 z^3 + \dfrac{2}{x^2}z = 0$ by the transformation $y = z + \dfrac{1}{x^2}$, i.e., a Bernoulli equation belonging to case i). □

EXAMPLE 5.8   Equation 1.41 of Kamke's collection is $y' + axy^3 + by^2 = 0$ with constants a and b, and has rational normal form

$$y' + \frac{(9a + 2b^2)^2 b^2}{729x^3}y^3 - \frac{6a + b^2}{3ax}y + 1 = 0, \tag{5.25}$$

i.e., it belongs to case ii). Its invariant $K = -\dfrac{729}{b^2}\dfrac{(3a + b^2)^3}{(9a + 2b^2)^2}$ entails that there is a symmetry generator

$$U = \frac{81x}{(9a + 2b^2)^{2/3}b^{2/3}}\left(\frac{6a + b^2}{ax}\partial_x + \frac{3y}{x}\partial_y\right).$$
□

5.3   Symmetries of Second Order Equations

In this section the objects of primary interest are not the various groups any more, but the differential equations and the symmetries attached to them. The link between these two concepts is provided by the tabulation of differential invariants given in the Section 5.1. Although it contains the complete

Symmetries of Differential Equations 211

information on all possible symmetries of any ode, the information given there is somehow organized by the wrong key for the applications that are the subject of this book. Here the symmetry problem occurs in the following form: for any given ode, determine its symmetries, and if there are any, utilize them for finding its closed form solutions.

Classification of Symmetries. As a first partial answer to the symmetry problem all possible symmetry types of a quasilinear ode of order two will be determined. This is an important constraint that will turn out to be of considerable help later on for identifying the symmetry class to which a given ode belongs.

An equation which is a function of the invariants of a certain group is *guaranteed* to allow a group of invariance transformations that *comprises* this group; it is possible, however, that its symmetry group is actually *larger*. It turns out that this does occur in a few cases. Taking these observations into account, the subsequent theorem describes the possible point symmetry groups for quasilinear ode's of order two. The explicit form of the ode's allowing the various symmetry groups which also follows from the listing in Section 5.1 is the subject of Chapter 6.

THEOREM 5.8 *(Lie 1883). Any symmetry algebra of a second order quasilinear ode is similar to one in canonical variables u and $v \equiv v(u)$ as given in the following listing. In addition the corresponding Janet basis for the determining system is given where $\alpha(u, v)$ and $\beta(u, v)$ are the coefficients of ∂_u and ∂_v respectively.*

One-parameter group

$\mathcal{S}_1^2 : \mathbf{g}_{27} = \{\partial_v\}$. *Janet basis* $\{\alpha, \beta_u, \beta_v\}$.

Two-parameter groups

$\mathcal{S}_{2,1}^2 : \mathbf{g}_{26} = \{\partial_u, \partial_v\}$. *Janet basis* $\{\alpha_u, \alpha_v, \beta_u, \beta_v\}$.

$\mathcal{S}_{2,2}^2 : \mathbf{g}_{25} = \{\partial_v, u\partial_u + v\partial_v\}$. *Janet basis* $\{\alpha_v, \beta_u, \beta_v - \frac{1}{v}\beta, \alpha_{uu}\}$.

Three-parameter groups

$\mathcal{S}_{3,1}^2 : \mathbf{g}_{10} = \{\partial_u + \partial_v, u\partial_u + v\partial_v, u^2\partial_u + v^2\partial_v\}$. *Janet basis*

$\{\alpha_v, \beta_u, \beta_v + \alpha_u + \frac{2}{u-v}(\beta - \alpha), \alpha_{uu} - \frac{2}{u-v}\alpha_u - \frac{2}{(u-v)^2}(\beta - \alpha)\}$.

$\mathcal{S}_{3,2}^2 : \mathbf{g}_{13} = \{\partial_u, 2u\partial_u + v\partial_v, u^2\partial_u + uv\partial_v\}$. *Janet basis*

$\{\alpha_u - \frac{2}{v}\beta, \alpha_v, \beta_v - \frac{1}{v}\beta, \beta_{uu}\}$.

$\mathcal{S}_{3,3}^2(c) : \mathbf{g}_7(c) = \{\partial_u, \partial_v, u\partial_u + cv\partial_v\}$, $c \neq 1$.

Janet basis $\alpha_v, \beta_u, \beta_v - c\alpha_u, \alpha_{uu}\}$.

$\mathcal{S}_{3,4}^2 : \mathbf{g}_{20}(r = 1) = \{\partial_u, \partial_v, u\partial_u + (u + v)\partial_v\}$.

Janet basis $\{\alpha_v, \beta_u - \alpha_u, \beta_v - \alpha_u, \alpha_{uu}\}$.

Eight-parameter group:

$\mathcal{S}_8^2 : \mathbf{g}_3 = \{\partial_u, \partial_v, u\partial_v, v\partial_v, u\partial_u, v\partial_u, u^2\partial_u + uv\partial_v, uv\partial_u + v^2\partial_v\}$.

Janet basis $\{\alpha_{vv}, \beta_{uu}, \beta_{uv}, \beta_{vv}, \alpha_{uuu}, \beta_{uuv}\}$.

This listing shows in particular that there does not exist any second order ode allowing a group of point symmetries with 4, 5, 6 or 7 parameters.

PROOF   At first invariant differential equations originating from the *invariants* listed in Section 5.1 will be considered. For the groups $\mathbf{g}_k$ with $k = 1$, 2, 3, 6, 8, 9, 11, 12, 14 and 24 an invariant of order two obviously does not exist, a second order ode with these symmetry groups is therefore excluded. For the groups with $k = 16, 18, 21, 22$ or 23 the parameter r is constrained to $r \geq 2$ as may be seen from the listing of Section 3.4. Therefore an invariant of order two is excluded for these groups as well. The same is true for $k = 19$ and $c \neq r$.

For the groups $\mathbf{g}_4$ and $\mathbf{g}_5$ the only invariant of order not higher than two has the form $\dfrac{y''}{y'}$, and there is no invariant of order one or zero, i.e., depending on y or y'. As a consequence, any quasilinear equation that may be constructed from it must be linear. By Lemma 4.7 its symmetry group is the eight-parameter projective group $\mathbf{g}_3$. A similar argument applies to the group $\mathbf{g}_{15}$ for $r = 2$ with the lowest invariant of nonvanishing order

$$D \equiv (\phi_1\phi_2' - \phi_1'\phi_2)y'' - (\phi_1\phi_3' - \phi_1'\phi_3)y' + (\phi_1'\phi_2'' - \phi_1''\phi_2)y$$

and the group $\mathbf{g}_{17}$ where D is a linear and homogeneous in $y, y', \ldots, y^{(r)}$ with constant coefficients.

For the remaining values $k = 7, 10, 13, 20$ with $r = 1$, 25, 26 and 27, equations may be constructed from the given invariants with the respective symmetry groups.

The only *lower equation* of order two is of the form $y'' = 0$ with the eight-parameter projective group $\mathbf{g}_3$ as symmetry group. This is also true for the remaining case of the group $\mathbf{g}_{19}$ with to $c = r$.   □

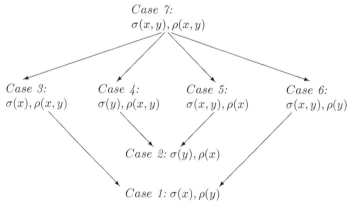

FIGURE 5.1: Arrows indicate a possible specializations. The notation *Case k* → *Case j* means that *Case j* is obtained by specialization from *Case k*.

Determining the Symmetry Class. In Theorem 5.8 the symmetry groups of a second order ode are characterized in terms of canonical variables. In applications, however, they occur in actual variables. In order to apply the

Symmetries of Differential Equations

results of this theorem, the relation between these two descriptions has to be found. In particular this applies to the Janet bases for the determining system because it can be determined algorithmically from the given ode. This is based on the following observation.

Take the Janet basis for any group defined by Theorem 5.8 in canonical variables u and v, and transform it into actual variables by means of (5.2). In general this change of variables will destroy the Janet basis property. In order to reestablish it, the algorithm *JanetBasis* has to be applied. During this process it may occur that a leading coefficient of an equation that is applied for reduction vanishes due to a special choice of the transformation (5.2). This has the consequence that alternatives may occur leading to different types of Janet bases. In order to obtain the complete answer, each of these alternatives has to be investigated separately. Finally all Janet bases have to be combined such that a minimal number of *generic cases* is retained, and all those that may be obtained from them by specialization are discarded. It turns out that up to a few exceptional cases all alternatives are due to vanishing first order partial derivatives of the transformation functions σ and ρ. Taking into account the constraint $\Delta \equiv \sigma_x \rho_y - \sigma_y \rho_x \neq 0$, there are altogether seven possible combinations. There exists a partial order between them described by the diagram shown in Figure 5.1. If there is only a single generic case, it must obviously be *Case 7*, all others may be obtained from it. Due to the size of the expressions involved, the details of the necessary calculations can only be given for the simplest groups.

In order to identify a particular group type of an ode from its Janet basis, the results of Chapter 3 relating various group properties to its determining system are applied. In general this proceeds in three steps.

▷ The first distinguishing property is the number of group parameters that is obvious from the Janet basis type.

▷ The second distinguishing property is the number of imprimitivity systems, it is purely geometric.

▷ Finally the algebraic structure of the corresponding Lie algebra expressed in terms of invariant properties of the coefficients of its characteristic polynomial as described in Theorem 3.7 is applied.

It is part of the problem to assure that a set of criteria is obtained that will lead to a unique answer for the symmetry type of the ode at hand. For second order quasilinear ode's the complete answer may be described as follows.

THEOREM 5.9 *(Schwarz 1996) The following criteria provide a decision procedure for the symmetry type of a second order ode if its Janet basis in a grlex term ordering with $\eta > \xi$, $y > x$ is given.*

One-parameter group

$\mathcal{S}_1^2$: *Group* $\mathbf{g}_{27}$, *Janet basis type* $\mathcal{J}_{1,1}^{(2,2)}$ *or* $\mathcal{J}_{1,2}^{(2,2)}$.

214

Two-parameter groups

$S_{2,1}^2$: *Group* $\mathbf{g}_{26}$, *Lie algebra* $\mathbf{l}_{2,2}$, *Janet basis type* $\mathcal{J}_{2,3}^{(2,2)}$.

$S_{2,2}^2$: *Group* $\mathbf{g}_{27}$, *Lie algebra* $\mathbf{l}_{2,1}$, *Janet basis type* $\mathcal{J}_{2,3}^{(2,2)}$.

Three-parameter groups

$S_{3,1}^2$: *Group* $\mathbf{g}_{10}$ *with two systems of imprimitivity, Lie algebra* $\mathbf{l}_{3,1}$, *Janet basis type* $\mathcal{J}_{3,6}^{(2,2)}$ *or* $\mathcal{J}_{3,7}^{(2,2)}$.

$S_{3,2}^2$: *Group* $\mathbf{g}_{13}$ *with one system of imprimitivity, Lie algebra* $\mathbf{l}_{3,1}$, *Janet basis type* $\mathcal{J}_{3,4}^{(2,2)}$, $\mathcal{J}_{3,6}^{(2,2)}$ *or* $\mathcal{J}_{3,7}^{(2,2)}$.

$S_{3,3}^2(c)$: *Group* $\mathbf{g}_7(c)$ *with two systems of imprimitivity, Lie algebra* $\mathbf{l}_{3,2}(c)$, *Janet basis type* $\mathcal{J}_{3,6}^{(2,2)}$ *or* $\mathcal{J}_{3,7}^{(2,2)}$.

$S_{3,4}^2$: *Group* $\mathbf{g}_{20}(r = 1)$ *with one system of imprimitivity, Lie algebra* $\mathbf{l}_{3,3}$, *Janet basis type* $\mathcal{J}_{3,6}^{(2,2)}$ *or* $\mathcal{J}_{3,7}^{(2,2)}$.

Eight-parameter group

S_8^2 : *Group* $\mathbf{g}_3$, *Janet basis type* $\{\xi_{yy}, \eta_{xx}, \eta_{xy}, \eta_{yy}, \xi_{xxx}, \xi_{xxy}\}$.

PROOF *One-parameter group.* A one-parameter symmetry is uniquely determined by the respective Janet basis type.

Two-parameter groups. The two groups are identified by their Lie algebra via Lemma 3.23.

Three-parameter groups. At first the groups are distinguished by the number of imprimitivity systems they allow by Theorem 3.21. The two pairs corresponding to one or two systems of imprimitivity are separated w.r.t. their Lie algebras by Theorem 3.24, 3.25 and 3.26.

The projective group. The eight-parameter symmetry is uniquly identified by its Janet basis type. ▯

EXAMPLE 5.9   Equation 6.90 of Kamke's collection

$$4x^2 y'' - x^4 y'^2 + 4y = 0 \qquad (5.26)$$

has the Janet basis

$$\eta + \frac{2y}{x}\xi = 0, \quad \xi_x - \frac{1}{x}\xi = 0, \quad \xi_y = 0 \qquad (5.27)$$

for its point symmetries. By Theorem 5.9 it follows that it has a one-parameter symmetry group. ▯

$$\textit{Symmetries of Differential Equations} \qquad\qquad 215$$

EXAMPLE 5.10 Equation 6.98 of Kamke's collection

$$x^4 y'' - x^2 y'^2 - x^3 y' + 4y^2 = 0 \qquad (5.28)$$

generates the type $\mathcal{J}_{1,2}^{(2,2)}$ Janet basis

$$\eta - \frac{2y}{x}\xi = 0, \quad \xi_x - \frac{1}{x}\xi = 0, \quad \xi_y = 0. \qquad (5.29)$$

Consequently, it has a one-parameter symmetry. The equation

$$y'' - \frac{2x(x-1)}{x+1}y'^2 - \frac{4(x-1)}{x+1}yy' + \frac{4x^2+5x+2}{x(x+1)^2} - \frac{2(x-1)}{x(x+1)}y^2 + \frac{2x+1}{x(x+1)^2}y = 0$$
$$(5.30)$$

generates the type $\mathcal{J}_{1,1}^{(2,2)}$ Janet basis $\{\xi = 0, \eta_x + \frac{1}{x}\eta = 0, \eta_y = 0\}$ and therefore also a one-parameter symmetry. The close relation between the two equations (5.28) and (5.30) will become clear later on. 〼

EXAMPLE 5.11 Equation 6.227 of Kamke's collection

$$(x - y')y'' + 4y'^2 = 0$$

generates the type $\mathcal{J}_{2,3}^{(2,2)}$ Janet basis

$$\xi_x - \frac{1}{x}\xi = 0, \quad \xi_y = 0, \quad \eta_x = 0, \quad \eta_y - \frac{1}{y}\eta = 0.$$

By Theorem 3.23 its Lie algebra is $l_{2,2}$, therefore by Theorem 5.9 its symmetry class is $\mathcal{S}_{2,1}^2$. 〼

EXAMPLE 5.12 Equation 6.159 of Kamke's collection

$$4y'' y - 3y'^2 - 12y^3 = 0$$

has the structure (5.34) with $\phi_1 = -\frac{15}{4}$, $\phi_2 = 0$, i.e., the projective group is excluded as symmetry group. The Janet basis

$$\xi_x + \frac{1}{2y}\eta = 0, \quad \xi_y = 0, \quad \eta_x = 0, \quad \eta_y - \frac{1}{y}\eta = 0$$

is of type $\mathcal{J}_{2,3}^{(2,2)}$. By Theorem 3.23 its Lie algebra is $l_{2,1}$, therefore by Theorem 5.9 its symmetry class is $\mathcal{S}_{2,2}^2$. 〼

EXAMPLE 5.13 The equation

$$x^2(x^2 y - 2)^2 y''^2 - 4x^4(x^2 y - 2)y'' y'^2 - 8x(x^2 y - 2)y'' y'$$

$$+ 4x^6 y'^4 + 24x^3 y'^3 + 24x^2 y'^2 y + 16y'^2 + 24xy' y^2 + 8y^3 = 0$$

216

has been written in rational form, therefore it is not quasilinear. Its Janet basis

$$\xi_y = 0, \quad \eta_x + \frac{2y}{x}\xi_x - \frac{x^2y+2}{x(x^2y-2)}\eta - \frac{4y^2}{x^2y-2}\xi = 0,$$

$$\eta_y + \xi_x - \frac{2x^2}{x^2y-2}\eta - \frac{3x^2y+2}{x(x^2y-2)}\xi = 0,$$

$$\xi_{xx} - \frac{2xy}{x^2y-2}\xi_x + \frac{4x}{(x^2y-2)^2}\eta + \frac{2y(x^2y+2)}{(x^2y-2)^2}\xi = 0$$

is of type $\mathcal{J}_{3,6}^{(2,2)}$. By Theorem 3.21 it allows two systems of imprimitivity. By Theorem 3.25 its Lie algebra type is $\mathfrak{l}_{3,1}$. Consequently by Theorem 5.9 its symmetry class is $\mathcal{S}_{3,1}^2$. ▯

EXAMPLE 5.14   Equation 6.133 of Kamke's collection

$$y''y + xy'' + y'^2 - y' = 0$$

generates the Janet basis

$$\xi_y + \xi_x - \frac{1}{x+y}\eta - \frac{1}{x+y}\xi = 0, \quad \eta_x - \xi_x + \frac{1}{x+y}\eta + \frac{1}{x+y}\xi = 0,$$

$$\eta_y + \xi_x - \frac{2}{x+y}\eta - \frac{2}{x+y}\xi = 0, \quad \xi_{xx} - \frac{3}{x+y}\xi_x + \frac{3}{(x+y)^2}\eta + \frac{3}{(x+y)^2}\xi = 0$$

of type $\mathcal{J}_{3,6}^{(2,2)}$. By Theorem 3.21 it allows a single system of imprimitivity. By Theorem 3.25 its Lie algebra type is $\mathfrak{l}_{3,1}$. Consequently by Theorem 5.9 its symmetry class is $\mathcal{S}_{3,2}^2$. ▯

EXAMPLE 5.15   The equation

$$y''y'yx^6 - 2y'^3x^6 + 2y'^2yx^5 + y^5 = 0 \tag{5.31}$$

generates the type $\mathcal{J}_{3,6}^{(2,2)}$ Janet basis

$$\xi_y = 0, \quad \eta_x = 0, \quad \eta_y - \frac{3}{2}\xi_x - \frac{2}{y}\eta + \frac{3}{x}\xi = 0, \quad \xi_{xx} - \frac{2}{x}\xi_x + \frac{2}{x^2}\xi = 0. \tag{5.32}$$

By Theorem 3.25 its Lie algebra type is $\mathfrak{l}_{3,2}(\frac{3}{2})$; therefore by Theorem 5.9 its symmetry class is $\mathcal{S}_{3,3}^2(\frac{3}{2})$. The equation

$$y'y'' + 2y'' - y'^4 - 12y'^3 - 54y'^2 - 108y' - 81 = 0$$

generates the Janet basis

$$\xi_x = 0, \quad \eta_x + 6\xi_y = 0, \quad \eta_y + 5\xi_y = 0, \quad \xi_{yy} = 0$$

of type $\mathcal{J}_{3,7}^{(2,2)}$. By a similar reasoning as in the preceding case the symmetry class is again $\mathcal{S}_{3,3}^2(\frac{3}{2})$. ▯

Symmetries of Differential Equations

EXAMPLE 5.16   Any equation of symmetry class $\mathcal{S}^2_{3,4}$ must contain a term proportional to an exponential in y'. An example is

$$y''y + y'^2 - \frac{1}{x}y'y - \frac{1}{2x} = 2x^2 \exp\left(-\frac{1 + 2y'y}{2x}\right) \tag{5.33}$$

with type $\mathcal{J}^{(2,2)}_{3,6}$ Janet basis

$$\xi_y = 0, \ \eta_x - \frac{x}{y}\xi_x - \frac{2x+1}{2xy}\xi = 0, \ \eta_y - \xi_x + \frac{1}{y}\eta - \frac{1}{x}\xi = 0, \ \xi_{xx} + \frac{1}{x}\xi_x - \frac{1}{x^2}\xi = 0.$$

By Theorem 3.25 its Lie algebra type is $l_{3,3}$, therefore by Theorem 5.9 its symmetry class is $\mathcal{S}^2_{3,3}(\frac{3}{2})$. □

Symmetry Type and Invariants. Tresse's invariants and their application to equivalence problems have been discussed in Section 4.3. In the theorem below it will be shown how they may be applied for determining the symmetry class of any second order quasilinear ode.

THEOREM 5.10   *(Tresse 1896) The following criteria provide a decision procedure for the type of symmetry group of a second order quasilinear ode in terms of its invariants under point transformations.*

i) *If there are three functionally independent absolute invariants the equation does not have a nontrivial symmetry.*

ii) *If there are two functionally independent absolute invariants the equation belongs to symmetry class $\mathcal{S}^2_1$.*

iii) *If there is a single functionally independent absolute invariant there is a two-parameter symmetry group. If for any two invariants ψ_1 and ψ_2 with weights r_1, s_1 and r_2, s_2, respectively, there holds*

$$2\begin{vmatrix} \Delta_x\psi_1 & \Delta_z\psi_1 \\ \Delta_x\psi_2 & \Delta_z\psi_2 \end{vmatrix} + \begin{vmatrix} \Delta_y\psi_1 & (3r_1 + 2s_1)\psi_1 \\ \Delta_y\psi_2 & (3r_2 + 2s_2)\psi_2 \end{vmatrix} = 0$$

the symmetry class is $\mathcal{S}^2_{2,1}$, and $\mathcal{S}^2_{2,2}$ otherwise.

iv) *If $\omega^4 H \neq 0$ and all absolute invariants are constant there is a three-parameter symmetry group. If $\omega^4 = 0$ the symmetry class is $\mathcal{S}^2_{3,2}$. If for any invariant ψ with weight r and s there holds*

$$\frac{\Delta_z\psi\Delta_x\psi}{\psi\Delta_y\psi} = \frac{32}{41}(3r + 2s)$$

the symmetry class is $\mathcal{S}^2_{3,4}$. If there holds

$$\Delta_x\psi = 0, \ \Delta_z\psi = 0, \ \frac{\Delta_y\psi}{\psi}\frac{1}{\sqrt[4]{\omega^4 H}} \neq \frac{1}{8}(3r + 2s)$$

the symmetry class is $\mathcal{S}^2_{3,1}$, otherwise the symmetry class is $\mathcal{S}^2_{3,3}$.

218

v) If $\omega^4 = 0$ and $H = 0$ the symmetry class is S_8^2.

The invariants used in this theorem have been defined in Theorem 4.24 on page 189. The proof may again be found in Tresse's thesis [181], page 83. In the following two examples the criteria of this theorem are applied for identifying the symmetry class of an ode. For comparison the same answer is obtained by the Janet basis analysis of Theorem 5.9

EXAMPLE 5.17 Consider the second order ode

$$8xy''y^6 - 9x^5y'^4 - 16xy'^2y^5 + 16y'y^6 = 0.$$

Its relative invariants up to order five are

$$\omega^4 = \frac{27x^4}{y^6}, \quad H = \frac{2187}{64}\frac{y'^8 x^{12}}{y^{18}}, \quad K = \frac{2187}{8}\frac{y'^7 x^{12}}{y^{18}},$$

$$H_{10} = \frac{98415}{64}\frac{y'^{11}x^{16}}{y^{24}}, \quad H_{01} = \frac{255879}{128}\frac{y'^{10}x^{16}}{y^{24}}.$$

All absolute invariants are constant, therefore the symmetry group has three parameters. Due to $\omega^4 \neq 0$,

$$\frac{\Delta_z H \Delta_x H}{H \Delta_y H} = \frac{80y'x^4 + 64y^5}{13y'^2x^4 + 8y^5} \neq \frac{256}{41}$$

and

$$\Delta_x H = \frac{98415}{64}(y'^2x^4 + \tfrac{4}{5}y^5)y'^9 x^{12}y^{12} \neq 0$$

by Theorem 5.10, *iv)* the symmetry class $S_{3,3}^2$ is identified. On the other hand the Janet basis for the determining system is

$$\xi_y = 0, \ \eta_x = 0, \ \eta_y - \frac{2}{3}\xi_x - \frac{2}{y}\eta + \frac{4}{3x}\xi = 0, \ \xi_{xx} - \frac{2}{x}\xi_x + \frac{2}{x^2}\xi = 0.$$

Applying Theorem 5.9 yields the same answer. ⬜

EXAMPLE 5.18 As a second example for applying Tresse's invariants consider the equation $y''y'^2 + y''y^2 + y^3 = 0$ with the absolute invariants

$$\hat{K} = \frac{67108864}{87890625}\frac{(y'^8 - \frac{21}{4}y'^6y^2 + \frac{7}{4}y'^4y^4 - \frac{23}{4}y'^2y^6 + \frac{9}{4}y^8)^8 y'^8}{(y'^4 - 2y'^2y^2 + \frac{1}{5}y^4)^{10}y^8(y'^2 + y'y + y^2)^6(y'^2 - y'y + y^2)^6},$$

$$\hat{H}_{10} = \frac{67108864}{225}\frac{(y'^8 - \frac{21}{4}y'^6y^2 + \frac{7}{4}y'^4y^4 - \frac{23}{4}y'^2y^6 + \frac{9}{4}y^8)^8 y'^8}{(y'^4 - 2y'^2y^2 + \frac{1}{5}y^4)^{10}y^8(y'^2 + y'y + y^2)^6(y'^2 - y'y + y^2)^6}.$$

The remaining invariants cannot be given here due to its size. As a consequence it is difficult to apply Theorem 5.10 for determining the symmetry

Symmetries of Differential Equations

class of this equation. The Janet basis $\{\xi_x = 0, \, \xi_y = 0, \, \eta_x = 0, \, \eta_y - \frac{1}{y}\eta = 0\}$ immediately yields the symmetry class $\mathcal{S}_{2,1}^2$.

If Tresse's invariants are computed for a large collection of ode's, the following general pattern appears. The calculations become extremely complex for small symmetry groups with less than three parameters. That means, despite the theoretical interest in his results, for any practical purpose it is advantageous to determine the symmetry class from the Janet basis by Theorem 5.9. This is even more true because the subsequent step in the solution algorithm, i.e., determining the canonical form, requires the knowledge of the Janet basis as well.

Symmetries of Lie's Equation. The case $\dfrac{\partial^4 \omega}{\partial y'^4} = 0$ has been excluded in Theorem 4.24. Quasilinear equations which are third order polynomials in y', are called *Lie's equation*. They deserve special attention due to its importance in various application fields and when the linearizability of second order ode's is discussed. Two of its symmetry classes may be identified by the conditions given below.

THEOREM 5.11   *Let a second order ode of the form*

$$y'' + Ay'^3 + By'^2 + Cy' + D = 0 \tag{5.34}$$

be given where $A \equiv A(x,y), \ldots, D \equiv D(x,y)$. If it allows a nontrivial point symmetry the following alternatives may occur.

i) *Its symmetry class is $\mathcal{S}_8^2$ iff the two relations $L_1 = 0$, $L_2 = 0$ are satisfied where*

$$L_1 \equiv D_{yy} + BD_y - AD_x + (B_y - 2A_x)D$$
$$+ \tfrac{1}{3}B_{xx} - \tfrac{2}{3}C_{xy} + \tfrac{1}{3}C(B_x - 2C_y),$$
$$L_2 \equiv 2AD_y + A_yD + \tfrac{1}{3}C_{yy} - \tfrac{2}{3}B_{xy} + A_{xx}$$
$$- \tfrac{1}{3}BC_y + \tfrac{2}{3}BB_x - A_xC - AC_x,$$

ii) *Its symmetry class is $\mathcal{S}_{3,2}^2$ if $L_1 \neq 0$ or $L_2 \neq 0$ and $\Phi = \Psi = 0$ where*

$$\Phi \equiv L_2(L_1L_{2,x} - L_{1,x}L_2) - L_1(L_1L_{2,y} - L_{1,y}L_2)$$
$$- AL_1^3 + BL_1^2L_2 - CL_1L_2^2 + DL_2^3,$$
$$\Psi \equiv L_2(L_1L_{2,yy} - L_{1,yy}L_2) - (2L_{2,y} + 3AL_1 - BL_2)(L_1L_{2,y} - L_{1,y}L_2)$$
$$- A^2L_1^3 + (AB + A_y)L_1^2L_2 - (AC - A_x + B_y)L_1L_2^2$$
$$+ (AD - \tfrac{1}{3}B_x + \tfrac{2}{3}C_y)L_2^3.$$

220

PROOF   The general expression for the transformation of a second order derivative $v''(u)$ by the transformation $u = \sigma(x,y)$, $v = \rho(x,y)$ is

$$v'' = \frac{1}{(\sigma_x + \sigma_y y')^3}\{(\sigma_x \rho_y - \sigma_y \rho_x)y'' + (\sigma_y \rho_{yy} - \sigma_{yy}\rho_y)y'^3$$
$$+[\sigma_x \rho_{yy} - \sigma_{yy}\rho_x + 2(\sigma_y \rho_{xy} - \sigma_{xy}\rho_y)]y'^2$$
$$+[\sigma_y \rho_{xx} - \sigma_{xx}\rho_y + 2(\sigma_x \rho_{xy} - \sigma_{xy}\rho_x)]y' + \sigma_x \rho_{xx} - \sigma_{xx}\rho_x\}.$$

It has the structure of Lie's equation. On the other hand, any equation with the structure (5.34) leads to a determining system of the form

$$\xi_{yy} - 2A\eta_y - B\xi_y + A\xi_x - A_y\eta - A_x\xi = 0,$$
$$\eta_{xx} - D\eta_y + C\eta_x + 2D\xi_x + D_y\eta + D_x\xi = 0,$$
$$\eta_{xy} - \tfrac{1}{2}\xi_{xx} + B\eta_x + \tfrac{3}{2}D\xi_y + \tfrac{1}{2}C\xi_x + \tfrac{1}{2}C_y\eta + \tfrac{1}{2}C_x\xi = 0,$$
$$\eta_{yy} - 2\xi_{xy} + B\eta_y + 3A\eta_x + 2C\xi_y + B_y\eta + B_x\xi = 0.$$

$$(5.35)$$

In order to transform it into a Janet basis, the algorithm *JanetBasis* described in Chapter 1 has to be applied. In the first step it yields the integrability conditions

$$\xi_{xxx} - 3D\xi_{xy} - (2B_x - C_y - 6AD)\eta_x$$
$$-(D_x - CD)\xi_y - (2C_x - 4D_y - 4BD + C^2)\xi_x$$
$$+(2B_yD - C_{xy} - C_yC + 2D_{yy} + 2D_yB)\eta$$
$$+(2B_xD - C_{xx} - C_xC + 2D_{xy} + 2D_xB)\xi = 0,$$

$$(5.36)$$

$$\xi_{xxy} - C\xi_{xy} - (\tfrac{2}{3}B_x - \tfrac{1}{3}C_y)\eta_y - (2A_x - 2AC)\eta_x$$
$$-(C_x - D_y - BD)\xi_y - (\tfrac{2}{3}B_x - \tfrac{1}{3}C_y - 3AD)\xi_x$$
$$+(A_yD - \tfrac{2}{3}B_{xy} + \tfrac{1}{3}C_{yy} + 2D_yA)\eta$$
$$+(A_xD - \tfrac{2}{3}B_{xx} + \tfrac{1}{3}C_{xy} + 2D_xA)\xi = 0.$$

$$(5.37)$$

If they are added to the original system (5.35) as the two highest equations, it has the same leading terms as the Janet basis for the projective symmetry group. In order to become a Janet basis for the group $\mathcal{S}_8^2$, the two integrability conditions

$$2L_2\eta_y + L_1\xi_y + L_2\xi_x + L_{2,y}\eta + L_{2,x}\xi = 0,$$
$$L_1\eta_y + L_2\eta_x + 2L_1\xi_x + L_{1,y}\eta + L_{1,x}\xi = 0$$

for this enlarged system are obtained. It is easy to see that they are a consequence of $L_1 = L_2 = 0$. This proves part i). The calculations for part ii) are similar, although they are more lengthy due to the larger coefficients.   ▯

The first part of this theorem has been known already to Lie [109]. He gave a rather implicit criterion for $\mathcal{S}_8^2$ symmetry in terms of a set of pde's from which

Symmetries of Differential Equations 221

the conditions $L_1 = L_2 = 0$ may be obtained as integrability conditions, see eqs. (3) and (4) in Lie [109], part II. Liouville [118] gave these constraints for the first time exlicitly in terms of the invariants L_1 and L_2. Tresse [181], Theorem VII on page 56, rederived and extended this result. Since then they have been rediscovered by various authors, Grissom [63] e. g. obtained them as conditions for linearizability of a second order ode. However, the above method of proving this theorem is more transparent and in addition it provides an explicit form of the Janet basis for any equation with S_8^2 symmetry. This remarkable result is formulated as follows.

COROLLARY 5.2 *For any second order equation (5.34) with S_8^2 symmetry the Janet basis in grlex term ordering with $\eta > \xi$ and $y > x$ is given by the union of the systems (5.35) and (5.36), (5.37).*

EXAMPLE 5.19 Equation 6.98 of Kamke's collection has been considered previously in Example 5.10. It has the structure (5.34). From its coefficients the functions

$$\phi_1 = \frac{8}{3}\frac{2x^2 - 3y}{x^6}, \quad \phi_2 = -\frac{4}{3x^5}$$

are obtained. Therefore the projective group is excluded. Because

$$\Phi = -\frac{400}{9}\frac{5x^4 + 2x^2 + x}{(x^2 + 1)^{10}}, \quad \Psi = -\frac{128}{81}\frac{x^2}{(x^2 + 1)^{11}}$$

the symmetry class $S_{3,2}^2$ is excluded as well. ⬚

EXAMPLE 5.20 Kamke's equation 6.133 has been considered in Example 6.11. By Theorem 5.11 with $\phi_{1,2} = \pm\dfrac{3}{(x+y)^3}$, the symmetry group of this equation cannot be the projective group. However both $\Phi = 0$ and $\Psi = 0$ and the above theorem yields the symmetry class $S_{3,2}^2$. ⬚

EXAMPLE 5.21 Equation Kamke 6.180 is

$$y''(y-1)x^2 - 2y'^2x^2 - 2y'(y-1)x - 2y(y-1)^2 = 0. \tag{5.38}$$

Its coefficients $A = 0$, $B = -\dfrac{2}{y-1}$, $C = -\dfrac{2}{x}$ and $D = -\dfrac{2y(y-1)}{x^2}$ satisfy the constraints $\phi_1 = \phi_2 = 0$ of Theorem 5.11, i.e., its symmetry group is the projective group. ⬚

A special case of Lie's equation is obtained for $A = B = C = 0$, sometimes it is called generalized Emden-Fowler equation. The subsequent theorem describes its partition into symmetry classes.

THEOREM 5.12 *The symmetry classes of the generalized Emden-Fowler equation $y'' + r(x, y) = 0$ may be described as follows.*

i) The symmetry class is S_8^2 if $r_{yy} = 0$.

ii) The symmetry class is $\mathcal{S}^2_{3,2}$ if $r_{yyy} \neq 0$ and

$$\left(\log r_{yyy} - \tfrac{6}{5} \log r_{yy} \right)_y = 0, \quad \left(\log r_{xyy} - \tfrac{6}{5} \log r_{yy} \right)_x = 0,$$

$$r_{xxyy} r_{yyy} r_{yy} - \tfrac{6}{5} r^2_{xyy} r_{yyy} - r^2_{yyy} r_{yy} r_y - 5 r_{yyy} r^2_{yy} r_y + \tfrac{25}{3} r^4_{yy} = 0.$$

iii) The symmetry class is $\mathcal{S}^2_{2,2}$ if $r_{yyy} = 0$ and

$$r_{xxxxyy} r^3_{yy} - \tfrac{32}{5} r_{xxxyy} r_{xyy} r^2_{yy} - \tfrac{43}{10} r^2_{xxyy} r^2_{yy} + \tfrac{594}{25} r_{xxyy} r^2_{xyy} r_{yy}$$

$$+ 5 r_{xxyy} r^3_{yy} r_y - 5 r_{xxy} r^4_{yy} - \tfrac{1782}{125} r^4_{xyy} - 10 r^2_{xyy} r^2_{yy} r_y$$

$$+ 10 r_{xyy} r_{xy} r^3_{yy} + 5 r^5_{yy} r - \tfrac{5}{2} r^4_{yy} r^2_y = 0.$$

iv) The symmetry class is $\mathcal{S}^2_1$ if r obeys a differential polynomial of order eight.

PROOF   The determining system for $y'' + r(x,y) = 0$ is

$$\xi_{yy} = 0, \quad \eta_{xx} - r\eta_y + 2r\xi_x + r_y \eta + r_x \xi = 0,$$
$$\eta_{xy} - \tfrac{1}{2}\xi_{xx} + \tfrac{3}{2}r\xi_y = 0, \quad \eta_{yy} - 2\xi_{xy} = 0 \tag{5.39}$$

with the integrability conditions

$$\xi_{xxx} - 3r\xi_{xy} - r_x\xi_y + 4r_y\xi_x + 2r_{yy}\eta + 2r_{xy}\xi = 0, \quad \xi_{xxy} + r_y\xi_y = 0,$$
$$r_{yy}\xi_y = 0, \quad 2r_{yy}\eta_y + 4r_{yy}\xi_x + 2r_{yyy}\eta + 2r_{xyy}\xi = 0.$$

The first two conditions, combined with (5.39), form a Janet basis for the symmetry class $\mathcal{S}^2_8$ if and only if $r_{yy} = 0$ such that the last two conditions are satisfied. This covers case *i*). If $r_{yy} \neq 0$, these latter conditions have to be added to the system. After two autoreduction steps it contains an equation with the leading term $\frac{r_{yyy}}{r_{yy}}\xi_x$ and a further branching occurs. If $r_{yyy} \neq 0$ a type $\mathcal{J}^{(2,2)}_{3,6}$ Janet basis is generated if the conditions *ii*) are satisfied. By Theorem 5.9 it determines uniquely the symmetry class $\mathcal{S}^2_{3,2}$. If $r_{yyy} = 0$, a type $\mathcal{J}^{(2,2)}_{2,3}$ is obtained under the constraints of case *iii*). By the criterion given in Theorem 5.9 the symmetry class $\mathcal{S}^2_{2,2}$ is identified. Finally it may belong to symmetry class $\mathcal{S}^2_1$ if r obeys a differential polynomial of order eight that is too complicated to be given here. $\square$

Lie Algebra of Symmetry Generators. The methods for solving differential equations given by Lie [113] require the explicit knowledge of the symmetry generators. By means of the algorithms of Chapter 2, all Liouvillian generators may be obtained from the Janet basis for its coefficients. Because the symmetry type is also known from the Janet basis, the only additional step is to generate a set of generators satisfying canonical commutation

Symmetries of Differential Equations 223

relations as it is described in Chapter 3 on page 116. These steps will be explained now by some examples.

EXAMPLE 5.22   The two equations 6.90 and 6.98 considered above have the generators $x\partial_x - 2y\partial_y$ and $x\partial_x + 2y\partial_y$ respectively. This shows the close connection between the two equations although this is not obvious from their appearance. $\Box$

EXAMPLE 5.23   Equation 6.228 has the Abelian two-parameter symmetry algebra with generators $U_1 = x\partial_x + y\partial_y$ and $U_2 = (x + y)\partial_y$. Example 5.12 with equation 6.159 has the two parameter algebra $U_1 = \partial_x$, $U_2 = x\partial_x - 2y\partial_y$ with commutator $[U_1, U_2] = U_1$ which is already in canonical form. $\Box$

EXAMPLE 5.24   Kamke's equation 6.133 of Example 5.14 in symmetry class $\mathcal{S}^2_{3,2}$ has generators

$$U_1 = \partial_x - \partial_y, \ U_2 = x\partial_x + y\partial_y,$$
$$U_3 = \left(x^2 - \tfrac{2}{3}xy - \tfrac{1}{3}y^2\right)\partial_x + \left(\tfrac{1}{3}x^2 + \tfrac{2}{3}xy - y^2\right)\partial_y.$$

In order to satisfy the commutation relations in canonical form, the linear combinations $\bar{U}_1 - \tfrac{3}{4}U_1$, $\bar{U}_2 = 2U_2$ and $\bar{U}_3 = U_3$ have to be formed. $\Box$

EXAMPLE 5.25   The first equation in symmetry class $\mathcal{S}^2_{3,3}$ considered in Example 5.15 has symmetry generators

$$U_1 = x\partial_x + \tfrac{3}{2}y\partial_y, \ U_2 = y^2\partial_y, \ U_3 = x^2\partial_x.$$

In order to obtain the canonical commutation relations of the $\mathfrak{l}_{3,2}$ Lie algebra the linear combinations $\bar{U}_1 = U_2$, $\bar{U}_2 = U_3$ and $\bar{U}_3 = -\tfrac{2}{3}U_1$ have to be formed. The second equation of this example in the same symmetry class has generators $U_1 = \partial_x$, $U_2 = \partial_y$ and $U_3 = y\partial_y - (6x + 5y)\partial_y$. A simple algebraic computation yields $\bar{U}_1 = -\tfrac{1}{3}U_1 + U_2$, $\bar{U}_2 = -\tfrac{1}{2}U_1 + U_2$ and $\bar{U}_3 = U_1 + U_2 - \tfrac{1}{3}U_3$. In either case there holds $[\bar{U}_1, \bar{U}_2] = 0$, $[\bar{U}_1, \bar{U}_3] = U_1$ and $[\bar{U}_2, \bar{U}_3] = \tfrac{2}{3}U_2$. $\Box$

EXAMPLE 5.26   Equation 6.180 has been considered above in Example 5.21. Its Janet basis in *grlex* term ordering is

$$\xi_{yy} + \frac{2}{y-1}\xi_y = 0,$$

$$\eta_{xx} + \frac{2y^2 - 2y}{x^2}\eta_y - \frac{2}{x}\eta_x - \frac{4y^2 - 4y}{x^2}\xi_x - \frac{4y - 2}{x^2}\eta + \frac{4y^2 - 4y}{x^3}\xi = 0,$$

$$\eta_{xy} - \frac{1}{2}\xi_{xx} - \frac{2}{y-1}\eta_x - \frac{3y^2 - 3y}{x^2}\xi_y - \frac{1}{x}\xi_x + \frac{1}{x^2}\xi = 0,$$

$$\eta_{yy} - 2\xi_{xy} - \frac{2}{y-1}\eta_y - \frac{4}{x}\xi_y + \frac{2}{y^2 - 2y + 1}\eta = 0,$$

$$\xi_{xxx} + \frac{6y^2 - 6y}{x^2}\xi_{xy} = 0, \quad \xi_{xxy} + \frac{2}{x}\xi_{xy} = 0$$

224

in accordance with Corollary 5.2. The special structure of this system allows it to be solved by an *ad hoc* method. It has a rational fundamental system leading to the vector fields

$$U_1 = x\partial_x, \; U_2 = y(y-1)\partial_y, \; U_3 = x^2\partial_x - x(y^2-1)\partial_y,$$

$$U_4 = x^2\partial_x - 2xy(y-1)\partial_y, \; U_5 = x^2(y-1)^2\partial_y,$$

$$U_6 = \frac{y}{x(y-1)}\partial_x - \frac{y^2}{x^2}\partial_y, \; U_7 = \frac{y-2}{y-1}\partial_x + \frac{2y}{x}\partial_y,$$

$$U_8 = \frac{y}{y-1}\partial_x - \frac{2y^2}{x}\partial_y. \qquad \qquad \square$$

5.4 Symmetries of Nonlinear Third Order Equations

The methods described in the preceding section may be extended to higher order equations more or less straightforwardly, although the larger number of symmetry types leads to more alternatives that have to be considered. For various reasons it appears to be appropriate to distinguish equations that are equivalent to a linear one from those that are not. These latter equations are simply called genuinely nonlinear third order equations; they are the subject of the current section whereas the linearizable equations are considered in Section 5.5.

Classification of Symmetries. The main result of this section, the equivalent of Theorem 5.8 for nonlinear third order equations, is given first.

THEOREM 5.13 *Any symmetry algebra of a genuinely nonlinear third order quasilinear ode is similar to one in canonical variables u and $v \equiv v(u)$ as given in the following listing. In addition the corresponding Janet basis is given where $\alpha(u,v)$ and $\beta(u,v)$ are the coefficients of ∂_u and ∂_v respectively.*
One-parameter group
  $\mathcal{S}_1^3 : \mathbf{g}_{27} = \{\partial_v\}$. *Janet basis* $\{\alpha, \beta_u, \beta_v\}$.
Two-parameter groups
  $\mathcal{S}_{2,1}^3 : \mathbf{g}_{26} = \{\partial_u, \partial_v\}$. *Janet basis* $\{\alpha_u, \alpha_v, \beta_u, \beta_v\}$.
  $\mathcal{S}_{2,2}^3 : \mathbf{g}_{25} = \{u\partial_u + v\partial_v, \partial_v\}$. *Janet basis* $\{\alpha_v, \beta_u, \beta_v - \frac{1}{v}\beta, \alpha_{uu}\}$.
  $\mathcal{S}_{2,3}^3 : \mathbf{g}_4 = \{v\partial_v, \partial_v\}$. *Janet basis* $\{\alpha, \beta_u, \beta_{vv}\}$.
  $\mathcal{S}_{2,4}^3 : \mathbf{g}_{15}(r=2) = \{\partial_v, u\partial_v\}$. *Janet basis* $\{\alpha, \beta_v, \beta_{uu}\}$.
Three-parameter groups
  $\mathcal{S}_{3,1}^3 : \mathbf{g}_{10} = \{\partial_u + \partial_v, u\partial_u + v\partial_v, u^2\partial_u + v^2\partial_v\}$. *Janet basis*
  $\{\alpha_v, \beta_u, \beta_v + \alpha_u + \frac{2}{u-v}(\beta - \alpha), \alpha_{uu} - \frac{2}{u-v}\alpha_u - \frac{2}{(u-v)^2}(\beta - \alpha)\}$.
  $\mathcal{S}_{3,2}^3 : \mathbf{g}_{13} = \{\partial_u, 2u\partial_u + v\partial_v, u^2\partial_u + uv\partial_v\}$. *Janet basis*
  $\{\alpha_u - \frac{2}{v}\beta, \alpha_v, \beta_v - \frac{1}{v}\beta, \beta_{uu}\}$.

$$\mathcal{S}_{3,3}^3(c) : \mathbf{g}_7 = \{\partial_u, \partial_v, u\partial_u + cv\partial_v\}, \ c \neq 0, 1.$$
Janet basis $\alpha_v, \beta_u, \beta_v - c\alpha_u, \alpha_{uu}\}.$
$$\mathcal{S}_{3,4}^3 : \mathbf{g}_{20}(r=1) = \{\partial_u, \partial_v, u\partial_u + (u+v)\partial_v\}. \ \text{Janet basis}$$
$\{\alpha_v, \beta_u - \alpha_u, \beta_v - \alpha_u, \alpha_{uu}\}.$
$$\mathcal{S}_{3,5}^3 : \mathbf{g}_5 = \{\partial_u, \partial_v, v\partial_v\}. \ \text{Janet basis } \{\alpha_u, \alpha_v, \beta_u, \beta_{vv}\}.$$
$$\mathcal{S}_{3,6}^3 : \mathbf{g}_{17}(l=1, \rho_1 = 1, \alpha_1 = 0) = \{\partial_u, \partial_v, u\partial_v\}. \ \text{Janet basis } \{\alpha_u, \alpha_v, \beta_v, \beta_{uu}\}.$$
$$\mathcal{S}_{3,7}^3 : \mathbf{g}_{17}(l=1, \rho_1 = 1, \alpha_1 = 1) = \{e^u\partial_v, ue^u\partial_v, \partial_u\}.$$
Janet basis $\{\alpha_u, \alpha_v, \beta_v, \beta_{uu} - 2\beta_u + \beta\}.$
$$\mathcal{S}_{3,8}^3 : \mathbf{g}_{17}(l=2, \rho_1 = \rho_2 = 0, \alpha_1 = 0, \alpha_2 = 1) = \{e^u\partial_v, \partial_v, \partial_u\}.$$
Janet basis $\{\alpha_u, \alpha_v, \beta_v, \beta_{uu} - \beta_u\}.$
$$\mathcal{S}_{3,9}^3(c) : \mathbf{g}_{17}(l=2, \rho_1 = \rho_2 = 0, \alpha_1 = 1, \alpha_2 = c) = \{e^u\partial_v, e^{cu}\partial_v, \partial_u\},$$
$c \neq 0, 1.$ Janet basis $\{\alpha_u, \alpha_v, \beta_v, \beta_{uu} - (a+1)\beta_u + a\beta\}.$
$$\mathcal{S}_{3,10}^3 : \mathbf{g}_{24} = \{\partial_u, \partial_v, u\partial_u + v\partial_v\}. \ \text{Janet basis } \{\alpha_v, \beta_u, \beta_v - \alpha_u, \alpha_{uu}\}.$$

Four-parameter groups

$$\mathcal{S}_{4,1}^3 : \mathbf{g}_6 = \{\partial_u, \partial_v, v\partial_v, u\partial_u\}. \ \text{Janet basis } \{\alpha_v, \beta_u, \alpha_{uu}, \beta_{vv}\}.$$
$$\mathcal{S}_{4,2}^3 : \mathbf{g}_{14} = \{\partial_u, u\partial_u, v\partial_v, u^2\partial_u + uv\partial_v\}.$$
Janet basis $\{\alpha_v, \beta_v - \frac{1}{v}\beta, \alpha_{uu} - \frac{2}{v}\beta_u, \beta_{uu}\}.$
$$\mathcal{S}_{4,3}^3(c) : \mathbf{g}_{19}(r=2) = \{\partial_u, \partial_v, u\partial_v, u\partial_u + cv\partial_v\}, \ c \neq 2$$
Janet basis $\{\alpha_v, \beta_v - c\alpha_u, \alpha_{uu}, \beta_{uu}\}.$
$$\mathcal{S}_{4,4}^3 : \mathbf{g}_{20}(r=2) = \{\partial_u, \partial_v, u\partial_v, u\partial_u + (u^2 + 2v)\partial_v\}.$$
Janet basis $\{\alpha_v, \beta_v - 2\alpha_u, \alpha_{uu}, \beta_{uu} - 2\alpha_u\}.$

Groups with six parameters

$$\mathcal{S}_6^3 : \mathbf{g}_{12} = \{\partial_u, \partial_v, u\partial_u, v\partial_v, u^2\partial_u, v^2\partial_v\}. \ \text{Janet basis } \{\alpha_v, \beta_u, \alpha_{uuu}, \beta_{vvv}\}.$$

PROOF At first invariant differential equations originating from the *invariants* listed in Section 5.1 will be considered. For the groups $\mathbf{g}_k$ with $k = 1, 2, 3, 11$ and 12 an invariant of order three obviously does not exist. A third order ode with these symmetry groups is therefore excluded. For the groups with $k = 21, 22$ or 23 the parameter r is constrained to $r \geq 2$ as it may be seen from the listing of Section 3.4. Therefore, an invariant of order three is excluded for these groups as well. The quasilinear equation $y'''y' - \frac{3}{2}y''^2 + r(x) = 0$ corresponding to $\mathbf{g}_8$ is invariant under the larger group $\mathbf{g}_{12} \supset \mathbf{g}_8$. Therefore its symmetry group is $\mathbf{g}_{12}$. A similar argument applies to group $\mathbf{g}_9$.

The group $\mathbf{g}_{15}$ for $r = 2$ allows the three lowest invariants $\Phi_1^{(15)} = x$,

$$\Phi_2^{(15)} = D \equiv \begin{vmatrix} \phi_1 & \phi_1' & \phi_1'' \\ \phi_2 & \phi_2' & \phi_2'' \\ y & y' & y'' \end{vmatrix} \quad \text{and} \quad \Phi_3^{(15)} = D_1 \equiv \begin{vmatrix} \phi_1 & \phi_1' & \phi_1''' \\ \phi_2 & \phi_2' & \phi_2''' \\ y & y' & y''' \end{vmatrix}$$

as it may be seen from the listing in Section 5.1. The special choice $\phi_1 = 1$ and $\phi_2 = x$ yields the simplification $\Phi_2^{(15)} = y''$ and $\Phi_3^{(15)} = y'''$ from which a nonlinear third order equation may be constructed. For $r = 3$, again $\Phi_1^{(15)} = x$ and

$$\Phi_2^{(15)} = \begin{vmatrix} \phi_1 & \phi_1' & \phi_1'' & \phi_1''' \\ \phi_2 & \phi_2' & \phi_2'' & \phi_2''' \\ y & y' & y'' & y''' \end{vmatrix}.$$

226

Choosing ϕ_1 and ϕ_2 as above and leaving ϕ_3 undetermined yields now $\Phi_2^{(15)} = \phi''y''' - \phi'''y''$. It is not possible to construct a nonlinear third order equation from it which is linear in y''' and is therefore excluded; see, however, Theorem 5.16.

The group $\mathbf{g}_{16}$ for $r = 2$ has the lowest invariants $\Phi_1^{(16)} = x$ and $\Phi_2^{(16)} = \frac{D_1}{D}$ with D and D_1 as above. Applying the same simplification leads to $\Phi_2^{(16)} = \frac{y'''}{y''}$. Again it is not possible to construct a nonlinear third order equation from it which is linear in y''' and is therefore excluded.

For the remaining values $k = 4, \ldots, 7, 9, 10, 12, 13, 14, 19, 20$ and $24, \ldots, 27$ equations may be constructed from the given invariants with the respective symmetry groups as listed above.

The groups $\mathbf{g}_{17}$ have at least three parameters. There exist exactly four of them for this minimal value; they have been listed in Section 3.4. Setting up the system (5.14) for these groups, the two lowest invariants of the form Φ, Φ' are obtained. For the groups $\mathcal{S}_{3,6}^3, \ldots, \mathcal{S}_{3,9}^3$ one gets $\Phi = v''$, $\Phi = v'' - 2v' + v$, $\Phi = v'' - v'$ and $\Phi = v'' - (\alpha + 1)v' + \alpha v$ respectively. In each case, a nonlinear third order equation that is linear in v''' may be constructed. Proceeding in a similar way for the seven types of four-parameter groups $\mathbf{g}_{17}$ listed in Section 3.4 yields lowest invariants of the form $v''' + c_1 v'' + c_2 v' + c_3 v$ with constants c_i from which a nonlinear equation, linear in v''', cannot be constructed. Therefore these groups are excluded.

The *lower equations* for $k = 1, \ldots, 10, 13, 14, 20, 23, \ldots, 27$ are excluded because its order is different from three. For $k = 19, 21$ and 22 the lower equation is linear, they are covered by Theorem 5.16 below. The lower equations for $\mathbf{g}_{11} \subset \mathbf{g}_{12}$ are already subsumed under the equation with $\mathbf{g}_{12}$ symmetry as listed above. $\Box$

In his book on differential equations Lie [113] gives in Chapter 22, §3 a complete listing of three parameter groups of a two-dimensional manifold. It is instructive to identify the symmetry groups of the above theorem in Lie's list. For symmetry classes $\mathcal{S}_{3,k}^3$, $k = 1, 2, \ldots, 6$ this is fairly obvious, the corresponding groups are 1), 2), 4), 8), 10) and 12) in his enumeration respectively. The symmetry type of class $\mathcal{S}_{3,7}^3$ is similar to Lie's group 9), the transformations functions are $\sigma = -x$, $\rho = ye^{-x}$; the symmetry type of class $\mathcal{S}_{3,8}^3$ is similar to Lie's group 11), the transformation functions are $\sigma = -\log x$, $\rho = \frac{y}{x}$. Finally the symmetry type of $\mathcal{S}_{3,9}^3$ is similar to Lie's group 5) with transformation functions $\sigma = \frac{1}{c-1}\log x$, $\rho = yx^{1(c-1)}$ and $\alpha = c$.

In the subsequent Section 5.5 it will be shown that equations with a five- or seven-parameter symmetry group are always linearizable, therefore they do not occur in the above theorem.

Determining the Symmetry Class. The preceding theorem describes the possible symmetry types of third order quasilinear ode's in canonical coordinates. Based on this result, the next theorem allows it to determine the

Symmetries of Differential Equations 227

symmetry type of any such equation in actual variables from the Janet basis of its determining system. In addition to the Janet bases that occurred for second order ode's in the previous section, various Janet bases of order four are required. They are given in Appendix B on page 380.

THEOREM 5.14   *The following criteria provide a decision procedure for determining the symmetry class of a third order ode if its Janet basis in a grlex term ordering with $\eta > \xi$, $y > x$ is given.*

One-parameter group

$\mathcal{S}_1^3$: *Janet basis of type $\mathcal{J}_{1,1}^{(2,2)}$ or $\mathcal{J}_{1,2}^{(2,2)}$.*

Two-parameter groups

$\mathcal{S}_{2,1}^3$: *Group $\mathbf{g}_{26}$ with a one-parameter system of imprimitivity, Lie algebra $\mathbf{l}_{2,2}$, Janet basis type $\mathcal{J}_{2,3}^{(2,2)}$.*

$\mathcal{S}_{2,2}^3$: *Group $\mathbf{g}_{27}$ with a one-parameter system of imprimitivity, Lie algebra $\mathbf{l}_{2,1}$, Janet basis type $\mathcal{J}_{2,3}^{(2,2)}$.*

$\mathcal{S}_{2,3}^3$: *Group $\mathbf{g}_4$ with two systems of imprimitivity, Lie algebra $\mathbf{l}_{2,1}$, Janet basis type $\mathcal{J}_{2,1}^{(2,2)}$, $\mathcal{J}_{2,2}^{(2,2)}$ or $\mathcal{J}_{2,4}^{(2,2)}$.*

$\mathcal{S}_{2,4}^3$: *Group $\mathbf{g}_{15}(r = 2)$ with one system of imprimitivity, Lie algebra $\mathbf{l}_{2,1}$, Janet basis type $\mathcal{J}_{2,1}^{(2,2)}$ or $\mathcal{J}_{2,5}^{(2,2)}$.*

Three-parameter groups

$\mathcal{S}_{3,1}^3$: *Group $\mathbf{g}_{10}$ with two systems of imprimitivity, Lie algebra $\mathbf{l}_{3,1}$, Janet basis type $\mathcal{J}_{3,6}^{(2,2)}$ or $\mathcal{J}_{3,7}^{(2,2)}$.*

$\mathcal{S}_{3,2}^3$: *Group $\mathbf{g}_{13}$ with one system of imprimitivity, Lie algebra $\mathbf{l}_{3,1}$, Janet basis type $\mathcal{J}_{3,4}^{(2,2)}$, $\mathcal{J}_{3,6}^{(2,2)}$ or $\mathcal{J}_{3,7}^{(2,2)}$.*

$\mathcal{S}_{3,3}^3(c)$: *Group $\mathbf{g}_7(c)$ with two systems of imprimitivity, Lie algebra $\mathbf{l}_{3,2}(c)$, Janet basis type $\mathcal{J}_{3,6}^{(2,2)}$ or $\mathcal{J}_{3,7}^{(2,2)}$.*

$\mathcal{S}_{3,4}^3$: *Group $\mathbf{g}_{20}(r = 1)$ with one system of imprimitivity, Lie algebra $\mathbf{l}_{3,3}$, Janet basis type $\mathcal{J}_{3,6}^{(2,2)}$ with $c_1 \neq 1$ or $\mathcal{J}_{3,7}^{(2,2)}$ with $c_1 \neq 0$.*

$\mathcal{S}_{3,5}^3$: *Group $\mathbf{g}_5$ with two systems of imprimitivity, Lie algebra $\mathbf{l}_{3,4}$, Janet basis type $\mathcal{J}_{3,4}^{(2,2)}$, $\mathcal{J}_{3,5}^{(2,2)}$, $\mathcal{J}_{3,6}^{(2,2)}$ or $\mathcal{J}_{3,7}^{(2,2)}$.*

$\mathcal{S}_{3,6}^3$: *Group $\mathbf{g}_{17}(l = 1, \rho_1 = 1, \alpha_1 = 0)$ with one system of imprimitivity, Lie algebra $\mathbf{l}_{3,5}$, Janet basis type $\mathcal{J}_{3,4}^{(2,2)}$, $\mathcal{J}_{3,6}^{(2,2)}$ or $\mathcal{J}_{3,7}^{(2,2)}$.*

$\mathcal{S}_{3,7}^3$: *Group* $\mathbf{g}_{17}(l = 1, \rho_1 = 1, \alpha_1 = 1)$ *with one system of imprimitivity, Lie algebra* $\mathbf{l}_{3,3}$, *Janet basis type* $\mathcal{J}_{3,4}^{(2,2)}$, $\mathcal{J}_{3,6}^{(2,2)}$ *with* $c_1 = 1$ *or* $\mathcal{J}_{3,7}^{(2,2)}$ *with* $c_1 = 0$.

$\mathcal{S}_{3,8}^3$: *Group* $\mathbf{g}_{17}(l = 2, \rho_1 = \rho_2 = 0, \alpha_1 = 0, \alpha_2 = 1)$ *with one system of imprimitivity, Lie algebra* $\mathbf{l}_{3,4}$, *Janet basis type* $\mathcal{J}_{3,4}^{(2,2)}$, $\mathcal{J}_{3,6}^{(2,2)}$ *or* $\mathcal{J}_{3,7}^{(2,2)}$.

$\mathcal{S}_{3,9}^3(c)$: *Group* $\mathbf{g}_{17}(l = 2, \rho_1 = \rho_2 = 0, \alpha_1 = 1, \alpha_2 = c)$ *with one system of imprimitivity, Lie algebra* $\mathbf{l}_{3,2}(c)$, *Janet basis type* $\mathcal{J}_{3,4}^{(2,2)}$, $\mathcal{J}_{3,6}^{(2,2)}$ *or* $\mathcal{J}_{3,7}^{(2,2)}$.

$\mathcal{S}_{3,10}^3$: *Group* $\mathbf{g}_{24}$ *with a one-parameter system of imprimitivity, Lie algebra* $\mathbf{l}_{3,2}(c = 1)$, *Janet basis type* $\mathcal{J}_{3,6}^{(2,2)}$.

Four-parameter groups

$\mathcal{S}_{4,1}^3$: *Group* $\mathbf{g}_6$ *with two systems of imprimitivity, Lie algebra* $\mathbf{l}_{4,10}$, *Janet basis type* $\mathcal{J}_{4,9}^{(2,2)}$, $\mathcal{J}_{4,10}^{(2,2)}$ *or* $\mathcal{J}_{4,14}^{(2,2)}$.

$\mathcal{S}_{4,2}^3$: *Group* $\mathbf{g}_{14}$ *with one system of imprimitivity, Lie algebra* $\mathbf{l}_{4,1}$, *Janet basis type* $\mathcal{J}_{4,9}^{(2,2)}$ *or* $\mathcal{J}_{4,12}^{(2,2)}$.

$\mathcal{S}_{4,3}^3(c)$: *Group* $\mathbf{g}_{19}(r = 2)$ *with one system of imprimitivity, Lie algebra* $\mathbf{l}_{4,5}(c)$, *Janet basis type* $\mathcal{J}_{4,9}^{(2,2)}$ *or* $\mathcal{J}_{4,14}^{(2,2)}$.

$\mathcal{S}_{4,4}^3$: *Group* $\mathbf{g}_{20}(r = 2)$ *with one system of imprimitivity, Lie algebra* $\mathbf{l}_{4,4}$, *Janet basis type* $\mathcal{J}_{4,9}^{(2,2)}$ *or* $\mathcal{J}_{4,14}^{(2,2)}$.

Six-parameter group

$\mathcal{S}_6^3$: *Janet basis of type* $\{\xi_y, \eta_x, \xi_{xxx}, \eta_{yyy}\}$, *type* $\{\xi_y, \eta_y, \xi_{xxx}, \eta_{xxx}\}$, *type* $\{\eta_x, \eta_y, \xi_{xy}, \xi_{xxx}, \xi_{yyy}\}$ *or type* $\{\eta_x, \eta_y, \xi_{yy}, \xi_{xxx}, \xi_{xxy}\}$.

PROOF *One-parameter group.* A one-parameter symmetry is uniquely determined by the Janet basis type.

Two-parameter groups. The first two symmetry classes are singled out by their one-parameter system of imprimitivity by Theorem 3.20. The further distinction between the two groups is obtained through their Lie algebra by Theorem 3.23. For Janet basis type $\mathcal{J}_{2,1}^{(2,2)}$ the symmetry class is uniquely identified by Theorem 3.20 through their systems of imprimitivity. If the Janet basis type is $\mathcal{J}_{2,2}^{(2,2)}$, $\mathcal{J}_{2,4}^{(2,2)}$ or $\mathcal{J}_{2,5}^{(2,2)}$ the symmetry class is unique.

Three-parameter groups. If there is a type $\mathcal{J}_{3,5}^{(2,2)}$ Janet basis, the symmetry class $\mathcal{S}_{3,5}^3$ is uniquely identified. For Janet basis type $\mathcal{J}_{3,4}^{(2,2)}$, by Theorem 3.21 again symmetry class $\mathcal{S}_{3,5}^3$ is uniquely identified by its two systems of imprimitivity. The remaining cases allow a single system of imprimitivity; the

Symmetries of Differential Equations 229

symmetry type is identified from its Lie algebra by means of Theorem 3.24. For Janet basis type $\mathcal{J}_{3,6}^{(2,2)}$, Theorem 3.21 gives the unique answer $\mathcal{S}_{3,10}^3$ if a one-parameter system of imprimitivity is found; otherwise it combines the symmetry classes into those allowing two systems of imprimitivity, i.e., $\mathcal{S}_{3,1}^3$, $\mathcal{S}_{3,3}^3(c)$ and $\mathcal{S}_{3,5}^3$, and the remaining ones. In the former case Theorem 3.25 leads to a unique identification by its Lie algebra. In the latter the same is true except for the two symmetry classes $\mathcal{S}_{3,4}^3$ and $\mathcal{S}_{3,7}^3$, both with a $\mathbf{l}_{3,3}$ Lie algebra. They are distinguished by their different connectivity properties which are identified by $c_1 \neq 1$ or $c_1 = 1$ respectively. For Janet basis type $\mathcal{J}_{3,7}^{(2,2)}$ the discussion is identical; Theorems 3.21 and 3.26 are applied now. The distinction between symmetry classes $\mathcal{S}_{3,4}^3$ and $\mathcal{S}_{3,7}^3$ is obtained by the condition $c_1 \neq 0$ or $c_1 = 0$ respectively.

Four-parameter groups. The symmetry classes $\mathcal{S}_{4,1}^3$ and $\mathcal{S}_{4,2}^3$ are uniquely identified from the Janet basis types $\mathcal{J}_{4,10}^{(2,2)}$ and $\mathcal{J}_{4,12}^{(2,2)}$. For Janet basis type $\mathcal{J}_{4,9}^{(2,2)}$, by Theorem 3.22, symmetry class $\mathcal{S}_{4,1}^3$ is uniquely identified by its two systems of imprimitivity. The remaining three symmetry types are distinguished by their Lie algebra applying Theorem 3.27. For Janet basis type $\mathcal{J}_{4,14}^{(2,2)}$, by Theorem 3.22, again symmetry class $\mathcal{S}_{4,1}^3$ is uniquely identified through its two systems of imprimitivity. The distinction between the two remaining cases is obtained from Theorem 3.28.

Six-parameter group. The only symmetry class $\mathcal{S}_6^3$ comprising six parameters is identified from any of the Janet bases of order six without constraints on its coefficients. ⬜

For later use some algebraic properties of the various symmetry algebras are given next.

COROLLARY 5.3   *All one- and two-dimensional symmetry algebras, all three-dimensional symmetry algebras except those of symmetry classes $\mathcal{S}_{3,1}^3$ and $\mathcal{S}_{3,2}^3$ that are simple and of type $\mathbf{l}_{3,1}$, and all four-dimensional symmetry algebras except that of symmetry class $\mathcal{S}_{4,1}^3$ of type $\mathbf{l}_{4,1}$ are solvable.*

PROOF   The solvable cases follow immediately from the Definition 3.6 of solvability. The nonsolvable algebra of $\mathcal{S}_{4,1}^3$ has the Levi decomposition $\mathcal{R} \bar{\oplus} \mathcal{S}$ with radical $\mathcal{R}$ generated by $\{y\partial_y\}$ and the simple Levi factor of type $\mathbf{l}_{3,1}$ generated by $\{\partial_x, x\partial_x + \frac{1}{2}y\partial_y, x^2\partial_x + xy\partial_y\}$. The six-dimensional algebra $\mathbf{g}_{12}$ is semi-simple with the direct-sum decomposition $\{\partial_u, u\partial_u, u^2\partial_u\} \oplus \{\partial_v, v\partial_v, v^2\partial_v\}$ into two simple type $\mathbf{l}_{3,1}$ algebras. ⬜

EXAMPLE 5.27   Equation 7.2 in Kamke's collection

$$x^2 y''' + xy''(y-1) + xy'^2 - y'(y-1) = 0$$

generates the type $\mathcal{J}_{1,2}^{(2,2)}$ Janet basis $\{\eta, \xi_x - \frac{1}{x}\xi, \xi_y\}$; consequently it belongs to symmetry class $\mathcal{S}_1^3$. ⬜

230

EXAMPLE 5.28   The equation

$$y'''y'y - 3y''^2y + y''y'(2xyy' + 3y' + 2y^2) - y'^3(y'^2 + 2xy' + 4y) = 0$$

generates the type $\mathcal{J}_{2,3}^{(2,2)}$ Janet basis $\{\xi_x + \frac{1}{y}\eta, \xi_y + \frac{1}{y}\xi, \eta_x, \eta_y\}$ with nonvanishing coefficients $a_1 = b_2 = \frac{1}{y}$. By Theorem 3.23 it generates a $l_{2,2}$ Lie algebra, therefore by Theorem 5.14 its symmetry class is $\mathcal{S}_{2,1}^3$.   ☐

EXAMPLE 5.29   Equation 7.7 from Kamke's collection $yy''' - y'y'' + y^2y' = 0$ generates the type $\mathcal{J}_{2,3}^{(2,2)}$ Janet basis $\{\xi_x + \frac{1}{y}\eta, \xi_y, \eta_x, \eta_y - \frac{1}{y}\eta\}$ with nonvanishing coefficients $a_1 = \frac{1}{y}$, $d_1 = -\frac{1}{y}$. By Theorem 3.23 it generates a $l_{2,1}$ Lie algebra; consequently, by Theorem 5.14 its symmetry class is $\mathcal{S}_{2,2}^3$.   ☐

EXAMPLE 5.30   The equation

$$y'''y'^3x^3 + x^3y''^3 - 3x(2y' + x)(xy' - 2)y'y'' - x^3(y + 1)y'^6 - 6xy'^4 + 8y'^3 = 0$$

generates the type $\mathcal{J}_{2,4}^{(2,2)}$ Janet basis; consequently it belongs to symmetry class $\mathcal{S}_{2,3}^3$.   ☐

EXAMPLE 5.31   The equation

$$y'''y'^2 - (3y' + \frac{4}{x^2})y''^2 + y'\left(y'^2 - \frac{6}{x}y' + \frac{4y}{x^3}\right)y'' + x^2y'^6 + \frac{2}{x}y'^4 - \frac{6}{x^2}y'^3 + \frac{4y}{x^4}y'^2\right) = 0$$

generates the type $\mathcal{J}_{2,5}^{(2,2)}$ Janet basis; consequently it belongs to symmetry class $\mathcal{S}_{2,4}^3$.   ☐

EXAMPLE 5.32   The equation

$$(y'''y' - 4y''^2)(y - x)^2 + 10y''y'(y' + 1)(y - x) - y'^2(10y'^2 + 3y' + 10) = 0$$

generates a type $\mathcal{J}_{3,6}^{(2,2)}$ Janet basis for its symmetry generators with nonvanishing coefficients $c_1 = 1$, $c_2 = -c_3 = \frac{2}{x - y}$, $d_1 = -\frac{2}{x - y}$ and $d_3 = -d_2 = \frac{2}{(x - y)^2}$. By Theorem 3.21 it allows two systems of imprimitivity. By Theorem 3.25, its Lie algebra is $l_{3,1}$. Consequently, by Theorem 5.14 its symmetry class is $\mathcal{S}_{3,1}^3$.   ☐

EXAMPLE 5.33   The equation

$$y'''y'y^6 - 3y''^2y^6 + \left(6y'y + \frac{2}{x^2}y' + \frac{3y^2}{x}\right)y''y'y^4 - \frac{1}{x^5}y'^5 - 2\left(3y + \frac{2}{x^2}\right)y'^4y^3 - \frac{6y^5}{x}y'^3 = 0$$

generates a type $\mathcal{J}_{3,7}^{(2,2)}$ Janet basis for its symmetry generators with nonvanishing coefficients $a_3 = -\frac{1}{x}$, $c_2 = -\frac{2}{y}$, $c_3 = -\frac{2}{x}$ and $d_1 = \frac{2}{y}$. By Theorem 3.22

Symmetries of Differential Equations 231

it allows a single system of imprimitivity, by Theorem 3.26 its Lie algebra is $l_{3,1}$. Consequently, by Theorem 5.14 its symmetry class is $\mathcal{S}^3_{3,2}$. $\square$

EXAMPLE 5.34   The equation

$$\left(y''' + \frac{3}{x}y''\right)\left(y'^3 - \frac{3}{x}y'^2y + \frac{3}{x^2}y'y^2 - \frac{1}{x^3}y^3\right) + \frac{1}{x^8} = 0$$

generates a type $\mathcal{J}^{(2,2)}_{3,6}$ Janet basis for its symmetry generators with coefficients $a_1 = 0$, $b_1 = \frac{y}{2x}$, $c_1 = -\frac{3}{2}$. By Theorem 3.21 it allows two systems of imprimitivity. By Theorem 3.25 its Lie algebra is $l_{3,2}(\frac{3}{2})$. Consequently, by Theorem 5.14 its symmetry class is $\mathcal{S}^3_{3,3}(\frac{3}{2})$. $\square$

EXAMPLE 5.35   The equation $y'''y'^2 - 3y''^2y' - y''^2 - y'^6e^{-2/y'} = 0$ generates a type $\mathcal{J}^{(2,2)}_{3,6}$ Janet basis for its symmetry generators with coefficients $a_1 = c_1 = -1$. By Theorem 3.21 it allows a single system of imprimitivity. By Theorem 3.25 its Lie algebra is $l_{3,3}$. Consequently, by Theorem 5.14 its symmetry class is $\mathcal{S}^3_{3,4}$. $\square$

EXAMPLE 5.36   The equation

$$x^2y'''y'y^4 - 4x^2y''^2y^4 + 10x^2y''y'^2y^3 - 10xy''y'y^4$$
$$-10x^2y'^4y^2 + 2x^2y'^4 + 20xy'^3y^3 - 10y'^2y^4 = 0$$

generates a type $\mathcal{J}^{(2,2)}_{3,6}$ Janet basis for its symmetry generators with nonvanishing coefficients $c_2 = -\frac{2}{y}$, $d_1 = -\frac{2}{x}$ and $d_3 = \frac{2}{x^2}$. By Theorem 3.21 it allows two systems of imprimitivity. By Theorem 3.25 its Lie algebra is $l_{3,4}$. Consequently, by Theorem 5.14 its symmetry class is $\mathcal{S}^3_{3,5}$. $\square$

EXAMPLE 5.37   Consider the following four equations generating type $\mathcal{J}^{(2,2)}_{3,4}$ Janet bases for its symmetry generators, the nonvanishing coefficients of which are also given.

$$x^{10}y'''y' + 2x^9y'''y' + x^8(6x - 1)y''^2 + 2x^7(9x - 2)y''y' + 4x^6(3x - 1)y'^2 - 1 = 0$$

with nonvanishing coefficients $a_3 = -\frac{2}{x}$ and $d_1 = \frac{2}{x}$.

$$x^{10}y'''y'' + 2x^8(x + 1)y'''y' + x^6y'''y + x^8(6x + 1)y''^2$$
$$+x^6(18x^2 + 16x + 1)y''y' + 6x^5y''y$$
$$+2x^4(x + 1)(6x^2 + 2x - 1)y'^2 + 3x^2(2x^2 - 1)y'y - y^2 - 1 = 0$$

with nonvanishing coefficients $a_3 = -\frac{2}{x}$, $d_1 = \frac{2}{x} + \frac{2}{x^2}$ and $d_2 = \frac{1}{x^4}$.

$$x^{10}y'''y'' + x^8(2x + 1)y'''y' + 6x^9y''^2 + x^6(18x^2 + 6x - 1)y''y'$$
$$+x^4(6x^2 - 1)(2x + 1)y'^2 - 1 = 0$$

232

with nonvanishing coefficients $a_3 = -\frac{2}{x}$ and $d_1 = \frac{2}{x} + \frac{1}{x^2}$.

$$x^{10}y'''y'' + 2x^8(x+2)y'''y' + 3x^6y'''y + 3x^8(2x+1)y''^2$$
$$+x^6(18x^2 + 36x + 11)y''y' + 6x^4(3x+1)y''y$$
$$+2x^4(x+2)(6x^2 + 6x - 1)y'^2 + 3x^2(6x^2 + 4x - 5)y'y - 9y^2 - 1 = 0$$

with nonvanishing coefficients $a_3 = -\frac{2}{x}$, $d_1 = \frac{2}{x} + \frac{4}{x^2}$ and $d_2 = \frac{2}{x^4}$. There are altogether six possible symmetry classes for the Janet basis type $\mathcal{J}_{3,4}^{(2,2)}$. By Theorem 3.21 it follows that in all four cases there is a single system of imprimitivity; consequently, symmetry class $\mathcal{S}_{3,5}^3$ is excluded. By Theorem 3.24 the Lie algebras $\mathfrak{l}_{3,5}$, $\mathfrak{l}_{3,3}$, $\mathfrak{l}_{3,4}$ and $\mathfrak{l}_{3,2}(\frac{1}{2})$ are identified. From the classification in Theorem 5.14 the symmetry classes are $\mathcal{S}_{3,6}^3$, $\mathcal{S}_{3,7}^3$, $\mathcal{S}_{3,8}^3$ and $\mathcal{S}_{3,9}^3$ respectively. ⬜

EXAMPLE 5.38   Consider the equation

$$x^2y'''y' + x^6y''^2y'^2 + x^2y''^2 + 4x^5y''y'^3 + 10xy''y' + 4x^4y'^4 + 10y'^2 = 0.$$

It generates a type $\mathcal{J}_{3,6}^{(2,2)}$ Janet basis for its symmetry generators with nonvanishing coefficients $c_1 = -1$, $c_3 = \frac{2}{x}$, $d_1 = -\frac{2}{x}$ and $d_3 = \frac{2}{x^2}$. By Theorem 3.21 it allows a one-parameter system of imprimitivity, by Theorem 5.14 its symmetry class $\mathcal{S}_{3,10}^3$ is identified. ⬜

EXAMPLE 5.39   Consider the equation

$$y'''y'y^3\left(y' + \frac{1}{x}y\right) - y''^2y^3\left(3y' + \frac{1}{2x}y\right)$$
$$-y''y'^2y^2\left(y' + \frac{11}{x}y\right) + 4xy'^6 + 16y'^5y + \frac{22}{x}y'^4y^2 = 0.$$

It generates a type $\mathcal{J}_{4,14}^{(2,2)}$ Janet basis for its symmetry generators with nonvanishing coefficients $a_1 = a_2 = a_3 = a_4 = 0$, $b_1 = \frac{y}{x}$ and $b_2 = -1$. By Theorem 3.22 it allows two systems of imprimitivity, therefore by Theorem 5.14 its symmetry class is $\mathcal{S}_{4,1}^3$. ⬜

EXAMPLE 5.40   Consider the equation

$$y''' - x^8y''^3 - 6x^6(2xy' + y)y''^2 - 3\left(16x^6y'^2 + 16x^5y'y + 4x^4y^2 - \frac{3}{x}\right)y''$$
$$-64x^5y'^3 - 96x^4y'^2y - 6\left(8x^3y^2 - \frac{3}{x^2}\right)y' - 8x^2y^3 + \frac{6}{x^3}y = 0$$

with Janet basis type $\mathcal{J}_{4,9}^{(2,2)}$ for its symmetry generators with nonvanishing coefficients $b_2 = -\frac{3}{2}$, b_4, c_2, c_4 and all $d_i's$. By Theorem 3.22 it allows a single system of imprimitivity, by Theorem 3.28 its Lie algebra is $\mathfrak{l}_{4,5}(\frac{3}{2})$. Consequently, by Theorem 5.14 its symmetry class is $\mathcal{S}_{4,3}^3(\frac{3}{2})$. ⬜

Symmetries of Differential Equations 233

Equations with a Six-Parameter Symmetry. Although equations with this symmetry type are covered by the above theorem, a different proceeding will be described next. It does not require the Janet basis for the determining system. Later on, for third order equations allowing a seven-parameter symmetry group, a similar method will be applied exclusively.

THEOREM 5.15   *If a quasilinear third order ode belongs to symmetry class $\mathcal{S}_6^3$, either of the following two cases applies where $A_j \equiv A_j(x,y)$ and $B_k \equiv B_k(x,y)$ for all j and k.*
i) The equation has the structure

$$(A_0 y'^2 + A_1 y' + 1)y''' - (3A_0 y' + \tfrac{3}{2}A_1)y''^2 + (A_2 y'^3 + A_3 y'^2 + A_4 y' + A_5)y''$$

$$+B_1 y'^6 + B_2 y'^5 + B_3 y'^4 + B_4 y'^3 + B_5 y'^2 + B_6 y' + B_7 = 0.$$

$$(5.40)$$

Necessary coefficient constraints for symmetry class $\mathcal{S}_6^3$ are as follows.

$$A_0 A_{1,x} - \tfrac{1}{2}A_0 A_1 A_5 + \tfrac{1}{3}A_0 A_4 - \tfrac{1}{4}A_{1,x}A_1^2 + \tfrac{1}{12}A_1^3 A_5 - \tfrac{1}{6}A_1 A_3 + \tfrac{1}{3}A_2 = 0,$$

$$A_0^2 A_5 + 3A_0 A_{1,y} - \tfrac{1}{2}A_0 A_1^2 A_5 - \tfrac{1}{2}A_0 A_1 A_4 + A_0 A_3 - \tfrac{3}{4}A_{1,y}A_1^2$$

$$+\tfrac{1}{4}A_1^3 A_4 - \tfrac{1}{2}A_1^2 A_3 + \tfrac{1}{2}A_1 A_2 = 0,$$

$$A_0^2 A_1 A_{5,x} - 2A_0^2 A_{5,y} - \tfrac{3}{2}A_0 A_{1,y}A_1 A_5 + A_0 A_{1,y}A_4 - \tfrac{1}{4}A_0 A_1^3 A_{5,x} + \tfrac{1}{4}A_0 A_1^3 A_5^2$$

$$-\tfrac{1}{2}A_0 A_1^2 A_{4,x} - \tfrac{1}{4}A_0 A_1^2 A_4 A_5 + A_0 A_1^2 A_{5,y} + A_0 A_1 A_{3,x} - \tfrac{1}{2}A_0 A_1 A_3 A_5$$

$$+\tfrac{1}{6}A_0 A_1 A_4^2 - 2A_0 A_{2,x} + A_0 A_2 A_5 + \tfrac{3}{8}A_{1,y}A_1^3 A_5 - \tfrac{1}{4}A_{1,y}A_1^2 A_4$$

$$+\tfrac{1}{8}A_1^4 A_{4,x} - \tfrac{1}{8}A_1^4 A_4 A_5 - \tfrac{1}{8}A_1^4 A_{5,y} - \tfrac{1}{4}A_1^3 A_{3,x} + \tfrac{5}{12}A_1^3 A_3 A_5 + \tfrac{1}{12}A_1^3 A_4^2$$

$$+\tfrac{1}{2}A_1^2 A_{2,x} - \tfrac{3}{4}A_1^2 A_2 A_5 - \tfrac{1}{3}A_1^2 A_3 A_4 + \tfrac{1}{2}A_1 A_2 A_4 + \tfrac{1}{6}A_1 A_3^2 - \tfrac{1}{3}A_2 A_3 = 0,$$

$$A_0^3 A_{5,x} - \tfrac{13}{4}A_0^2 A_1^2 A_{5,x} - \tfrac{1}{2}A_0^2 A_1 A_{4,x} + \tfrac{13}{2}A_0^2 A_1 A_{5,y} + A_0^2 A_{3,x}$$

$$-A_0^2 A_{4,y} + \tfrac{1}{6}A_0^2 A_4^2 + \tfrac{3}{2}A_0 A_{1,y}^2 + \tfrac{33}{8}A_0 A_{1,y}A_1^2 A_5 - \tfrac{7}{8}A_0 A_{1,y}A_1 A_4$$

$$+3A_0 A_{1,y}A_3 + \tfrac{3}{4}A_0 A_1^4 A_{5,x} - \tfrac{17}{24}A_0 A_1^4 A_5^2 + \tfrac{13}{8}A_0 A_1^3 A_{4,x} + \tfrac{19}{24}A_0 A_1^3 A_4 A_5$$

$$-\tfrac{25}{8}A_0 A_1^3 A_{5,y} - \tfrac{13}{4}A_0 A_1^2 A_{3,x} + \tfrac{7}{6}A_0 A_1^2 A_3 A_5 + \tfrac{1}{4}A_0 A_1^2 A_{4,y} - \tfrac{13}{24}A_0 A_1^2 A_4^2$$

$$+\tfrac{11}{2}A_0 A_1 A_{2,x} - \tfrac{31}{12}A_0 A_1 A_2 A_5 + \tfrac{1}{2}A_0 A_1 A_{3,y} - \tfrac{1}{3}A_0 A_1 A_3 A_4 - A_0 A_{2,y}$$

$$+\tfrac{1}{4}A_{1,y}A_1 A_2 - \tfrac{3}{8}A_1^5 A_{4,x} + \tfrac{17}{48}A_1^5 A_4 A_5 + \tfrac{3}{4}A_1^5 A_{5,y} + \tfrac{3}{4}A_1^4 A_{3,x}$$

$$-\tfrac{29}{24}A_1^4 A_3 A_5 - \tfrac{13}{48}A_1^4 A_4^2 - \tfrac{11}{8}A_1^3 A_{2,x} + \tfrac{13}{6}A_1^3 A_2 A_5 - \tfrac{1}{8}A_1^3 A_{3,y}$$

$$+\tfrac{31}{24}A_1^3 A_3 A_4 + \tfrac{1}{4}A_1^2 A_{2,y} - \tfrac{43}{24}A_1^2 A_2 A_4 - A_1^2 A_3^2 + \tfrac{25}{12}A_1 A_2 A_3 - \tfrac{5}{6}A_2^2 = 0,$$

$$B_5 - \tfrac{5}{2}B_6 A_1 - A_0 B_7 + 4A_1^2 B_7 + \tfrac{1}{2}A_1 A_{5,x} + \tfrac{1}{2}A_1 A_5^2 - \tfrac{1}{3}A_4 A_5 - A_{5,y} = 0,$$

$$B_4 - 2B_6 A_0 - 2B_6 A_1^2 + 4A_0 A_1 B_7 + \tfrac{1}{3}A_0 A_{5,x} + \tfrac{2}{9}A_0 A_5^2 + 4A_1^3 B_7$$

$$+\tfrac{2}{3}A_1^2 A_{5,x} + \tfrac{4}{9}A_1^2 A_5^2 + \tfrac{1}{6}A_1 A_{4,x} - \tfrac{3}{2}A_1 A_{5,y} - \tfrac{1}{3}A_3 A_5 - \tfrac{1}{3}A_{4,y} - \tfrac{1}{9}A_4^2 = 0,$$

$$B_3 - 3B_6 A_0 A_1 - \tfrac{1}{2} B_6 A_1^3 + A_0^2 B_7 + 7 A_0 A_1^2 B_7 + \tfrac{13}{12} A_0 A_1 A_{5,x} + \tfrac{2}{3} A_0 A_1 A_5^2$$
$$+ \tfrac{1}{6} A_0 A_{4,x} - \tfrac{2}{9} A_0 A_4 A_5 - \tfrac{5}{3} A_0 A_{5,y} + A_1^4 B_7 + \tfrac{1}{6} A_1^3 A_{5,x} + \tfrac{1}{9} A_1^3 A_5^2$$
$$+ \tfrac{1}{8} A_1^2 A_{4,x} - \tfrac{11}{24} A_1^2 A_{5,y} + \tfrac{1}{12} A_1 A_{3,x} - \tfrac{1}{18} A_1 A_3 A_5 - \tfrac{1}{3} A_1 A_{4,y}$$
$$+ \tfrac{1}{18} A_1 A_4^2 - \tfrac{2}{9} A_2 A_5 - \tfrac{1}{6} A_{3,y} - \tfrac{2}{9} A_3 A_4 = 0,$$

$$B_2 A_0 - \tfrac{1}{4} B_2 A_1^2 - B_6 A_0^3 - \tfrac{3}{4} B_6 A_0^2 A_1^2 + \tfrac{1}{4} B_6 A_0 A_1^4 + 4 A_0^3 A_1 B_7 + A_0^2 A_1^3 B_7$$
$$+ \tfrac{1}{2} A_0^2 A_1 A_{4,x} - \tfrac{1}{2} A_0^2 A_1 A_{5,y} - \tfrac{1}{3} A_0^2 A_{3,x} - \tfrac{1}{12} A_0^2 A_4^2 + \tfrac{3}{4} A_0 A_{1,y}^2$$
$$- \tfrac{3}{16} A_0 A_{1,y} A_1^2 A_5 - \tfrac{3}{4} A_0 A_{1,y} A_1 A_4 + \tfrac{1}{3} A_0 A_{1,y} A_3 - \tfrac{1}{2} A_0 A_1^5 B_7$$
$$- \tfrac{5}{144} A_0 A_1^4 A_5^2 + \tfrac{1}{24} A_0 A_1^3 A_{4,x} + \tfrac{31}{144} A_0 A_1^3 A_4 A_5 - \tfrac{1}{24} A_0 A_1^3 A_{5,y}$$
$$- \tfrac{1}{4} A_0 A_1^2 A_{3,x} - \tfrac{1}{72} A_0 A_1^2 A_3 A_5 + \tfrac{1}{48} A_0 A_1^2 A_4^2 + A_0 A_1 A_{2,x}$$
$$- \tfrac{43}{72} A_0 A_1 A_2 A_5 - \tfrac{1}{3} A_0 A_1 A_{3,y} - \tfrac{1}{9} A_0 A_1 A_3 A_4 + \tfrac{1}{3} A_0 A_{2,y} - \tfrac{1}{6} A_0 A_2 A_4$$
$$- \tfrac{3}{16} A_{1,y}^2 A_1^2 + \tfrac{1}{32} A_{1,y} A_1^4 A_5 + \tfrac{7}{32} A_{1,y} A_1^3 A_4 - \tfrac{7}{48} A_{1,y} A_1^2 A_3 + \tfrac{1}{8} A_{1,y} A_1 A_2$$
$$- \tfrac{1}{24} A_1^5 A_{4,x} - \tfrac{1}{96} A_1^5 A_4 A_5 + \tfrac{1}{24} A_1^5 A_{5,y} + \tfrac{1}{12} A_1^4 A_{3,x} - \tfrac{5}{144} A_1^4 A_3 A_5$$
$$- \tfrac{5}{96} A_1^4 A_4^2 - \tfrac{1}{4} A_1^3 A_{2,x} + \tfrac{25}{144} A_1^3 A_2 A_5 + \tfrac{1}{12} A_1^3 A_{3,y} + \tfrac{5}{36} A_1^3 A_3 A_4$$
$$- \tfrac{1}{12} A_1^2 A_{2,y} - \tfrac{11}{144} A_1^2 A_2 A_4 - \tfrac{1}{72} A_1^2 A_3^2 - \tfrac{11}{72} A_1 A_2 A_3 + \tfrac{13}{36} A_2^2 = 0,$$

$$B_1 - \tfrac{1}{2} B_2 A_1 + \tfrac{1}{2} B_6 A_0 A_1^3 + A_0^3 B_7 - A_0^2 A_1^2 B_7 + \tfrac{1}{6} A_0^2 A_{4,x} - \tfrac{1}{2} A_0^2 A_{5,y}$$
$$- \tfrac{1}{8} A_0 A_{1,y} A_1 A_5 + \tfrac{1}{12} A_0 A_{1,y} A_4 - A_0 A_1^4 B_7 - \tfrac{3}{16} A_0 A_1^3 A_{5,x}$$
$$- \tfrac{13}{144} A_0 A_1^3 A_5^2 - \tfrac{1}{6} A_0 A_1^2 A_{4,x} - \tfrac{1}{48} A_0 A_1^2 A_4 A_5 + \tfrac{13}{24} A_0 A_1^2 A_{5,y}$$
$$+ \tfrac{5}{72} A_0 A_1 A_3 A_5 + \tfrac{1}{6} A_0 A_1 A_{4,y} + \tfrac{1}{72} A_0 A_1 A_4^2 + \tfrac{1}{6} A_0 A_{2,x} - \tfrac{5}{36} A_0 A_2 A_5$$

$$- \tfrac{1}{6} A_0 A_{3,y} + \tfrac{1}{32} A_{1,y} A_1^3 A_5 - \tfrac{1}{48} A_{1,y} A_1^2 A_4 + \tfrac{1}{96} A_1^4 A_{4,x} - \tfrac{1}{96} A_1^4 A_4 A_5$$
$$- \tfrac{1}{96} A_1^4 A_{5,y} - \tfrac{1}{48} A_1^3 A_{3,x} + \tfrac{5}{144} A_1^3 A_3 A_5 + \tfrac{1}{144} A_1^3 A_4^2 - \tfrac{1}{24} A_1^2 A_{2,x}$$
$$+ \tfrac{1}{12} A_1^2 A_{3,y} - \tfrac{1}{18} A_1^2 A_3 A_4 + \tfrac{11}{72} A_1 A_2 A_4 + \tfrac{5}{72} A_1 A_3^2 - \tfrac{5}{36} A_2 A_3 = 0.$$

ii) The equation has the structure

$$(A_0 y'^2 + y') y''' - (3 A_0 y' + \tfrac{3}{2}) y''^2 + (A_2 y'^3 + A_3 y'^2) y''$$
$$+ B_1 y'^6 + B_2 y'^5 + B_3 y'^4 + B_4 y'^3 + B_5 y'^2 = 0. \tag{5.41}$$

Its symmetry class is $\mathcal{S}_6^3$ if and only if its coefficients satisfy the following

Symmetries of Differential Equations 235

constraints.

$$B_1 - B_2 A_0 + B_3 A_0^2 - B_4 A_0^3 + B_5 A_0^4 - \tfrac{1}{6} A_3^2 A_0^2 + \tfrac{1}{3} A_3 A_2 A_0 - \tfrac{1}{6} A_2^2 = 0,$$

$$A_{0,x} + \tfrac{1}{3} A_3 = 0, \quad A_{0,y} + \tfrac{1}{3} A_2 = 0,$$

$$A_{2,y} - 2A_{2,x} A_0 - 3B_2 + 6B_3 A_0 - 8B_4 A_0^2 + 8B_5 A_0^3 - \tfrac{4}{3} A_3^2 A_0 + \tfrac{4}{3} A_3 A_2 = 0,$$

$$A_{3,x} - B_4 + 4B_5 A_0 = 0, \quad A_{3,y} - A_{2,x} = 0,$$

$$B_{4,x} - 3B_{5,y} - B_{5,x} A_0 - \tfrac{2}{3} B_5 A_3 = 0, \quad B_{3,x} - B_{4,y} - 2B_{5,y} A_0 - \tfrac{4}{3} B_5 A_2 = 0,$$

$$A_{2,x,x} - B_{4,y} + 4B_{5,y} A_0 - \tfrac{4}{3} B_5 A_2 = 0,$$

$$B_{2,xx} - \tfrac{1}{3} B_{4,yy} - \tfrac{8}{3} B_{5,yy} A_0 + \tfrac{4}{9} B_{4,y} A_3 - \tfrac{4}{9} B_{5,y} A_3 A_0 - \tfrac{20}{9} B_{5,y} A_2$$

$$- \tfrac{10}{9} A_{2,x} B_4 + \tfrac{8}{9} A_{2,x} B_5 A_0 - \tfrac{4}{3} B_2 B_5 + \tfrac{2}{3} B_3 B_4$$

$$- \tfrac{8}{9} B_4^2 A_0 + \tfrac{8}{9} B_4 B_5 A_0^2 - \tfrac{4}{27} B_4 A_3^2 + \tfrac{8}{27} B_5 A_3 A_2 = 0.$$

The proof of this theorem is similar to that of Theorem 5.11. Starting from a generic equation with structure (5.40) or (5.41) that is obtained from the canonical form $v'''v' - \tfrac{3}{2}v''^2 = 0$ (see Theorem 6.16) by a general point transformation, a Janet basis involving the undetermined functions A_j and B_k is generated. The above constraints assure that this Janet basis corresponds to a six-parameter symmetry. The intermediate calculations are very lengthy and are therefore omitted. The advantage of these criteria is that they are very fast.

EXAMPLE 5.41   Consider the equation

$$2x^3 y''' y'^2 y - 2xy''' y^3 - 6x^3 y''^2 y' y + 6x^3 y'' y'^3 - 6x^2 y'' y'^2 y$$

$$+6xy'' y' y^2 - 6y'' y^3 + 6x^2 y'^4 - 12xy'^3 y + 6y'^2 y^2 = 0$$

with the structure (5.40). Its coefficients

$$A_0 = -\frac{x^2}{y^2}, \ A_1 = 0, \ A_2 = -\frac{3x^2}{y^3}, \ A_3 = \frac{3x}{y^2}, \ A_4 = -\frac{3}{y}, \ A_5 = \frac{3}{x},$$

$$B_1 = B_2 = 0, \ B_3 = -\frac{3x}{y^3}, \ B_4 = \frac{6}{y^2}, \ B_5 = -\frac{3}{xy}, \ B_6 = B_7 = 0$$

satisfy the necessary constraints of Theorem 5.15, case i). Furthermore, the Janet basis

$$\eta_x - \frac{y^2}{x^2} \xi_y = 0, \quad \eta_y - \xi_x - \tfrac{1}{y}\eta + \tfrac{1}{x}\xi = 0,$$

$$\xi_{y,y} - \frac{x^2}{y^2} \xi_{x,x} + \tfrac{1}{y}\xi_y + \frac{x}{y^2}\xi_x - \frac{1}{y^2}\xi = 0, \quad \xi_{xxx} = 0, \quad \xi_{xxy} - \tfrac{1}{x}\xi_{x,y} = 0$$

is generated. Consequently, it belongs to symmetry class $\mathcal{S}_6^3$. If the above ode is slightly changed, e. g. by adding the term y, i.e., $B_7 = y$, the symmetry

236

is destroyed. This may be discovered by testing the coefficient constraints of Theorem 5.15 which is very fast, avoiding in this way the much more elaborate Janet basis calculation. ⬜

EXAMPLE 5.42   Consider the equation

$$\left[\frac{x}{(x-1)y}y'^2 + y' \right] y''' - \left[\frac{3x}{(x-1)y}y' + \frac{3}{2} \right] y''^2$$

$$+ \left[\frac{3x}{(x-1)y^2}y'^3 + \frac{3}{(x-1)^2y}y'^2 \right] y'' + \frac{3x(x-2)}{(x-1)^2y^2}y'^4$$

$$+ \frac{2(x^2 - 3x + 3)}{(x-1)^2y}y'^3 + \frac{x^2 - 4x + 6}{2(x-1)^2}y'^2 = 0$$

with the structure (5.41). Its coefficients

$$A_0 = \frac{x}{xy - y}, \quad A_2 \frac{3x}{xy^2 - y^2}, \quad A_3 = \frac{3}{x^2y - 2xy + y}, \quad B_1 = B_2 = 0,$$

$$B_3 = \frac{3x^2 - 6x}{x^2y^2 - 2xy^2 + y^2}, \quad B_4 = \frac{2x^2 - 6x + 6}{x^2y - 2xy + y}, \quad B_5 = \frac{x^2 - 4x + 6}{2x^2 - 4x + 2}$$

satisfy the constraints of Theorem 5.15, case ii). Consequently, it belongs to symmetry class $\mathcal{S}_6^3$. ⬜

5.5   Symmetries of Linearizable Equations

For order higher than two, linearizable equations comprise more than a single equivalence class as it has been shown in Theorem 4.9. Moreover, they combine into several symmetry classes as it will be seen in this section. Because the methods for solving linear equations are different from the nonlinear ones, a special section is devoted to them. At first equations of third order are considered; the results for them are more detailed than for arbitrary order.

Classification of Symmetries. At first a complete survey of all possible symmetry types is provided.

THEOREM 5.16   *Any symmetry algebra of a third order linearizable ode is similar to one in canonical variables u and $v \equiv v(u)$ as given in the following listing. In addition the corresponding Janet basis is given where $\alpha(u, v)$ and $\beta(u, v)$ are the coefficients of ∂_u and ∂_v respectively.*
Four-parameter symmetry
$\quad \mathcal{S}_{4,5}^3 : \mathbf{g}_{16}(r = 3) = \{\partial_v, u\partial_v, \phi(u)\partial_v, v\partial_v\}.$

$\quad$ *Janet basis* $\left\{ \alpha, \beta_{uv}, \beta_{vv}, \beta_{uuu} - \frac{\phi'''}{\phi''}\beta_{uu} \right\}.$
Five-parameter symmetry

$$\mathcal{S}_{5,1}^3 : \mathbf{g}_{18}(l = 2, \rho_1 = 1, \rho_2 = 0, \alpha_1 = 0, \alpha_2 = 1) = \{\partial_v, u\partial_v, e^u\partial_v, v\partial_v, \partial_u\}.$$

Janet basis: $\{\alpha_u, \alpha_v, \beta_{uv}, \beta_{vv}, \beta_{uuu} - \beta_{uu}\}.$

$$\mathcal{S}_{5,2}^3(a) \equiv \mathbf{g}_{18}(l = 3, \rho_1 = \rho_2 = \rho_3 = 0, \alpha_1 = 0, \alpha_2 = 1, \alpha_3 = a) =$$
$$\{\partial_v, e^u\partial_v, e^{au}\partial_v, v\partial_v, \partial_u\}, \ a \neq 0, 1.$$

Janet basis: $\{\alpha_u, \alpha_v, \beta_{uv}, \beta_{vv}, \beta_{uuu} - (a + 1)\beta_{uu} + a\beta_u\}.$

Seven-parameter symmetry

$$\mathcal{S}_7^3 : \mathbf{g}_{23}(r = 3) = \{\partial_v, u\partial_v, u^2\partial_v, v\partial_v, \partial_u, u\partial_u, u^2\partial_u + 2uv\partial_v\}.$$

Janet basis: $\{\alpha_v, \beta_{uv} - \alpha_{uu}, \beta_{vv}, \alpha_{uuu}, \beta_{uuu}\}.$

PROOF   The equivalence classes of third order equations with a linear representative have been determined in Theorem 4.9. The canonical equation $v''' = 0$ for $\mathcal{E}_1^3$ generates the Janet basis of the seven-parameter group $\mathbf{g}_{23}(r = 3)$ given in the listing on page 142. It corresponds to symmetry class $\mathcal{S}_7^3$. The canonical equation $v''' + v = 0$ for $\mathcal{E}_2^3$ has been shown to be equivalent to $\bar{v}''' - (a + 1)\bar{v}'' + a\bar{v} = 0$ with $a = \frac{1}{2}(1 \pm i\sqrt{2})$ in Lemma 4.3. Similarly, the canonical equations $v''' + cv' + v = 0$ for $c \neq 0$, $c \neq -\frac{3}{2}\sqrt[3]{2}$ of $\mathcal{E}_3^3(c)$ have been shown to be equivalent to equations of the same form if c and a are related by (4.32). Equations of this latter form generate the Janet basis $\{\alpha_u, \alpha_v, \beta_{uv}, \beta_{vv}, \beta_{uuu} - (a + 1)\beta_{uu} + a\beta_u\}$. From the listing on page 142 the group $\mathbf{g}_{18}(l = 3, \rho_1 = \rho_2 = \rho_3 = 0, \alpha_1 = 0, \alpha_2 = 1, \alpha_3 = a)$ is identified as a symmetry group. It corresponds to symmetry class $\mathcal{S}_{5,2}^3(a)$. If $c = -\frac{3}{2}\sqrt[3]{2}$ in the canonical form of $\mathcal{E}_3^3(c)$, it is equivalent to $v''' - v'' = 0$ with Janet basis $\{\alpha_u, \alpha_v, \beta_{uv}, \beta_{vv}, \beta_{uuu} - \beta_{uu}\}$. From the listing on page 142 group $\mathbf{g}_{18}(l = 2, \rho_1 = 1, \rho_2 = 0, \alpha_1 = 0, \alpha_2 = 1)$ is identified as a symmetry group corresponding to symmetry class $\mathcal{S}_{5,1}^3$. Finally the canonical representative $v''' + Q(x)v = 0$ of $\mathcal{E}_4^3(Q)$ generates the Janet basis $\{\alpha, \beta_{uv}, \beta_{vv}, \beta_{uuu} + Q\beta\}$. From the listing on page 142 the group $\mathbf{g}_{16}(r = 3)$ for symmetry class $\mathcal{S}_{4,5}^3$ is obtained.   ⬚

In Section 4.1 on page 172 the equivalence classes of linearizable third order equations have been completely identified. The subsequent corollary combines these results with the symmetry classes obtained in the preceding theorem.

COROLLARY 5.4   *The symmetry classes of linear third order equations are composed of equivalence classes as follows.*

$$\mathcal{S}_7^3 \equiv \mathcal{E}_1^3, \ \ \mathcal{S}_{5,1}^3 \equiv \mathcal{E}_3^3(-\tfrac{3}{2}\sqrt{2}),$$

$$\mathcal{S}_{5,2}^3(a) \equiv \mathcal{E}_2^3 \cup \mathcal{E}_3^3(c) \ \text{ with } \ c \neq -\tfrac{3}{2}\sqrt{2}, \ \mathcal{S}_{4,5}^3 \equiv \mathcal{E}_4^3(Q).$$

The constants a and c for the five-parameter symmetries are related by (4.32).

The symmetries identified above will be applied in the next chapter for solving the corresponding equations. To this end the algebraic structure of these symmetry algebras is described next.

238

Corollary 5.5   *The symmetry algebras of the four- and five-dimensional symmetries are solvable. The seven-dimensional algebra has the Levi decomposition $\mathcal{R} \rtimes \mathcal{S}$ with radical $\mathcal{R} = \{\partial_v, u\partial_v, u^2\partial_v, v\partial_v\}$ of type $\mathfrak{l}_{4,6}(a = b = 1)$ and the simple Levi factor $\mathcal{S} = \{\partial_u, u\partial_u + v\partial_v, u^2\partial_u + 2uv\partial_v\}$ of type $\mathfrak{l}_{3,1}$.*

Proof   The solvable cases follow immediately from the Definition 3.6 of solvability, the seven-dimensional algebra has been considered before in Example 3.6. $\qquad\square$

Determining the Symmetry Class. Based on the above classification of symmetries, the subsequent theorem provides algorithmic means for determining the symmetry class of any given quasilinear linearizable third order ode. The Janet basis types required for its identification are listed in Appendix B on page 380.

Theorem 5.17   *The following criteria provide a decision procedure for identifying the symmetry class of a linearizable third order ode if its Janet basis in grlex term ordering with $\eta > \xi$, $y > x$ is given.*

Four-parameter symmetry

$\mathcal{S}^3_{4,5}$: *Janet basis of type $\mathcal{J}^{(2,2)}_{4,2}$, type $\mathcal{J}^{(2,2)}_{4,17}$ or type $\mathcal{J}^{(2,2)}_{4,19}$.*

Five-parameter symmetries

$\mathcal{S}^3_{5,1}$ *and* $\mathcal{S}^3_{5,2}(a)$: *Three Janet basis types may occur. In all three cases the parameter a determining the symmetry type is a solution of the equation*

$$P^2(a^2 - a + 1)^3 + Q^2 R(a - \tfrac{1}{2})^2(a+1)^2(a-2)^2) = 0. \tag{5.42}$$

If $a \neq 0$ the symmetry class is $\mathcal{S}^3_{5,2}(a)$ and $\mathcal{S}^3_{5,1}$ otherwise. Its coefficients P, Q and R for the various cases are given below.

i) Janet basis type $\mathcal{J}^{(2,2)}_{5,1}$.

$$P = -\tfrac{9}{2}\big(a_{1,x}a_1 + \tfrac{1}{3}a_{1,x}e_1 + c_{3,x} - a_1^3 - \tfrac{2}{3}a_1^2 e_1 - 2a_1 c_3$$
$$- \tfrac{1}{3}a_1 e_1^2 + a_1 e_3 + c_1 e_2 - \tfrac{2}{27}e_1^3 + \tfrac{1}{3}e_1 e_3 - e_4\big),$$
$$Q = a_{1,x} - 2a_1^2 - a_1 e_1 - 3c_3 - \tfrac{1}{3}e_1^2 + e_3, \quad R = 3Q.$$

ii) Janet basis type $\mathcal{J}^{(2,2)}_{5,2}$.

$$P = -\tfrac{9}{2}\big(b_{1,y}b_1 + \tfrac{1}{3}b_{1,y}e_1 + d_{2,y} - b_1^3 - \tfrac{2}{3}b_1^2 e_1 - 2b_1 d_2$$
$$- \tfrac{1}{3}b_1 e_1^2 + b_1 e_2 + c_1 e_3 - \tfrac{2}{27}e_1^3 + \tfrac{1}{3}e_1 e_2 - e_5\big),$$
$$Q = b_{1,y} - 2b_1^2 - b_1 e_1 - 3d_2 - \tfrac{1}{3}e_1^2 + e_2, \quad R = 3Q.$$

$$\text{Symmetries of Differential Equations} \qquad 239$$

iii) Janet basis type $\mathcal{J}_{5,3}^{(2,2)}$.

$$P = \tfrac{9}{2}\left(a_{3,x}a_3c_1^3 - \tfrac{1}{3}a_{3,x}c_1^2 e_1 + a_{3,x}c_1 c_3 + c_{5,x} + a_2 a_3^2 c_1^3 - \tfrac{1}{3}a_2 a_3 c_1^2 e_1\right.$$
$$+ a_2 a_3 c_1 c_3 - a_2 c_5 + 2a_3^3 c_1^4 - a_3^2 c_1^3 e_1 + 3a_3^2 c_1^2 c_3 + a_3 c_1^3 e_2$$
$$+ \tfrac{1}{3}a_3 c_1^2 e_1^2 - a_3 c_1^2 e_3 + 2a_3 c_1 c_5 + a_3 c_4 - \tfrac{1}{3}c_1^2 e_1 e_2$$
$$\left. + c_1 c_3 e_2 - \tfrac{2}{27}c_1 e_1^3 + \tfrac{1}{3}c_2 e_1 e_3 - c_1 e_5 - e_4\right),$$
$$Q = a_{3,x}c_1^2 + a_2 a_3 c_1^2 + 3a_3^2 c_1^3 - a_3 c_1^2 e_1 + 3a_3 c_1 c_3 + c_1^2 e_2$$
$$+ \tfrac{1}{3}c_1 e_1^2 - c_1 e_3 + 3c_5, \qquad R = \tfrac{3}{c_1}Q.$$

Seven-parameter symmetry

$\mathcal{S}_7^3$: *Janet basis of type* $\{\xi_y, \eta_{xy}, \eta_{yy}, \xi_{xxx}, \eta_{xxx}\}$, *type* $\{\eta_x, \xi_{xx}, \eta_{yy}, \xi_{xyy}, \xi_{yyy}\}$ *or type* $\{\eta_y, \xi_{yy}, \eta_{xx}, \xi_{xxx}, \xi_{xxy}\}$.

PROOF A Janet basis of type $\mathcal{J}_{4,2}^{(2,2)}$, $\mathcal{J}_{4,17}^{(2,2)}$ or $\mathcal{J}_{4,19}^{(2,2)}$ identifies the symmetry class $\mathcal{S}_{4,5}^3$ uniquely because these Janet basis types do not occur for any other four-parameter symmetry as it may be seen from Theorem 5.14. If the equations between the generic coefficients of a Janet basis of type $\mathcal{J}_{5,1}^{(2,2)}$, $\mathcal{J}_{5,2}^{(2,2)}$ or $\mathcal{J}_{5,3}^{(2,2)}$ and the transformation functions $\sigma(x,y)$ and $\rho(x,y)$ are transformed into a Janet basis, one of the equations is (5.42) from which the value of a and thereby the symmetry class $\mathcal{S}_{5,2}^3(a)$ may be obtained if $a \neq 0$, or $\mathcal{S}_{5,1}^3$ if $a = 0$. A Janet basis allowing a seven-parameter symmetry group identifies uniquely symmetry class $\mathcal{S}_7^3$. ▯

In the first three examples equations in symmetry class $\mathcal{S}_{4,5}^3$ are discussed. As it will turn out in the next chapter, they are all equivalent to each other.

EXAMPLE 5.43 The equation

$$y'''y'^2 - 6y''y'y + \frac{3}{x}y''y'^2 + 6y'^3 - \frac{6}{x}y'^2 y - \frac{8x}{2x-1}y'y^2 = 0$$

generates the type $\mathcal{J}_{4,2}^{(2,2)}$ Janet basis

$$\xi = 0, \quad \eta_{xy} - \frac{2}{y}\eta_x = 0,$$
$$\eta_{yy} - \frac{2}{y}\eta_y + \frac{2}{y^2}\eta = 0, \quad \eta_{xxx} + \frac{3}{x}\eta_{xx} - \frac{8x}{2x-1}\eta_x = 0.$$

By case $a)$ of the above theorem it belongs to symmetry class $\mathcal{S}_{4,5}^3$. ▯

EXAMPLE 5.44 The equation

$$(y'''y' - 3y''^2)y^3(y - \tfrac{1}{2}) - 3y''y'^2 y^2(y - \tfrac{1}{2})$$
$$-y'^5 x(8y^3 - 6y + 3) + y'^4 y(4y^3 - 6y + 3) = 0$$

240

generates the type $\mathcal{J}_{4,19}^{(2,2)}$ Janet basis

$$\eta = 0, \quad \xi_{xx} = 0, \quad \xi_{xy} = 0,$$

$$\xi_{yyy} - \frac{3}{y}\xi_{yy} - \frac{4y^3 - 6y + 3}{y^3(y - \frac{1}{2})}\xi_y - \frac{x(8y^3 - 6y + 3)}{y^3(y - \frac{1}{2})}\xi_y + \frac{8y^3 - 6y + 3}{y^3(y - \frac{1}{2})}\xi = 0.$$

By case b) of the above theorem it belongs to symmetry class $\mathcal{S}_{4,5}^3$. ☐

EXAMPLE 5.45   The equation

$$y'''(y' + 1)(y + x - \tfrac{1}{2}) - 3y''^2(y + x - \tfrac{1}{2}) - 4y'^5 x + 4y'^4(y - 4x)$$

$$+8y'^3(2y - 3x) + 8y'^2(3y - 2x) + 4y'(4y - x) + 4y = 0$$

generates the type $\mathcal{J}_{4,17}^{(2,2)}$ Janet basis

$$\eta + \xi = 0, \quad \xi_{xy} - \xi_{xx} = 0, \quad \xi_{yy} - \xi_{xx} = 0,$$

$$\xi_{xxx} - \frac{4y}{x + y - \frac{1}{2}}\xi_y - \frac{4x}{x + y - \frac{1}{2}}\xi_x + \frac{4}{x + y - \frac{1}{2}}\xi = 0.$$

By case c) of the above theorem it belongs to symmetry class $\mathcal{S}_{4,5}^3$. ☐

EXAMPLE 5.46   The equation

$$y'''y'y + \frac{1}{x}y'''y^2 - 3y''^2 y - xy''y'^2 y + 3y''y'^2 - 2y''y'y^2 - \frac{12}{x}y''y'y$$

$$-\frac{1}{x}y''y^3 - \frac{3}{x^2}y''y^2 + 2xy'^4 + 3y'^3 y + \frac{15}{x}y'^3 - \frac{1}{x}y'^2 y^2$$

$$-\frac{3}{x^2}y'^2 y - \frac{3}{x^2}y'y^3 - \frac{9}{x^3}y'y^2 - \frac{1}{x^3}y^4 - \frac{3}{x^4}y^3 = 0$$

generates the type $\mathcal{J}_{5,3}^{2,2}$ Janet basis

$$\eta_x + \frac{y}{x}\xi_x + \frac{1}{x}\eta = 0, \qquad \eta_y + \frac{y}{x}\xi_y + \frac{1}{x}\xi = 0,$$

$$\xi_{xy} - \frac{x}{y}\xi_{xx} - \frac{1}{x}\xi_y + \frac{3}{y}\xi_x + \frac{1}{y^2}\eta - \frac{2}{xy}\xi = 0,$$

$$\xi_{yy} - \frac{x^2}{y^2}\xi_{xx} + \frac{3}{y}\xi_y + \frac{3x}{y^2}\xi_x + \frac{2x}{y^3}\eta - \frac{1}{y^2}\xi = 0,$$

$$\xi_{xxx} - \frac{xy + 6}{x}\xi_{xx} + \frac{xy^2 + 3y}{x^3}\xi_y + \frac{3xy + 15}{x^2}\xi_x + \frac{2xy + 9}{x^2 y}\eta - \frac{xy + 6}{x^3}\xi = 0.$$

By the above theorem, a five-parameter symmetry group is assured. From the Janet basis coefficients $P = -\frac{1}{3}xy^2$, $Q = -\frac{1}{3}xy$ and $R = xy$ are obtained. Substituting these values into (5.42) leads to $a^2(a - 1) = 0$ with the zero $a = 0$ corresponding to the symmetry class $\mathcal{S}_{5,1}^3$. ☐

$$\text{Symmetries of Differential Equations} \qquad 241$$

EXAMPLE 5.47   The equation

$$y'''y'y^3 - 3y''^2y^3 - 4y''y'^2y^3 - 3y''y'^2y^2 + 3xy'^5y^2$$
$$+8xy'^5y + 6xy'^5 - 3y'^4y^3 - 8y'^4y^2 - 6y'^4y = 0$$

generates the type $\mathcal{J}_{5,2}^{2,2}$ Janet basis

$$\eta_x = 0, \quad \eta_y = 0, \quad \xi_{xx} = 0, \quad \xi_{xy} + \frac{1}{y^2}\eta = 0,$$
$$\xi_{yyy} - \frac{4y+3}{y}\xi_{yy} + \frac{3y^2+8y+6}{y^2}\xi_y + \frac{3xy^2+8xy+6x}{y^3}\xi_x$$
$$+\frac{3xy^2+16xy+18x}{y^4}\eta - \frac{3y^2+8y+6}{y^3}\xi = 0.$$

The Janet basis coefficients yield $P = -\frac{10}{3}$, $Q = -\frac{7}{3}xy$ and $R = -7$. With these values (5.42) leads to

$$(a+2)(a+\tfrac{1}{2})(a-\tfrac{1}{3})(a-\tfrac{2}{3})(a-\tfrac{3}{2})(a-3) = 0.$$

The proper choice for a will be determined in the subsequent chapter. □

Third Order Equations with Maximal Symmetry. Similar to second order equations with projective symmetry, third order equations with symmetry groups comprising seven parameters deserve special attention. Due to the fact that their canonical form $v''' = 0$ does not contain parameters or undetermined functions, equations with this symmetry type have a special structure from which the $\mathcal{S}_7^3$ symmetry may be identified without generating a Janet basis. The answer has been given essentially by Zorawski [192], a student of Lie at the end of the 19th century. As usual, $\sigma(x,y)$ and $\rho(x,y)$ are the transformation functions from a canonical form to the actual variables x and y.

THEOREM 5.18   (*Zorawski 1897, Neumer 1929*) *A quasilinear third order ode belongs to symmetry class $\mathcal{S}_7^3$ if and only if either of the following two cases applies where $A_j \equiv A_j(x,y)$ and $B_k \equiv B_k(x,y)$ for all j and k.*
a) If $\sigma_x \neq 0$ the equation has the structure

$$(A_0y'+1)y''' - 3A_0y''^2 + (A_1y'^2 + A_2y' + A_3)y''$$
$$+B_1y'^5 + B_2y'^4 + B_3y'^3 + B_4y'^2 + B_5y' + B_6 = 0. \tag{5.43}$$

The coefficients satisfy the constraints

$$A_{0,xx} - \tfrac{1}{6}A_{0,x}A_3 - \tfrac{1}{6}A_0A_{3,x} + \tfrac{1}{6}A_{2,x} - \tfrac{1}{6}A_{3,y} = 0,$$
$$A_{0,y} - A_{0,x}A_0 + \tfrac{1}{6}A_0^2A_3 - \tfrac{1}{6}A_0A_2 + \tfrac{1}{6}A_1 = 0,$$

$$B_{5,x} + \tfrac{2}{3}B_5 A_3 - 2B_{6,y} - 3B_{6,x}A_0 - 5B_6 A_{0,x} - 2B_6 A_0 A_3$$

$$- \tfrac{2}{3}B_6 A_2 - \tfrac{1}{3}A_{3,xx} - \tfrac{2}{3}A_{3,x}A_3 - \tfrac{4}{27}A_3^3 = 0,$$

$$B_{5,y} - 2B_5 A_{0,x} + \tfrac{2}{3}B_5 A_0 A_3 - 7B_{6,y}A_0 + 2B_{6,x}A_0^2 + 5B_6 A_{0,x}A_0$$

$$- \tfrac{7}{6}B_6 A_0^2 A_3 - \tfrac{3}{2}B_6 A_0 A_2 + \tfrac{5}{6}B_6 A_1 + \tfrac{4}{3}A_{0,x}A_{3,x}$$

$$+ \tfrac{11}{18}A_{0,x}A_3^2 + \tfrac{1}{3}A_0 A_{3,xx} - \tfrac{1}{18}A_0 A_{3,x}A_3 - \tfrac{4}{27}A_0 A_3^3$$

$$+ \tfrac{1}{3}A_{2,xx} + \tfrac{1}{18}A_{2,x}A_3 - \tfrac{4}{3}A_{3,x,y} - \tfrac{13}{18}A_{3,y}A_3 = 0,$$

$$B_4 - 4B_5 A_0 + 10B_6 A_0^2 + A_0 A_{3,x} + \tfrac{2}{3}A_0 A_3^2 - \tfrac{1}{3}A_2 A_3 - A_{3,y} = 0,$$

$$B_3 - 6B_5 A_0^2 + 20B_6 A_0^3 + \tfrac{7}{3}A_0^2 A_{3,x} + \tfrac{23}{18}A_0^2 A_3^2 + \tfrac{1}{3}A_0 A_{2,x}$$

$$- \tfrac{5}{18}A_0 A_2 A_3 - \tfrac{7}{3}A_0 A_{3,y} - \tfrac{5}{18}A_1 A_3 - \tfrac{1}{3}A_{2,y} - \tfrac{1}{9}A_2^2 = 0,$$

$$B_2 - 4B_5 A_0^3 + 15B_6 A_0^4 + \tfrac{11}{6}A_0^3 A_{3,x} + \tfrac{17}{18}A_0^3 A_3^2 + \tfrac{1}{2}A_0^2 A_{2,x}$$

$$- \tfrac{5}{36}A_0^2 A_2 A_3 - \tfrac{11}{6}A_0^2 A_{3,y} + \tfrac{1}{6}A_0 A_{1,x} - \tfrac{1}{6}A_0 A_1 A_3$$

$$- \tfrac{1}{2}A_0 A_{2,y} - \tfrac{1}{36}A_0 A_2^2 - \tfrac{1}{6}A_{1,y} - \tfrac{7}{36}A_1 A_2 = 0,$$

$$B_1 - B_5 A_0^4 + 4B_6 A_0^5 + \tfrac{1}{2}A_0^4 A_{3,x} + \tfrac{1}{4}A_0^4 A_3^2 + \tfrac{1}{6}A_0^3 A_{2,x}$$

$$- \tfrac{1}{36}A_0^3 A_2 A_3 - \tfrac{1}{2}A_0^3 A_{3,y} + \tfrac{1}{6}A_0^2 A_{1,x} - \tfrac{1}{18}A_0^2 A_1 A_3$$

$$- \tfrac{1}{6}A_0^2 A_{2,y} - \tfrac{1}{6}A_0 A_{1,y} - \tfrac{1}{36}A_0 A_1 A_2 - \tfrac{1}{12}A_1^2 = 0.$$

b) If $\sigma_x = 0$ the equation has the structure

$$y'y''' - 3y''^2 + (A_1 y'^2 + A_2 y')y'' + B_1 y'^5 + B_2 y'^4 + B_3 y'^3 + B_4 y'^2 = 0. \quad (5.44)$$

The coefficients satisfy the constraints

$$B_4 + \tfrac{1}{3}A_{2,x} + \tfrac{1}{9}A_2^2 = 0, \quad B_3 + A_{1,x} + \tfrac{1}{3}A_1 A_2 = 0,$$

$$A_{2,y} - A_{1,x} = 0, \quad A_{1,xy} + \tfrac{2}{3}A_1 A_{1,x} + B_{2,x} = 0,$$

$$A_{1,yy} + 3B_{2,y} - 6B_{1,x} + 2A_1 A_{1,y} + \tfrac{4}{9}A_1^3 + 2(A_1 B_2 - A_2 B_1) = 0.$$

PROOF   If the expressions for the derivatives up to third order (5.3), (5.5) and (5.6) (see also (B.6), (B.7)) are substituted into the canonical form $v''' = 0$, and the resulting expression is separated w.r.t. derivatives of y, the structure (5.43) or (5.44) follows. For any equation obtained in this way, equating the coefficients of the monomials in the derivatives to the coefficients A_j and B_k of (5.43) or (5.44) leads to a system of equations that is obeyed by the transformation functions. The above constraints are obtained by transforming this system into a Janet basis. If they are satisfied it is guaranteed that a general solution involving seven constants does exist.   ⧠

In the subsequent chapter the systems of pde's determining the transformation functions to canonical form will be studied in more detail. In particular, its relation to certain invariants will become apparent.

Symmetries of Differential Equations   243

Neumer (1929) has actually given more complete results on equations with the structure (5.43) or (5.44). On pages 20-23 he determines criteria for identifying a four-, five- or seven-dimensional symmetry in terms of its coefficients. However, in the former two cases they are considerably more complicated, and the Janet bases are required anyhow for determining the transformation functions to the equivalent linear equation. Therefore they are not given here. A special case of this type of results for identifying an enlarged symmetry group of a linear equation is discussed in the Exercise 5.8.

EXAMPLE 5.48   Equation 7.8 from Kamke's collection

$$y'''y^2 - \tfrac{9}{2}y''y'y + \tfrac{15}{4}y'^3 = 0$$

has structure (5.43) with nonvanishing coefficients $A_2 = -\dfrac{9}{2y}$ and $B_3 = \dfrac{15}{4y^2}$. They satisfy the necessary constraints of Theorem 5.18, case a); consequently, this equation belongs to symmetry class $\mathcal{S}_7^3$. Alternatively, the symmetry class may be read off from its Janet basis

$$\xi_y = 0, \quad \eta_{xy} - \xi_{xx} - \frac{3}{2y}\eta_x = 0,$$

$$\eta_{yy} - \frac{3}{2y}\eta_y + \frac{3}{2y^2}\eta = 0, \quad \xi_{xxx} = 0, \quad \eta_{xxx} = 0. \qquad \square$$

EXAMPLE 5.49   Consider the equation

$$y'''y'y + \tfrac{1}{x}y'''y^2 - 3y''^2y + 3y''y'^2 - \tfrac{12}{x}y''y'y - \tfrac{3}{x^2}y''y^2$$
$$+ \tfrac{15}{x}y'^3 - \tfrac{3}{x^2}y'^2y - \tfrac{9}{x^3}y'y^2 - \tfrac{3}{x^4}y^3 = 0$$

with the structure (5.43). Its coefficients

$$A_0 = \tfrac{x}{y}, \ A_1 = \frac{3x}{y^2}, \ A_2 = \frac{-12}{y}, \ A_3 = \frac{-3}{x},$$
$$B_1 = B_2 = 0, \ B_3 = \frac{15}{y^2}, \ B_4 = \frac{-3}{xy} \ B_5 = \frac{-9}{x^2}, \ B_6 = \frac{-3y}{x^3}$$

satisfy the necessary constraints of Theorem 5.18, case a); consequently it belongs to symmetry class $\mathcal{S}_7^3$. $\qquad\square$

EXAMPLE 5.50   Consider the equation

$$y'''y' - 3y''^2 + \tfrac{9}{y}y''y'^2 + \frac{3x^2 - 6x + 6}{x(x-1)}y''y' - \frac{6x}{y^3(x-1)}y'^5 - \frac{18}{y^2}y'^4$$
$$- \frac{9x^2 - 18x + 18}{xy(x-1)}y'^3 - \frac{x^3 - 3x^2 + 6x - 6}{x^2(x-1)}y'^2 = 0$$

244

with the structure (5.44). Its coefficients

$$A_1 = \frac{9}{y}, \ A_2 = \frac{3x^2 - 6x + 6}{x(x-1)}, \ B_1 = \frac{-6x}{y^3(x-1)}, \ B_2 = \frac{-18}{y^2},$$
$$B_3 = \frac{-9x^2 + 18x - 18}{xy(x-1)}, \ B_4 = \frac{-x^3 + 3x^2 - 6x + 6}{x^2(x-1)}$$

satisfy the constraints of Theorem 5.18, case *b*); consequently, its symmetry class is $\mathcal{S}_7^3$. □

Linear Equations of Any Order. The special structure of these equations entails that their symmetries may be described in much more detail than for a general quasilinear equation. Amazingly they have been studied only recently by Krause and Michel [95, 96], and Mahomed and Leach [129]. Let a linear homogeneous equation of *n*-th order be given in rational normal form by

$$L \equiv y^{(n)} + q_2 y^{(n-2)} + q_3 y^{(n-3)} + \ldots + q_{n-1} y' + q_n y = 0. \tag{5.45}$$

The following theorem describes the possible symmetries of these equations.

THEOREM 5.19   *(Mahomed and Leach 1990, Krause and Michel 1988, 1991) The symmetry algebra of a linear nth order differential equation (5.45) with fundamental system $y_1, y_2, \ldots, y_n$ has one of three possible forms.*

i) A $(n+4)$-parameter symmetry algebra $\{y_1 \partial_y, y_2 \partial_y, \ldots, y_n \partial_y, y \partial_y\} \rtimes sl_2$. This case applies if and only if (5.45) is equivalent to $v^{(n)} = 0$.

ii) A $(n+2)$-parameter symmetry algebra $\{\partial_x, y_1 \partial_y, y_2 \partial_y, \ldots, y_n \partial_y, y \partial_y\}$. This case applies if and only if (5.45) is equivalent to a linear equation with constant coefficients.

iii) A $(n+1)$-parameter symmetry algebra $\{y_1 \partial_y, y_2 \partial_y, \ldots, y_n \partial_y, y \partial_y\}$.

It cannot have a $n+3$-dimensional symmetry algebra.

The proof of this theorem is based on a direct computation of the vector fields generating the symmetries. The details may be found in the articles by Mahomed [129] or Krause and Michel [96]. If an equation has maximal symmetry, the transformation to $v^{(n)} = 0$ is actually achieved by the Laguerre-Forsyth transformation. The special case of third order equations has been treated above in Theorem 5.17.

Linear equations with maximal symmetry group have a very special structure, in fact they may be parametrized by a single independent coefficient as the following result shows.

THEOREM 5.20   *(Mahomed and Leach 1990) For linear equations (5.45) with maximal symmetry, the coefficient p of $y^{(n-2)}$ determines the equation* *completely. For n not higher than five these equations are*

$$y''' + py' + \tfrac{1}{2}p'y = 0,$$

$$y^{(4)} + py'' + p'y' + \left(\tfrac{3}{10}p'' + \tfrac{9}{100}p^2\right)y = 0,$$

$$y^{(5)} + py''' + \tfrac{3}{2}p'y'' + \left(\tfrac{9}{10}p'' + \tfrac{4}{25}p^2\right)y' + \left(\tfrac{1}{5}p''' + \tfrac{4}{25}pp'\right)y = 0.$$

In the article by Mahomed [129] these equations are given up to order eight. The severe constraints on the coefficients of an equation with maximal symmetry described in this theorem have their counterpart in a very special structure of the solutions of such equations as it is shown next.

THEOREM 5.21   *(Krause and Michel 1988) The linear equations with maximal symmetry of Theorem 5.20 are the symmetric power of the second order source equation $M \equiv z'' + \dfrac{1}{\binom{n+1}{3}}pz = 0$, i.e., $L \equiv M^{\circledS n-1}$; a fundamental system therefore has the form $y_k = z_1^k z_2^{n-1-k}$ for $k = 0, \ldots, n-1$ where z_1, z_2 is a fundamental system for M.*

This result reduces the problem of solving a n-th order equation with maximal symmetry to a second order problem.

EXAMPLE 5.51   The equation

$$y^{(4)} - \left(\frac{5}{2} + \frac{20}{x^2}\right)y'' + \frac{40}{x^3}y' + \left(\frac{9}{16} + \frac{9}{x^2}\right)y = 0 \tag{5.46}$$

has a maximal, i.e., an eight-parameter symmetry group. The source equation $z'' - \left(\tfrac{1}{4} + \tfrac{2}{x^2}\right)z = 0$ has the two independent solutions $z_{1,2} = \left(1 \pm \tfrac{2}{x}\right)\exp\left(\mp\tfrac{x}{2}\right)$. By the above theorem a fundamental system for (5.46) is

$$y_1 = \frac{(x-2)^3}{x^2}\exp\left(\frac{3x}{2}\right), \quad y_2 = \frac{(x+2)(x-2)^2}{x^3}\exp\left(\frac{x}{2}\right),$$

$$y_3 = \frac{(x+2)^2(x-2)}{x^3}\exp\left(-\frac{x}{2}\right), \quad y_4 = \frac{(x+2)^3}{x^3}\exp\left(-\frac{3x}{2}\right).$$

It is left as Exercise 5.9 to obtain the same answer from the Loewy decomposition of (5.46). ▯

246

Exercises

EXERCISE 5.1   Prove the relation between $U^{(n)}$ and D given in Lemma 5.2.

EXERCISE 5.2   Let $y'' + r(x)y = 0$ be a second order linear ode in rational normal form with the general solution $y = C_1 y_1 + C_2 y_2$. Determine its eight symmetry generators in terms of this fundamental system.

EXERCISE 5.3   Show that any linear homogeneous ode of order n for $y \equiv y(x)$ with fundamental system $y_1, y_2, \ldots, y_n$ has $n + 1$ symmetry generators $y_k(x)\partial_y$, $k = 1, \ldots, n$, and $y\partial_y$.

EXERCISE 5.4   The form of the two equations (2.67) is constrained by the requirements that the complete system for w is linear and homogeneous and that no new integrability conditions are introduced. Determine their most general form in conformity with these requirements.

EXERCISE 5.5 $\mathcal{S}_{3,1}^3, \ldots, \mathcal{S}_{3,10}^3$ of Theorem 5.13 coincide with the listing given in Lie [113], pages 501-502. Why is this list only partially included in Theorem 5.8 ?

EXERCISE 5.6   Determine the structure invariance group for the canonical form equation $v''' + v'r\left(\frac{v''}{v'}, u\right) = 0$ of the symmetry class $\mathcal{S}_{2,3}^3$.

EXERCISE 5.7   The same for $v''' + r(u, v'') = 0$ and symmetry class $\mathcal{S}_{2,4}^3$.

EXERCISE 5.8   Show that $y''' + A(x)y' + B(x)y = 0$ allows a five-parameter symmetry group if and only if one of the following two conditions is satisfied.

$$6\frac{\Theta_3''}{\Theta_3} - 7\left(\frac{\Theta_3'}{\Theta_3}\right)^2 - 9A = 0 \quad \text{or} \quad \frac{\Theta_3'''}{\Theta_3} - 4\frac{\Theta_3'' \Theta_3'}{\Theta_3 \Theta_3} + \frac{28}{9}\left(\frac{\Theta_3'}{\Theta_3}\right)^3 + A\frac{\Theta_3'}{\Theta_3} - \frac{3}{2}A' = 0$$

where Θ_3 is defined in Theorem 4.9.

EXERCISE 5.9   Determine the fundamental system of (5.46) from its Loewy decomposition.

Chapter 6

Transformation to Canonical Form

It is assumed now that a differential equation of order two or three allows a nontrivial symmetry of a known type or, if its order is one, at least a single symmetry generator is explicitly known. This knowledge is utilized for transforming it into a canonical form corresponding to its symmetry class whenever this is possible. To this end a system of equations is constructed the solutions of which are the desired transformation functions. These equations must be such that algorithms are available for determining its solutions in well-defined function fields, e. g. rational or Liouvillian functions. A complete understanding of these equations is achieved by determining for each symmetry class the structure invariance group of its canonical form equations, and the differential invariants that may be constructed from the transformation functions.

6.1   First Order Equations

Any first order equation $y' + r(x, y) = 0$ is equivalent to the canonical form $v' = 0$ according to Theorem 4.10. However, in general there is no algorithm available for determining the transformation functions that actually generate it. Furthermore, it is not guaranteed that a symmetry generator may be explicitly obtained despite the fact that infinitely many do exist as has been explained in Section 5.2. As a consequence, in order to obtain a manageable problem it must be assumed that at least a single symmetry generator is a priori known. That means, the first order ode *together* with one or more symmetry generators is the input to the problem at issue. The following theorem, which is due to Lie [113], Kapitel 6, § 5, describes how a canonical form for the given ode may be obtained if a single nontrivial symmetry generator is known.

THEOREM 6.1   *(Lie 1881) Let the first order ode $y' + r(x, y) = 0$ allow the symmetry generator $U = \xi(x, y)\partial_x + \eta(x, y)\partial_y$. If canonical variables u and v may be determined such that it represents the translation ∂_v, the ode assumes the form $v' + s(u) = 0$ in these variables.*

248

PROOF   The first prolongation of the canonical generator ∂_v allows the invariant $\Phi(u, v')$. From this the given form of the canonical ode follows immediately.   ⧠

The crucial step in this theorem is to solve the ode $\dfrac{dx}{\xi(x, y)} = \dfrac{dy}{\eta(x, y)}$ as required by Lemma 3.4. If it fails, a canonical form may not be obtained. In this case the symmetry may be applied to generate an integrating factor as will be explained in Section 7.1.

EXAMPLE 6.1   Equation 1.15 from the collection by Kamke

$$y' + y^2 - 2x^2 y + x^4 - 2x - 1 = 0$$

allows the symmetry generator $U = \partial_x + 2x\partial_y$. By Lemma 3.4 the transformation function $\sigma(x, y)$ to canonical variables is obtained by solving $\dfrac{dy}{dx} = 2x$ with the result $u \equiv \sigma(x, y) = y - x^2$, then $v \equiv \rho(x, y)$ follows from

$$\rho = \int \frac{dy}{\sqrt{y - u}}\bigg|_{u=\sigma} = \sqrt{y - u}\,\bigg|_{u=\sigma} = x.$$

The inverse functions $x = v$, $y = u + v^2$ yield $y' = 2v + \dfrac{1}{v'}$ and finally lead to the canonical form $v' + \dfrac{1}{u^2 - 1} = 0$.   ⧠

If two generators corresponding to the groups $\mathbf{g}_{26}$ or $\mathbf{g}_{25}$ are known, the transformation to canonical form is always possible as is shown next.

THEOREM 6.2   *Let the first order ode $y' + r(x, y) = 0$ allow two symmetry generators $U_i = \xi_i(x, y)\partial_x + \eta_i(x, y)\partial_y$ for $i = 1, 2$. If they generate the group $\mathbf{g}_{25}$ or $\mathbf{g}_{26}$, a Liouvillian transformation according to Lemma 3.5 transforms it to $v' + a = 0$ where a is a constant.*

PROOF   The canonical form is an immediate consequence of the invariants given in the listing in Chapter 5.1 on page 205.   ⧠

The constant a introduced in this theorem is *not* an integration constant. Its value is uniquely tied up with the particular canonical form. The system of equations for the transformation functions given in Lemma 3.5, and the solution algorithms described in Lemma 2.9 and Lemma 2.10, imply that knowing two symmetry generators for a first order ode $y' + r(x, y) = 0$ *guarantees* that a transformation to canonical form may be achieved by Liouvillian functions which may always be determined.

EXAMPLE 6.2   The equation $y' + \dfrac{(x - y)^2 + 1}{(x - y)^2 - 1} = 0$ allows the two symmetry generators $U_1 = \partial_x + \partial_y$ and $U_2 = y\partial_x + x\partial_y$ of a type $\mathbf{g}_{25}$ group. Its transformation to canonical form has been obtained in Example 3.28 with the

result $x \equiv \phi(u, v) = u - v + \frac{1}{2v}$ and $y \equiv \psi(u, v) = u - v - \frac{1}{2v}$. It yields
$y' = \dfrac{2v^2 + (1 - 2v^2)v'}{2v^2 - (1 + 2v^2)v'}$ and finally $v' = \frac{2}{3}$. ⬚

6.2   Second Order Equations

Let a second order ode be given in actual variables x and $y \equiv y(x)$. In general it will be assumed that it is polynomial in the derivatives, and rational in the dependent and the independent variables. In addition the dependence on y'' is assumed to be linear if it is not stated otherwise explicitly. For any equation of this kind with a nontrivial symmetry new variables u and $v(u)$ are searched for such that the equation is transformed to a canonical form corresponding to its symmetry class. This process consists of two steps.

i) A system of equations is constructed that is obeyed by the desired transformation functions.

ii) Solutions of these equations in well-defined function fields have to be determined.

A priori there is no guarantee that the desired transformation functions may be found. However, due to the special structure of these equations, by the methods that have been described in Chapter 2 it is assured that large classes of transformation functions may be obtained algorithmically. A complete listing of possible equation types and the corresponding classes of solutions that may be obtained for them is given next, it is a partial summary of the results of this section. In order to arrive at these results, Lemma 2.9, 2.10 and 2.11 are applied expressing a solution in terms of integrals.

THEOREM 6.3   *In order to transform a second order quasilinear ode with a nontrivial symmetry to canonical form, the type of equations and the function fields for which solution algorithms are available may be described as follows.*

S_1^2: *First order ode, in general no algorithm available.*

$S_{2,1}^2$: *Third order system*
$$z_{xx} + a_1 z_y + a_2 z_x = 0, \ z_{xy} + b_1 z_y + b_2 z_x = 0, \ z_{yy} + c_1 z_y + c_2 z_y = 0. \quad (6.1)$$
Hyperexponential solutions may be determined.

$S_{2,2}^2$: *Liouvillian functions explicitly known.*

$S_{3,1}^2$: *Partial Riccati-like system*
$$z_x + a_1 z^2 + a_2 z + a_3 = 0, \ z_y + b_1 z^2 + b_2 z + b_3 = 0.$$
Liouvillian functions in terms of integrals over rational solutions for z.

250

$S^2_{3,2}$: *Partial Riccati-like system*

$$z_{xx} + a_1 z_x^2 + a_2 z_x + a_3 = 0, \; z_y + b_1 z_x + b_2 = 0.$$

Liouvillian functions in terms of integrals over rational solutions for z.

$S^2_{3,3}$ and $S^2_{3,4}$: *Liouvillian functions over the base field explicitly known.*

S^2_8: *Third order system (6.1) with coefficients that are rational solutions of partial Riccati-like systems*

$$z_{1,x} + z_1^2 + a_1 z_1 + a_2 z_2 + a_3 = 0, \; z_{1,y} + z_1 z_2 + b_1 z_1 + b_2 z_2 + b_3 = 0,$$
$$z_{2,x} + z_1 z_2 + c_1 z_1 + c_2 z_2 + c_3 = 0, \; z_{2,y} + z_2^2 + d_1 z_1 + d_2 z_2 + d_3 = 0.$$

Liouvillian solutions determined by first order Loewy components of (6.1).

These results show that for an equation in symmetry classes $S^2_{2,2}$, $S^2_{3,3}$ or $S^2_{3,4}$ a canonical form may *always* be obtained from its Janet basis coefficients. If it belongs to symmetry class S^2_2, $S^2_{3,1}$, $S^2_{3,2}$ or S^2_8 a canonical form may be obtained for large classes of transformation functions. An equation in symmetry class S^2_1 in general may not be transformed to canonical form.

The details of these proceedings will be described now for the various symmetry classes one after another, proofs are given for the individual cases. At first the corresponding structure invariance groups will be determined, and after that its differential invariants up to second order. As usual the notation $\Delta = \sigma_x \rho_y - \sigma_y \rho_x$ is applied.

Canonical Forms and Structure Invariance Groups. In general the canonical forms of the invariant equations for the various symmetry classes are not unique. For the special case of quasilinear equations this freedom is completely described in Theorem 6.4 below. The structure invariance groups given there have an important meaning for the respective differential equation. In the first place they describe the degree of arbitrariness for the transformation functions to canonical form. This is an important information for solving the systems of pde's describing these transformations as it will be seen later on in this chapter. Secondly, knowing the freedom that is allowed for the canonical form is a necessary prerequirement in order to obtain definite statements on the existence of exact solutions and for designing solution algorithms.

THEOREM 6.4 *In canonical variables u and $v \equiv v(u)$ the second order quasilinear equations with nontrivial symmetries have the following canonical forms and corresponding structure invariance groups; f, g and r are undetermined functions of their respective arguments, a, c and a_i for all i are constants.*

One-parameter symmetry group

S^2_1: $v'' + r(u, v') = 0$ *allows the transformations* $u = f(x)$, $v = g(x) + cy$.

Two-parameter symmetry groups

$\mathcal{S}^2_{2,1}$: $v'' + r(v') = 0$ *allows* $u = a_1 x + a_2 y + a_3$, $v = a_4 x + a_5 y + a_6$.

$\mathcal{S}^2_{2,2}$: $v'' u + r(v') = 0$ *allows* $u = a_1 x$, $v = a_2 x + a_3 y + a_4$.

Three-parameter symmetry groups

$\mathcal{S}^2_{3,1}$: $v''(u - v) + 2v'(v' + a\sqrt{v'} + 1) = 0$ *allows*

$$u = \frac{a_1 + (a_2 - a_1 a_3)x}{1 - a_3 x}, \quad v = \frac{a_1 + (a_2 - a_1 a_3)y}{1 - a_3 y}.$$

$\mathcal{S}^2_{3,2}$: $v'' v^3 + a = 0$, $a \neq 0$, *allows*

$$u = \frac{a_1(x + a_3)}{1 + a_4(x + a_3)}, \quad v = \frac{a_2 y}{1 + a_4(x + a_3)}.$$

$\mathcal{S}^2_{3,3}(c)$: $v'' + a v'^{(c-2)/(c-1)} = 0$, $c \neq 0, 1$, *allows*
$$u = a_1 x + a_2, \quad v = a_3 y + a_4.$$

$\mathcal{S}^2_{3,4}$: $v'' - a e^{-v'} = 0$ *allows* $u = a_1 x + a_2$, $v = a_3 x + a_1 y + a_4$.

Eight-parameter symmetry group

$\mathcal{S}^2_8$: $v'' = 0$, *projective group of the plane.*

PROOF   A general point transformation $u = \sigma(x, y)$ and $v = \rho(x, y)$ with $y \equiv y(x)$ generates the transformation $v' = \dfrac{\rho_x + \rho_y y'}{\sigma_x + \sigma_y y'}$ of the first derivative. For $\mathcal{S}^2_1$, in order to avoid any dependence on y to be generated via $r(u, v')$, $\sigma_y = \rho_{xy} = \rho_{yy} = 0$ is required. This yields the above structure with

$$v' = \frac{1}{f'}(g' + cy'), \quad v'' = \frac{1}{f'^3}(cf'y'' - cf''y' + f'g'' - f''g').$$

The expression for v'' shows that no further constraints are necessary.

For the two parameter groups, any dependence on x and y in the transformed function r is avoided if

$$x = a_1 u + a_2 v + a_3, \quad y = a_4 u + a_5 v + a_6$$

where $a_1, \ldots, a_6$ are constant. The second derivative is transformed according to

$$y'' = \frac{a_1 a_5 - a_2 a_4}{(a_1 + a_2 v')^3} v''.$$

For $\mathcal{S}^2_{2,1}$ this assures the desired structure. For $\mathcal{S}^2_{2,2}$ the second derivative must be proportional to the independent variable. This requires in addition $a_2 = a_3 = 0$.

252

For $S_{3,1}^2$ the transformed first derivative must be proportional to y'; this requires $\sigma_y = \rho_x = 0$. The condition for the invariance of the second order invariant leads to a fairly complicated system of pde's for σ and ρ. A Janet basis for the full system is

$$\sigma_y = 0, \ \rho_x = 0, \ \rho_y \sigma_x - \left(\frac{\rho - \sigma}{x - y}\right)^2 = 0, \ \sigma_{xx} + \frac{2\sigma_x^2}{\rho - \sigma} + \frac{2\sigma_x}{x - y} = 0.$$

The transformation functions given above, containing three constants, give its general solution.

For $S_{3,2}^2$ the transformed second derivative must be proportional to y'' and must be independent of x and y'. This is assured if

$$\sigma_y = 0, \ \rho_{yy} = 0, \ 2\sigma_x \rho_{xy} - \sigma_{xx}\rho_y = 0, \ \sigma_x \rho_{xx} - \sigma_{xx}\rho_x = 0$$

holds. Then $v'' = \frac{\rho_y}{\sigma_x^2} y''$. In the transformed equation the coefficient of the second derivative must be independent of x and proportional to y^3. This requires in addition

$$\left(\frac{\rho_y \rho^3}{\sigma_x}\right)_x = 0, \ \left(\frac{\rho_y \rho^3}{\sigma_x}\right)_y \frac{\sigma_x^2}{\rho_y \rho^3} - \frac{3}{y} = 0.$$

The combined constraints may be transformed into the Janet basis

$$\sigma_y = 0, \ y\rho_y - \rho = 0, \ \sigma_{xx}\rho - 2\rho_x\sigma_x = 0, \ \rho_{xx}\rho - 2\rho_x^2 = 0$$

with the general solution

$$\sigma = \frac{C_1 x + C_2}{C_3(C_3 x + C_4)}, \ \rho = \frac{y}{C_3 x + C_4}.$$

By a suitable change of the integration constants C_k the above expressions for the transformation functions are obtained.

For the group $S_{3,3}^2$ the transformed first derivative must be proportional to y'; this requires $\sigma_y = \rho_x = 0$. The transformed second derivative cannot contain a term proportional to a power of y'; this requires $\sigma_{xx} = \rho_{yy} = 0$. The general solution of these equations represent the transformation functions given in the above theorem.

For the group $S_{3,4}^2$ the transformed first derivative must have the form $-y' + constant$; this requires

$$\sigma_y = 0, \ \rho_y - \sigma_x = 0, \ \sigma_x \rho_{xy} - \sigma_{xy}\rho_x = 0, \ \sigma_x \rho_{xx} - \sigma_{xx}\rho_x = 0$$

or the equivalent Janet basis $\sigma_y = 0, \ \sigma_{xx} = 0, \ \rho_y - \sigma_x = 0, \ \rho_{xx} = 0$ with the solution given above.

Finally the structure invariance group of $v'' = 0$ is obviously identical to its symmetry group S_8^2, i.e., the eight parameter projective group of the plane.

□

Transformation to Canonical Form 253

For later use the Lie algebras corresponding to the two- and three-parameter symmetries of the preceding theorem are given explicitly next.

COROLLARY 6.1

Two-parameter symmetry groups

$$\mathcal{S}_{2,1}^2: \{\partial_x, \partial_y, x\partial_x, y\partial_x, x\partial_y, y\partial_y\}. \qquad \mathcal{S}_{2,2}^2: \{\partial_y, x\partial_x, x\partial_y, y\partial_y\}$$

Three-parameter symmetry groups

$$\mathcal{S}_{3,1}^2: \{\partial_x + \partial_y, x\partial_x + y\partial_y, x^2\partial_x + y^2\partial_y\}. \qquad \mathcal{S}_{3,2}^2: \{\partial_x, x\partial_x, y\partial_y, x^2\partial_x + xy\partial_y\}.$$

$$\mathcal{S}_{3,3}^2(c): \{\partial_x, \partial_y, x\partial_x, y\partial_y\}. \qquad \mathcal{S}_{3,4}^2: \{\partial_x, \partial_y, x\partial_x + y\partial_y, x\partial_y\}.$$

The proof is a straightforward application of the definition of the infinitesimal generators given in Chapter 3.

A symmetry class may comprise only a single equivalence class. This is true e. g. for $\mathcal{S}_8^2$. Because its canonical representative $v'' = 0$ does not contain any undetermined function or parameter, this symmetry class is at the same time an equivalence class as has been mentioned before. In general, however, a symmetry class may comprise more than a single equivalence class. In these cases there remains the question of how the individual equivalence classes contained in it may be uniquely parametrized. For the symmetry classes corresponding to three-parameter groups the answer is given next.

COROLLARY 6.2   *The symmetry class $\mathcal{S}_{3,1}^2$ decomposes into equivalence classes which are uniquely determined by the value of the parameter a in the canonical form equation. The remaining symmetry classes $\mathcal{S}_{3,2}^2$, $\mathcal{S}_{3,3}^2(c)$ and $\mathcal{S}_{3,4}^2$ comprise a single equivalence class.*

PROOF   The structure invariance group of $\mathcal{S}_{3,1}^2$ is identical to the symmetry group. Therefore it cannot transform equations with different values of a into each other. In the remaining cases, canonical equations with parameter values a and b are transformed into each other if the group parameters of the respective structure invariance groups are determined by

$$\frac{b}{a} = \left(\frac{a_1}{a_2^2}\right)^2, \text{ and } \frac{b}{a} = \left(\frac{a_1^c}{a_3}\right)^{1/(c-1)} \text{ or } \frac{b}{a} = \frac{1}{a_1}\exp\left(-\frac{a_3}{a_1}\right).$$

That is, in all three cases there is a one-parameter set of transformations by which this transformation may be achieved. ⬚

For two-parameter symmetries the situation is more complicated. A partial answer may be obtained from the following example dealing with symmetry class $\mathcal{S}_{2,2}^2$.

EXAMPLE 6.3   Applying a general transformation of the structure invariance group to the canonical form $uv'' + r(v') = 0$ of symmetry class $\mathcal{S}_{2,2}^2$ yields

254

$xy'' + s(y') = 0$ with $s(y') = \frac{1}{a_3} r(\frac{a_2}{a_1} + \frac{a_3}{a_1} y')$. Its action consists essentially of a linear transformation of the argument of r. □

EXAMPLE 6.4   Applying a general transformation of the structure invariance group to the canonical form $v''v^3 + a = 0$ for $\mathcal{S}_{3,2}^2$ symmetry yields $y''y^3 + b = 0$ with $b = a \frac{a_1^2}{a_2^4}$, i.e., any value for the constant b may be achieved, e. g. $b = 1$. □

The parameters and undetermined functions occurring in the canonical forms of Theorem 6.4 have to obey certain constraints in order to guarantee the respective symmetry type. They are a consequence of the definition that the symmetry type of an equation is determined by the *maximal* group leaving it invariant. Obvious examples of these constraints are $a \neq 0$ in the canonical forms of $\mathcal{S}_{3,2}^2$, $\mathcal{S}_{3,3}^2$ and $\mathcal{S}_{3,4}^2$, or $\frac{dr(v')}{dv'} \neq 0$ in the canonical forms of $\mathcal{S}_{2,1}^2$ and $\mathcal{S}_{2,2}^2$ assuring that the projective symmetry is excluded. For the applications in this book, however, this has no importance because the symmetry type of an equation is determined by its Janet basis for the symmetries which always gives the correct answer for a given equation in actual variables. These questions are discussed in more detail in Exercise 6.1

The discussion in the remaining part of this chapter is based on a series of articles by Lie [109], especially part III. As Lie mentions at the beginning of page 371, he outlines the ample calculations only schematically. Here all necessary steps and case distinctions are described in full detail such that the theorems given below readily may be applied for solving concrete examples.

The differential equations satisfied by the transformation functions to canonical form have to obey severe constraints which are a consequence of the following result due to Lie [109], part III, page 379ff.

THEOREM 6.5   *(Lie 1883) The differential equations satisfied by the transformation functions to canonical form allow the structure invariance group of the respective symmetry class as a symmetry group.*

Lie's proof uses the fact that the determining systems for the actual differential equation and the corresponding equation in canonical form are related by the desired transformation functions as well. Knowing their symmetry group and its differential invariants yields a better understanding of their structure. On page 382, part III of [109] the following remarkable result is proved.

THEOREM 6.6   *(Lie 1883) Let I_k be the invariants for a particular symmetry type, and $\sigma(x, y)$ and $\rho(x, y)$ be the transformation functions to canonical form. There are relations of the form $I_k = B_k(x, y)$ with known functions B_k.*

Transformation to Canonical Form 255

After these preliminary remarks the equations determining the transformation functions are determined explicitly.

One-Parameter Symmetry. This is the simplest type of nontrivial invariance that may occur for any ode. The freedom involved in the canonical form transformation, i.e., two unspecified functions of a single variable and a constant, correspond to the respective quantities generating the structure invariance group.

THEOREM 6.7   *If a second order ode belongs to symmetry class $\mathcal{S}_1^2$, two types of Janet bases may occur; f, g, h and k are undetermined functions of its argument.*

i) *Janet basis of type $\mathcal{J}_{1,1}^{(2,2)}$: $\sigma = f(x)$, $\rho = \int \exp\left(F(x,y)\right)dy + g(x)$ where $F(x,y) = \oint a(x,y)dx + b(x,y)dy$.*

ii) *Janet basis of type $\mathcal{J}_{1,2}^{(2,2)}$: $\sigma = h(\phi)$ where $\phi(x,y) = C$ is the solution of $y' + a = 0$, $\rho = C \int \exp\left(G(x,y)\right)dx + k(y)$ where C is a constant, and $G(x,y) = \oint b(x,y)dx + c(x,y)dy$.*

PROOF   In case *i)* the Janet basis

$$\xi = 0, \quad \eta_x + (\log \rho_y)_x \eta = 0, \quad \eta_y + (\log \rho_y)_y \eta = 0$$

yields the relations $(\log \rho_y)_x = a$, $(\log \rho_y)_y = b$ and $\sigma_y = 0$ with the solution given above. In case *ii)* the Janet basis

$$\eta + \frac{\sigma_x}{\sigma_y}\xi = 0, \quad \xi_x + \left(\log \frac{\Delta}{\sigma_y}\right)_x \xi = 0, \xi_y + \left(\log \frac{\Delta}{\sigma_y}\right)_y \xi = 0$$

leads to the system

$$\frac{\sigma_x}{\sigma_y} = a, \quad \left(\log \frac{\Delta}{\sigma_y}\right)_x = b, \quad \left(\log \frac{\Delta}{\sigma_y}\right)_y = c$$

for σ and ρ. A few simplifications lead to

$$\sigma_y - \frac{1}{a}\sigma_x = 0, \quad \rho_y - \frac{1}{a}\rho_x = \frac{1}{a} \exp G(x,y)$$

with G as defined above. By Theorem 2.19 the invariant $\phi(x,y)$ for the first equation is obtained. By Theorem 2.20 the function $w(x,y,\rho)$ obeys the homogeneous equation $w_y - \frac{1}{a}w_x + \frac{1}{a}e^G w_\rho = 0$. The variable change $\bar{y} = \phi(x,y)$ yields $w_x - e^G w_\rho = 0$ with the solution ρ as given above.   ⬚

The arbitrariness due to the undetermined elements is obviously a consequence of the structure invariance group given in Theorem 6.4.

The crucial step of this proceeding is to solve the ode $y' + a(x,y) = 0$ in case *ii)*. If a solution cannot be found it fails. Consequently, the examples of

256

second order equations with a one-parameter symmetry given in the literature generate a Janet basis with a simple expression for the coefficient $a(x, y)$, e. g. linear in y. Many equations are in the form $y'' + r(y, y') = 0$. Interchanging the dependent and the independent variable with each other by $x = v$, $y = u$ and applying (B.18) yields $v'' - v'^3 r\left(u, \frac{1}{v'}\right) = 0$. This case is discussed in more detail in Exercise 6.4.

EXAMPLE 6.5   Equation 6.90 of Kamke's collection has been shown in 5.9 to belong to symmetry class $\mathcal{S}_1^2$. Its type $\mathcal{J}_{1,2}^{(2,2)}$ Janet basis yields $a = \frac{2y}{x}$, $b = -\frac{1}{x}$ and $c = 0$. According to case ii) $y' + \frac{2}{x}y = 0$ is obtained with the solution $x^2 y = C$. It leads to $\phi = x^2 y$ and $e^G = C\frac{1}{x}$. Special solutions therefore are $\sigma = x^2 y$ and $\rho = \log x$. Inverting these relations with the result $x = e^v$ and $y = ue^{-2v}$ leads to the canonical form

$$v'' + u(u + 5)v'^3 - (u + 3)v'^2 + \tfrac{1}{4}v' = 0. \qquad \square$$

EXAMPLE 6.6   Equation 6.98 has been considered before in Example 5.10. Its type $\mathcal{J}_{1,2}^{(2,2)}$ Janet basis differs from the preceding case only by the value of a; now it is $a = -\frac{2y}{x}$. A similar calculation as above yields $\sigma = \frac{y}{x^2}$, $\rho = \log x$ and $x = e^v$, $y = ue^{2v}$ with the canonical form

$$v'' + 2(2u - 1)v'^2 + v' = 0. \qquad (6.2)$$

The second equation considered in Example 5.10 had a type $\mathcal{J}_{1,1}^{(2,2)}$ Janet basis with $a = \frac{1}{x}$ and $b = 0$. Therefore case i) of the above theorem applies. A few simple calculations yield the general transformation $\sigma = f(x)$ and $\rho = Cxy + g(x)$. With the special choice $f = \frac{1}{x+1}$, $C = 1$ and $g = 0$ the variable transformations

$$u = \frac{1}{x+1}, \quad v = xy \text{ and } x = \frac{1-u}{u}, \quad y = \frac{uv}{1-u}$$

are obtained. It transforms equation (5.30) into the canonical form, too. $\square$

Two-Parameter Symmetries. According to Corollary 6.1 the structure invariance group for the symmetry class $\mathcal{S}_{2,1}^2$ is generated by the six operators $\{\partial_x, \partial_y, x\partial_x, y\partial_x, x\partial_y, y\partial_y\}$. Its invariants up to second order are given next.

LEMMA 6.1   *The pde's for the canonical form transformations $\sigma(x, y)$ and $\rho(x, y)$ for the symmetry class $\mathcal{S}_{2,1}^2$ allow the following invariants w.r.t. the invariance group generated by $\{\partial_\sigma, \partial_\rho, \sigma\partial_\sigma, \rho\partial_\sigma, \sigma\partial_\rho, \rho\partial_\rho\}$. There are no first order invariants, and there are six second order invariants*

$$I_1 = \frac{\sigma_y \rho_{yy} - \sigma_{yy}\rho_y}{\sigma_x \rho_y - \sigma_y \rho_x}, \quad I_2 = \frac{\sigma_x \rho_{yy} - \sigma_{yy}\rho_x}{\sigma_x \rho_y - \sigma_y \rho_x}, \quad I_3 = \frac{\sigma_y \rho_{xy} - \sigma_{xy}\rho_y}{\sigma_x \rho_y - \sigma_y \rho_x},$$

$$I_4 = \frac{\sigma_y \rho_{xx} - \sigma_{xx}\rho_y}{\sigma_x \rho_y - \sigma_y \rho_x}, \quad I_5 = \frac{\sigma_x \rho_{xy} - \sigma_{xy}\rho_x}{\sigma_x \rho_y - \sigma_y \rho_x}, \quad I_6 = \frac{\sigma_x \rho_{xx} - \sigma_{xx}\rho_x}{\sigma_x \rho_y - \sigma_y \rho_x}.$$

$$\text{PROOF} \quad \text{The system of pde's determining any invariant } \Phi \text{ of order not}$$

PROOF   The system of pde's determining any invariant Φ of order not higher than two comprises $\Phi_\sigma = \Phi_\rho = 0$ and

$$\sigma_x \Phi_{\sigma_x} + \sigma_y \Phi_{\sigma_y} + \sigma_{xx} \Phi_{\sigma_{xx}} + \sigma_{xy} \Phi_{\sigma_{xy}} + \sigma_{yy} \Phi_{\sigma_{yy}} = 0,$$
$$\rho_x \Phi_{\sigma_x} + \rho_y \Phi_{\sigma_y} + \rho_{xx} \Phi_{\sigma_{xx}} + \rho_{xy} \Phi_{\sigma_{xy}} + \rho_{yy} \Phi_{\sigma_{yy}} = 0,$$
$$\sigma_x \Phi_{\rho_x} + \sigma_y \Phi_{\rho_y} + \sigma_{xx} \Phi_{\rho_{xx}} + \sigma_{xy} \Phi_{\rho_{xy}} + \sigma_{yy} \Phi_{\rho_{yy}} = 0,$$
$$\rho_x \Phi_{\rho_x} + \rho_y \Phi_{\rho_y} + \rho_{xx} \Phi_{\rho_{xx}} + \rho_{xy} \Phi_{\rho_{xy}} + \rho_{yy} \Phi_{\rho_{yy}} = 0.$$

The subsystem containing only derivatives w.r.t. σ, ρ and its first order derivatives contains six equations in six indeterminates with a nonvanishing coefficient determinant $\sigma_x \rho_y - \sigma_y \rho_x$. Consequently, a first order invariant does not exist. The full system for the second order invariants comprises six equations in twelve indeterminates σ, σ_x, σ_y, σ_{xx}, σ_{xy} and σ_{yy}, and the corresponding derivatives of ρ, i.e., six second order invariants may exist. They may be obtained by the methods described in Chapter 2. $\quad\square$

THEOREM 6.8   If a second order ode has a Janet basis of type $\mathcal{J}_{2,3}^{(2,2)}$ and $a_1 = b_2$ and $c_1 = d_2$, it belongs to symmetry class $\mathcal{S}_{2,1}^2$. The transformation function σ is determined by

$$\sigma_{xx} - c_2 \sigma_y - a_2 \sigma_x = 0, \ \sigma_{xy} - c_1 \sigma_y - a_1 \sigma_x = 0, \ \sigma_{yy} - d_1 \sigma_y - b_1 \sigma_x = 0. \quad (6.3)$$

There is an identical set of equations for ρ. The general solution of the full system contains six constants.

PROOF   From the Janet basis given in equation (11) of Schwarz [165], the coefficients $a_1, a_2, \ldots, d_2$ may be expressed in terms of the transformation functions σ and ρ as

$$\begin{aligned}
\sigma_y \rho_{xy} - \sigma_{xy} \rho_y &= a_1 \Delta, & \sigma_y \rho_{xx} - \sigma_{xx} \rho_y &= a_2 \Delta, \\
\sigma_y \rho_{yy} - \sigma_{yy} \rho_y &= b_1 \Delta, & \sigma_y \rho_{xy} - \sigma_{xy} \rho_y &= b_2 \Delta, \\
\sigma_x \rho_{xy} - \sigma_{xy} \rho_x &= -c_1 \Delta, & \sigma_x \rho_{xx} - \sigma_{xx} \rho_x &= -c_2 \Delta, \\
\sigma_x \rho_{yy} - \sigma_{yy} \rho_x &= -d_1 \Delta, & \sigma_x \rho_{xy} - \sigma_{xy} \rho_x &= -d_2 \Delta
\end{aligned} \quad (6.4)$$

where as usual $\Delta = \sigma_x \rho_y - \sigma_y \rho_x$. This system may be transformed into a Janet basis in grlex term ordering with $\rho > \sigma > d_2 > d_1 > \ldots > a_2 > a_1$ and $y > x$ with the result

$$b_2 - a_1 = 0, \ \ d_2 - c_1 = 0,$$
$$a_{2,y} - a_{1,x} + c_2 b_1 - c_1 a_1 = 0, \ b_{1,x} - a_{1,y} + d_1 a_1 - b_1 c_1 + b_1 a_2 - a_1^2 = 0,$$
$$c_{2,y} - c_{1,x} + d_1 c_2 - c_2 a_1 - c_1^2 + c_1 a_2 = 0, \ d_{1,x} - c_{1,y} + b_1 c_2 - c_1 a_1 = 0,$$
$$\sigma_{xx} - c_2 \sigma_y - a_2 \sigma_x = 0, \ \sigma_{xy} - c_1 \sigma_y - a_1 \sigma_x = 0, \ \sigma_{yy} - d_1 \sigma_y - b_1 \sigma_x = 0,$$
$$\rho_{xx} - c_2 \rho_y - a_2 \rho_x = 0, \ \rho_{xy} - c_1 \rho_y - a_1 \rho_x = 0, \ \rho_{yy} - d_1 \rho_y - b_1 \rho_x = 0.$$
$$(6.5)$$

258

The lower equations not involving σ and ρ represent the coherence conditions for the Janet basis coefficients. The upper half of this Janet basis represents the two identical linear systems for σ and ρ respectively as given above. □

EXAMPLE 6.7   The symmetry type of equation 6.227 has been identified as $\mathcal{S}_{2,1}^2$ in Example 5.11. By Theorem 6.8, case $i)$ the type $\mathcal{J}_{3,2}^{(1,2)}$ Janet basis for the transformation function σ is

$$\left\{\sigma_{xx} + \frac{1}{x}\sigma_x = 0, \ \sigma_{xy} = 0, \ \sigma_{yy} + \frac{1}{y}\sigma_y = 0\right\},$$

and an identical one for ρ. It is not completely reducible; its Loewy factor comprises only $\{\sigma_x, \sigma_y\}$. Dividing it out yields a type $\mathcal{J}_{2,3}^{(2,2)}$ Janet basis for the quotient in terms of $\sigma_1 \equiv \sigma_x$ and $\sigma_2 \equiv \sigma_y$. It is completely reducible and decomposes into type $\mathcal{J}_{1,1}^{(2,2)}$ and type $\mathcal{J}_{1,2}^{(2,2)}$ Janet bases according to

$$\left\{\sigma_{1,x} + \tfrac{1}{x}\sigma_1 = 0, \ \sigma_{1,y} = 0, \ \sigma_{2,x} = 0, \ \sigma_{2,y} + \tfrac{1}{y}\sigma_2 = 0\right\}$$
$$= Lclm\left(\left\{\sigma_1 = 0, \ \sigma_{2,x} = 0, \ \sigma_{2,y} + \tfrac{1}{y}\sigma_2 = 0\right\},\right.$$
$$\left.\left\{\sigma_2 + \tfrac{x}{y}\sigma_1 = 0, \ \sigma_{1,x} + \tfrac{1}{x}\sigma_1 = 0, \ \sigma_{1,y} = 0\right\}\right)$$

from which the special solutions $\sigma_1 = 0$, $\sigma_2 = \frac{1}{y}$ and $\sigma_1 = \frac{1}{x}$, $\sigma_2 = -\frac{1}{y}$, and finally $\sigma = \log y$ and $\sigma = \log \frac{x}{y}$ are obtained. A possible choice of the transformation functions to new variables $u \equiv \sigma$ and $v \equiv \rho$ therefore is $\sigma = \log y$ and $\rho = \log \frac{x}{y}$ with the inverse $x = e^{u+v}$, $y = e^u$. It yields the canonical form $v'' + \dfrac{v'^3 + 5v'^2 + 8v' + 4}{v'} = 0$. □

As has been shown in Theorem 6.4, the canonical form for the symmetry classes corresponding to two-parameter groups is highly non-unique. This arbitrariness is discussed next.

EXAMPLE 6.8   The preceding example is continued. A fundamental system for the transformation functions σ and ρ has been shown to be $\{1, \log x, \log y\}$. Any linear combination leading to a nonvanishing functional determinant is appropriate, although it is not a priori clear whether there is a particularly favorable one for the subsequent solution procedure. The linear combinations $u = c_1 \log x + c_2 \log y$, $v = c_3 \log x + c_4 \log y$ with $c_1 c_4 - c_2 c_3 \neq 0$ lead to the most general transformation $e^u = x^{c_1}y^{c_2}$ and $e^v = x^{c_3}y^{c_4}$. The special choice $c_1 = c_4 = 0$, $c_2 = c_3 = 1$ leads to $v'' + \dfrac{v'(v'+1)^2}{v'-1} = 0$ for which $r(v')$ decomposes into linear factors. The transformation between the two canonical forms is discussed in Exercise 6.2. □

The structure invariance group for the symmetry class $\mathcal{S}_{2,2}^2$ is generated by the four operators $\{\partial_y, x\partial_x, x\partial_y, y\partial_y\}$ according to Corollary 6.1. Its invariants are similar to the preceding case.

$$\text{TRANSFORMATION TO CANONICAL FORM} \qquad 259$$

LEMMA 6.2   *The pde's for the canonical form transformations $\sigma(x,y)$ and $\rho(x,y)$ for the symmetry class $\mathcal{S}_{2,2}^2$ allow the following invariants w.r.t. the invariance group that is generated by $\{\partial_\rho, \sigma\partial_\sigma, \sigma\partial_\rho, \rho\partial_\rho\}$. There are two first order invariants $K_1 = \frac{\sigma_x}{\sigma}$ and $K_2 = \frac{\sigma_y}{\sigma}$. If $\sigma_x \neq 0$, $\rho_x \neq 0$ and $\rho_y \neq 0$, there are the same six second order invariants that have been determined in Lemma 6.1*

PROOF   The system of pde's determining the invariants Φ of order not higher than two comprises $\Phi_\rho = 0$ and

$$\sigma\Phi_\sigma + \sigma_x\Phi_{\sigma_x} + \sigma_y\Phi_{\sigma_y} + \sigma_{xx}\Phi_{\sigma_{xx}} + \sigma_{xy}\Phi_{\sigma_{xy}} + \sigma_{yy}\Phi_{\sigma_{yy}} = 0,$$

$$\sigma_x\Phi_{\rho_x} + \sigma_y\Phi_{\rho_y} + \sigma_{xx}\Phi_{\rho_{xx}} + \sigma_{xy}\Phi_{\rho_{xy}} + \sigma_{yy}\Phi_{\rho_{yy}} = 0,$$

$$\rho_x\Phi_{\rho_x} + \rho_y\Phi_{\rho_y} + \rho_{xx}\Phi_{\rho_{xx}} + \rho_{xy}\Phi_{\rho_{xy}} + \rho_{yy}\Phi_{\rho_{yy}} = 0.$$

The subsystem containing only derivatives w.r.t. σ, ρ and its first derivatives comprises four independent equations in six variables; i.e., there are two first order invariants. The four independent equations in twelve variables of the full system allow 12-4-2=6 second order invariants.   ⬚

In Exercise 6.3 the Janet basis coefficients of the two parameter symmetries will be determined in terms of the respective invariants.

THEOREM 6.9   *If a second order ode has a Janet basis of type $\mathcal{J}_{2,3}^{(2,2)}$ and belongs to symmetry class $\mathcal{S}_{2,2}^2$, the transformation function σ is*

$$\sigma = \exp\left[\oint (c_1 - d_2)dx + (b_2 - a_1)dy \right]. \tag{6.6}$$

For the other transformation function ρ two alternatives may occur.

a) If $c_1 \neq d_2$, ρ is explicitly given by

$$\rho = \oint \rho_1 \frac{c_2 dx + d_2 dy}{d_2 - c_1} - \sigma \oint \frac{\rho_1}{\sigma} \frac{c_2 dx + c_1 dy}{d_2 - c_1} \tag{6.7}$$

where

$$\rho_1 = \exp\left[\oint \left(\frac{a_1 - b_2}{c_1 - d_2}c_2 + c_1 \right) dx + \left(\frac{a_1 - b_2}{c_1 - d_2}c_1 + b_2 - a_1 + d_1 \right) dy \right].$$

b) If $c_1 = d_2$, ρ is explicitly given by

$$\rho = \oint \rho_1 \frac{a_1 dx - b_1 dy}{a_1 - b_2} - \sigma \oint \frac{\rho_1}{\sigma} \frac{b_2 dx + b_1 dy}{a_1 - b_2} \tag{6.8}$$

where

$$\rho_1 = \exp\left(\oint a_2 dx + b_2 dy \right).$$

260

PROOF From the Janet basis given in equation (10) of Schwarz [165], the coefficients $a_1, a_2, \ldots, d_2$ may be expressed in terms of the transformation functions σ and ρ as

$$(\sigma_y \rho_{xy} - \sigma_{xy} \rho_y)\sigma + \sigma_y \Delta = -a_1 \sigma \Delta, \quad (\sigma_y \rho_{xx} - \sigma_{xx} \rho_y)\sigma + \sigma_x \Delta = -a_2 \sigma \Delta,$$
$$\sigma_y \rho_{yy} - \sigma_{yy} \rho_y = -b_1 \Delta, \quad \sigma_y \rho_{xy} - \sigma_{xy} \rho_y = -b_2 \Delta,$$
$$\sigma_x \rho_{xy} - \sigma_{xy} \rho_x = c_1 \Delta, \quad \sigma_x \rho_{xx} - \sigma_{xx} \rho_x = c_2 \Delta,$$
$$(\sigma_x \rho_{yy} - \sigma_{yy} \rho_x)\sigma - \sigma_y \Delta = d_1 \sigma \Delta, \quad (\sigma_x \rho_{xy} - \sigma_{xy} \rho_x)\sigma - \sigma_x \Delta = d_2 \sigma \Delta.$$

(6.9)

In the same term ordering as system (6.5) above the Janet basis

$$a_{2,y} - a_{1,x} + c_2 b_1 - c_1 b_2 = 0, \quad b_{1,x} - a_{1,y} + a_1 d_1 - b_1 c_1 - a_1 b_2 + a_2 b_1 = 0,$$
$$b_{2,x} - a_{1,x} + a_1 d_2 - c_1 b_2 = 0, \quad b_{2,y} - a_{1,y} + b_1 d_2 - b_2 d_1 + a_1 d_1 - c_1 b_2 = 0,$$
$$c_{2,y} - c_{1,x} - c_1 d_2 + c_2 d_1 - a_1 c_2 + c_1 a_2 = 0, \quad d_{1,x} - c_{1,y} - a_1 d_2 + b_1 c_2 = 0,$$
$$d_{2,x} - c_{1,x} - a_2 d_2 + b_2 c_2 - a_1 c_2 + c_1 a_2 = 0, \quad d_{2,y} - c_{1,y} - a_1 d_2 + c_1 b_2 = 0,$$
$$\sigma_x + (d_2 - c_1)\sigma = 0, \quad \sigma_y + (a_1 - b_2)\sigma = 0,$$
$$\rho_{xx} - c_2 \rho_y + (d_2 - c_1 - a_2)\rho_x = 0,$$
$$\rho_{xy} - c_1 \rho_y - b_2 \rho_x = 0,$$
$$\rho_{yy} + (a_1 - d_1 - b_2)\rho_y - c\rho_x = 0$$

is obtained. The eight lowest equations, i.e., the first four lines, are the coherence conditions for the coefficients. The two subsequent equations determine σ in terms of the path integral (6.6). The three remaining equations for ρ allow the solutions $\rho = 1$ and $\rho = \sigma$. This follows from the fact that they may be reduced to zero by both $\{\rho_x, \rho_y\}$ and also by $\{\rho_x + (d_2 - c_1)\rho, \rho_y + (a_1 - b_2)\rho\}$, i.e., the same system that determines σ. If $c_1 \neq d_2$, dividing out the least common left multiple

$$Lclm\big(\{\rho_x, \rho_y\}, \{\rho_x + (d_2 - c_1)\rho, \rho_y + (a_1 - b_2)\rho\}\big)$$
$$= \left\{ \rho_1 \equiv \rho_y + \frac{a_1 - b_2}{c_1 - d_2} \rho_x, \rho_2 \equiv \rho_{xx} + \left(\frac{a_1 - b_2}{c_1 - d_2} c_2 + d_2 - c_1 - a_2 \right) \rho_x \right\}$$

(6.10)

yields the exact quotient

$$\left\{ \rho_2 - c_2 \rho_1, \ \rho_{1,x} - \left(\frac{a_1 - b_2}{c_1 - d_2} c_2 + c_1 \right) \rho_1, \ \rho_{1,y} - \left(\frac{a_1 - b_2}{c_1 - d_2} c_1 + b_2 - a_1 + d_1 \right) \rho_1 \right\}$$

from which ρ_1 and ρ_2 may be be determined. Consequently, the second line of (6.10) forms an inhomogeneous system of pde's with known solution of the homogeneous part and known right hand sides. It has been shown in Excercise 5 of Chapter 2, see also Appendix A, how to express a special solution in terms of integrals. Applying the results obtained there leads to the expression (6.7). If $c_1 = d_2$ it follows $c_1 = c_2 = d_2 = 0$. The least common left multiple now is

$$Lclm\big(\{\rho_x, \rho_y\}, \{\rho_x, \rho_y + (a_1 - b_2)\rho\}\big) =$$
$$\{\rho_1 \equiv \rho_x, \rho_2 \equiv \rho_{yy} + (a_1 - b_2 - d_1)\rho_y\}$$

(6.11)

Transformation to Canonical Form 261

and yields the exact quotient $\{\rho_2 - b_1\rho_1, \rho_{1,x} - a_2\rho_1, \rho_{1,y} - b_2\rho_1\}$. Applying the results of Exercise 5 in Chapter 2 leads to the representation (6.8). □

This result for the symmetry class $S^2_{2,2}$ is more explicit than for symmetry class $S^2_{2,1}$ because the transformation functions are obtained in closed form without solving any equation. Consequently, whenever a second order quasilinear equation allows this symmetry type, it is *guaranteed* that it may be transformed into canonical form by a Liouvillian transformation.

EXAMPLE 6.9   It has been shown in Example 5.12 that Equation 6.159 of Kamke's collection belongs to symmetry class $S^2_{2,2}$. The integral (6.6) of Theorem 6.9 yields $\sigma = \dfrac{1}{\sqrt{y}}$. Because $c_1 = d_2 = 0$, the integral (6.8) yields $\rho = x$. The inverse of this transformation $x = v$, $y = \dfrac{1}{u^2}$ yields the canonical form $v''u - \frac{3}{2}v'(v'^2 - 1) = 0$. □

Three-Parameter Symmetries. The general proceeding in this subsection is the same as in the preceding one, i.e., at first the invariants of the structure invariance groups for the various symmetry classes are determined. Subsequently the equations for the transformation functions to canonical form are obtained. They turn out to be considerably more complicated than for the two-parameter symmetries.

LEMMA 6.3   *The pde's for the canonical form transformations $\sigma(x,y)$ and $\rho(x,y)$ for the symmetry type $S^2_{3,1}$ allow the following invariants w.r.t. the invariance group that is generated by $\{\partial_\sigma + \partial_\rho, \sigma\partial_\sigma + \rho\partial_\rho, \sigma^2\partial_\sigma + \rho^2\partial_\rho\}$. If $\sigma_x \neq 0$ and $\rho_x \neq 0$ there are three first order invariants $K_1 = \dfrac{\sigma_y}{\sigma_x}$, $K_2 = \dfrac{\rho_y}{\rho_x}$ and $K_3 = \dfrac{\sigma_x\rho_x}{(\rho - \sigma)^2}$, and six second order invariants*

$$I_1 = \frac{\sigma_{xx}}{\sigma_x} + \frac{2\sigma_x}{\rho - \sigma}, \quad I_2 = \frac{\sigma_{xy}}{\sigma_x} + \frac{2\sigma_y}{\rho - \sigma}, \quad I_3 = \frac{\sigma_{yy}}{\sigma_y} + \frac{2\sigma_y}{\rho - \sigma},$$

$$I_4 = \frac{\rho_{xx}}{\rho_x} - \frac{2\rho_x}{\rho - \sigma}, \quad I_5 = \frac{\rho_{xy}}{\rho_x} - \frac{2\rho_y}{\rho - \sigma}, \quad I_6 = \frac{\rho_{yy}}{\rho_y} - \frac{2\rho_y}{\rho - \sigma}.$$

PROOF   The system of pde's determining any invariant Φ of order not higher than two comprises $\Phi_\sigma + \Phi_\rho = 0$ and

$$\sigma\Phi_\sigma + \rho\Phi_\rho + \sigma_x\Phi_{\sigma_x} + \sigma_y\Phi_{\sigma_y} + \rho_x\Phi_{\rho_x} + \rho_y\Phi_{\rho_y}$$
$$+\sigma_{xx}\Phi_{\sigma_{xx}} + \sigma_{xy}\Phi_{\sigma_{xy}} + \sigma_{yy}\Phi_{\sigma_{yy}} + \rho_{xx}\Phi_{\rho_{xx}} + \rho_{xy}\Phi_{\rho_{xy}} + \rho_{yy}\Phi_{\rho_{yy}} = 0,$$
$$\sigma^2\Phi_\sigma + \rho^2\Phi_\rho + 2\sigma\sigma_x\Phi_{\sigma_x} + 2\sigma\sigma_y\Phi_{\sigma_y} + 2\rho\rho_x\Phi_{\rho_x} + 2\rho\rho_y\Phi_{\rho_y}$$
$$+2(\sigma\sigma_{xx} + \sigma_x^2)\Phi_{\sigma_{xx}} + 2(\sigma\sigma_{xy} + \sigma_x\sigma_y)\Phi_{\sigma_{xy}} + 2(\sigma\sigma_{yy} + \sigma_y^2)\Phi_{\sigma_{yy}}$$
$$+2(\rho\rho_{xx} + \rho_x^2)\Phi_{\rho_{xx}} + 2(\rho\rho_{xy} + \rho_x\rho_y)\Phi_{\rho_{xy}} + 2(\rho\rho_{yy} + \rho_y^2)\Phi_{\rho_{yy}} = 0.$$

262

The subsystem containing only derivatives w.r.t. σ, ρ and its first derivatives comprises three independent equations in six variables, i.e., there are three first order invariants. The three independent equations in twelve variables involving second derivatives allow 12-3-3=6 second order invariants. $\quad\Box$

THEOREM 6.10 *If a second order ode belongs to symmetry class $\mathcal{S}_{3,1}^2$ two types of Janet bases may occur. The transformation functions σ and ρ are determined by the following systems of equations.*

a) *Janet basis type $\mathcal{J}_{3,6}^{(2,2)}$: Three cases have to be distinguished. If $b_1 \neq 0$ define $z^2 \equiv a_1 b_1 + 1$ and*

$$P(x,y) \equiv \frac{1}{(a_1 b_1 + 1)b_1} \big[(z+1)(a_1 b_1 + 1)b_3 - z(a_3 b_1 + c_3 - b_2)b_1 \big],$$

$$Q(x,y) \equiv \frac{b_1}{4z^2} \big[z^2(a_{3,x} - a_1 d_3 - d_2) + a_3(a_3 b_1 + 2c_3 - 3b_2) \big.$$
$$\big. - a_1(c_3 - b_2)^2 - a_1 a_3 b_1 b_2 \big].$$
$$(6.12)$$

R is determined from

$$R_x + R^2 - PR + Q = 0,$$
$$R_y + \frac{z+1}{b_1}R^2 - \Big[\frac{z+1}{b_1}P + \Big(\frac{z+1}{b_1} \Big)_x \Big] R + \frac{z-1}{b_1}Q = 0.$$
$$(6.13)$$

The system for σ and ρ is

$$\sigma_y - \frac{z+1}{b_1}\sigma_x = 0, \quad \sigma_{xx} + (2R - P)\sigma_x = 0, \quad \rho = \frac{1}{R}\sigma_x + \sigma = 0.$$
$$(6.14)$$

If $b_1 = 0$ define

$$P \equiv \frac{1}{2}(a_{3,x} - a_1 d_3 - a_1 c_3^2 + 2a_3 c_3 - d_2), \quad Q \equiv a_1 c_3 + c_2 - a_3. \quad (6.15)$$

R is either determined from

$$R_x + P = 0, \quad R_y + R^2 - QR - \tfrac{1}{2}a_1 P = 0 \qquad (6.16)$$

and the system for $\sigma(x,y)$ and $\rho(x,y)$ is

$$\sigma_x = 0, \quad \sigma_{yy} + (2R - Q)\sigma_y = 0, \quad \rho = \frac{1}{R}\sigma_y + \sigma. \qquad (6.17)$$

Or R is determined from

$$R_x + R^2 - c_3 R = 0, \quad R_y + \tfrac{1}{2}a_1 R^2 - (a_1 c_3 - a_3)R + P = 0 \qquad (6.18)$$

and the system for σ and ρ is

$$\sigma_y + \tfrac{1}{2}a_1\sigma_x = 0, \quad \sigma_{xx} + (2R - c_3)\sigma_x = 0, \quad \rho = \frac{1}{R}\sigma_x + \sigma = 0. \qquad (6.19)$$

Transformation to Canonical Form

b) *Janet basis type* $\mathcal{J}_{3,7}^{(2,2)}$: *Defining* b_1 *and* B *by*

$$b_1 \equiv -c^2, \quad B \equiv \exp\left(2 \oint a_3 dx + \frac{c_x + a_3 c}{c^2} dy \right),$$

at first R *and* S *may be determined from*

$$R_x + R^2 - (a_2 c + a_3)R + B = 0,$$
$$R_y - \tfrac{1}{c}R^2 - \frac{c_x - (a_2 c + a_3)c}{c^2}R - \tfrac{1}{c}B = 0,$$
$$S_x - S^2 + (a_2 c - a_3)S - B = 0,$$
$$S_y + \tfrac{1}{c}S^2 - \frac{c_x + (a_2 c - a_3)c}{c^2}S + \tfrac{1}{c}B = 0.$$

$$(6.20)$$

Then the system for σ *and* ρ *is*

$$\sigma_y - \tfrac{1}{c}\sigma_x = 0, \quad \sigma_{xx} + (2R - a_2 c - a_3)\sigma_x = 0,$$
$$\rho_y + \tfrac{1}{c}\rho_x = 0, \quad \rho_{xx} - (2S - a_2 c + a_3)\rho_x = 0.$$

$$(6.21)$$

PROOF   At first case *a)* will be considered. If $b_1 \neq 0$, the Janet basis for σ and ρ in *grlex* ordering with $\rho > \sigma$, $y > x$ is

$$\sigma_y - \frac{z+1}{b_1}\sigma_x = 0, \quad \rho_y + \frac{z-1}{b_1}\rho_x = 0,$$
$$\sigma_{xx}(\rho - \sigma) + 2\sigma_x^2 - P\sigma_x(\rho - \sigma) = 0, \quad \rho_x\sigma_x - Q(\rho - \sigma)^2 = 0.$$

The new function $R \equiv \dfrac{\sigma_x}{\rho - \sigma}$ is introduced with the result

$$\sigma_x - R(\rho - \sigma) = 0, \quad \sigma_y - \frac{z+1}{b_1}R(\rho - \sigma) = 0,$$
$$\rho_x - \frac{Q}{P}(\rho - \sigma) = 0, \quad \rho_y - \frac{z-1}{b_1}\frac{Q}{R}(\rho - \sigma) = 0.$$

P, Q and z are defined as above. The integrability conditions of this system are the two first equations of (6.14). If a *lex* term order with $\rho > \sigma > R$ is applied, the complete system (6.14) is obtained.

If $b_1 = 0$, $\sigma_x = 0$, $\rho_x \neq 0$, the Janet basis for σ and ρ in *grlex* term ordering with $\rho > \sigma$, $x > y$ is

$$\sigma_x = 0, \quad \sigma_{yy}(\rho - \sigma) + 2\sigma_y^2 - Q\sigma_y(\rho - \sigma) = 0,$$
$$\rho_x\sigma_y - P(\rho - \sigma)^2 = 0, \quad \rho_y\sigma_y + \tfrac{1}{2}a_1 P(\rho - \sigma)^2 = 0$$

where P and Q are defined by (6.15). The new function $R = \dfrac{\sigma_y}{\rho - \sigma}$ is introduced. It yields the system

$$\sigma_x = 0, \quad \sigma_{yy} + (2R - Q)\sigma_y = 0, \quad \rho_y R + \tfrac{1}{2}a_1 P(\rho - \sigma) = 0, \quad \rho_x R - P(\rho - \sigma) = 0.$$

264

Its integrability conditions are the equations (6.16). In order to show this, the relations $Q_x + 2P = 0$, $P_x = c_3 P$ and $P_y = c_2 P$ are useful.

If $b_1 = 0$, $\sigma_x \neq 0$ and $\rho_x = 0$, the Janet basis for σ and ρ in the same term ordering as above is

$$\sigma_y + \tfrac{1}{2}a_1\sigma_x = 0, \quad \sigma_{xx}(\rho - \sigma) + 2\sigma_x^2 - c_3\sigma_x(\rho - \sigma) = 0,$$

$$\rho_x = 0, \quad \rho_y\sigma_x - P(\rho - \sigma)^2 = 0.$$

Introducing now $R = \dfrac{\sigma_x}{\rho - \sigma}$ as a new function, the system

$$\sigma_x - R(\rho - \sigma) = 0, \ \sigma_y + \tfrac{1}{2}a_1 R(\rho - \sigma) = 0, \ \rho_x = 0, \ \rho_y R - P(\rho - \sigma) = 0$$

for σ and ρ is obtained. A similar reasoning to the above leads to equations (6.18) and (6.19).

In case b), the type $\mathcal{J}_{3,7}^{(2,2)}$ Janet basis for σ and ρ is

$$\sigma_y - \tfrac{1}{c}\sigma_x = 0, \quad \rho_y + \tfrac{1}{c}\rho_x = 0,$$

$$\sigma_{xx}(\rho - \sigma) + 2\sigma_x^2 - (a_2 c + a_3)\sigma_x(\rho - \sigma) = 0,$$

$$\rho_{xx}(\rho - \sigma) - 2\rho_x^2 + (a_2 c - a_3)\rho_x(\rho - \sigma) = 0.$$

Defining $R \equiv \dfrac{\sigma_x}{\rho - \sigma}$ and $S \equiv \dfrac{\rho_x}{\rho - \sigma}$ leads to the equations

$$R_x + R^2 + RS - (a_2 c + a_3)R = 0, \ R_y - \tfrac{1}{c}R^2 - \tfrac{1}{c}RS - \tfrac{1}{c^2}[c_x - (a_2 c + a_3)c]R = 0,$$

$$S_x - S^2 - RS + (a_2 c - a_3)S = 0, \ S_y + \tfrac{1}{c}S^2 + \tfrac{1}{c}RS - \tfrac{1}{c^2}[c_x + (a_2 c - a_3)c]S = 0.$$

Combining the two equations determining the x-derivatives and the y-derivatives respectively leads to

$$(RS)_x - 2a_3 RS = 0, \quad (RS)_y - \frac{2}{c^2}(c_x + a_3 c)RS = 0$$

from which the above expression for B follows. $\qquad\qquad\qquad\qquad\qquad$ □

EXAMPLE 6.10   The equation considered in Example 5.13 in symmetry class $S_{3,1}^2$ is continued. Due to $b_1 \neq 0$ the first alternative of case a) in the above theorem applies. The Riccati system (6.13) has the form

$$R_x + R^2 + \frac{2x}{x^2 - \dfrac{2}{y}}R - \frac{\dfrac{2}{y}}{\left(x^2 - \dfrac{2}{y}\right)^2} = 0,$$

$$R_y + \frac{x}{y}R^2 + \left(\frac{2}{y - \dfrac{2}{x^2}} - \frac{1}{y}\right)R = 0$$

with the rational solution $R = \dfrac{-xy}{x^2y - 2} + \dfrac{y}{C + xy}$. Substituting the special solution $\dfrac{-xy}{x^2y - 2}$ corresponding to $C \to \infty$ into (6.14) yields

$$\sigma_{xx} = 0, \ \sigma_y - \frac{x}{y}\sigma_x = 0, \ \rho + (x - \frac{2}{xy})\sigma_x - \sigma = 0.$$

The Janet basis for σ is completely reducible, it may be written as

$$Lclm\left(\{\sigma_x - \frac{1}{x}\sigma, \sigma_y - \frac{1}{y}\sigma\}, \{\sigma_x, \sigma_y\}\right).$$

The non-constant solution $\sigma = xy$ yields $\rho = \frac{2}{x}$. The corresponding transformation to canonical variables u and v is $x = \frac{2}{v}$, $y = \frac{1}{2}uv$. Substitution into the original equation finally yields the equation in canonical form

$$v''^2(u - v)^2 + 4v''v'(v' + 1)(u - v) + 4v'^4 + 4v'^3 + 4v'^2 = 0. \qquad \square$$

LEMMA 6.4   *The pde's for the canonical form transformations $\sigma(x, y)$ and $\rho(x, y)$ for the symmetry class $\mathcal{S}^2_{3,2}$ allow the following invariants w.r.t. the invariance group that is generated by $\{\partial_\sigma, \sigma\partial_\sigma, \rho\partial_\rho, \sigma^2\partial_\sigma + \sigma\rho\partial_\rho\}$. If $\sigma_x \neq 0$ and $\rho_x \neq 0$ there are two first order invariants $K_1 = \dfrac{\sigma_y}{\sigma_x}$, $K_2 = \dfrac{\rho_y}{\rho} - \dfrac{\rho_x}{\rho}\dfrac{\sigma_y}{\sigma_x}$, and six second order invariants*

$$I_1 = \frac{\sigma_{xx}}{\sigma_x} - 2\frac{\rho_x}{\rho}, \quad I_2 = \frac{\sigma_{xy}}{\sigma_y} - 2\frac{\rho_x}{\rho}, \quad I_3 = \frac{\sigma_{yy}}{\sigma_y} - 2\frac{\rho_y}{\rho},$$

$$I_4 = \frac{\rho_{xx}}{\rho} - \frac{\rho_x}{\rho}\frac{\sigma_{xx}}{\sigma_x}, \quad I_5 = \frac{\rho_{xy}}{\rho} - \frac{1}{2}\frac{\rho_x}{\rho}\frac{\sigma_{xy}}{\sigma_x} - \frac{1}{2}\frac{\rho_y}{\rho}\frac{\sigma_{xy}}{\sigma_y}, \quad I_6 = \frac{\rho_{yy}}{\rho} - \frac{\rho_y}{\rho}\frac{\sigma_{yy}}{\sigma_y}.$$

PROOF   The system of pde's determining any invariant Φ of order not higher than two comprises $\Phi_\sigma = 0$ and

$$\sigma_x\Phi_{\sigma_x} + \sigma_y\Phi_{\sigma_y} + \sigma_{xx}\Phi_{\sigma_{xx}} + \sigma_{xy}\Phi_{\sigma_{xy}} + \sigma_{yy}\Phi_{\sigma_{yy}} = 0,$$

$$\rho\Phi_\rho + \rho_x\Phi_{\rho_x} + \rho_y\Phi_{\rho_y} + \rho_{xx}\Phi_{\rho_{xx}} + \rho_{xy}\Phi_{\rho_{xy}} + \rho_{yy}\Phi_{\rho_{yy}} = 0,$$

$$\sigma\rho\Phi_\rho + 2\sigma\sigma_x\Phi_{\sigma_x} + 2\sigma\sigma_y\Phi_{\sigma_y} + (\rho\sigma_x + \sigma\rho_x)\Phi_{\rho_x} + (\rho\sigma_y + \sigma\rho_y)\Phi_{\rho_y}$$

$$+2(\sigma\sigma_{xx} + \sigma_x^2)\Phi_{\sigma_{xx}} + 2(\sigma\sigma_{xy} + \sigma_x\sigma_y)\Phi_{\sigma_{xy}} + 2(\sigma\sigma_{yy} + \sigma_y^2)\Phi_{\sigma_{yy}}$$

$$+(\rho\sigma_{xx} + \sigma\rho_{xx} + 2\sigma_x\rho_x)\Phi_{\rho_{xx}} + (\rho\sigma_{xy} + \sigma\rho_{xy} + \sigma_x\rho_y + \sigma_y\rho_x)\Phi_{\rho_{xy}}$$

$$+(\rho\sigma_{yy} + \sigma\rho_{yy} + 2\sigma_y\rho_y)\Phi_{\rho_{yy}} = 0.$$

The number of invariants is determined in the same way as in the proof of Lemma 6.2. $\qquad \square$

The four parameters of the structure invariance group for the symmetry class $\mathcal{S}^2_{3,2}$ are obtained by combining the integration constant of a second order Riccati-like system and a second order Janet basis for σ or ρ respectively.

266

THEOREM 6.11   *If a second order ode belongs to symmetry class $\mathcal{S}_{3,2}^2$, three types of Janet bases may occur. The transformation functions σ and ρ are determined by the following systems of equations.*

a) *Janet basis of type $\mathcal{J}_{3,4}^{(2,2)}$.*

$$R_{xx} + R_x^2 - a_3 R_x - \tfrac{1}{4}(c_{3,x} + c_3^2 - a_3 c_3 - d_2) = 0, \ R_y - \tfrac{1}{2}a_2 = 0,$$

$$\rho_x + R_x\rho = 0, \ \rho_y + R_y\rho = 0, \ \sigma_y = 0, \ \sigma_{xx} - (2R_x - a_3)\sigma_x = 0.$$
$$\tag{6.22}$$

b) *Janet basis of type $\mathcal{J}_{3,7}^{(2,2)}$.*

$$R_{yy} + R_y^2 - c_2 R_y - \tfrac{1}{4}(a_{2,y} + a_2^2 - a_2 c_2 - d_3) = 0, \ R_x - \tfrac{1}{2}c_3 = 0,$$

$$\rho_x + R_x\rho = 0, \ \rho_y + R_y\rho = 0, \ \sigma_x = 0, \ \sigma_{yy} + (2R_y - c_2)\sigma_y = 0.$$
$$\tag{6.23}$$

c) *Janet basis of type $\mathcal{J}_{3,6}^{(2,2)}$.*

$$R_{xx} + R_x^2 + a_1 b_3 R_x - \frac{Q}{P} = 0, \ R_y + a_1 R_x - \tfrac{1}{2}P = 0,$$

$$\rho_x + R_x\rho = 0, \ \rho_y + R_y\rho = 0, \ \sigma_y + a_1\sigma_x = 0, \ \sigma_{xx} + (2R_x + a_1 b_3)\sigma_x = 0$$
$$\tag{6.24}$$

where $P(x,y) \equiv a_1 c_3 - a_1 b_2 - a_3 \equiv -2\dfrac{\Delta}{\sigma_x \rho} \neq 0$ and

$$Q(x,y) = \tfrac{1}{4}\big[(b_1 a_3 + c_3 - b_2)(a_{3,x} - d_2) + a_1(c_3 - b_2)^3$$

$$-d_3 P(x,y) + 2 b_3 P(x,y)^2 + a_3(b_2 c_3 - a_3 b_1 b_2 - c_3^2)\big].$$

PROOF   The proof will be given in detail for case a) for Janet basis type $\mathcal{J}_{3,4}^{(2,2)}$. If its coefficients are compared to the corresponding expressions in terms of the transformation functions (see also Theorem 5.8), the following system of pde's is obtained.

$$(\log \rho^2)_y = a_2, \ \left(\log \frac{\sigma_x}{\rho^2}\right)_x = a_3, \ \left(\log \frac{\rho_y}{\rho}\right)_y = c_1, \ c_2 = 0, \ \left(\log \frac{\rho_y}{\rho}\right)_x = c_3,$$

$$\left(\log \frac{\rho_y^2}{\sigma_x}\right) = d_1, \ (\log \rho_x)_y (\log \sigma_x)_x - (\log \rho_{xx})_y + 3(\log \rho)_x \left(\log \frac{\sigma_x}{\rho_x}\right)_x = d_2,$$

$$\frac{\rho_x}{\rho_y}\left[\left(\log \frac{\rho_{xx}}{\sigma_{xx}}\right)_x + 3 \left(\log \frac{\sigma_x}{\rho_x}\right)_x \left(\log \frac{\sigma_x}{\rho}\right)_x\right] = d_3.$$

$$\textit{Transformation to Canonical Form} \qquad 267$$

From this system of equations a Janet basis in total degree, then lexicographic term order $\rho > \sigma > d_3 > d_2 > \ldots > a_3 > a_2$ is generated.

$$d_1 - 2c_3 - a_3 = 0, \; a_{2,x} - a_2 c_3 = 0, \; a_{2,y} - a_2 c_2 = 0, \; a_{3,y} - c_3 a_2 = 0,$$

$$c_{3,y} - c_{2,x} = 0, \; d_{3,y} - d_{2,x} + d_3 c_2 - 2d_3 a_2 - d_2 c_3 + 2d_2 a_3 = 0,$$

$$c_{2,xx} - d_{2,y} - 2c_{3,x} a_2 + 2c_{2,x} c_3 - c_{2,x} a_3 - 3c_3^2 a_2 + 2a_2 a_3 c_3 = 0,$$

$$c_{3,xx} - d_{2,x} + 2c_{3,x} c_3 - 3c_{3,x} a_3 - a_{3,x} c_3 - 2d_3 a_2 + 2d_2 a_3 - 2c_3^2 a_3 + 2c_3 a_3^2 = 0,$$

$$\sigma_y = 0, \; \sigma_{xx}\rho - 2\sigma_x \rho_x - \sigma_x \rho a_3 = 0,$$

$$\rho_y + \tfrac{1}{2}\rho a_2 = 0, \; \rho_{xx}\rho - 2\rho_x^2 - \rho_x \rho a_3 + \tfrac{1}{4}\rho^2(c_{3,x} - d_2 + c_3^2 - a_3 c_3) = 0.$$

Introducing the new function $R \equiv -\log\rho$ into the last four equations yields system (6.22). For case $b)$ the four highest equations of the Janet basis for the full system expressing σ and ρ in terms of the coefficients $a_2, \ldots, d_2, d_3$ are

$$\sigma_x = 0, \; \rho_x + \tfrac{1}{2}c_3\rho = 0, \; \sigma_{yy}\rho - 2\rho_y \sigma_y - c_2 \sigma_y \rho = 0,$$

$$\rho_{yy}\rho - 2\rho_y^2 - \rho_y \rho c_2 + \tfrac{1}{4}(a_{2,y} + a_2^2 - d_3 - a_2 c_2)\rho^2 = 0.$$

Substituting $R \equiv -\log\rho$ yields (6.24). Finally for case $c)$ a fairly complicated system of pde's is obtained for σ and ρ. Introducing again $R \equiv -\log\rho$, it may be written in the form (6.23). $\quad\square$

The equations for R in the systems (6.22), (6.23) and (6.24) are of the form (2.58) on page 89. The solution procedure described there is always applied.

EXAMPLE 6.11 The symmetry class of equation 6.133 of Kamke's collection has been identified as $\mathcal{S}_{3,2}^2$ in Example 5.14. Because its Janet basis is of type $\mathcal{J}_{3,6}^{(2,2)}$, case $c)$ of the above theorem applies. The system (6.24) is

$$R_{xx} + R_x^2 + \frac{1}{x+y}R_x - \frac{\tfrac{1}{4}}{(x+y)^2} = 0, \; R_y + R_x + \frac{1}{x+y} = 0.$$

Solving the first equation leads to the system

$$R_x = \frac{1}{x - y + C} - \frac{\tfrac{1}{2}}{x + y}, \quad R_y = -\frac{1}{x - y + C} - \frac{\tfrac{1}{2}}{x + y}$$

for R with the integration constant $C \equiv C(y)$. Because the equation for $C(y)$ obtained from (2.60) is linear, the special solution corresponding to $C \to \infty$ may be chosen and the system for σ becomes $\{\sigma_y + \sigma_x, \sigma_{xx}\}$. It is completely reducible and may be written as

$$Lclm\left(\left\{\sigma_x - \frac{1}{x - y + C_1}\sigma, \sigma_y + \frac{1}{x - y + C_1}\sigma\right\}\right) =$$

$$Lclm\left(\{\sigma_x, \sigma_y\}, \left\{\sigma_x + \frac{1}{x - y}\sigma, \sigma_y - \frac{1}{x - y}\sigma\right\}\right) = 0.$$

268

The two components of the latter representation correspond to the values $C_1 \to \infty$ and $C_1 = 0$ respectively. A special solution is $\sigma = x - y$. To the chosen value of C there corresponds the system $R_x = R_y = \dfrac{1}{2(x+y)}$. A special solution is $R = \frac{1}{2}\log(x+y)$ and consequently $\rho = \sqrt{x+y}$. The inverse of the transformation

$$u = x - y, \quad v = \sqrt{x+y} \quad \text{is} \quad x = \tfrac{1}{2}(v^2 + u), \quad y = \tfrac{1}{2}(v^2 - u),$$

yields the canonical form $v''v^3 + \frac{1}{4} = 0$ for equation 6.133 of Kamke's collection. The transformations of the structure invariance group may be applied to generate any value for the constant term. Choosing $a_1 = 2$, $a_2 = 1$ and $a_3 = a_4$ for example yields $v''v^3 + 1 = 0$. $\quad\square$

For both symmetry classes $\mathcal{S}_{3,3}^2(c)$ and $\mathcal{S}_{3,4}^2$ the four-parameter structure invariance originates from the second order Janet bases for the transformation functions σ and ρ as it is shown in the next lemma. Moreover, the same remarks apply as for symmetry class $\mathcal{S}_{2,1}^2$ on page 261, i.e., no equations have to be solved in order to determine the transformation functions.

LEMMA 6.5   The pde's for the canonical form transformations $\sigma(x,y)$ and $\rho(x,y)$ for symmetry class $\mathcal{S}_{3,3}^2(c)$ allow the following invariants w.r.t. the invariance group that is generated by $\{\partial_\sigma, \partial_\rho, \sigma\partial_\sigma, \rho\partial_\rho\}$. If all first derivatives are different from zero there are two first order invariants $K_1 = \dfrac{\sigma_y}{\sigma_x}$, $K_2 = \dfrac{\rho_y}{\rho_x}$ and six second order invariants

$$I_1 = \frac{\sigma_{xx}}{\sigma_x}, \; I_2 = \frac{\sigma_{xy}}{\sigma_x}, \; I_3 = \frac{\sigma_{yy}}{\sigma_y}, \; I_4 = \frac{\rho_{xx}}{\rho_x}, \; I_5 = \frac{\rho_{xy}}{\rho_x}, \; I_6 = \frac{\rho_{yy}}{\rho_y}.$$

PROOF   The system of pde's determining any invariant Φ of order not higher than two comprises $\Phi_\sigma = \Phi_\rho = 0$ and

$$\sigma_x \Phi_{\sigma_x} + \sigma_y \Phi_{\sigma_y} + \sigma_{xx} \Phi_{\sigma_{xx}} + \sigma_{xy} \Phi_{\sigma_{xy}} + \sigma_{yy} \Phi_{\sigma_{yy}} = 0,$$

$$\rho_x \Phi_{\rho_x} + \rho_y \Phi_{\rho_y} + \rho_{xx} \Phi_{\rho_{xx}} + \rho_{xy} \Phi_{\rho_{xy}} + \rho_{yy} \Phi_{\rho_{yy}} = 0.$$

The number of invariants is determined in the same way as in the proof of Lemma 6.2. $\quad\square$

For the symmetry class $\mathcal{S}_{3,3}^2(c)$ the parameter c has been determined in Theorem 5.8 and may be considered as already known.

THEOREM 6.12   If a second order ode belongs to symmetry class $\mathcal{S}_{3,3}^2(c)$ with $c \neq 1$, two types of Janet bases may occur. The transformation functions σ and ρ are determined by the following systems of pde's.

a) Janet basis of type $\mathcal{J}_{3,6}^{(2,2)}$ and $\sigma_x\rho_y - c\sigma_y\rho_x \neq 0$. There are two alternatives. If $b_1 = 0$ the system is either

$$\sigma_x = 0, \; \sigma_{yy} - \left(a_1 c_3 + c_2 + \tfrac{a_3}{c}\right)\sigma_y = 0,$$
$$\rho_y + \frac{ca_1}{c-1}\rho_x = 0, \; \rho_{xx} + cc_3\rho_x = 0 \tag{6.25}$$

or

$$\sigma_y - \frac{a_1}{c-1}\sigma_x = 0, \ \sigma_{xx} - d_1\sigma_x = 0,$$

$$\rho_x = 0, \ \rho_{yy} - (a_1c_3 + c_2 + ca_3)\rho_y = 0. \tag{6.26}$$

If $b_1 \neq 0$ and $c + 1 \neq 0$, the system is

$$\sigma_y - \frac{c_1 + c}{(c+1)b_1}\sigma_x = 0, \ \sigma_{xx} + \frac{R(x,y)}{T(x,y)}\sigma_x = 0,$$

$$\rho_y - \frac{cc_1 + 1}{(c+1)b_1}\rho_x = 0, \ \rho_{xx} + \frac{S(x,y)}{T(x,y)}\rho_x = 0 \tag{6.27}$$

where

$$R(x,y) \equiv c^2\big[b_1(a_3b_1 + c_3 - b_2 - a_1b_3) - b_3\big]$$

$$+ c\big[b_1(c_1 + 2)(a_3b_1 - a_1b_3 + c_3 - b_2) - (2c_1 + 1)b_3\big]$$

$$+ (c_1 + 1)\big[b_1(a_3b_1 - a_1b_3 + c_3 - b_2) - c_1b_3\big],$$

$$S(x,y) \equiv c^3b_1(a_3b_1 + c_3 - b_2)$$

$$+ c^2\big[b_1(c_1 + 2)(a_3b_1 + c_3 - b_2) - (a_1b_1 + 1)c_1b_3\big]$$

$$+ c\big[b_1(c_1 + 1)(a_3b_1 + c_3 - b_2) - (a_1b_1c_1 + 1)b_3\big],$$

$$T(x,y) = b_1\big[c_1 + a_1b_1 + c(c+2)(a_1b_1 + 1)\big].$$

b) *Janet basis of type $\mathcal{J}_{3,7}^{(2,2)}$ and $\sigma_x\rho_y - c\sigma_y\rho_x = 0$. The transformation functions σ and ρ are determined by*

$$\sigma_y - \frac{c+1}{cc_1}\sigma_x = 0, \ \sigma_{xx} + \frac{a_2c_1 - (c+1)a_3}{c+1}\sigma_x = 0,$$

$$\rho_y - \frac{c+1}{c_1}\rho_x = 0, \ \rho_{xx} + \frac{a_2c_1 + (c+1)a_3}{c+1}\rho_x = 0. \tag{6.28}$$

PROOF   The original expressions of the Janet basis coefficients in terms of σ and ρ are too voluminous to be given here explicitly. If they are transformed into a Janet basis in total degree, then lexicographic term order with $\rho > \sigma$ and $y > x$, the above linear system for σ and ρ is obtained.   □

EXAMPLE 6.12   In Chapter 5, Example 5.15, two equations in symmetry class $\mathcal{S}_{3,3}^2(c)$ have been considered. For the first equation, due to $b_1 = 0$ the first alternative of case a) applies. It is not a priori clear which choice of c leads to the correct canonical form. For the first alternative, $c = \frac{2}{3}$ and the system

$$\sigma_x = 0, \ \sigma_{yy} + \frac{2}{y}\sigma_y = 0, \ \rho_y = 0, \ \rho_{xx} = 0$$

is obtained. Its general solution is $\sigma = \frac{C_1}{y} + C_2$, $\rho = \frac{C_3}{x} + C_4$. The special choice $C_1 = C_3 = 1$, $C_2 = C_4 = 0$ yields the canonical form $v'' + v'^4 = 0$. The

270

second alternative leads to $y'y'' - 1 = 0$. Therefore the former choice is the correct one. For the second equation c is determined by $c^2 - \frac{13}{6}c + 1 = 0$ with the same solutions as above. The value $c = \frac{2}{3}$ leads to the Janet basis

$$\sigma_y - \tfrac{1}{2}\sigma_x = 0, \ \sigma_{xx} = 0, \ \rho_y - \tfrac{1}{3}\rho_x = 0, \ \rho_{xx} = 0.$$

A special solution is $\sigma = 2x+y$, $\rho = 3x+y$. Substituting the inverse $x = v-u$ and $y = 3u - 2v$ into the latter equation of Example 5.15 yields $v'' + v'^4 = 0$, i.e., the same canonical form as the preceding equation. ⬚

LEMMA 6.6   *The pde's for the canonical form transformations $\sigma(x,y)$ and $\rho(x,y)$ for the symmetry class $\mathcal{S}_{3,4}^2$ allow the following invariants w.r.t. the invariance group that is generated by $\{\partial_\sigma, \partial_\rho, \sigma\partial_\sigma + \rho\partial_\rho, \sigma\partial_\rho\}$. If $\sigma_x \neq 0$ there are two first order invariants $K_1 = \dfrac{\sigma_y}{\sigma_x}$, $K_2 = \dfrac{\sigma_x\rho_y - \sigma_y\rho_x}{\sigma_x^2}$. If $\sigma_x \neq 0$ and $\sigma_y \neq 0$ there are six second order invariants*

$$I_1 = \frac{\sigma_{xx}}{\sigma_x}, \ I_2 = \frac{\sigma_{xy}}{\sigma_x}, \ I_3 = \frac{\sigma_{yy}}{\sigma_y},$$

$$I_4 = \frac{\sigma_{xx}\rho_x - \sigma_x\rho_{xx}}{\sigma_x^2}, \ I_5 = \frac{\sigma_{yy}\rho_y - \sigma_y\rho_{yy}}{\sigma_y^2}, \ I_6 = \frac{\sigma_{xy}\rho_x - \sigma_x\rho_{xy}}{\sigma_x^2}.$$

PROOF   The system of pde's determining any invariant Φ of order not higher than two comprises $\Phi_\sigma = \Phi_\rho = 0$ and

$$\sigma_x\Phi_{\sigma_x} + \sigma_y\Phi_{\sigma_y} + \rho_x\Phi_{\rho_x} + \rho_y\Phi_{\rho_y} + \sigma_{xx}\Phi_{\sigma_{xx}} + \sigma_{xy}\Phi_{\sigma_{xy}}$$

$$+\sigma_{yy}\Phi_{\sigma_{yy}} + \rho_{xx}\Phi_{\rho_{xx}} + \rho_{xy}\Phi_{\rho_{xy}} + \rho_{yy}\Phi_{\rho_{yy}} = 0,$$

$$\sigma_x\Phi_{\rho_x} + \sigma_y\Phi_{\rho_y} + \sigma_{xx}\Phi_{\rho_{xx}} + \sigma_{xy}\Phi_{\rho_{xy}} + \sigma_{yy}\Phi_{\rho_{yy}} = 0.$$

The number of invariants is determined in the same way as in the proof of Lemma 6.2. ⬚

THEOREM 6.13   *If a second order ode belongs to symmetry class $\mathcal{S}_{3,4}^2$ two types of Janet bases may occur.*

a) *If the Janet basis is of type $\mathcal{J}_{3,6}^{(2,2)}$ three cases have to be distinguished. If $b_1 \neq 0$, $c_1 + 1 \neq 0$ and $T(x,y) \equiv 2a_1b_1 + c_1 + 1 \neq 0$, the transformation functions σ and ρ are determined by*

$$\sigma_y + \frac{2a_1}{c_1+1}\sigma_x = 0, \quad \sigma_{xx} + \frac{R(x,y)}{T(x,y)}\sigma_x = 0,$$

$$\rho_y - \frac{c_1+1}{2b_1}\rho_x - \frac{c_1-1}{2b_1}\sigma_x = 0, \quad \rho_{xx} + \frac{R(x,y)}{T(x,y)}\rho_x + \frac{S(x,y)}{b_1T(x,y)}\sigma_x = 0$$

$$(6.29)$$

where

$$R(x,y) \equiv (c_1 + 1)(a_3b_1 + c_3 - 4) - (c_1 - 1)a_1b_3,$$

$$S(x,y) \equiv (c_1 + 1)[(a_3b_1 + c_3 - b_2)b_1 + b_3] - (c_1 - 3)a_1b_1b_3.$$

$$\text{Transformation to Canonical Form} \qquad 271$$

If $b_1 \neq 0$, $c_1 + 1 = T = 0$ the system for σ and ρ is

$$\sigma_y = 0, \quad \sigma_{xx} - (c_3 - b_2)\sigma_x = 0,$$

$$\rho_y + \frac{1}{b_1}\sigma_x = 0, \quad \rho_{xx} + (c_3 - b_2)\rho_x + \left(c_3 - b_2 + \frac{f}{b_1}\right)\sigma_x = 0. \tag{6.30}$$

If $b_1 = c_1 + 1 = T = 0$ there is always $a_1 \neq 0$, the system for σ and ρ is

$$\sigma_x = 0, \quad \sigma_{yy} - (a_1 c_3 + a_3 - c_2)\sigma_y = 0,$$

$$\rho_x + \frac{1}{a_1}\sigma_y = 0, \quad \rho_{yy} - (a_1 c_3 + a_3 - c_2)\rho_y - \left(a_1 c_3 + a_3 - \frac{a_2}{a_1}\right)\sigma_y = 0. \tag{6.31}$$

b) If the Janet basis is of type $\mathcal{J}_{3,7}^{(2,2)}$ there holds $c_1 \neq 0$. The system for σ and ρ is

$$\sigma_y - \frac{2}{c_1}\sigma_x = 0, \quad \sigma_{xx} + \left(\tfrac{1}{2}a_2 c_1 - a_3\right)\sigma_x = 0,$$

$$\rho_y - \frac{2}{c_1}\rho_x - \frac{2}{c_1}\sigma_x = 0, \quad \rho_{xx}\left(\tfrac{1}{2}a_2 c_1 - a_3\right)\rho_x + \tfrac{1}{2}a_2 c_1 \sigma_x = 0. \tag{6.32}$$

PROOF   At first case a) is considered. In terms of σ and ρ the coefficients a_1, b_1 and c_1 are

$$a_1 = -\frac{\sigma_y^2}{\sigma_x \sigma_y - \Delta}, \quad b_1 = \frac{\sigma_x^2}{\sigma_x \sigma_y - \Delta}, \quad c_1 = \frac{\sigma_x \sigma_y + \Delta}{\sigma_x \sigma_y - \Delta}$$

where as usual $\Delta = \sigma_x \rho_y - \sigma_y \rho_x$. It follows that

$$c_1 + 1 = \frac{2\sigma_x \sigma_y}{\sigma_x \sigma_y - \Delta}, \quad T(x, y) = -\frac{2\sigma_x \sigma_y \Delta}{\sigma_x \sigma_y - \Delta}.$$

The three alternatives correspond to $\sigma_x \neq 0$ and $\sigma_y \neq 0$, $\sigma_x \neq 0$, $\sigma_y = 0$ and $\sigma_x = 0$, $\sigma_y \neq 0$ respectively. In the latter case there follows $a_1 \neq 0$. In case b) there holds $c_1 = 2\frac{\sigma_x}{\sigma_y}$. Consequently, $c_1 = 0$ entails $\sigma_x = 0$ which is not possible due to the constraint $\sigma_x \sigma_y = \Delta \neq 0$, i.e., $p \neq 0$ is assured. ⬚

EXAMPLE 6.13   An equation with this symmetry type has been considered in Example 5.16. From the type $\mathcal{J}_{3,6}^{(2,2)}$ Janet basis given there it follows $a_1 = 0$, $b_1 = -\frac{x}{y}$ and $c_1 = -1$. Therefore the second alternative of case a) applies. It yields

$$\sigma_y = 0, \quad \sigma_{xx} + \frac{1}{x}\sigma_x = 0, \quad \rho_y - \frac{y}{x}\sigma_x = 0, \quad \rho_{xx} - \frac{1}{x}\rho_x + \frac{1}{x}\sigma_x = 0.$$

A non-constant solution for σ is $\sigma = x^2$. Substituting it in the equations for ρ yields the independent solution $\rho = x + y^2$. ⬚

272

The Symmetry Class of the Projective Group. The largest group of point symmetries that any second order ode may allow has been shown to be the eight-parameter projective group of $\mathbb{R}^2$. Because there is no degree of freedom in the canonical form $v'' = 0$ corresponding to this symmetry, its structure invariance group is identical to the symmetry group itself. The invariants of order two of the projective group and its largest subgroup are given in the following lemma. The notation for the invariants is chosen such that its relation to the differential equations to be discussed subsequently is obvious.

LEMMA 6.7 *(Lie 1883) The projective group of the plane with generators*

$$\{\partial_\sigma, \partial_\rho, \sigma\partial_\rho, \rho\partial_\rho, \sigma\partial_\sigma, \rho\partial_\sigma, \sigma^2\partial_\sigma + \sigma\rho\partial_\rho, \sigma\rho\partial_\sigma + \rho^2\partial_\rho\} \tag{6.33}$$

does not have any first order invariants. A fundamental system of four second order invariants is

$$\frac{\sigma_y\rho_{yy} - \sigma_{yy}\rho_y}{\sigma_x\rho_y - \sigma_y\rho_x} \equiv A, \quad \frac{\sigma_x\rho_{yy} - \sigma_{yy}\rho_x}{\sigma_x\rho_y - \sigma_y\rho_x} + 2\frac{\sigma_y\rho_{xy} - \sigma_{xy}\rho_y}{\sigma_x\rho_y - \sigma_y\rho_x} \equiv B,$$

$$\frac{\sigma_y\rho_{xx} - \sigma_{xx}\rho_y}{\sigma_x\rho_y - \sigma_y\rho_x} + 2\frac{\sigma_x\rho_{xy} - \sigma_{xy}\rho_x}{\sigma_x\rho_y - \sigma_y\rho_x} \equiv C, \quad \frac{\sigma_x\rho_{xx} - \sigma_{xx}\rho_x}{\sigma_x\rho_y - \sigma_y\rho_x} \equiv D. \tag{6.34}$$

The largest subgroup of the projective group is generated by

$$\{\partial_\sigma, \partial_\rho, \sigma\partial_\rho, \rho\partial_\rho, \sigma\partial_\sigma, \rho\partial_\sigma\},$$

it does not have any first order invariant either. A fundamental system of six second order invariants is

$$\frac{\sigma_y\rho_{yy} - \sigma_{yy}\rho_y}{\sigma_x\rho_y - \sigma_y\rho_x} \equiv A, \quad \frac{\sigma_x\rho_{yy} - \sigma_{yy}\rho_x}{\sigma_x\rho_y - \sigma_y\rho_x} \equiv B_1, \quad \frac{\sigma_y\rho_{xy} - \sigma_{xy}\rho_y}{\sigma_x\rho_y - \sigma_y\rho_x} \equiv B_2,$$

$$\frac{\sigma_y\rho_{xx} - \sigma_{xx}\rho_y}{\sigma_x\rho_y - \sigma_y\rho_x} \equiv C_1, \quad \frac{\sigma_x\rho_{xy} - \sigma_{xy}\rho_x}{\sigma_x\rho_y - \sigma_y\rho_x} \equiv C_2, \quad \frac{\sigma_x\rho_{xx} - \sigma_{xx}\rho_x}{\sigma_x\rho_y - \sigma_y\rho_x} \equiv D.$$

The proof may be found in the above quoted article by Lie [109], part III, §1.

THEOREM 6.14 *(Lie 1883) Any second order ode*

$$y'' + A(x,y)y'^3 + B(x,y)y'^2 + C(x,y)y' + D(x,y) = 0 \tag{6.35}$$

with projective symmetry group is similar to the equation $v''(u) = 0$. The transformation functions $u = \sigma(x,y)$ and $v = \rho(x,y)$ are solutions of

$$\sigma_{xx} - D\sigma_y + (C - 2C_2)\sigma_x = 0,$$
$$\sigma_{xy} - C_2\sigma_y + B_2\sigma_x = 0, \tag{6.36}$$
$$\sigma_{yy} - (B - 2B_2)\sigma_y + A\sigma_x = 0$$

Transformation to Canonical Form

273

and an identical system for ρ such that $\sigma_x \rho_y - \sigma_y \rho_x \neq 0$. The coefficients $B_2 = b$ and $C_2 = -a$ are determined by the Riccati system

$$
\begin{aligned}
a_x + a^2 + Ca - Db + D_y + BD &= 0, \\
a_y + ab - \tfrac{1}{3}B_x + \tfrac{2}{3}C_y + AD &= 0, \\
b_x + ab - \tfrac{2}{3}B_x + \tfrac{1}{3}C_y + AD &= 0, \\
b_y + b^2 - Bb + Aa - A_x + AC &= 0
\end{aligned}
\tag{6.37}
$$

of the form (2.65). Its rational solutions may be determined by the algorithm described in Chapter 2. Substituting them into (6.36) generates Janet bases of type $\mathcal{J}_{3,2}^{(1,2)}$ for σ and ρ. They may be decomposed by the algorithms described in Chapter 2 as well.

PROOF   A general point transformation $u = \sigma(x,y)$, $v = \rho(x,y)$ of $v''(u) = 0$ generates an equation of the form (6.35) where the coefficients A, B, C and D are given by the corresponding expressions in Lemma 6.7. These equations may be considered as a system of pde's expressing σ and ρ in terms of the coefficients of (6.35). Furthermore, there are the relations expressing B_1, B_2, C_1 and C_2 from the second part of this lemma in terms of σ and ρ. By means of the obvious relations $B = B_1 + 2B_2$ and $C = C_1 + 2C_2$, B_1 and C_1 are expressed in terms of B_2 and C_2, and the resulting system is transformed into a Janet basis with lexicographical term ordering $\rho > \sigma > B_2 > C_2 > A > \ldots > D$. The full Janet basis comprises twelve equations, the upper half of which is given by (6.36), and an identical system for ρ. The two lowest equations are the constraints for $A, \ldots, D$ of Theorem 5.11 guaranteeing the projective symmetry. In between there is the system

$$
\begin{aligned}
C_{2,x} - C_2^2 + DB_2 + CC_2 - D_y - BD &= 0, \\
C_{2,y} + B_2C_2 + \tfrac{1}{3}B_x - \tfrac{2}{3}C_y - AD &= 0, \\
B_{2,x} - B_2C_2 - \tfrac{2}{3}B_x + \tfrac{1}{3}C_y + AD &= 0, \\
B_{2,y} + B_2^2 - BB_2 - AC_2 - A_x + AC &= 0.
\end{aligned}
\tag{6.38}
$$

Combined with the two lowest equations they may be considered as coherence conditions for the linear homogeneous system (6.36). Equations (6.38) express the functions B_2 and C_2 in terms of the known coefficients A, B, C and D. Substituting $C_2 = -a$ and $B_2 = b$ yields the Riccati system (6.37) for a and b.   □

The above decomposition of the coefficients of equation (6.35) into B_1, B_2, C_1 and C_2 is not obvious. It seems to be more natural to start with the coefficients A, B, C and D, expressed in terms of σ and ρ, compare them to the actual coefficients in terms of x and y and generate a Janet basis from it. *In principle* this leads to the same answer. However, the resulting system contains several fairly complicated equations for which a solution algorithm does not seem to exist. This is particularly obvious if B_2

274

and C_2 are substituted by their definition in terms of σ and ρ into the system (6.36). Lemma 6.7 shows clearly the group theoretical origin of the proceeding described in the above proof, it follows closely the discussion by Lie [109], part III, pages 364-370.

The existence of the crucial relations (6.38) may be understood as follows. In addition to the six second order invariants the group (6.34) has eight independent third order invariants. Due to the fact that the twelve first derivatives of the second order invariants are also invariants of third order, four relations must exist between them. For example, by differentiation and some rearrangement the difference $C_{2,x} - D_y$ may be written as

$$C_{2,x} - D_y = \frac{2}{\Delta}\left(\sigma_{xx}\rho_{xy} - \sigma_{xy}\rho_{xx}\right) - \frac{\Delta_x}{\Delta}C_2 + \frac{\Delta_y}{\Delta}D$$

with $\Delta = \sigma_x\rho_y - \sigma_y\rho_x$. Using $\frac{\Delta_x}{\Delta} = C_2 - C_1$ and $\frac{\Delta_y}{\Delta} = B_1 - B_2$ and applying some simplifications, the relation

$$C_{2,x} - D_y = (B_1 + B_2)D - (C_1 + C_2)C_2$$

is obtained which is the first equation of (6.38). Due to the group structure they are obtained as integrability conditions during the Janet basis algorithms.

EXAMPLE 6.14   Consider equation 6.180

$$x^2(y-1)y'' - 2x^2y'^2 - 2x(y-1)y' - 2y(y-1)^2 = 0$$

of the collection by Kamke with $A = 0$, $B = -\frac{2}{y-1}$, $C = -\frac{2}{x}$ and $D = -\frac{2y(y-1)}{x^2}$. By Theorem 5.34 it belongs to symmetry class $\mathcal{S}_8^2$. The system (6.37) for a and b is

$$a_x + a^2 - \frac{2}{x}a + \frac{2y(y-1)}{x^2}b + \frac{2}{x^2} = 0, \quad a_y + ab = 0,$$
$$b_x + ab = 0, \quad b_y + b^2 + \frac{2}{y-1}b = 0.$$

Its general solution is rational and may be written as

$$a = \frac{C_2\dfrac{y-1}{y} + \dfrac{2x(y-1)}{y}}{C_1 + C_2\dfrac{x(y-1)}{y} + \dfrac{x^2(y-1)}{y}} + \frac{1}{x},$$

$$b = \frac{C_2\dfrac{x}{y^2} + \dfrac{x^2}{y^2}}{C_1 + C_2\dfrac{x(y-1)}{y} + \dfrac{x^2(y-1)}{y}} - \frac{1}{y(y-1)}$$

Transformation to Canonical Form

where C_1 and C_2 are the integration constants. For the values $C_1 \to \infty$ and $C_2 = 0$ the special solution $a = 0$ and $b = -\dfrac{1}{y(y-1)}$ is obtained. They yield the Janet basis

$$\sigma_{xx} + \frac{2y(y-1)}{x^2}\sigma_y - \frac{2}{x}\sigma_x = 0, \ \sigma_{xy} - \frac{1}{y(y-1)}\sigma_x = 0, \ \sigma_{yy} + \frac{2}{y}\sigma_y = 0$$

for σ and an identical one for ρ. It is completely reducible with the Loewy decomposition

$$Lclm\left(\sigma_x - \frac{(C_2+2x)(y-1)}{C_1y + (C_2+x)x(y-1)}\sigma, \ \sigma_y - \frac{(C_2+x)x}{C_1y^2 + (C_2+x)xy(y-1)}\sigma\right) = 0.$$

Choosing $C_1 \to \infty$, $C_2 = 0$ or $C_1 = C_2 = 0$ or $C_1 = 0$, $C_2 \to \infty$ leads to the representation

$$Lclm\left(\{\sigma_x, \ \sigma_y\}, \{\sigma_x - \frac{2}{x}\sigma, \ \sigma_y - \frac{1}{y^2-y}\sigma\}, \{\sigma_x - \frac{1}{x}\sigma, \ \sigma_y - \frac{1}{y^2-y}\sigma\}\right) = 0$$

from which the fundamental system $\left\{1, \ \dfrac{x(y-1)}{y}, \ \dfrac{x^2(y-1)}{y}\right\}$ is obtained. Therefore possible transformation functions are

$$u = \frac{x(y-1)}{y}, \ v = \frac{x^2(y-1)}{y} \quad \text{with inverse} \ x = \frac{v}{u}, \ y = \frac{v}{v-u^2}.$$

The projective symmetry entails that any transformation

$$\bar{\sigma} = \frac{a_1\sigma + a_2\rho + a_3}{a_7\sigma + a_8\rho + a_9}, \quad \bar{\rho} = \frac{a_4\sigma + a_5\rho + a_6}{a_7\sigma + a_8\rho + a_9}$$

generates the same equation 6.180 from the canonical form $v''(u) = 0$. For example, choosing $a_k = 1$ for $k = 1,3,5,\ldots,9$ and $a_2 = a_4 = 2$ yields the transformation

$$u \equiv \bar{\sigma} = \frac{x(2x+1)(y-1)}{y}, \quad v \equiv \bar{\rho} = \frac{x(x+2)(y-1)}{y}$$

and its reverse

$$x \equiv \phi = \frac{v-2u}{u-2v}, \ y \equiv \psi = \frac{3(v-2u)}{u^2 - 4uv - 6u + 4v^2 + 3v}.$$

If it is applied to equation 6.180 it generates

$$-\frac{(2u-v)^3(u-2v)^6}{3[(4v+3)v + u^2 - 2(2v+3)u]^3(uv'-v)^3}v'' = 0,$$

i.e., up to a lower order factor the canonical form $v'' = 0$. ⬚

276

EXAMPLE 6.15   As a second example consider equation 6.124

$$yy'' - 3y'^2 + 3yy' - y^2 = 0$$

of the collection by Kamke where $A = 0$, $B = -\frac{3}{y}$, $C = 3$ and $D = -y$. By Theorem 5.34 it belongs to symmetry class $\mathcal{S}_8^2$. The system (6.37) is

$$a_x + a^2 + 3a + by + 2 = 0, \ a_y + ab = 0, \ b_x + ab = 0, \ b_y + b^2 + \frac{3}{y}b = 0.$$

Its rational solutions are $a = -1$, $b = 0$ or $a = -2$, $b = 0$ or $a = 0$, $b = -\frac{2}{y}$. Choosing the first alternative yields $B_2 = 0$, $C_2 = 1$ and leads to the Janet basis

$$\sigma_{xx} + y\sigma_y + \sigma_x = 0, \ \sigma_{xy} - \sigma_y = 0, \ \sigma_{yy} + \frac{3}{y}\sigma_y = 0.$$

It is completely reducible and may be represented as

$$Lclm\left(\{\sigma_x, \ \sigma_y\}, \ \{\sigma_x + \sigma, \ \sigma_y\}, \{\sigma_x - \sigma, \ \sigma_y + \frac{2}{y}\sigma\}\right) = 0.$$

There is an identical one for ρ. They yield the fundamental system $\{1, \ e^{-x}, \ \frac{e^x}{y^2}\}$. A possible choice for the transformation functions is

$$u = e^{-x}, \ v = \frac{e^x}{y^2} \ \text{ with inverse } \ x = \log\frac{1}{u}, \ y = \frac{1}{\sqrt{uv}}.$$

It is instructive to evaluate the same example after a transformation to a new set of variables has been performed. Introducing $\bar{x}$ and $\bar{y}$ by $x = \frac{\bar{y}}{\bar{x}}$, $y = \frac{1}{\bar{x}}$ and renaming them x and y, the equation 6.124 becomes

$$y'' - \frac{1}{x}y'^3 + \frac{3}{x^2}y'^2y - \frac{3}{x}y'^2 - \frac{3}{x^3}y'y^2 + \frac{6}{x^2}y'y - \frac{3}{x}y' + \frac{1}{x^4}y^3 - \frac{3}{x^3}y^2 + \frac{3}{x^2}y = 0.$$

There is a third order equation for the x-dependence of w now with the Loewy decomposition

$$Lclm\left[w_x, w_{xx} + \left(\frac{4}{x} - \frac{2x-y}{P}\right)w_x + \left(\frac{3}{xy} - \frac{1}{x^2} + \frac{2y}{x^3} - \frac{y^2}{x^4} - \frac{3x-2y}{yP}\right)w\right] = 0$$

where $P \equiv x^2 - xy + \frac{1}{3}y^2$. The first order factor yields the Janet basis

$$\sigma_{xxx} + \frac{6x^2 - 9xy + 4y^2}{3x^3 - 3x^2y + xy^2}\sigma_{xx} - \frac{6x^4 - 12x^3y + 10x^2y^2 - 5xy^3 + y^4}{3x^6 - 3x^5y + x^4y^2}\sigma_x = 0,$$

$$\sigma_y - \frac{\frac{1}{3}x^4}{x^2y - xy^2 + \frac{1}{3}y^3}\sigma_{x,x} + \frac{\frac{1}{3}x^3 - \frac{2}{3}x^2y + \frac{1}{3}xy^2}{x^2y - xy^2 + \frac{1}{3}y^3}\sigma_x = 0.$$

Transformation to Canonical Form 277

It is completely reducible with the representation

$$Lclm\left(\left\{\sigma_{xx} - \frac{1}{x-y}\sigma_x + \frac{3x^2y - 3xy^2 + y^3}{x^5 - x^4y}\sigma,\right.\right.$$

$$\left.\left.\sigma_y - \frac{x}{x-y}\sigma_x + \frac{1}{x-y}\sigma\right\}, \{\sigma_x, \sigma_y\}\right) = 0. \qquad \Box$$

In some applications there occur equations of the general form (5.34) with $A = 0$. This may be due to the fact that the transformation to canonical form is achieved by a so-called *fiber-preserving* transformation, i.e., the transformation function σ is independent of y. This specialization leads to a considerable simplification of the the procedure described in the above Theorem 6.14. The details are given in the subsequent corollary.

COROLLARY 6.3 *(Hsu and Kamran 1989) A second order ode of the form $y'' + B(x, y)y'^2 + C(x, y)y' + D(x, y) = 0$ is similar to $v''(u) = 0$ by a fiber-preserving point transformation if and only if its coefficients satisfy*

$$C_y - 2B_x = 0, \quad \left(C_x + \tfrac{1}{2}C^2 - 2D_y - 2BD\right)_y = 0.$$

The transformation function $u = \sigma(x)$ and $v = \rho(x, y)$ are determined by

$$\rho_{xx} - D\rho_y - \frac{\sigma_{xx}}{\sigma_x}\rho_x = 0, \quad \rho_{xy} - \tfrac{1}{2}C\rho_y - \tfrac{1}{2}\frac{\sigma_{xx}}{\sigma_x}\rho_x = 0, \quad \rho_{yy} - B\rho_y = 0,$$

$$\left(\frac{\sigma_{xx}}{\sigma_x}\right)_x - \tfrac{1}{2}\left(\frac{\sigma_{xx}}{\sigma_x}\right)^2 + C_x + \tfrac{1}{2}C^2 - 2D_y - 2BD = 0.$$

PROOF Applying a general fiber-preserving transformation $u = \sigma(x)$ and $v = \rho(x, y)$ to $v'' = 0$ yields

$$\sigma_x\rho_y y'' + \sigma_x\rho_{yy}y'^2 + (2\sigma_x\rho_{xy} - \sigma_{xx}\rho_y)y' + \sigma_x\rho_{xx} - \sigma_{xx}\rho_x = 0$$

from which the relations

$$\frac{\rho_{yy}}{\rho_y} = B, \quad 2\frac{\rho_{xy}}{\rho_y} - \frac{\sigma_{xx}}{\sigma_x} = C, \quad \frac{\rho_x}{\rho_y}\left(\frac{\rho_{xx}}{\rho_x} - \frac{\sigma_{xx}}{\sigma_x}\right) = D$$

follow. Transforming this system of pde's into a Janet basis in lexicographic term ordering with $\rho > \sigma > B > C > D$ and $y > x$ yields the above equations for σ and ρ. The two lowest equations in the Janet basis are the coefficient constraints. $\qquad \Box$

It should be noticed that the Riccati equation for $\frac{\sigma_{xx}}{\sigma_x}$ does not depend on y due to the second coefficient constraint. In this form the above corollary solves the equivalence problem for second order equations with a maximal number of fiber-preserving symmetries *constructively*, i.e., for all equations that occur Liouvillian solutions may be obtained by the algorithms described in Chapter 2.

Canonical Form Transformation and Infinitesimal Generators. In his main publication on solving differential equations Lie determines at first the

278

symmetry generators for a given second order ode explicitly, and then applies them for determining its canonical form. The following theorem is based on Lie [113], Kapitel 18, pages 412-425 and Kapitel 23, pages 503-531.

THEOREM 6.15   (Lie 1891) Let $\xi_i \partial_x + \eta_i \partial_y$, $i = 1, \ldots, r$, $r = 1, 2$ or 3, be the infinitesimal generators for a second order ode obeying canonical commutation relations. In terms of these generators, the transformation functions σ and ρ are determined by the following systems of equations where $\Delta_{ij} = \xi_i \eta_j - \xi_j \eta_i$.

$$\mathcal{S}_1^2 : \quad \xi_1 \sigma_x + \eta_1 \sigma_y = 0, \quad \xi_1 \rho_x + \eta_1 \rho_y = 1.$$

$$\mathcal{S}_{2,1}^2 : \quad \sigma_x = \frac{\eta_2}{\Delta_{12}}, \quad \sigma_y = -\frac{\xi_2}{\Delta_{12}}, \quad \rho_x = -\frac{\eta_1}{\Delta_{12}}, \quad \rho_y = \frac{\xi_1}{\Delta_{12}}.$$

$$\mathcal{S}_{2,2}^2 : \quad \sigma_x + \frac{\eta_1}{\Delta_{12}}\sigma = 0, \quad \sigma_y - \frac{\xi_1}{\Delta_{12}}\sigma = 0, \quad \rho_x + \frac{\eta_1}{\Delta_{12}}\rho = \frac{\eta_2}{\Delta_{12}}, \quad \rho_y - \frac{\xi_1}{\Delta_{12}}\rho = -\frac{\xi_2}{\Delta_{12}}.$$

$$\mathcal{S}_{3,1}^2 : \quad \sigma + \rho = \frac{\Delta_{13}}{\Delta_{12}}, \quad \sigma\rho = \frac{\Delta_{23}}{\Delta_{12}}.$$

$$\mathcal{S}_{3,2}^2 : \quad \sigma^2 = \frac{\Delta_{13}}{\Delta_{12}}, \quad \rho_x + \frac{\eta_1}{\Delta_{12}}\rho = 0, \quad \rho_y - \frac{\xi_1}{\Delta_{12}}\rho = 0.$$

$$\mathcal{S}_{3,3}^2 : \quad \sigma = -\frac{\Delta_{23}}{\Delta_{12}}, \quad c\rho = \frac{\Delta_{13}}{\Delta_{12}}. \qquad \mathcal{S}_{3,4}^2 : \quad \sigma = \frac{\Delta_{13}}{\Delta_{12}}, \quad \rho = -\frac{\Delta_{13} + \Delta_{23}}{\Delta_{12}}.$$

PROOF   For a one-parameter symmetry the solution is given by Lemma 3.4 on page 130. For two-parameter symmetries the proof has been given in Lemma 3.5 on page 132. For symmetry class $\mathcal{S}_{2,1}^2$ corresponding to group $\mathbf{g}_{26}$, the system

$$\xi_1 \sigma_x + \eta_1 \sigma_y = 1, \quad \xi_2 \sigma_x + \eta_2 \sigma_y = 0, \quad \xi_1 \rho_x + \eta_1 \rho_y = 0, \quad \xi_2 \rho_x + \eta_2 \rho_y = 1$$

is obtained; and for symmetry class $\mathcal{S}_{2,2}^2$ corresponding to group $\mathbf{g}_{25}$ it is

$$\xi_1 \sigma_x + \eta_1 \sigma_y = 0, \quad \xi_2 \sigma_x + \eta_2 \sigma_y = \sigma, \quad \xi_1 \rho_x + \eta_1 \rho_y = 1, \quad \xi_2 \rho_x + \eta_2 \rho_y = \rho.$$

In both cases the above result follows by algebraic elimination of the first order derivatives. The transformstion functions are obtained in terms of path integrals.

For the three-parameter groups, due to $N = 3$ in Lemma 2.12, two algebraic relations for σ and ρ are obtained. For the symmetry $\mathcal{S}_{3,1}^2$ the systems for σ and ρ are

$$\xi_1 \sigma_x + \eta_1 \sigma_y = 1, \quad \xi_2 \sigma_x + \eta_2 \sigma_y = \sigma, \quad \xi_3 \rho_x + \eta_3 \rho_y = \sigma^2,$$
$$\xi_1 \rho_x + \eta_1 \rho_y = 1, \quad \xi_2 \rho_x + \eta_2 \rho_y = \rho, \quad \xi_3 \rho_x + \eta_3 \rho_y = \rho^2$$

leading to

$$\Delta_{12}\sigma^2 - \Delta_{13}\sigma + \Delta_{23} = 0, \quad \Delta_{12}\rho^2 - \Delta_{13}\rho + \Delta_{23} = 0$$

with the solution as given above. For the symmetry class $\mathcal{S}_{3,2}^2$ the systems for σ and ρ are

$$\xi_1\sigma_x + \eta_1\sigma_y = 1, \quad \xi_2\sigma_x + \eta_2\sigma_y = 2\sigma, \quad \xi_3\rho_x + \eta_3\rho_y = \sigma^2,$$

$$\xi_1\rho_x + \eta_1\rho_y = 0, \quad \xi_2\rho_x + \eta_2\rho_y = \rho, \quad \xi_3\rho_x + \eta_3\rho_y = \sigma\rho^2.$$

The two algebraic relations

$$\Delta_{12}\sigma^2 + 2\Delta_{23}\sigma = 0, \quad \Delta_{12}\sigma\rho - \Delta_{13}\rho = 0$$

yield $\sigma = \frac{\Delta_{13}}{\Delta_{12}}$ and $\sigma^2 = \frac{\Delta_{23}}{\Delta_{12}}$. The second function ρ is not determined by these relations. Therefore, the first two equations for ρ have to be solved for its first derivatives with the above result. For symmetry class $\mathcal{S}_{3,3}^2$ the system for σ and ρ is

$$\xi_1\sigma_x + \eta_1\sigma_y = 1, \quad \xi_2\sigma_x + \eta_2\sigma_y = 0, \quad \xi_3\rho_x + \eta_3\rho_y = \sigma,$$

$$\xi_1\rho_x + \eta_1\rho_y = 0, \quad \xi_2\rho_x + \eta_2\rho_y = \rho, \quad \xi_3\rho_x + \eta_3\rho_y = c\rho^2$$

leading to $\Delta_{12}\sigma + \Delta_{23} = 0$ and $c\Delta_{12} - \Delta_{13} = 0$. Finally, for $\mathcal{S}_{3,4}^2$ the system

$$\xi_1\sigma_x + \eta_1\sigma_y = 0, \quad \xi_2\sigma_x + \eta_2\sigma_y = 0, \quad \xi_3\rho_x + \eta_3\rho_y = \sigma,$$

$$\xi_1\rho_x + \eta_1\rho_y = 1, \quad \xi_2\rho_x + \eta_2\rho_y = 0, \quad \xi_3\rho_x + \eta_3\rho_y = \sigma + \rho^2$$

leads to $\Delta_{12}\sigma - \Delta_{13} = 0$ and $\Delta_{12}(\sigma + \rho) + \Delta_{23} = 0$ from which the above relations follow immediately. ⬜

This result shows clearly the decreasing complexity of finding the canonical variables with increasing number of symmetries. For the full problem, however, it must be taken into account that symmetry generators have to be determined first.

6.3 Nonlinear Third Order Equations

For second order equations there is only a single equivalence class with a *linear* canonical representative which may be chosen as $v'' = 0$. This equivalence class makes up the full symmetry class $\mathcal{S}_8^2$. The situation is different for equations of order higher than two. There are infinitely many equivalence classes with linear canonical representatives that may be combined to form

280

several symmetry classes. As it has been shown in Chapter 2, there are special solution procedures for linear equations that do not exist for nonlinear ones. Therefore, it appears reasonable to consider the two cases separately. The subject of this Section 6.3 is third order equations that are *genuinely* nonlinear, i.e., they are not equivalent to a linear equation. These latter equations will be treated in the subsequent Section 6.4.

Canonical Forms and Structure Invariance Groups. Similar to equations of order two, the tabulation of the groups of the plane including its invariants allows determining the possible forms of invariant equations. They are described in the subsequent theorem together with its structure invariance groups.

THEOREM 6.16   *In canonical variables u and $v \equiv v(u)$ the third order quasilinear equations with nontrivial symmetries have the following canonical forms and corresponding structure invariance groups of point transformations $u = \sigma(x,y)$ and $v = \rho(x,y)$; f, g and r are undetermined functions of the respective arguments, a, a_i for all i and c are constants.*

   One-parameter symmetry group

$\mathcal{S}_1^3$: $v''' + r(u, v', v'') = 0$ *allows the pseudogroup $u = f(x)$, $v = g(x) + cy$.*

   Two-parameter symmetry groups

$\mathcal{S}_{2,1}^3$: $v''' + r(v'', v') = 0$ *allows $u = a_1 x + a_2 y + a_3$, $v = a_4 x + a_5 y + a_6$.*

$\mathcal{S}_{2,2}^3$: $u^2 v''' + r(uv'', v') = 0$ *allows $u = a_1 x$, $v = a_2 x + a_3 y + a_4$.*

$\mathcal{S}_{2,3}^3$: $v''' + v' r\left(\dfrac{v''}{v'}, u\right) = 0$ *allows $u = f(x)$, $v = a_1 y + a_2$.*

$\mathcal{S}_{2,4}^3$: $v''' + r(v'', u) = 0$ *allows*

$$
\sigma = \frac{(x + a_1)a_2}{1 - (x + a_1)a_2 a_3}, \qquad \rho = \frac{a_3 y}{1 - (x + a_1)a_2 a_4} + f(x).
$$

   Three-parameter symmetry groups

$\mathcal{S}_{3,1}^3$:
$$
\Phi_1^{(10)} = \frac{[(u - v)v'' + 2v'(v' + 1)]^2}{v'^3},
$$

$$
\Phi_2^{(10)} = \frac{(u - v)^2 v''' + 6(u - v)(v' + 1)v''}{v'^2} + \frac{6(v'^2 + 4v' + 1)}{v'}.
$$

$\omega\left(\Phi_1^{(10)}, \Phi_2^{(10)}\right) = 0$ *allows*

$$
u = \frac{a_1(x + a_2)}{1 - a_1 a_3(x + a_2)}, \qquad v = \frac{a_1(y + a_2)}{1 - a_1 a_3(y + a_2)}.
$$

Transformation to Canonical Form

$\mathcal{S}_{3,2}^3$: $v'''v^5 + 3v''v'v^4 + r(v''v^3) = 0$ *allows*

$$u = \frac{a_1(x + a_3)}{1 + a_4(x + a_3)}, \quad v = \frac{a_2 y}{1 + a_4(x + a_3)}.$$

$\mathcal{S}_{3,3}^3(c)$: $v''' + v'^{(c-3)/(c-1)} r\left(v''v'^{(2-c)/(c-1)}\right) = 0$ *allows* $u = a_1 x + a_2$, $v = a_3 y + a_4$.

$\mathcal{S}_{3,4}^3$: $v'''e^{2v'} + r(v''e^{v'}) = 0$ *allows* $u = a_1 x + a_2$, $v = a_3 x + a_1 y + a_4$.

$\mathcal{S}_{3,5}^3$: $v''' + v'r\left(\frac{v''}{v'}\right) = 0$ *allows* $u = a_1 x + a_2$, $v = a_3 y + a_4$.

$\mathcal{S}_{3,6}^3$: $v''' + r(v'') = 0$ *allows* $u = a_1 x + a_2$, $v = a_3 y + a_4 x^2 + a_5 x + a_6$.

$\mathcal{S}_{3,7}^3$: $v''' - 2v'' + v' + r(v'' - 2v' + v) = 0$ *allows* $u = x + a_1$,
$\qquad v = a_2 y + a_3 + a_4 e^x + a_5 x e^x$.

$\mathcal{S}_{3,8}^3$: $v''' - v'' + r(v'' - v') = 0$ *allows* $u = x + a_1$, $v = a_2 y + a_3 e^x + a_4 x + a_5$.

$\mathcal{S}_{3,9}^3$: $v''' - (\alpha + 1)v'' + \alpha v' + r[v'' - (\alpha + 1)v' + \alpha v] = 0$ *allows* $u = x + a_1$,
$\qquad v = a_2 y + a_3 + a_4 e^x + a_5 e^{\alpha x}$.

$\mathcal{S}_{3,10}^3$: $v''' + v''^2 r(v') = 0$ *allows* $u = a_1 x + a_2 y + a_3$, $v = a_4 x + a_5 y + a_6$.

Four-parameter symmetry groups

$\mathcal{S}_{4,1}^3$: $v'''v' + av''^2 = 0$ *with* $a \neq 0, -\frac{3}{2}$ *allows* $u = a_1 x + a_2$, $v = a_3 y + a_4$.

$\mathcal{S}_{4,2}^3$: $v'''v + 3v''v' + av''\sqrt{v''v} = 0$ *allows* $u = \frac{a_1 x}{1 - a_2 x} + a_3$, $v = \frac{a_4 y}{1 - a_2 x}$.

$\mathcal{S}_{4,3}^3(c)$: $v''' + av''^{(c-3)/(c-2)} = 0$ *allows* $u = a_1 x + a_2$, $v = a_3 x + a_4 y + a_5$.

$\mathcal{S}_{4,4}^3$: $v''' + ae^{-\frac{1}{2}v''} = 0$ *allows* $u = a_1 x + a_2$, $v = a_1^2 y + a_3 x^2 + a_4 x + a_5$.

Six-parameter symmetry group

$\mathcal{S}_6^3$: $v'''v' - \frac{3}{2}v''^2 = 0$ *allows* $u = a_1 + a_2 x + a_3 x^2$, $v = a_4 + a_5 y + a_6 y^2$.

The proof is similar to that for Theorem 6.4 and is not given. A few symmetry types are considered in the exercises.

One- and Two-Parameter Symmetries. The symmetry class $\mathcal{S}_1^3$ and the first two symmetry classes $\mathcal{S}_{2,1}^3$ and $\mathcal{S}_{2,2}^3$ described in Theorem 5.13 correspond to the same groups as the respective symmetry classes for second order equations. Its transformation to canonical form is therefore known by Theorems 6.8 and 6.9.

282

EXAMPLE 6.16   The equation considered in Example 5.27 in symmetry class $\mathcal{S}_1^3$ generates a type $\mathcal{J}_{1,2}^{(2,2)}$ Janet basis. Therefore case *b*) of Theorem 6.7 applies. The equation $\dfrac{dy}{dx} = 0$ yields $\sigma = y$, and the path integral leads to $\rho = -\log x$. Substituting these two functions the canonical form

$$v'''v' - 3v''^2 - (u-4)v''v'^2 + (2u-4)v'^4 + v'^3 = 0$$

is obtained.   □

EXAMPLE 6.17   The equation considered in Example 5.28 in symmetry class $\mathcal{S}_{2,1}^3$ generates a type $\mathcal{J}_{2,3}^{(2,2)}$ Janet basis. Therefore case *b*) of Theorem 6.8 applies. It yields the system $\sigma_{xx} = 0$, $\sigma_{xy} - \frac{1}{y}\sigma_x = 0$, $\sigma_{yy} = 0$ and an identical system for ρ with a solution basis $\{1, y, xy\}$. The choice $\sigma = y$ and $\rho = xy$ leads to $x = \frac{v}{u}$, $y = u$ and the canonical equation $v''' + 2v''v' + 1 = 0$.   □

EXAMPLE 6.18   The equation considered in Example 5.29 in symmetry class $\mathcal{S}_{2,2}^3$ generates a type $\mathcal{J}_{2,3}^{(2,2)}$ Janet basis. The expressions (6.6) and (6.8) of Theorem 6.9 yield $\sigma = \frac{1}{y}$ and $\rho = x$ from which the canonical form

$$u^2 v''' - \frac{3}{v}u^2 v''^2 - 5uv'' - v'^3 - 4v' = 0$$

is obtained.   □

The remaining symmetry classes $\mathcal{S}_{2,3}^3$ and $\mathcal{S}_{2,4}^3$ of this subsection are treated in the following two theorems.

THEOREM 6.17   *If a third order ode belongs to symmetry class $\mathcal{S}_{2,3}^3$, the transformation functions σ and ρ may be determined as follows; C_i are constants.*

a) *Janet basis type $\mathcal{J}_{2,2}^{(2,2)}$: $\sigma = f(x)$, f is an undetermined function of x, ρ obeys*

$$\rho_x = 0, \quad \rho_{yy} - b_1 \rho_y = 0. \tag{6.39}$$

b) *Janet basis type $\mathcal{J}_{2,1}^{(2,2)}$: σ is the same as in the preceding case, ρ obeys*

$$\rho_y + a_1 \rho_x = 0, \quad \rho_{xx} + \left(2\frac{a_{1,x}}{a_1} - b_1\right)\rho_x = 0. \tag{6.40}$$

c) *Janet basis of type $\mathcal{J}_{2,5}^{(2,2)}$: $\sigma(x,y) = f(\varphi)$, $\varphi(x,y) = C_0$ is the integral of*

$$\frac{dy}{dx} = -a_1, \tag{6.41}$$

ρ obeys

$$\rho_x = 0, \quad \rho_{yy} + \left(2\frac{a_{1,y}}{a_1} - c_1\right)\rho_y = 0. \tag{6.42}$$

Transformation to Canonical Form 283

d) *Janet basis type* $\mathcal{J}_{2,4}^{(2,2)}$: σ *is the same as in the preceding case,* ρ *obeys*

$$\rho_y + b_1\rho_x = 0, \quad \rho_{xx} + \left[\frac{2(a_1b_1)_x}{a_1b_1 + 1} - c_1\right]\rho_x = 0. \tag{6.43}$$

PROOF   In all four cases the equations for ρ are obtained as part of the Janet basis expressing the transformation functions in terms of the actual coefficients. In case a) and b) σ obeys $\sigma_y = 0$; in case c) and d) it obeys $\sigma_y - \frac{1}{a_1}\sigma_x = 0$ from which the given expressions follow from Corollary 2.7 on page 78. ▯

EXAMPLE 6.19   The equation considered in Example 5.30 in symmetry class $\mathcal{S}_{2,3}^3$ generates a type $\mathcal{J}_{2,4}^{(2,2)}$ Janet basis. The expressions (6.6) and (6.8) of Theorem 6.9 yield $\sigma = \frac{1}{y}$ and $\rho = x$ from which the canonical form

$$v''' + v'\left[\left(\frac{v''}{v'}\right)^3 + u + 1\right] = 0$$

is obtained. ▯

THEOREM 6.18   *If a third order ode belongs to symmetry class* $\mathcal{S}_{2,4}^3$, *three types of Janet bases may occur;* C_i *are constants.*

a) *Janet basis type* $\mathcal{J}_{2,1}^{(2,2)}$: $\sigma = C_1 \int \exp\left(\int S(x)dx\right)dx + C_2$, $S \neq 0$ *is a solution of*

$$S_x - \tfrac{1}{2}S^2 + b_{1,x} + \tfrac{1}{2}b_1^2 - 2b_2 = 0, \quad S_y = 0 \tag{6.44}$$

or $\sigma = C_1 x + C_2$ *if* $S = 0$.

$$\rho = C_3 \exp\left(\tfrac{1}{2}\int(S + b_1)dx - \int a_2 dy\right)\int \exp\left(\int a_2 dy\right)dy + f(x)$$

where $f(x)$ *is an undetermined function of its argument.*

b) *Janet basis type* $\mathcal{J}_{2,5}^{(2,2)}$: $\sigma = C_1 \exp\left(\int R(y)dy\right) + C_2$, $R \neq 0$ *a solution of*

$$R_y - \tfrac{1}{2}R^2 + c_{1,y} + \tfrac{1}{2}c_1^2 - 2c_2 = 0, \quad R_x = 0 \tag{6.45}$$

or $\sigma = C_1 y + C_2$ *if* $R = 0$.

$$\rho = C_3 \exp\left(\tfrac{1}{2}\int(R + c_1)dy - \int b_1 dx\right)\int \exp\left(\int b_1 dx\right)dx + g(y)$$

where $g(y)$ *is an undetermined function of its argument.*

c) *Janet basis type* $\mathcal{J}_{2,4}^{(2,2)}$: σ *obeys the system* $\sigma_y + b_1\sigma_x = 0$, $\sigma_{xx} - S\sigma_x = 0$ *where* S *is a solution of*

$$S_x - \tfrac{1}{2}S^2 + c_{1,x} + \tfrac{1}{2}c_1^2 - 2c_2 = 0. \tag{6.46}$$

284

If σ is known, $\Delta \equiv \sigma_x \rho_y - \sigma_y \rho_x$ is determined from

$$\Delta_x - \left(\tfrac{3}{2} \tfrac{\sigma_{xx}}{\sigma_x} + a_{2,x} b_1 + \tfrac{1}{2} c_1 \right) \Delta = 0,$$

$$\Delta_y - \left(\tfrac{3}{2} \tfrac{\sigma_{yy}}{\sigma_y} + \tfrac{1}{2} b_{1,x} + \tfrac{1}{2} a_2 b_{1,y} + b_2 - \tfrac{1}{2} b_1 c_1 \right) \Delta = 0$$

and, finally,

$$\rho = \int \frac{\Delta(\phi(\bar{x}, y), y)}{\sigma_x(\phi(\bar{x}, y), y)} dy \, \Big|_{\bar{x} = \sigma(x, y)} \tag{6.47}$$

where $\phi(\bar{x}, y)$ is defined as the inverse of $\bar{x} = \sigma(x, y)$.

PROOF   In case $a)$ for σ and ρ the Janet basis

$$\sigma_y = 0, \quad \rho_{xy} - \tfrac{1}{2} \left(\tfrac{\sigma_{xx}}{\sigma_x} + b_1 \right) \rho_y = 0, \quad \rho_{yy} - a_2 \rho_y = 0,$$

$$\sigma_{xxx} \sigma_x - \tfrac{3}{2} \sigma_{xx}^2 + (b_{1,x} + \tfrac{1}{2} b_1^2 - 2b_2) \sigma_x^2 = 0$$

is obtained. Defining $S \equiv \tfrac{\sigma_{xx}}{\sigma_x}$, the last equation becomes the Riccati equation for S given above with the general solution $S \equiv S(x, C_0)$, C_0 the integration constant. From the third equation $\rho = \bar{C}_3(x) \int \exp\left(\int a_2 dy \right) dy + f(x)$ is obtained. Substituting the derivatives ρ_y and ρ_{xy} into the second equation yields the linear first order equation $\bar{C}_{3,x} + \left(\int a_{2,x} dy - \tfrac{1}{2} S - \tfrac{1}{2} b_1 \right) \bar{C}_3 = 0$ for $\bar{C}_3$. Consequently,

$$\bar{C}_3 = C_3 \exp \left(\tfrac{1}{2} \int (S + b_1) dx - \int a_{2,x} dy dx \right)$$

with C_3 constant. Substituting this value into the expression for ρ completes case $a)$. In case $b)$ the equations for σ and ρ are

$$\sigma_x = 0, \quad \rho_{xx} - b_1 \rho_x = 0, \quad \rho_{xy} - \tfrac{1}{2} \left(\tfrac{\sigma_{yy}}{\sigma_y} + c_1 \right) \rho_x,$$

$$\sigma_{yyy} \sigma_y - \tfrac{3}{2} \sigma_{yy}^2 + (c_{1,y} + \tfrac{1}{2} c_1^2 - 2c_2) \sigma_y^2 = 0.$$

A similar calculation as in case $a)$ leads to the above result. In case $c)$ the equations for σ are

$$\sigma_y + b_1 \sigma_x = 0, \quad \sigma_{xxx} \sigma_x - \tfrac{3}{2} \sigma_{xx}^2 + (c_{1,x} - 2c_2 + \tfrac{1}{2} c_1^2) \sigma_x^2 = 0$$

from which the given representation follows if $S \equiv \tfrac{\sigma_{xx}}{\sigma_x}$ is defined. The system for ρ is

$$\rho_{xy} + b_1 \rho_{xx} - \tfrac{1}{2} \left(\tfrac{\sigma_{xx}}{\sigma_x} + 2a_{2,x} b_1 + c_1 \right) \rho_y - \tfrac{1}{2} b_1 \left(\tfrac{\sigma_{xx}}{\sigma_x} + c_1 \right) \rho_x = 0,$$

$$\rho_{yy} - b_1^2 \rho_{xx} + b_1 \left(\tfrac{\sigma_{xx}}{\sigma_x} - a_{2,y} + a_{2,x} b_1 + c_1 + a_2 b_2 \right) \rho_y$$

$$+ b_1^2 \left(\tfrac{\sigma_{xx}}{\sigma_x} + c_1 + a_2 b_2 \right) \rho_x = 0.$$

Defining Δ as above and expressing the derivatives of ρ in terms of Δ, the above system for Δ is obtained. If σ and Δ are known, ρ may be determined from it applying part $ii)$ of Corollary 2.7. □

EXAMPLE 6.20   The equation considered in Example 5.31 in symmetry class $\mathcal{S}_{2,4}^3$ generates a type $\mathcal{J}_{2,5}^{(2,2)}$ Janet basis. Consequently, case $b)$ applies. $R = 0$ leads to the special solution $\sigma = y$. The corresponding value for ρ is $\frac{1}{x}$. From these transformation functions the canonical form $v''' + uv''^2 + v'' + 1 = 0$ follows. □

Three-Parameter Symmetries. The first four symmetry classes in this subsection are $\mathcal{S}_{3,1}^3, \dots, \mathcal{S}_{3,4}^3$. They correspond to the respective symmetry classes of second order equations, and it has been described in Theorems 6.10 to 6.13 how the transformation functions to canonical form are determined. These results may be applied to third order equations without any change.

EXAMPLE 6.21   The canonical form transformation of the ode in symmetry class $\mathcal{S}_{3,1}^3$ considered in Example 5.32 is covered by Theorem 6.10, case $a)$. Because $b_1 = 0$, P and Q are determined from (6.15) with the result $P = \frac{1}{(x-y)^2}$, $Q = \frac{2}{x-y}$. R is obtained from (6.16) with the result $R = \frac{x}{y(x-y)}$. The system for σ is $\sigma_x = 0$, $\sigma_{yy} + \frac{2}{y}\sigma_y = 0$ with the non-constant solution $\sigma = \frac{1}{y}$ from which $\rho = \frac{1}{x}$ follows. With these transformation functions the canonical form

$$v''' + \frac{6v''(v'+1)}{u-v} + \frac{6v'}{(u-v)^2}(v'^2 + 4v' + 1)$$

$$+ \frac{v'^2}{(u-v)^2}\left[\frac{(v''(u-v) + 2v'(v'+1))^2}{v'^3} + 1\right] = 0$$

is obtained. □

EXAMPLE 6.22   The canonical form transformation of the ode in symmetry class $\mathcal{S}_{3,2}^3$ considered in Example 5.33 is covered by Theorem 6.11, case $b)$. At first $R_{yy} + R_y^2 + \frac{2}{y}R_y = 0$ yields the special solution $R_y = -\frac{1}{y}$. With $R_x = -\frac{1}{x}$ the system $\rho_x - \frac{1}{x}\rho =$, $\rho_y - \frac{1}{y}\rho =$ is obtained with the solution $\rho = xy$. Finally, $\sigma_x = \sigma_{yy} = 0$ with the special solution $\sigma = y$ leads to the canonical form $v'''v^5 + 3v''v'v^4 + 2v''v^2 + 1 = 0$. □

EXAMPLE 6.23   The canonical form transformation of the ode in symmetry class $\mathcal{S}_{3,3}^3(c)$ considered in Example 5.34 is covered by Theorem 6.12. Because $b_1 = \frac{y}{2x} \neq 0$ of its type $\mathcal{J}_{3,6}^{(2,2)}$ Janet basis, the last case of $a)$ applies. The systems for σ and ρ are $\sigma_y = 0$, $\sigma_{xx} + \frac{2}{x}\sigma_x = 0$ and $\rho_y + \frac{x}{y}\rho_x = 0$, $\rho_{xx} + \frac{x}{y}\rho_x = 0$ from which the transformation functions $\sigma = \frac{1}{x}$ and $\rho = \frac{y}{x}$ are obtained. They yield the canonical form $v'''v'^3 + 1 = 0$. □

286

EXAMPLE 6.24   The canonical form transformation of the ode in symmetry class $S_{3,4}^3$ considered in Example 5.35 is covered by Theorem 6.13, case a). Due to $b_1 = c_1 = -1$, the system for σ is $\sigma_x = \sigma_{yy} = 0$ with the special solution $\sigma = y$. The remaining equations $\rho_x = \sigma_y$, $\rho_{yy} = 0$ yield $\rho = x$ and the the canonical form $v''' + v''^2 + e^{-2v'} = 0$ follows. $\Box$

The remaining six three-parameter symmetries $S_{3,5}^3, \ldots, S_{3,10}^3$ do not have a counterpart for second order equations and are treated next. For the symmetry class $S_{3,9}^3(c)$ the parameter c has been determined in Theorem 5.14 in terms of Janet basis coefficients; i.e., the symmetry class may be considered as known. The proof of Theorem 6.19 to 6.24 is essentially identical in all cases, the given equations are immediately obtained as part of a Janet basis for the pde's expressing the general determining system in terms of the respective Janet basis coefficients.

THEOREM 6.19   *If a third order ode belongs to symmetry class $S_{3,5}^3$, the transformation functions σ and ρ may be determined by the following systems of pde's.*

a) *Janet basis type $\mathcal{J}_{3,5}^{(2,2)}$:*

$$\sigma_y = 0, \quad \sigma_{xx} - a_2\sigma_x = 0, \quad \rho_x = 0, \quad \rho_{yy} - d_1\rho_y = 0. \qquad (6.48)$$

b) *Janet basis type $\mathcal{J}_{3,4}^{(2,2)}$:*

$$\sigma_y = 0, \quad \sigma_{xx} - a_3\sigma_x = 0, \quad \rho_y + c_1\rho_x = 0, \quad \rho_{xx} - (2c_3 - 2a_3 - d_1)\rho_x = 0. \qquad (6.49)$$

c) *Janet basis type $\mathcal{J}_{3,7}^{(2,2)}$: There holds always $c_1 \neq 0$.*

$$\sigma_y - \frac{1}{c_1}\sigma_x = 0 = 0, \quad \sigma_{xx} - a_3\sigma_x = 0, \quad \rho_x = 0, \quad \rho_{yy} - (2a_2 - 2c_2 - d_1)\rho_y = 0. \qquad (6.50)$$

d) *Janet basis type $\mathcal{J}_{3,6}^{(2,2)}$: If $b_1 = 0$, the system is*

$$\sigma_x = 0, \quad \sigma_{yy} - c_2\sigma_y = 0, \quad \rho_y + a_1\rho_x = 0, \quad \rho_{xx} - d_1\rho_x = 0 \qquad (6.51)$$

and if $b_1 \neq 0$

$$\sigma_y - \frac{1}{b_1}\sigma_x = 0, \quad \sigma_{xx} - \frac{b_3}{b_1}\sigma_x = 0,$$
$$\rho_y + a_1\rho_x = 0, \quad \rho_{xx} + (2a_1b_3 - 2a_3b_1 - d_1)\rho_x = 0. \qquad (6.52)$$

EXAMPLE 6.25   The Janet basis coefficients given in Example 5.36 lead to the systems $\sigma_x = 0$, $\sigma_{yy} + \frac{2}{y}\sigma_y = 0$ and $\rho_y = 0$, $\rho_{xx} + \frac{2}{x}\rho_x = 0$. A functionally

Transformation to Canonical Form 287

independent set of solutions is $\sigma = \frac{1}{y}$, $\rho = \frac{1}{x}$. It yields the canonical form $v'''v' + v''^2 - 2v'^2 = 0$. $\square$

THEOREM 6.20   *If a second order ode belongs to symmetry class $\mathcal{S}_{3,6}^3$, the transformation functions σ and ρ are determined by the following systems of pde's.*

a) *Janet basis type* $\mathcal{J}_{3,4}^{(2,2)}$:

$$\sigma_y = 0, \quad \sigma_{xx} - a_3\sigma_x = 0, \quad \rho_{xy} - c_3\rho_y = 0, \quad \rho_{yy} - c_2\rho_y = 0,$$

$$\rho_{xxx} - 3a_3\rho_{xx} - d_3\rho_y - (a_{3,x} - 2a_3^2)\rho_x = 0.$$

b) *Janet basis type* $\mathcal{J}_{3,7}^{(2,2)}$:

$$\sigma_x = 0, \quad \sigma_{yy} - c_2\sigma_y = 0, \quad \rho_{xx} - a_3\rho_x = 0, \quad \rho_{xy} - \tfrac{1}{2}(c_2 + d_1)\rho_x = 0,$$

$$\rho_{yyy} - 3c_2\rho_{yy} + (d_{1,y} + \tfrac{3}{2}c_2^2 + \tfrac{1}{2}d_1^2 - 2d_3)\rho_y - d_2\rho_x = 0.$$

c) *Janet basis type* $\mathcal{J}_{3,6}^{(2,2)}$: *There holds always $a_1 \neq 0$.*

$$\sigma_y + a_1\sigma_x = 0, \quad \sigma_{xx} + a_1b_3\sigma_x = 0,$$

$$\rho_{xy} + a_1\rho_{xx} - (a_1b_3 - a_3b_1 + b_2)\rho_y - a_3\rho_x = 0,$$

$$\rho_{yy} - a_1^2\rho_{xx} + (a_1^2b_3 + a_3 - c_2)\rho_y + (a_1a_3 - a_2)\rho_x = 0,$$

$$\rho_{xxx} + 3a_1b_3\rho_{xx} - (b_{3,x} + 3a_1b_3^2 + b_2b_3 + 2\tfrac{a_3}{a_1}b_3 + \tfrac{d_3}{a_1})\rho_y$$
$$-(2a_3b_3 + d_3)\rho_x = 0.$$

An example for a canonical form transformation in symmetry class $\mathcal{S}_{3,6}^3$ is given following Theorem 6.23 in conjunction with examples for symmetry classes $\mathcal{S}_{3,7}^3$, $\mathcal{S}_{3,8}^3$ and $\mathcal{S}_{3,9}^3(c)$ which are treated in the next three theorems.

THEOREM 6.21   *If a second order ode belongs to symmetry class $\mathcal{S}_{3,7}^3$, the transformation functions σ and ρ are determined by the following systems of pde's.*

a) *Janet basis type* $\mathcal{J}_{3,4}^{(2,2)}$:

$$\sigma_y = 0, \quad \sigma_x = c_3 - \tfrac{1}{2}(a_3 + d_1), \quad \rho_{xy} - c_3\rho_y = 0, \quad \rho_{yy} - c_2\rho_y = 0,$$

$$\rho_{xxx} + (d_1 - 2a_3 - 2c_3)\rho_{xx} - d_3\rho_y$$

$$-(a_{3,x} - \tfrac{5}{4}a_3^2 - \tfrac{1}{4}d_1^2 + d_1c_3 + \tfrac{1}{2}a_3d_1 - c_3^2 - a_3c_3)\rho_x = 0.$$

b) *Janet basis type* $\mathcal{J}_{3,7}^{(2,2)}$:

$$\sigma_x = 0, \ \sigma_y = a_2 - \tfrac{1}{2}(c_1 + d_1), \ \rho_{xx} - a_3\rho_x = 0, \ \ \rho_{xy} - a_2\rho_x = 0,$$

$$\rho_{yyy} - (2a_2 + 2c_2 - d_1)\rho_{yy} + (d_{1,y} + a_2^2 + a_2 c_2 - a_2 d_1$$
$$+ \tfrac{3}{4}c_2^2 - \tfrac{1}{2}c_2 d_1 + \tfrac{3}{4}d_1^2 - 2d_3)\rho_y - d_2\rho_x = 0.$$

c) *Janet basis type* $\mathcal{J}_{3,6}^{(2,2)}$: *There holds always* $a_1^2 b_3 - a_1 d_1 + 2a_3 \neq 0$ *and* $a_1 \neq 0$.

$$\sigma_x = a_1 b_3 - 2a_3 b_1 - d_1, \ \sigma_y = a_1 d_1 - a_1^2 b_3 - 2a_3,$$

$$\rho_{xy} + a_1\rho_{xx} - (a_1 b_3 - a_3 b_1 + b_2)\rho_y - a_3\rho_x = 0,$$

$$\rho_{yy} - a_1^2\rho_{xx} + (a_1^2 b_3 + a_3 - c_2)\rho_y + (a_1 a_3 - a_2)\rho_x = 0,$$

$$\rho_{xxx} + \frac{P_1(x,y)}{Q(x,y)}\rho_{xx} + \frac{P_2(x,y)}{Q(x,y)}\rho_y + \frac{P_3(x,y)}{4Q(x,y)}\rho_x = 0$$

where

$$P_1(x,y) \equiv 2a_1^3 b_3^2 - a_1^2 b_3 d_1 + 2a_1 a_3 b_3 - a_1 d_1^2 + 4a_3^2 b_1 + 4a_3 d_1, \quad (6.53)$$

$$P_2(x,y) \equiv -b_{3,x} a_1^2 b_3 + b_{3,x} a_1 d_1 - 2b_{3,x} a_3 - 2a_1^3 b_3^3 - a_1^2 b_2 b_3^2$$
$$+ a_1^2 b_3^2 d_1 - 4a_1 a_3 b_3^2 + a_1 b_2 b_3 d_1 + a_1 b_3 d_1^2 - a_1 b_3 d_3 \quad (6.54)$$
$$+ 2a_3 b_1 d_3 - 2a_3 b_2 b_3 - 2a_3 b_3 d_1 + d_1 d_3,$$

$$P_3(x,y) \equiv a_1^4 b_3^3 - 3a_1^3 b_3^2 d_1 - 2a_1^2 a_3 b_3^2 + 3a_1^2 b_3 d_1^2 - 4a_1^2 b_3 d_3$$
$$- 4a_1 a_3 b_3 d_1 - a_1 d_1^3 + 4a_1 d_1 d_3 + 8a_3^3 b_1^2$$
$$+ 12a_3^2 b_1 d_1 - 4a_3^2 b_3 + 6a_3 d_1^2 - 8a_3 d_3,$$

$$Q(x,y) \equiv a_1^2 b_3 - a_1 d_1 + 2a_3. \quad (6.55)$$

THEOREM 6.22 *If a second order ode belongs to symmetry class* $\mathcal{S}_{3,8}^3$, *the transformation functions* σ *and* ρ *are determined by the following systems of pde's.*

a) *Janet basis type* $\mathcal{J}_{3,4}^{(2,2)}$:

$$\sigma_x = 2c_3 - a_3 - d_1, \ \sigma_y = 0, \ \rho_{xy} - c_3\rho_y = 0, \ \rho_{yy} - c_2\rho_y = 0,$$

$$\rho_{xxx} + (d_1 - 2a_3 - 2c_3)\rho_{xx} - d_3\rho_y$$
$$- (a_{3,x} - a_3^2 + a_3 d_1 - 2a_3 c_3)\rho_x = 0.$$

Transformation to Canonical Form

b) Janet basis type $\mathcal{J}_{3,7}^{(2,2)}$:

$$\sigma_x = 0, \ \sigma_y = 2a_2 - c_2 - d_1, \ \rho_{xx} - a_3\rho_x = 0, \ \rho_{xy} - a_2\rho_x = 0,$$

$$\rho_{yyy} - (2a_2 + 2c_2 - d_1)\rho_{yy} + (d_{1,y} - 2a_2^2 + 4a_2c_2$$

$$+ 2a_2d_1 - 2c_2d_1 - 2d_3)\rho_y - d_2\rho_x = 0.$$

c) Janet basis type $\mathcal{J}_{3,6}^{(2,2)}$: There holds always $a_1 \neq 0$.

$$\sigma_x = a_1b_3 - 2a_3b_1 - d_1, \ \sigma_y = a_1d_1 - a_1^2b_3 - 2a_3,$$

$$\rho_{xy} + a_1\rho_{xx} - (a_1b_3 - a_3b_1 + b_2)\rho_y - a_3\rho_x = 0,$$

$$\rho_{yy} - a_1^2\rho_{xx} + (a_1^2b_3 + a_3 - c_2)\rho_y + (a_1a_3 - a_2)\rho_x = 0,$$

$$\rho_{xxx} + (2a_1b_3 + d_1 - 2\frac{a_3}{a_1})\rho_{xx}$$

$$- (b_{3,x} + b_3d_1 + 2a_1b_3^2 + b_2b_3 + \frac{d_3}{a_1})\rho_y - 2a_3b_3\rho_x = 0.$$

THEOREM 6.23 *If a second order ode belongs to symmetry class $\mathcal{S}_{3,9}^3(c)$ the transformation functions σ and ρ are determined by the following systems of pde's.*

a) Janet basis type $\mathcal{J}_{3,4}^{(2,2)}$:

$$\sigma_x = \frac{1}{c+1}(2c_3 - a_3 - d_1), \ \sigma_y = 0, \ \rho_{xy} - c_3\rho_y = 0, \ \rho_{yy} - c_2\rho_y = 0,$$

$$\rho_{xxx} + (d_1 - 2a_3 - 2c_3)\rho_{xx} - d_3\rho_y + \Big[a_{3,x} - a_3^2 - \frac{c}{(c+1)^2}(a_3^2 + d_1^2)$$

$$- \frac{c^2+1}{(c+1)^2}a_3(2c_3 - d_1) - \frac{4c}{(c+1)^2}c_3(c_3 - d_1)\Big]\rho_x = 0.$$

b) Janet basis type $\mathcal{J}_{3,7}^{(2,2)}$:

$$\sigma_x = 0, \ \sigma_y = \frac{1}{c+1}(2a_2 - c_2 - d_1), \ \rho_{xx} - a_3\rho_x = 0, \ \rho_{xy} - a_2\rho_x = 0,$$

$$\rho_{yyy} + (d_1 - 2a_2 - 2c_2)\rho_{yy} + \Big[c_{2,y} - c_2^2 - \frac{c}{(c+1)^2}(c_2^2 + d_1^2)$$

$$- \frac{c^2+1}{(c+1)^2}c_2(2a_2 - d_1) - \frac{4c}{(c+1)^2}a_2(a_2 - d_1)\Big]\rho_x = 0.$$

c) Janet basis type $\mathcal{J}_{3,6}^{(2,2)}$: There holds always $a_1 \neq 0$.

$$\sigma_x = \frac{1}{c+1}\Big(a_1b_3 - d_1 + 2\frac{a_3}{a_1}\Big), \ \sigma_y = \frac{1}{c+1}(a_1d_1 - a_1^2b_3 - 2a_3),$$

$$\rho_{xy} + a_1\rho_{xx} - (a_1b_3 - a_3b_1 + b_2)\rho_y - a_3\rho_x = 0,$$

$$\rho_{yy} - a_1^2\rho_{xx} + (a_1^2b_3 + a_3 - c_2)\rho_y + (a_1a_3 - a_2)\rho_x = 0,$$

$$\rho_{xxx} + (2a_1b_3 + d_1 - 2\tfrac{a_3}{a_1})\rho_{xx} - (b_{3,x} + 2a_1b_3^2 + b_2b_3 + b_3d_1 + \tfrac{d_3}{a_1})\rho_y$$

$$- \Big[d_3 + 2\frac{c^2+1}{(c+1)^2}a_3b_3 - \frac{c}{(c+1)^2}(a_1b_3 - d_1)^2$$

$$- \frac{4c}{(c+1)^2}\frac{a_3}{a_1^2}(a_3 - a_1d_1)\Big]\rho_x = 0.$$

EXAMPLE 6.26   The type $\mathcal{J}_{3,4}^{(2,2)}$ Janet bases for the ode's that have been considered in Example 5.37 are dealt with in case $a)$ of Theorems 6.20 to 6.23. In the first case, for σ the equations $\sigma_y = 0$, $\sigma_{xx} - \tfrac{2}{x}\sigma_x = 0$ are obtained with the fundamental system $\{1, \tfrac{1}{x}\}$. For the remaining three cases σ obeys $\sigma_y = 0$, $\sigma_x = \tfrac{1}{x^2}$ with the special solution $\sigma = -\tfrac{1}{x}$. For all four cases the systems for ρ comprise the equations $\rho_{xy} = 0$ and $\rho_{yy} = 0$. In addition there is a third order equation

$$\rho_{xxx} + \tfrac{6}{x}\rho_{xx} + \tfrac{6}{x^2}\rho_x = 0, \ \ \rho_{xxx} + \Big(\tfrac{6}{x} + \tfrac{2}{x^2}\Big)\rho_{xx} + \Big(\tfrac{6}{x^2} + \tfrac{4}{x^3} + \tfrac{3}{x^4}\Big)\rho_x = 0,$$

$$\rho_{xxx} + \Big(\tfrac{6}{x} + \tfrac{1}{x^2}\Big)\rho_{xx} - \Big(\tfrac{2}{x} - \tfrac{2}{x^3}\Big)\rho_x = 0$$

or

$$\rho_{xxx} + \Big(\tfrac{6}{x} + \tfrac{4}{x^2}\Big)\rho_{xx} - \Big(\tfrac{2}{x^2} - \tfrac{8}{x^3} + \tfrac{6}{x^4}\Big)\rho_x = 0$$

It is not necessary to obtain a fundamental system for ρ, rather it is sufficient to find a single solution such that σ and ρ are functionally independent. It turns out that $\rho = y$ together with $\sigma = -\tfrac{1}{x}$ have this property and transform the ode's of Example 5.37 into a canonical form. In all four cases it has the general structure $\tfrac{dz}{dx}z + z^2 + 1 = 0$ where z is defined by $z \equiv v''$, $z \equiv v'' - 2v' + v$, $z \equiv v'' - v'$ or $z \equiv v'' - 4v' + 3v$. $\Box$

THEOREM 6.24   If a second order ode belongs to symmetry class $\mathcal{S}_{3,10}^3$ the transformation function σ is determined by the system of pde's

$$\sigma_{xx} - b_3\sigma_y - (b_2 - c_3)\sigma_x = 0, \ \ \sigma_{xy} - b_2\sigma_y - a_3\sigma_x = 0,$$
$$\sigma_{yy} - (a_3 + c_2)\sigma_y - a_2\sigma_x = 0. \tag{6.56}$$

The function ρ satisfies an identical system.

EXAMPLE 6.27   The Janet basis coefficients given in Example 5.38 lead to the system $\sigma_{xx} + \tfrac{2}{x}\sigma_x = 0$, $\sigma_{xy} = 0$, $\sigma_{yy} = 0$ for σ and an identical one for ρ with the solution basis $\{1, \tfrac{1}{x}, y\}$. According to the preceding theorem a proper choice for the transformation functions is $\sigma = \tfrac{1}{x}$, $\rho = y$. It yields the canonical form $v'''v' + v''^2(v'^2 + 1) = 0$. $\Box$

Transformation to Canonical Form

Four-Parameter Symmetries. They occur for the first time for equations of order three. For the symmetry class $\mathcal{S}_{4,3}^3(c)$ the parameter c has been determined in Theorem 5.14 in terms of Janet basis coefficients; i.e., the symmetry class may be considered as known.

THEOREM 6.25   If a second order ode belongs to symmetry class $\mathcal{S}_{4,1}^3$ the transformation functions σ and ρ are determined by the following systems of pde's.

a) Janet basis type $\mathcal{J}_{4,10}^{(2,2)}$:

$$\sigma_x = 0, \quad \sigma_{yy} - d_1\sigma_y = 0, \quad \rho_y = 0, \quad \rho_{xx} - c_2\rho_x = 0. \tag{6.57}$$

b) Janet basis type $\mathcal{J}_{4,9}^{(2,2)}$:

$$\sigma_y + b_1\sigma_x = 0, \quad \sigma_{xx} + (b_1d_2 - c_2)\sigma_x, \quad \rho_y = 0, \quad \rho_{xx} - c_2\rho_x = 0. \tag{6.58}$$

c) Janet basis type $\mathcal{J}_{4,14}^{(2,2)}$: Two cases have to be distinguished. If $a_1 = 0$

$$\sigma_y - \frac{1}{b_1}\sigma_x = 0, \quad \sigma_{xx} - c_2\sigma_x = 0, \quad \rho_x = 0, \quad \rho_{yy} - (2b_3 - c_1)\rho_y = 0 \tag{6.59}$$

or an identical system with σ and ρ exchanged; and if $a_1 \neq 0$

$$\sigma_y - \frac{1}{\sqrt{-a_1}}\sigma_x = 0, \quad \sigma_{xx} + \left(b_4 + a_1e_2 - d_2\sqrt{-a_1}\right)\sigma_x = 0,$$
$$\rho_y + \frac{1}{\sqrt{-a_1}}\rho_x = 0, \quad \rho_{xx} + \left(b_4 + a_1e_2 + d_2\sqrt{-a_1}\right)\rho_x = 0. \tag{6.60}$$

PROOF   The given equations are immediately obtained as part of a Janet basis for the pde's expressing the general determining system in terms of the respective Janet basis coefficients. In case c) for $a_1 = 0$ it cannot be decided a priori which alternative applies. If $u = \sigma$ and $v = \rho$, as obtained by the above system, does not yield the canonical form, the proper transformation is $u = \rho$ and $v = \sigma$. The algebraic function that may occur for $a_1 \neq 0$ does not prevent the solution because only integrations are involved. ⬜

EXAMPLE 6.28   The coefficients of the type $\mathcal{J}_{4,14}^{(2,2)}$ Janet basis in Example 5.39 single out the first alternative of case c). The system $\sigma_y - \frac{y}{x}\sigma_x = 0$, $\sigma_{xx} = 0$ yields $\sigma = xy$, and the system $\rho_x = 0$, $\rho_{yy} + \frac{2}{y}\rho_y = 0$ yields $\rho = \frac{1}{y}$. Substitution into the equation of Example 5.39 yields the canonical form $v'''v' - \frac{1}{2}v''^2 = 0$. Exchanging σ and ρ, the canonical form $v'''v' - \frac{5}{2}v''^2 = 0$ is obtained. ⬜

292

THEOREM 6.26   *If a second order belongs to symmetry class $S^3_{4,2}$ the transformation functions σ and ρ are determined by the following systems of pde's.*

a) *Janet basis type* $\mathcal{J}^{(2,2)}_{4,12}$:

$$R_x + \tfrac{1}{2}d_2 = 0, \ R_{yy} + R_y^2 - d_1 R_y - \tfrac{1}{4}c_1 d_2 = 0,$$

$$\rho_x + R_x \rho = 0, \ \rho_y + R_y \rho = 0, \ \sigma_x = 0, \ \sigma_{yy} - (2R_y + d_1)\sigma_y = 0.$$
$$(6.61)$$

b) *Janet basis type* $\mathcal{J}^{(2,2)}_{4,9}$:

$$R_y + b_1 R_x + \tfrac{1}{2}b_1^2 d_2 + a_4 - b_1 c_2 - \tfrac{1}{2}c_1 = 0,$$

$$R_{xx} + R_x^2 + (b_1 d_2 - c_2)R_x + \tfrac{1}{4}d_2(b_1^2 d_2 + 2a_4 - 2b_1 c_2 - c_1) = 0,$$

$$\rho_x + R_x \rho = 0, \ \rho_y + R_y \rho = 0,$$

$$\sigma_y + b_1 \sigma_x = 0, \ \sigma_{xx} + (2R_x + b_1 d_2 - c_2)\sigma_x = 0.$$
$$(6.62)$$

PROOF   In case a) the system of pde's expressing the transformation functions in terms of the Janet basis coefficients comprises the following equations containing σ and ρ.

$$\sigma_x = 0, \ \sigma_{yy} - (2\tfrac{\rho_y}{\rho} + d_1)\sigma_y = 0,$$

$$\rho_x + \tfrac{1}{2}d_2\rho = 0, \ \rho_{yy}\rho - 2\rho_y^2 - d_1\rho_y\rho + \tfrac{1}{4}c_1 d_2 = 0.$$

Introducing the new function R by $\log \rho = -R$ leads to the equations given above. In case b) the equations obtained in the first place are

$$\sigma_y + b_1 \sigma_x = 0, \ \sigma_{xx} - (2\tfrac{\rho_x}{\rho} - b_1 d_2 c_2)\sigma_x = 0,$$

$$\rho_y + b_1 \rho_x + (b_1 c_2 + \tfrac{1}{2}c_1 - a_4 - \tfrac{1}{2}b_1^2 d_2)\rho = 0,$$

$$\rho_{xx}\rho - 2\rho_x^2 + (b_1 d_2 - c_2)\rho_x \rho$$

$$-(\tfrac{1}{2}a_4 d_2 + \tfrac{1}{4}a_4 d_2 + \tfrac{1}{4}b_1^2 b_2^2 - \tfrac{1}{2}b_1 c_2 d_2 - \tfrac{1}{4}c_1 d_2)\rho^2 = 0.$$

The same substitution $\log \rho = -R$ yields the equations for case b).   ☐

THEOREM 6.27   *If a second order ode belongs to symmetry class $S^3_{4,3}(c)$ the transformation functions σ and ρ are determined by the following systems of pde's.*

a) *Janet basis type* $\mathcal{J}^{(2,2)}_{4,9}$:

$$\sigma_y = 0, \ \sigma_{xx} - c_2 \sigma_x = 0, \ \rho_{xx} - \frac{d_2}{b_2 + 2}\rho_y - c_2 \rho_x = 0,$$
$$(6.63)$$

$$\rho_{xy} + (b_2 c_2 - b_4)\rho_y = 0, \ \rho_{yy} - b_3 \rho_y = 0.$$

Transformation to Canonical Form

b) *Janet basis type* $\mathcal{J}_{4,14}^{(2,2)}$:

$$\sigma_y - \frac{b_2+1}{a_2}\sigma_x = 0, \ \sigma_{xx} + \left[(b_2+1)(a_3+d_1 - 2\frac{a_4}{a_2} - \frac{c_1}{a_2}) + b_2c_2\right]\sigma_x = 0,$$
$$(6.64)$$

$$\rho_{xx} + \frac{1}{2b_2+1}\left[2a_2(b_2+1)(a_3+d_1) - (2b_2+1)(2a_4+c_1) + 2a_2b_2c_2\right]\rho_y$$
$$\frac{1}{(b_2+1)(2b_2+1)}\left[(b_2+1)(2b_4+d_1-a_3) + a_2b_2(b_3+d_2+e_1)\right]\rho_x = 0,$$

$$\rho_{xy} - \frac{b_1+1}{a_2}\rho_{xx} + \frac{1}{2a_2}\left[2a_4(b_2+1) - a_2(a_3+2b_4+c_2+d_1)\right]\rho_y$$
$$+ \frac{1}{2a_2}(b_2+1)(a_3+c_2+d_1)\rho_x = 0,$$

$$\rho_{yy} + \left[\tfrac{1}{2}b_2(b_3+d_2+e_1) - b_3 - \frac{a_2e_2}{(b_1+1)(2b_2+1)}\right]\rho_y + \frac{e_2}{2b_2+1}\rho_x = 0.$$
$$(6.65)$$

PROOF   The given equations are immediately obtained as part of a Janet basis for the pde's expressing the general determining system in terms of the respective Janet basis coefficients. In case a) the constraint $c \neq -2$ assures that the denominator $b_2 + 2 \neq 0$, see Theorem 5.13. □

EXAMPLE 6.29   The type $\mathcal{J}_{4,9}^{(2,2)}$ Janet basis in Example 5.40 singles out case a). The system $\sigma_y = 0$, $\sigma_{xx} + \frac{2}{x}\sigma_x = 0$ yields the non-constant solution $\sigma = \frac{1}{x}$. The system

$$\rho_{xx} - \frac{2y}{x^2}\rho_y + \frac{2}{x}\rho_x = 0, \ \rho_{xy} - \frac{1}{x}\rho_y = 0, \ \rho_{yy} = 0$$

yields the independent solution $\rho = xy$. Substitution into the equation of Example 5.40 leads to the canonical form $v''' + v''^3 = 0$. □

THEOREM 6.28   *If a second order ode belongs to symmetry class* $\mathcal{S}_{4,4}^3$ *the transformation functions* σ *and* ρ *are determined by the following systems of pde's.*

a) *Janet basis type* $\mathcal{J}_{4,9}^{(2,2)}$: *There holds always* $d_2 \neq 0$.

$$\sigma_y = 0, \ \sigma_{xx} - c_2\sigma_x = 0,$$
$$(6.66)$$
$$\rho_y = -\frac{2\sigma_x^2}{d_2}, \ \rho_{xxx} - 3c_2\rho_{xx} + (2c_2^2 - c_4)\rho_x = 2\left(c_2 - \frac{d_4}{d_2}\right)\sigma_x^2.$$

b) *Janet basis type* $\mathcal{J}_{4,14}^{(2,2)}$: *There holds always* $e_2 \neq 0$. *Two cases have to be distinguished. If* $b_1 = 0$ *the system is*

$$\sigma_x = 0, \ \sigma_{yy} - (b_3+d_2)\sigma_y = 0, \ \rho_x = -\frac{1}{e_2}\sigma_y^2,$$
$$\rho_{yyy} - 3(b_3+d_2)\rho_{yy} - \left(b_{3,y} + \tfrac{1}{2}e_4\right.$$
$$(6.67)$$
$$\left. -2b_3^2 - 3b_3d_2 + b_4e_2 - 3d_2^2\right)\rho_y = \left(2d_2 + \frac{e_3}{e_2}\right)\sigma_y^2$$

294

and if $b_1 \neq 0$

$$\sigma_y - \frac{3}{2b_1}\sigma_x = 0, \quad \sigma_{xx} - \left(\frac{4}{9}b_1^2 e_2 + 3\frac{a_4}{b_1} + \frac{3c_1}{2b_1}\right)\sigma_x = 0,$$

$$\rho_y - \frac{3}{2b_1}e_2\rho_x = \left(\frac{3}{2b_1}\right)^3\sigma_x^2, \tag{6.68}$$

$$\rho_{xxx} - \frac{1}{b_1}\left(9a_4 + \frac{4}{3}b_1^3 e_2 + \frac{9}{2}c_1\right)\rho_{xx}$$

$$+\frac{1}{b_1^2}\left(\frac{3}{2}c_{1,x}b_1 + 18a_4^2 + 8a_4 b_1^3 e_2 + \frac{63}{4}a_4 c_1 + \frac{64}{81}b_1^6 e_2^2\right.$$

$$\left.+\frac{8}{3}b_1^3 c_1 e_2 + \frac{2}{3}b_1^3 c_3 + 2b_1^2 c_1 d_2 - b_1^2 c_4 + \frac{3}{2}b_1 b_4 c_1 + \frac{9}{2}c_1^2\right)\rho_x$$

$$= \frac{\sigma_x^2}{b_1^4 e_2}\left(-\frac{27}{8}c_{1,x}b_1 + \frac{243}{16}a_4 c_1 + 3b_1^3 c_1 e_2 - \frac{3}{2}b_1^3 c_3 - \frac{9}{2}b_1^2 c_1 d_2 - \frac{27}{8}b_1 b_4 c_1\right). \tag{6.69}$$

PROOF   The given equations are immediately obtained as part of a Janet basis for the pde's expressing the general determining system in terms of the respective Janet basis coefficients. $\square$

Six-Parameter Symmetry. The only symmetry class is $\mathcal{S}_6^3$ with canonical form $v'''v' - \frac{3}{2}v''^2 = 0$. It is treated along similar lines as the projective symmetry $\mathcal{S}_8^2$ of second order equations. At first the invariants are determined from which the coefficients of an equation with this symmetry may be constructed.

LEMMA 6.8   *The differential invariants up to order three for the group $\mathcal{S}_6^3$ generated by the vector fields $\{\partial_\sigma, \partial_\rho, \sigma\partial_\sigma, \rho\partial_\rho, \sigma^2\partial_\sigma, \rho^2\partial_\rho\}$ and the transformation $u = \sigma(x,y)$, $v = \rho(x,y)$ may be described as follows.*

a) *For $\sigma_x \neq 0$ and $\rho_x \neq 0$ there are two first order invariants, four second order invariants*

$$I_1 = \frac{\sigma_y}{\sigma_x}, \quad I_2 = I_{1,x}, \quad I_3 = I_{1,y}, \quad K_1 = \frac{\rho_y}{\rho_x}, \quad K_2 = K_{1,x}, \quad K_3 = K_{1,y}$$

and eight third order invariants

$$I_4 = \frac{\sigma_{xxx}}{\sigma_x} - \frac{3}{2}\left(\frac{\sigma_{xx}}{\sigma_x}\right)^2, I_5 = \frac{\sigma_{yyy}}{\sigma_y} - \frac{3}{2}\left(\frac{\sigma_{yy}}{\sigma_y}\right)^2, I_6 = I_{2,x}, I_7 = I_{2,y},$$

$$K_4 = \frac{\rho_{xxx}}{\rho_x} - \frac{3}{2}\left(\frac{\rho_{xx}}{\rho_x}\right)^2, K_5 = \frac{\rho_{yyy}}{\rho_y} - \frac{3}{2}\left(\frac{\rho_{yy}}{\rho_y}\right)^2, K_6 = K_{2,x}, K_7 = K_{2,y}.$$

b) *For $\sigma_x = 0$, $\rho_x \neq 0$ there is no invariant involving derivatives of σ up to order two and one invariant*

$$J_1 = \frac{\sigma_{yyy}}{\sigma_y} - \frac{3}{2}\left(\frac{\sigma_{yy}}{\sigma_y}\right)^2$$

of order three with $J_{1,x} = 0$. The invariants $K_1, \ldots, K_7$ involving ρ are the same as in the preceding case.

Transformation to Canonical Form

c) *For* $\sigma_x \neq 0$, $\rho_x = 0$ *the invariants* $I_1, \ldots, I_7$ *involving* σ *are the same as in case a). There is no invariant involving derivatives of* ρ *up to order two and one invariant*

$$L_1 = \frac{\rho_{yyy}}{\rho_y} - \frac{3}{2}\left(\frac{\rho_{yy}}{\rho_y}\right)^2$$

of order three with $L_{1,x} = 0$.

PROOF   The prolongations of the group generators up to order three are

$$\partial_\sigma, \quad \sigma_x\partial_{\sigma_x} + \sigma_y\partial_{\sigma_y} + \sigma_{xx}\partial_{\sigma_{xx}} + \sigma_{xy}\partial_{\sigma_{xy}} + \sigma_{yy}\partial_{\sigma_{yy}}$$

$$+\sigma_{xxx}\partial_{\sigma_{xxx}} + \sigma_{xxy}\partial_{\sigma_{xxy}} + \sigma_{xyy}\partial_{\sigma_{xyy}} + \sigma_{yyy}\partial_{\sigma_{yyy}},$$

$$\sigma_x^2\partial_{\sigma_{xx}} + \sigma_x\sigma_y\partial_{\sigma_{xy}} + \sigma_y^2\partial_{\sigma_{yy}} + 3\sigma_x\sigma_{xx}\partial_{\sigma_{xxx}}$$

$$+(2\sigma_x\sigma_{xy} + \sigma_{xx}\sigma_y)\partial_{\sigma_{xxy}} + (2\sigma_{xy}\sigma_y + \sigma_x\sigma_{yy})\partial_{\sigma_{xyy}} + 3\sigma_y\sigma_{yy}\partial_{\sigma_{yyy}}$$

and the same set with σ replaced by ρ. Due to the occurrence ∂_σ and ∂_ρ, any invariant is independent of σ and ρ. There are two equations involving the four first order derivatives. Consequently, there are two first order invariants. Up to order two there are twelve variables involved leading to six invariants altogether. Up to order three there are twenty variables involved leading to fourteen invariants up to this order. If the respective numbers of lower order invariants are taken into account the above figures are obtained. The functional independence is guaranteed due to the fact that each invariant depends on a different set of variables. The actual determination of the invariants is achieved by the methods described in Section 5 of Chapter 1. If $\sigma_x = 0$ and $\rho_x \neq 0$, the three prolongations involving derivatives of σ are

$$\partial_\sigma, \quad \sigma_y\partial_{\sigma_y} + \sigma_{yy}\partial_{\sigma_{yy}} + \sigma_{yyy}\partial_{\sigma_{yyy}}, \quad \sigma_y\partial_{\sigma_{yy}} + 3\sigma_{yy}\partial_{\sigma_{yyy}}$$

with the obvious solution J_1. If $\sigma_x \neq 0$ and $\rho_x = 0$, the same prolongations involving derivatives of ρ are obtained with the solution L_1. $\square$

LEMMA 6.9   *The coefficients of an equation with* $\mathcal{S}_6^3$ *symmetry may be expressed in terms of the invariants introduced in the above lemma.*

a) *For equation (5.40),* $\sigma_x \neq 0$ *and* $\rho_x \neq 0$, *the expressions for the coefficients are:*

$$A_0 = I_1K_1, \quad A_1 = I_1 + K_1, \quad A_2 = 3\frac{I_3K_1^2 - I_1^2K_3}{I_1 - K_1},$$

$$A_3 = 3\frac{K_1^2I_2 - I_1^2K_2 + 2(K_1I_3 - I_1K_3)}{I_1 - K_1},$$

$$A_4 = 3\frac{I_3 - K_3 + 2(K_1I_2 - I_1K_2)}{I_1 - K_1}, \quad A_5 = 3\frac{I_2 - K_2}{I_1 - K_1},$$

$$B_1 = I_1^2K_1^2\frac{I_5 - K_5}{I_1 - K_1},$$

$$B_2 = A_0(A_0 + A_1^2)B_6 - 2A_0A_1(2A_0 + A_1)B_7,$$

$$B_3 = A_1(3A_0 + \tfrac{1}{2}A_1^2)B_6 - (A_0^2 + 7A_0A_1^2 + A_1^4)B_7,$$

$$B_4 = 2(A_0 + A_1^2)(B_6 - 2A_1B_7), \quad B_5 = \tfrac{5}{2}A_1B_6 + (A_0 - 4A_1^2)B_7,$$

$$B_6 = \frac{3(I_6 - K_6) + 2(K_1I_4 - I_1K_4) + I_1I_4 - K_1K_4}{I_1 - K_1}, \quad B_7 = \frac{I_4 - K_4}{I_1 - K_1}.$$

b) *For equation (5.41), $\sigma_x = 0$ and $\rho_x \neq 0$ the expressions for the coefficients are*

$$A_0 = K_1, \quad A_2 = -3K_3, \quad A_3 = 3K_2, \quad B_1 = K_1^2(J_1 - K_5),$$

$$B_2 = 2J_1K_1 - 3K_1^3K_4 - 3K_1^2K_6 + \tfrac{3}{2}K_1K_2^2 - K_1K_5$$

$$-3K_1K_7 + 3K_2K_3 + \tfrac{3}{2}\frac{K_3^2}{K_1},$$

$$B_3 = J_1 - 6K_1^2K_4 - 6K_1K_6 + 3K_2^2 - 3K_7,$$

$$B_4 = -4K_1K_4 - 3K_6, \quad B_5 = -K_4.$$

c) *For equation (5.41), $\sigma_x \neq 0$ and $\rho_x = 0$ the expressions for the coefficients are*

$$A_0 = I_1, \quad A_2 = -3I_3, \quad A_3 = -3I_2, \quad B_1 = I_1^2(L_1 - I_5),$$

$$B_2 = 2L_1I_1 - 3I_1^3I_4 - 3I_1^2I_6 + \tfrac{3}{2}I_1I_2^2 - I_1I_5$$

$$-3I_1I_7 + 3I_2I_3 + \tfrac{3}{2}\frac{I_3^2}{I_1},$$

$$B_3 = L_1 - 6I_1^2I_4 - 6I_1I_6 + 3I_2^2 - 3I_7,$$

$$B_4 = -4I_1I_4 - 3I_6, \quad B_5 = -I_4.$$

The proof is essentially the same as for Lemma 6.9 except that the canonical form now is $v'''v' - \tfrac{3}{2}v''^2 = 0$. Therefore it is omitted.

THEOREM 6.29   *The transformation of an equation in symmetry class $\mathcal{S}_6^3$ to canonical form may be determined as follows.*

a) *If it has the form (5.40) the transformation functions σ and ρ are determined by the following system of pde's.*

$$S_x - \tfrac{1}{2}S^2 - I_4 = 0, \quad S_y - \tfrac{1}{2}I_1S^2 - I_{1,x}S - I_1I_4 - I_{1,xx} = 0,$$
$$\sigma_y - I_1\sigma_x = 0, \quad \sigma_{xx} - S\sigma_x = 0, \tag{6.70}$$

$$R_x - \tfrac{1}{2}R^2 - K_4 = 0, \quad R_y - \tfrac{1}{2}K_1R^2 - K_{1,x}R - K_1K_4 - K_{1,xx} = 0,$$
$$\rho_y - K_1\rho_x = 0, \quad \rho_{xx} - R\rho_x = 0. \tag{6.71}$$

The required invariants may be expressed in terms of the coefficients of (5.40) as follows.

$$I_1 = \tfrac{1}{2}(A_1 + Z), \quad I_4 = \tfrac{1}{2}B_6 - \tfrac{1}{2}(3A_1 - Z)B_7 - \tfrac{3}{2}\frac{Z_{xx}}{Z}.$$

$$\text{Transformation to Canonical Form} \qquad 297$$

K_1 and K_4 are obtained from the corresponding I's by replacing $Z \equiv \sqrt{A_1^2 - 4A_0}$ with $-Z$. Z is contained in the same field as the transformation functions.

b) If it has the form (5.41), two cases have to be distinguished. Either the transformation function σ is determined by the system

$$S_x = 0, \quad S_y - \tfrac{1}{2}S^2 - J_1 = 0, \quad \sigma_x = 0, \quad \sigma_{yy} - S\sigma_y = 0 \qquad (6.72)$$

where

$$J_1 = B_3 - 6A_0^2 B_5 + 6A_0 A_{0,xx} - 3A_{0,x}^2 + 3A_{0,xy},$$

and the transformation function ρ is obtained from system (6.71) where now $K_1 = A_0$, $K_4 = -B_5$. Or σ is determined by the system (6.70) and ρ is determined by the system

$$R_x = 0, \quad R_y - \tfrac{1}{2}R^2 - J_1 = 0, \quad \rho_x = 0, \quad \rho_{yy} - R\rho_y = 0 \qquad (6.73)$$

with J_1 as above.

PROOF   Case a). The first equation of (6.70) for $S \equiv \frac{\sigma_{xx}}{\sigma_x}$ is essentially the definition of I_4 in Lemma 6.8. The second equation involving the y-derivative of S is obtained in the same way as in Theorem 6.33. The equations for σ are essentially the definitions of I_1 and S. Similar arguments lead to the system (6.71) for $R \equiv \frac{\rho_{xx}}{\rho_x}$.

Case b). The equation for $S \equiv \frac{\sigma_{yy}}{\sigma_y}$ in (6.72) is essentially the definition of J_1 in Lemma 6.8.

In either case, the expressions for the invariants are obtained by elimination from the relations given in Lemma 6.10. Writing Z as $Z = \sqrt{A_1^2 - 4A_0} = \frac{\Delta^2}{\sigma_x^2 \rho_x^2}$, $\Delta = \sigma_x \rho_y - \sigma_y \rho_x$, shows that it is contained in the field determined by the transformation functions. ☐

EXAMPLE 6.30   An equation in symmetry class $\mathcal{S}_6^3$, case a), has been considered in Example 5.41 of the preceding chapter. From the coefficients given there, the invariants $I_1 = \frac{x}{y}$, $K_1 = -\frac{x}{y}$, $I_4 = K_4 = 0$ are obtained. Consequently, the equations for S are $S_x - \tfrac{1}{2}S^2 = 0$, $S_y - \frac{x}{2y}S^2 - \tfrac{1}{y}S = 0$ with the general solution $S = -\frac{2y}{xy + C}$. Choosing $C \to \infty$ yields $S = 0$ and finally $\sigma = xy$. The system for R is $R_x - \tfrac{1}{2}R^2 = 0$, $R_y + \frac{x}{2y}R^2 + \tfrac{1}{y}R = 0$ with the general solution $R = -\frac{2}{Cy + x}$. Choosing now $C = 0$ yields $R = -\frac{2}{x}$ and $\rho_y + \frac{x}{y}\rho_x = 0$, $\rho_{xx} + \frac{2}{x}\rho_x = 0$ with the non-constant solution $\rho = \frac{y}{x}$. ☐

EXAMPLE 6.31   An equation in symmetry class $\mathcal{S}_6^3$, case b), has been considered in Example 5.42 of the preceding chapter. From the coefficients given

298

there, the invariants

$$J_1 = 0, \quad K_1 = \frac{x}{(x-1)y}, \quad K_4 = -\frac{1}{2}\frac{x^2 - 4x + 6}{(x-1)^2}$$

are obtained. Consequently, the equations for S and σ are the same as in Example 6.39, the special solution $\sigma = \frac{1}{y}$ is chosen again. R is determined by the system

$$R_x - \frac{1}{2}R^2 + \frac{1}{2}\frac{x^2 - 4x + 6}{(x-1)^2} = 0,$$

$$R_y - \frac{1}{2}\frac{x}{(x-1)y}R^2 + \frac{1}{(x-1)^2 y}R + \frac{1}{2}\frac{(x-2)(x^2 - 2x + 2)}{(x-1)^3 y} = 0.$$

It has the two rational solutions $\frac{x^2 - 2x + 2}{x(x-1)}$ and $-\frac{x-2}{x-1}$. Choosing the first alternative, the system

$$\rho_y - \frac{x}{xy - y}\rho_x = 0, \quad \rho_{xx} - \frac{x^2 - 2x + 2}{x^2 - x}\rho_x = 0$$

for ρ is obtained with the non-constant solution $\rho = \frac{y}{x}e^x$. $\quad\square$

6.4 Linearizable Third Order Equations

The distinctive feature of the equations covered by this section is the fact that they allow canonical forms that are linear. Consequently, the solution procedures described in Chapter 2 for linear differential equations apply as soon as the transformation to canonical form has been found. At first the canonical forms and its structure invariance groups are described, after that the transformation to these canonical forms for the various symmetry classes is considered.

Canonical Forms and Structure Invariance Groups. The subsequent theorem describes canoncial forms and structure invariance groups for the four symmetry types that occur in this section.

THEOREM 6.30   *In canonical variables u and $v \equiv v(u)$ the third order equations that are equivalent to a linear one and allow nontrivial symmetries have the following canonical forms and corresponding structure invariance groups of point transformations $u = \sigma(x,y)$ and $v = \rho(x,y)$; a_i for all i are constant.*

Four-parameter symmetry group

$$S_{4,5}^3: \ v''' + P(u)v' + Q(u)v = 0 \ allows \ u = \frac{a_1 x + a_2}{a_3 x + a_4}, \quad v = \frac{y}{(a_3 x + a_4)^2}.$$

Transformation to Canonical Form

Five-parameter symmetry groups

$\mathcal{S}_{5,1}^3$: $v''' - v'' = 0$ allows $u = x + a_1$, $v = a_2y + a_3x + a_4e^x + a_5$.

$\mathcal{S}_{5,2}^3(a)$: $v''' - (a+1)v'' + av' = 0$ allows $u = x + a_1$, $v = a_2y + a_3e^x + a_4a_1^ae^{ax} + a_5$. In addition five transformations $u = \alpha x$, $v = e^{\beta u}v$ with $\alpha = \frac{1}{a}$, $\beta = 0$; $\alpha = \beta = -1$; $\alpha = \beta = \frac{1}{a-1}$; $\alpha = \frac{1}{a}$, $\beta = 1$ and $\alpha = \frac{1}{a-1}$, $\beta = \frac{a}{a-1}$.

Seven-parameter symmetry groups

$\mathcal{S}_7^3$: $v''' = 0$ allows $u = \dfrac{a_1 + a_2x}{x + a_3}$, $v = \dfrac{a_4 + a_5x + a_6x^2 + a_7y^2}{x + a_3}$.

PROOF   The structure invariance group for the symmetry class $\mathcal{S}_{4,5}^3$ has been given in Theorem 4.3. For the symmetry classes $\mathcal{S}_{5,1}^3$ and $\mathcal{S}_7^3$ they are identical to the respective symmetry groups as given in Theorem 5.16 because there are no parameters or undetermined functions involved. $\square$

Four-Parameter Symmetry. There is only a single symmetry class, the generic linear homogeneous third order ode belongs to it. The subsequent theorem describes how to transform a general equation in this symmetry class to a linear equation in rational normal form.

THEOREM 6.31   If a third order ode belongs to the symmetry class $\mathcal{S}_{4,5}^3$ three types of Janet bases may occur. The transformation functions σ and ρ to rational normal form $v''' + P(u)v' + Q(u)v = 0$ are determined as follows; f, g and h are undetermined functions of its arguments; $A \equiv P(\sigma)$ and $B \equiv Q(\sigma)$.

a) Janet basis of type $\mathcal{J}_{4,2}^{(2,2)}$: $\sigma = f(x)$, the system for ρ is

$$\rho_{xy} - \left(\frac{\sigma_{xx}}{\sigma_x} + \frac{1}{3}c_1 \right)\rho_y = 0, \quad \rho_{yy} - b_2\rho_y = 0,$$

$$\rho_{xxx} - 3\frac{\sigma_{xx}}{\sigma_x}\rho_{xx} + c_2\rho_y$$

$$- \left[\frac{3}{2}\left(\frac{\sigma_{xx}}{\sigma_x} \right)^2 + \frac{1}{2}c_{1,x} + \frac{1}{6}c_1^2 - \frac{1}{2}c_3 + \frac{3}{2}\sigma_x^2 A \right]\rho_x + \sigma_x^3 B\rho = 0. \quad (6.74)$$

A and B are determined through

$$2\left(\frac{\sigma_{xx}}{\sigma_x} \right)_x - \left(\frac{\sigma_{xx}}{\sigma_x} \right)^2 + c_{1,x} + \frac{1}{3}c_1^2 - c_3 + \sigma_x^2 A = 0,$$

$$c_{1,xx} + 2c_1c_{1,x} - 3c_{3,x} - 6b_2c_2 + \tfrac{4}{9}c_1^3 - 2c_1c_3 + 6c_4 + 3\sigma_x^2 A_x - 6\sigma_x^3 B = 0. \quad (6.75)$$

300

b) Janet basis of type $\mathcal{J}_{4,19}^{(2,2)}$: $\sigma = g(y)$, the system for ρ is

$$\rho_{xx} - b_2\rho_x = 0, \quad \rho_{xy} - \left(\frac{\sigma_{yy}}{\sigma_y} + \tfrac{1}{3}c_1 \right) \rho_x = 0,$$

$$\rho_{yy} - 3\frac{\sigma_{yy}}{\sigma_y}\rho_{yy} - \left[\tfrac{3}{2} \left(\frac{\sigma_{yy}}{\sigma_y} \right)^2 + \tfrac{1}{2}c_{1,y} + \tfrac{1}{6}c_1^2 - \tfrac{1}{3}c_2 + \tfrac{3}{2}\sigma_y^2 A \right] \rho_y$$

$$+ c_3\rho_x + \sigma_y^3 B\rho = 0. \quad (6.76)$$

A and B are determined through

$$2 \left(\frac{\sigma_{yy}}{\sigma_y} \right)_y - \left(\frac{\sigma_{yy}}{\sigma_y} \right)^2 + c_{1,y} + \tfrac{1}{3}c_1^2 - c_2 + \sigma_y^2 A = 0,$$

$$c_{1,yy} + 2c_1 c_{1,y} - 3c_{2,y} - 6b_2 c_3 + \tfrac{4}{9}c_1^3 - 2c_1 c_2 + 6c_4 + 3\sigma_y^2 A_y - 6\sigma_y^3 B = 0. \tag{6.77}$$

c) Janet basis of type $\mathcal{J}_{4,17}^{(2,2)}$: $\sigma = h(\phi)$ where $\phi(x, y) = C$ is the solution of $y' - \frac{1}{b_1} = 0$, the system for ρ is

$$\rho_{xy} + b_1\rho_{xx} - \left(\frac{\sigma_{xx}}{\sigma_x} - b_2 + \tfrac{1}{3}d_1 \right) \rho_y - \left(b_1\frac{\sigma_{xx}}{\sigma_x} + \tfrac{1}{3}b_1 d_1 \right) \rho_x = 0,$$

$$\rho_{yy} - b_1^2\rho_{xx} + \left(2b_1\frac{\sigma_{xx}}{\sigma_x} + \tfrac{2}{3}b_1 d_1 + c_2 - 2b_3 \right) \rho_y$$

$$+ \left(2b_1^2\frac{\sigma_{xx}}{\sigma_x} + \tfrac{2}{3}b_1^2 d_1 - b_1 b_3 \right) \rho_x = 0,$$

$$\rho_{xxx} - 3\frac{\sigma_{xx}}{\sigma_x}\rho_{xx} + d_2\rho_y - \left[\tfrac{3}{2} \left(\frac{\sigma_{xx}}{\sigma_x} \right)^2 + \tfrac{1}{2}d_{1,x} \right.$$

$$\left. + \tfrac{1}{6}d_1^2 + \tfrac{3}{2}b_1 d_2 - \tfrac{1}{2}d_3 + \tfrac{3}{2}\sigma_x^2 A \right] \rho_x + \sigma_x^3 B\rho = 0. \quad (6.78)$$

A and B are determined through

$$2 \left(\frac{\sigma_{xx}}{\sigma_x} \right)_x - \left(\frac{\sigma_{xx}}{\sigma_x} \right)^2 + d_{1,x} + \tfrac{1}{3}d_1^2 - d_3 + b_1 d_2 + \sigma_x^2 A = 0,$$

$$d_{1,xx} + 2d_1 d_{1,x} - 3d_{3,x} + 3b_1 d_{2,x} + 2d_{1,x}d_1 + 6d_4 - 2d_1 d_3$$

$$+2b_1 d_1 d_2 - 6b_2 d_2 - 3b_1 b_2 d_2 + \tfrac{4}{9}d_1^3 + 3\sigma_x^2 A_x - 6\sigma_x^3 B = 0. \tag{6.79}$$

PROOF   In case a) the given equations are obtained from a Janet basis relating σ and ρ to the coefficients of the given type $\mathcal{J}_{4,2}^{(2,2)}$ for the symmetry generators. $A(x) \equiv P(\sigma)$ and $B(x) \equiv Q(\sigma)$ are the transformed coefficient functions P and Q. Their actual shape is determined by the choice of σ.

Transformation to Canonical Form 301

This freedom is allowed by the structure invariance group for the canonical form according to Theorem 6.30. The remaining four constants occur in the solution for ρ. The discussion for case b) is similar. □

EXAMPLE 6.32   The equation considered in Example 5.43 generates the type $\mathcal{J}_{4,2}^{(2,2)}$ Janet basis, therefore case a) of the above theorem applies. If $\sigma = x$ is chosen, the system (6.74) yields the independent solution $\rho = \frac{x}{y}$ and the canonical form

$$v''' - \frac{8u}{2u - 1}v' + \frac{8}{2u - 1}v = 0. \tag{6.80}$$

Choosing $\sigma = \frac{1}{x}$ yields $\rho = \frac{1}{xy}$, and the canonical form

$$v''' + \frac{8}{u^4(u - 2)}v' - \frac{8}{u^5(u - 2)}v = 0 \tag{6.81}$$

is obtained. □

EXAMPLE 6.33   The equation considered in Example 5.44 generates the type $\mathcal{J}_{4,19}^{(2,2)}$ Janet basis, therefore case b) of the above theorem applies. If $\sigma = y$ is chosen, the system (6.74) yields the independent solution $\rho = \frac{x}{y}$ and the canonical form (6.80). Choosing $\sigma = \frac{1}{y}$ instead yields $\rho = \frac{1}{xy}$ with the canonical form (6.81). □

EXAMPLE 6.34   The equation considered in Example 5.45 generates the type $\mathcal{J}_{4,17}^{(2,2)}$ Janet basis, therefore case c) of the above theorem applies. The first order equation for σ is $y' + 1 = 0$; i.e., $\phi(x, y) = x + y$. Choosing $\sigma = x + y$ yields a system for ρ with the fundamental system $\{0, x, y\}$. Consequently, any linear combination $\rho = c_1 x + c_2 y$ with $c_1 \neq c_2$ yields a possible transformation to a canonical form which turns out to be (6.80). □

Five-Parameter Symmetries. The parameter a that defines the symmetry class $\mathcal{S}_{5,2}^3(a)$ has been determined in Theorem 5.17. The expressions for P, Q and R occurring in the following theorem are also given there. The Janet bases of order five that are required are listed on page 382.

THEOREM 6.32   *If a third order ode belongs to the symmetry class $\mathcal{S}_{5,1}^3$ or $\mathcal{S}_{5,2}^3(a)$, three types of Janet bases may occur. The systems of pde's for the transformation functions σ and ρ are as follows.*

a) *Janet basis type $\mathcal{J}_{5,1}^{(2,2)}$:*

$$\sigma_x = \frac{P}{Q} \frac{a^2 - a + 1}{(a - \frac{1}{2})(a + 1)(a - 2)}, \quad \sigma_y = 0, \tag{6.82}$$

302

$$\rho_{xy} - \tfrac{1}{3}\big[(a+1)\sigma_x + 3a_1 + e_1\big]\rho_y = 0, \quad \rho_{yy} - c_1\rho_y = 0,$$

$$\rho_{xxx} - \big[(a+1)\sigma_x + 3a_1\big]\rho_{xx} + e_2\rho_y \tag{6.83}$$

$$+\big[a\sigma_x^2 + (a+1)a_1\sigma_x + 2a_1^2 - a_{1,x}\big]\rho_x = 0.$$

b) *Janet basis type* $\mathcal{J}_{5,2}^{(2,2)}$:

$$\sigma_x = 0, \quad \sigma_y = \frac{P}{Q}\frac{a^2 - a + 1}{(a - \tfrac{1}{2})(a+1)(a-2)}, \tag{6.84}$$

$$\rho_{xx} - c_1\rho_x = 0, \quad \rho_{xy} - \tfrac{1}{3}\big[(a+1)\sigma_y + 3b_1 + e_1\big]\rho_x = 0,$$

$$\rho_{yyy} - \big[(a+1)\sigma_y + 3b_1\big]\rho_{yy}$$

$$+\big[a\sigma_y^2 + (a+1)b_1\sigma_y + 2b_1^2 - b_{1,y}\big]\rho_y + e_3\rho_x = 0.$$

c) *Janet basis type* $\mathcal{J}_{5,3}^{(2,2)}$:

$$\sigma_x = \frac{P}{Q}\frac{a^2 - a + 1}{(a - \tfrac{1}{2})(a+1)(a-2)}, \quad \sigma_y = -c_1\sigma_x, \tag{6.85}$$

$$\rho_{xy} + c_1\rho_{xx} - \tfrac{1}{3}\big[(a+1)\sigma_x + e_1 + 3a_2\big]\rho_y$$

$$-\tfrac{1}{3}\big[c_1e_1 - 3c_1^2a_3 + c_1(a+1)\sigma_x\big]\rho_x = 0,$$

$$\rho_{yy} - c_1^2\rho_{xx} + \big[\tfrac{2}{3}(a+1)c_1\sigma_x + \tfrac{2}{3}e_1c_1 - c_3 - c_1^2a_3 - b_2\big]\rho_y$$

$$+\big[\tfrac{2}{3}(a+1)c_1\sigma_x + \tfrac{2}{3}e_1c_1 - c_3 - 2c_1^2a_3\big]c_1\rho_x = 0,$$

$$\rho_{xxx} - \big[(a+1)\sigma_x - 3c_1a_3\big]\rho_{xx} + e_2\rho_y$$

$$+\big[a\sigma_x^2 - (a+1)c_1a_3\sigma_x + (a_{3,x} + e_2 + a_3a_2 + 3c_1a_3^2)c_1\big]\rho_x = 0. \tag{6.86}$$

EXAMPLE 6.35   From the Janet basis coefficients in Example 5.46, the system for σ is $\sigma_x = y$, $\sigma_y = x$, i.e., $\sigma = xy + C$. For ρ the system

$$\rho_{xy} - \frac{x}{y}\rho_{xx} + \frac{1}{x}\rho_y - \frac{2}{y}\rho_x = 0,$$

$$\rho_{yy} - \frac{x^2}{y^2}\rho_{xx} + \frac{1}{y}\rho_y - \frac{x}{y^2}\rho_x = 0, \rho_{xxx} - y\rho_{xx} + \frac{xy^2 + 3y}{x^3}\rho_y - \frac{xy + 3}{x^2}\rho_x = 0$$

is obtained with the three rational solutions $\rho = 1$, $\rho = xy$ and $\rho = \frac{y}{x}$. The last alternative yields a transformation with nonvanishing functional determinant; i.e., the desired transformation is $x = \sqrt{\frac{u}{v}}$, $y = \sqrt{uv}$, and the canonical form is $v''' - v'' = 0$. ⬜

Transformation to Canonical Form

EXAMPLE 6.36   Choosing $a = 3$ from Example 5.47, case $b)$ of the above theorem leads to $\sigma_x = 0$, $\sigma_y = 1$, i.e., $\sigma = y + C$. The system

$$\rho_{xx} = 0, \quad \rho_{xy} + \frac{1}{y}\rho_x = 0,$$

$$\rho_{yyy} - 4\rho_{yy} + 3\rho_y + \frac{3xy^2 + 8xy + 6x}{y^3}\rho_x = 0$$

has the rational solution $\rho = \frac{x}{y}$. Together with the special solution $\sigma = y$, it yields the canonical form $v''' - 4v'' + 3v' = 0$ for the ode of Example 5.47. □

Maximal Symmetry Group with Seven Parameters.   The canonical form corresponding to this group does not contain any unspecified elements; its structure invariance group is therefore identical to its symmetry group. The proceeding is similar to the second order equations with projective symmetry. This case has been considered in detail by Zorawski [192].

LEMMA 6.10   *(Zorawski 1897) The differential invariants up to order three for the group $\mathcal{S}_7^3$ generated by $\{\partial_\sigma, \partial_\rho, \sigma\partial_\sigma, \rho\partial_\rho, \sigma\partial_\rho, \sigma^2\partial_\sigma, \sigma^2\partial_\sigma + 2\sigma\rho\partial_\rho\}$ with $\sigma \equiv \sigma(x, y)$ and $\rho \equiv \rho(x, y)$ may be described as follows.*

a) *For $\sigma_x \neq 0$ there is a single first order invariant $I_1 = \frac{\sigma_y}{\sigma_x}$, there are four second order invariants $I_2 = I_{1,x}$, $I_3 = I_{1,y}$,*

$$I_4 = \frac{\sigma_x\rho_{xy} - \sigma_{xy}\rho_x + \sigma_{xx}\rho_y - \sigma_y\rho_{xx}}{\sigma_x\rho_y - \sigma_y\rho_x} - 2\frac{\sigma_{xx}}{\sigma_x},$$

$$I_5 = \frac{\sigma_x\rho_{yy} - \sigma_{yy}\rho_x + \sigma_{xy}\rho_y - \sigma_y\rho_{xy}}{\sigma_x\rho_y - \sigma_y\rho_x} - 2\frac{\sigma_{xy}}{\sigma_x}$$

and there are eight third order invariants

$$I_6 = \frac{\sigma_{xxx}}{\sigma_x} - \frac{3}{2}\left(\frac{\sigma_{xx}}{\sigma_x}\right)^2, \quad I_7 = \frac{\sigma_x\rho_{xxx} - \sigma_{xxx}\rho_x}{\sigma_x\rho_y - \sigma_y\rho_x} - 3\frac{\sigma_x\rho_{xx} - \sigma_{xx}\rho_x}{\sigma_x\rho_y - \sigma_y\rho_x}\frac{\sigma_{xx}}{\sigma_x},$$

$$I_8 = I_{4,x}, \quad I_9 = I_{4,y}, \quad I_{10} = I_{5,y}, \quad I_{11} = I_{2,x}, \quad I_{12} = I_{2,y}, \quad I_{13} = I_{3,y}.$$

b) *For $\sigma_x = 0$ there are no first order invariants, there are two second order invariants*

$$J_1 = \frac{\rho_{xx}}{\rho_x}, \quad J_2 = \frac{\rho_{xy}}{\rho_x} - \frac{\sigma_{yy}}{\sigma_y}$$

and there are five third order invariants

$$J_3 = \frac{\sigma_{yyy}}{\sigma_y} - \frac{3}{2}\left(\frac{\sigma_{yy}}{\sigma_y}\right)^2, \quad J_5 = J_{1,x}, \quad J_6 = J_{1,y}, \quad J_7 = J_{2,y},$$

$$J_4 = \frac{\rho_{yyy}\sigma_y - \sigma_{yyy}\rho_y}{\sigma_y\rho_x} - 3\sigma_{yy}\frac{\rho_{yy}\sigma_y - \sigma_{yy}\rho_y}{\sigma_y^2\rho_x}.$$

PROOF   The prolongations of the group generators up to order three are, after some reductions have been performed, ∂_σ, ∂_ρ,

$$\sigma_x \partial_{\sigma_x} + \sigma_y \partial_{\sigma_y} + \sigma_{xx} \partial_{\sigma_{xx}} + \sigma_{xy} \partial_{\sigma_{xy}} + \sigma_{yy} \partial_{\sigma_{yy}}$$
$$+\sigma_{xxx} \partial_{\sigma_{xxx}} + \sigma_{xxy} \partial_{\sigma_{xxy}} + \sigma_{xyy} \partial_{\sigma_{xyy}} + \sigma_{yyy} \partial_{\sigma_{yyy}},$$
$$\rho_x \partial_{\rho_x} + \rho_y \partial_{\rho_y} + \rho_{xx} \partial_{\rho_{xx}} + \rho_{xy} \partial_{\rho_{xy}} + \rho_{yy} \partial_{\rho_{yy}}$$
$$+\rho_{xxx} \partial_{\rho_{xxx}} + \rho_{xxy} \partial_{\rho_{xxy}} + \rho_{xyy} \partial_{\rho_{xyy}} + \rho_{yyy} \partial_{\rho_{yyy}},$$
$$\sigma_x \partial_{\rho_x} + \sigma_y \partial_{\rho_y} + \sigma_{xx} \partial_{\rho_{xx}} + \sigma_{xy} \partial_{\rho_{xy}} + \sigma_{yy} \partial_{\rho_{yy}}$$
$$+\sigma_{xxx} \partial_{\rho_{xxx}} + \sigma_{xxy} \partial_{\rho_{xxy}} + \sigma_{xyy} \partial_{\rho_{xyy}} + \sigma_{yyy} \partial_{\rho_{yyy}},$$
$$\sigma_x^2 \partial_{\rho_{xx}} + \sigma_x \sigma_y \partial_{\rho_{xy}} + \sigma_y^2 \partial_{\rho_{yy}} + 3\sigma_x \sigma_{xx} \rho_{xxx}$$
$$+(2\sigma_x \sigma_{xy} + \sigma_{xx} \sigma_y)\partial_{\rho_{xxy}} + (2\sigma_{xy} \sigma_y + \sigma_x \sigma_{yy})\partial_{\rho_{xyy}} + 3\sigma_y \sigma_{yy} \partial_{\rho_{yyy}},$$
$$\rho_x \sigma \partial_{\rho_x} + \rho_y \sigma \partial_{\rho_y} + \sigma_x^2 \partial_{\sigma_{xx}} + 2\sigma_x \rho_x \partial_{\rho_{xx}} + \sigma_x \sigma_y \partial_{\sigma_{xy}} + (\rho_x \sigma_y + \rho_y \sigma_x)\partial_{\rho_{xy}}$$
$$+\sigma_y^2 \partial_{\sigma_{yy}} + 2\rho_y \sigma_y \partial_{\rho_{yy}} + 3\sigma_{xx} \sigma_x \partial_{\sigma_{xxx}} + 3(\rho_{xx} \sigma_x + \rho_x \sigma_{xx})\partial_{\rho_{xxx}}$$
$$+(2\sigma_{xy} \sigma_x + \sigma_{xx} \sigma_y)\partial_{\sigma_{xxy}} + (2\sigma_{xy} \sigma_y + \sigma_x \sigma_{yy})\partial_{\sigma_{xyy}}$$
$$+(2\rho_{xy} \sigma_x + \rho_{xxy} \sigma + \rho_{xx} \sigma_y + 2\rho_x \sigma_{xy} + \rho_y \sigma_{xx})\partial_{\rho_{xxy}}$$
$$+(2\rho_{xy} \sigma_y + \sigma_x \sigma_{yy} + \rho_{yy} \sigma_x + 2\rho_y \sigma_{xy})\partial_{\rho_{xyy}} + 3\sigma_{yy} \sigma_y \partial_{\sigma_{yyy}}$$
$$+3(\rho_{yy} \sigma_y + \rho_y \sigma_{yy})\partial_{\rho_{yyy}}.$$

They generate a system of linear homogeneous pde's in the derivatives of σ and ρ for the desired invariants. There are 4, 10 or 18 derivatives up to order 1, 2 or 3 respectively. If all derivatives are nonvanishing, the above system has rank three w.r.t. to derivatives of first order, and rank five w.r.t. derivatives of order higher than one. From these figures the number of invariants in case $a)$ follows. Their explicit form is obtained by applying the methods described in Section 5 of Chapter 2. The occurrence of the derivatives of the various orders guarantees their functional independence. If some partial derivatives vanish as it occurs in case $b)$, the corresponding terms in the system for the invariants disappear. □

LEMMA 6.11   (*Zorawski 1897*) *The coefficients of an equation with S_7^3 symmetry may be expressed in terms of the invariants introduced in the above lemma.*

a) *For equation (5.43) the expressions for the coefficients in terms of the I's are:*

$$A_0 = I_1, \quad A_1 = 3I_1 I_5 - 6I_3, \quad A_2 = 3I_1 I_4 + 3I_5 - 6I_2, \quad A_3 = 3I_4,$$

$$B_1 = I_1^5 I_7 + I_1^4(2I_6 + I_{11} + I_4^2) + I_1^3(I_{12} + I_8 + I_4 I_5 - I_2 I_4)$$
$$+ I_1^2(I_{13} + I_3^2 - I_3 I_4 - I_2 I_5) - I_1(I_{10} + 3I_3 I_5 - I_2 I_3) + 3I_3^2,$$

$$B_2 = 5I_1^4 I_7 + I_1^3(8I_6 + 5I_{11} + 5I_4^2) + I_1^2(5I_{12} + 3I_8 + 5I_4I_5 - 5I_2I_4)$$

$$+ I_1(2I_{13} - 2I_9 + 2I_5^2 - 5I_3I_4 - 5I_2I_5 + I_2^2) - I_{10} - 3I_3I_5 + 7I_2I_3,$$

$$B_3 = 10I_1^3 I_7 + 2I_1^2(6I_6 + 5I_{11} + 5I_4^2) + I_1(7I_{12} + 2I_8 + 7I_4I_5 - 10I_2I_4)$$

$$+ I_{13} - 2I_9 + I_5^2 - 4I_3I_4 - 4I_2I_5 + 4I_2^2,$$

$$B_4 = 10I_1^2 I_7 + I_1(8I_6 + 9I_{11} + 9I_4^2) + 3I_{12} + 3I_4I_5 - 6I_2I_4,$$

$$B_5 = 5I_1I_7 + 2I_6 + 3I_{11} + 3I_4^2, \quad B_6 = I_7.$$

b) *For equation (5.44) the expressions for the coefficients in terms of the J's are:*

$$A_1 = 3J_2, \quad A_2 = 3J_1, \quad B_1 = -J_4, \quad B_2 = -3(J_2^2 + J_7 + J_3),$$

$$B_3 = -3(J_1J_2 + J_6), \quad B_4 = -J_1^2 - J_5.$$

PROOF   If the canonical form equation $v'''(u) = 0$ is transformed by a general transformation $u = \sigma(x, y)$ and $v = \rho(x, y)$, the coefficients A_j and B_k of (5.43) or (5.44) are obtained in terms of σ, ρ and its derivatives depending on whether $\sigma_x \neq 0$ or $\sigma_x = 0$. On the other hand, the invariants have been expressed in terms of them in Lemma 6.10. If σ, ρ and its derivatives are eliminated between the two sets of equations, the above expressions for the coefficients are obtained.   □

THEOREM 6.33   *The transformation of an equation with S_7^3 symmetry to canonical form is determined by a system of partial differential equations the general solution of which contains seven constants.*

a) *For equation (5.43), the transformation function σ is determined by the system*

$$S_x - \tfrac{1}{2}S^2 - I_6 = 0, \quad S_y - \tfrac{1}{2}I_1S^2 - I_{1,x}S - I_1I_6 - I_{1,xx} = 0, \quad (6.87)$$

$$\sigma_y - I_1\sigma_x = 0, \quad \sigma_{xx} - S\sigma_x = 0. \quad (6.88)$$

The transformation function ρ is explicitly given by

$$\rho = -\oint \frac{1}{2}I_7 \left(\frac{\sigma}{\sigma_x}\right)^2 \rho_1 dx$$

$$+ \left[\frac{1}{2}\left(\frac{\sigma}{\sigma_x}\right)^2 (\rho_{1,xx} - I_1I_7\rho_1) - \frac{\sigma}{\sigma_x}\left(\frac{1}{2}\frac{\sigma}{\sigma_x}\frac{\sigma_{xx}}{\sigma_x} + 1\right)\rho_{1,x} + \rho_1\right] dy$$

$$+ \sigma \oint I_7\frac{\sigma}{\sigma_x^2} dx + \frac{\sigma}{\sigma_x^2}\left[\rho_{1,xx} - \left(\frac{\sigma_{xx}}{\sigma_x} + \frac{\sigma_x}{\sigma}\right)\rho_{1,x} + I_1I_7\rho_1\right] dy$$

$$- \sigma^2 \oint \frac{1}{2\sigma_x^2}\left[I_7\rho_1 dx + \left(\rho_{1,xx} - \frac{\sigma_{xx}}{\sigma_x}\rho_{1,x} + I_1I_7\rho_1\right)\right] dy$$

$$(6.89)$$

306

where

$$\rho_1 = \exp \oint \left(I_4 + \frac{\sigma_{xx}}{\sigma_x} \right) dx + \left(I_5 + I_{1,x} + I_1 \frac{\sigma_{xx}}{\sigma_x} \right) dy.$$

The invariants occurring in these equations are expressed in terms of the coefficients of (5.43) by

$$I_1 = A_0, \quad I_4 = \tfrac{1}{3} A_3, \quad I_5 = 2A_{0,x} + \tfrac{1}{3}(A_2 - A_0 A_3),$$
$$I_6 = -\tfrac{1}{2} A_{3,x} + \tfrac{1}{2} B_5 - \tfrac{1}{6} A_3^2 - \tfrac{5}{2} A_0 B_6, \quad I_7 = B_6.$$

b) *For equation (5.44), the transformation function σ is determined by the system*

$$S_x = 0, \quad S_y - \tfrac{1}{2} S^2 - J_3 = 0, \quad \sigma_x = 0, \quad \sigma_{yy} - S\sigma_y = 0. \qquad (6.90)$$

The transformation function ρ is explicitly given by

$$\rho = \oint \left[-\frac{1}{2} \left(\frac{\sigma}{\sigma_y} \right)^2 \rho_{1,yy} + \frac{\sigma}{\sigma_y} \left(\frac{1}{2} \frac{\sigma}{\sigma_y} \frac{\sigma_{yy}}{\sigma_y} + 1 \right) \rho_{1,y} - \rho_1 \right] dx$$

$$- \tfrac{1}{2} J_4 \left(\frac{\sigma}{\sigma_y} \right)^2 \rho_1 dy$$

$$+ \sigma \oint \frac{\sigma}{\sigma_y^2} \left[\rho_{1,yy} - \left(\frac{\sigma_{yy}}{\sigma_y} + \frac{\sigma_y}{\sigma} \right) \rho_{1,y} \right] dx + J_4 \frac{\sigma}{\sigma_y^2} \rho_1 dy$$

$$- \sigma^2 \oint \frac{1}{2\sigma_y^2} \left[\left(\rho_{1,yy} - \frac{\sigma_{yy}}{\sigma_y} \rho_{1,y} \right) dx + J_4 \rho_1 dy \right]$$

$$(6.91)$$

where

$$\rho_1 = \exp \oint J_2 dx + \left(J_2 + \frac{\sigma_{yy}}{\sigma_y} \right) dy.$$

The required invariants may be expressed in terms of the coefficients as

$$J_1 = \tfrac{1}{3} A_2, \quad J_2 = \tfrac{1}{3} A_1, \quad J_3 = -\tfrac{1}{2} B_2 - \tfrac{1}{2} A_{1,y} - \tfrac{1}{6} A_1^2, \quad J_{3,x} = 0, \quad J_4 = -B_1.$$

PROOF   Case a). The first equation (6.87) for $S \equiv \frac{\sigma_{xx}}{\sigma_x}$ is essentially the definition of I_6 in Lemma 6.10. In order to obtain the second equation, the relation $\frac{\sigma_y}{\sigma_x} = I_1$ is differentiated twice w.r.t. x. In the resulting expression, $\left(\frac{\sigma_{xx}}{\sigma_x} \right)_x$ is substituted by means of the first equation of (6.87), and the obvious relation $\left(\frac{\sigma_{xx}}{\sigma_x} \right)_y = \frac{\sigma_{xxy}}{\sigma_x} - \frac{\sigma_{xx}}{\sigma_x} \frac{\sigma_{xy}}{\sigma_x}$ is used to eliminate $\frac{\sigma_{xxy}}{\sigma_x}$.

Eliminating the ρ-derivatives from the expressions for I_4, I_5 and I_7 in Lemma 6.10 and using the equations (6.87) for eliminating derivatives of $\frac{\sigma_{xx}}{\sigma_x}$,

Transformation to Canonical Form

the linear homogeneous system

$$\rho_{xy} - I_1\rho_{xx} - \left(I_4 + \frac{\sigma_{xx}}{\sigma_x} \right)\rho_y + \left(I_1 I_4 - I_{1,x} + I_1 \frac{\sigma_{xx}}{\sigma_x} \right)\rho_x = 0,$$

$$\rho_{yy} - I_1^2\rho_{xx} - \left(I_1 I_4 + I_5 + I_{1,x} + 2I_1\frac{\sigma_{xx}}{\sigma_x} \right)\rho_y$$

$$+ \left(I_1^2 I_4 + I_1 I_5 - I_{1,y} + 2I_1^2\frac{\sigma_{xx}}{\sigma_x} \right)\rho_x = 0,$$

$$\rho_{xxx} - 3\frac{\sigma_{xx}}{\sigma_x}\rho_{xx} - I_7\rho_y + \left[I_1 I_7 - I_6 + \frac{3}{2}\left(\frac{\sigma_{xx}}{\sigma_x} \right)^2 \right]\rho_x = 0 \qquad (6.92)$$

for ρ is obtained. It has the solutions $\rho = 1$, $\rho = \sigma$ and $\rho = \sigma^2$ as it may be verified by substitution. Consequently, the right factor

$$Lclm\left(\{\rho_x, \rho_y\}, \{\rho_x - \tfrac{\sigma_x}{\sigma}\rho, \rho_y - \tfrac{\sigma_y}{\sigma}\rho\}, \{\rho_x - 2\tfrac{\sigma_x}{\sigma}\rho, \rho_y - 2\tfrac{\sigma_y}{\sigma}\rho\} \right) =$$

$$\left\{ \rho_y - I_1\rho_x \equiv \rho_1, \rho_{xxx} - 3\tfrac{\sigma_{xx}}{\sigma_x}\rho_{xx} - \left(I_6 - \tfrac{3}{2}\tfrac{\sigma_{xx}^2}{\sigma_x^2} \right)\rho_x \equiv \rho_2 \right\}$$

$$(6.93)$$

may be divided out yielding the exact quotient

$$\left\{ \rho_2 - I_7\rho_1, \ \rho_{1,x} - \left(I_4 + \tfrac{\sigma_{xx}}{\sigma_x} \right)\rho_1, \ \rho_{1,y} - (I_5 + I_{1,x} + I_1\tfrac{\sigma_{xx}}{\sigma_x})\rho_1 \right\}$$

from which ρ_1 and ρ_2 may be determined by integration. Substituting it into the right hand sides of the second line of (6.93) yields an inhomogeneous system for ρ with known solution of the homogeneous part. It has been shown in Exercise 6 of Chapter 2, see also Appendix A, how a special solution for it may be obtained in terms of integrals. Applying these results leads to the representation (6.89) for ρ.

Case b). The function σ follows from the expression for the invariant J_3 in Lemma 6.10, the substitution $\frac{\sigma_{yy}}{\sigma_y} = -\frac{1}{2}S$ and two subsequent integrations. The expressions for J_1, J_2 and J_4 in the same lemma yield the system

$$\rho_{xx} - J_1\rho_x = 0, \quad \rho_{xy} - \left(J_2 + \frac{\sigma_{yy}}{\sigma_y} \right)\rho_x = 0,$$

$$\rho_{yyy} - 3\frac{\sigma_{yy}}{\sigma_y}\rho_{yy} - \left(\frac{\sigma_{yyy}}{\sigma_y} - 3\frac{\sigma_{yy}^2}{\sigma_y^2} \right)\rho_y - J_4\rho_x = 0$$

for ρ with the three solutions $\rho = 1$, $\rho = \sigma$ and $\rho = \sigma^2$, see Lie [109], part III, page 409. Consequently, the right factor

$$\left\{ \rho_x \equiv \rho_1, \rho_{yyy} - 3\frac{\sigma_{yy}}{\sigma_y}\rho_{yy} - \left(\frac{\sigma_{yyy}}{\sigma_y} - 3\frac{\sigma_{yy}^2}{\sigma_y^2} \right)\rho_y \equiv \rho_2 \right\}$$

may be divided out yielding the exact quotient

$$\left\{ \rho_2 - J_4\rho_1, \ \rho_{1,x} - J_1\rho_1, \ \rho_{1,y} - \left(J_2 + \frac{\sigma_{yy}}{\sigma_y} \right)\rho_1 \right\}.$$

308

By similar methods as in case *a)* the representation (6.91) is obtained.

The required invariants are obtained by elimination from the expressions for $A_0, \ldots, A_3$, B_6 and B_7 in Lemma 6.11. ▯

In either case, determining σ requires solving a first order partial Riccati-like system and integrations. Although the expressions (6.89) and (6.91) for ρ appear somewhat awkward, they are in fact very simple. Containing only arithmetic and integrations, they are *guaranteed* to be Liouvillian over the field determined by σ and ρ_1.

EXAMPLE 6.37   Equation 7.8 of Kamke's collection in symmetry class $\mathcal{S}_7^3$ has been considered in Example 5.48 of the preceding chapter. From its coefficients the invariants $I_1 = I_4 = I_6 = I_7 = 0$ and $I_5 = -\frac{3}{2y}$ are obtained. They yield $S_x - \frac{1}{2}S^2 = 0$, $S_y = 0$ for S. The special solution $S = 0$ leads to $\sigma_y = \sigma_{xx} = 0$ with the non-constant solution $\sigma = x$. Substituting it together with $\rho_1 = \frac{1}{y\sqrt{y}}$ into (6.89) yields $\rho = \frac{2}{\sqrt{y}} - 2$. ▯

EXAMPLE 6.38   An equation in symmetry class $\mathcal{S}_7^3$, case a), has been considered in Example 5.49 of the preceding chapter. From the coefficients given there the invariants

$$I_1 = \frac{x}{y}, \quad I_4 = -\frac{1}{x}, \quad I_5 = -\frac{1}{y}, \quad I_6 = 0, \quad I_7 = -\frac{3y}{x^3}$$

are obtained.   Substituting I_1 and I_6 into (6.87) yields $S_x - \frac{1}{2}S^2 = 0$, $S_y + \frac{x}{2y}S^2 - \frac{1}{y}S = 0$ with the only rational solution $S = 0$. It leads to $\sigma_y - \frac{x}{y}\sigma_x = 0$, $\sigma_{xx} = 0$ a special solution of which is $\sigma = xy$. Substituting this choice of σ and the corresponding value $\rho_1 = \frac{1}{x}$ into (6.89) finally yields $\rho = \frac{y}{x}$. ▯

EXAMPLE 6.39   An equation in symmetry class $\mathcal{S}_7^3$, case b), has been considered in Example 5.50 of the preceding chapter. From the coefficients given there the invariants

$$J_1 = \frac{x^2 - 2x + 2}{x(x-1)}, \quad J_2 = \frac{3}{y}, \quad J_3 = 0, \quad J_4 = -\frac{6x}{y^3(x-1)}$$

are obtained. Substituting J_3 into (6.90) yields $S_y - \frac{1}{2}S^2 = 0$ with the solution $S = -\frac{2}{y+C}$. Choosing $C = 0$ leads to $\sigma_{yy} + \frac{2}{y}\sigma_y = 0$ a special solution of which is $\sigma = \frac{1}{y}$. Substituting this choice of σ and the corresponding value $\rho_1 = \frac{(x-1)y}{x^2}e^x$ into (6.91) finally yields $\rho = \frac{y}{x}e^x$. ▯

Exercises

EXERCISE 6.1 Determine the constraints for the parameters and undetermined functions occurring in the canonical forms of Theorem 6.4 guaranteeing the correct symmetry type. Show that these constraints are related to the structure of the symmetry algebras.

EXERCISE 6.2 Determine the transformation between the two canonical forms given in Examples 6.7 and 6.8.

EXERCISE 6.3 Express the Janet basis coefficients of the two parameter symmetry groups in terms of the invariants determined in Lemma 6.1 and 6.2. Show that this is an application of Theorem 6.6.

EXERCISE 6.4 Apply Theorem 6.7 to equation $y'' + r(y, y') = 0$ in order to generate the most general canonical form as given in Theorem 6.4.

EXERCISE 6.5 Consider again the equation discussed in Examples 5.15 and 6.12. Assume now that only a two-parameter group of type $\mathbf{g}_{26}$ is known. Determine the canonical form under this assumption and discuss the result.

EXERCISE 6.6 Consider again equation 6.133 from Kamke's collection discussed in Example 6.11. Assume now that only a two-parameter group of type $\mathbf{g}_{25}$ is known. Determine the canonical form under this assumption and discuss the result.

EXERCISE 6.7 The number of subgroups that may be applied for generating a canonical form of a second order equation is maximal for symmetry class $\mathcal{S}_8^2$. Reconsider equation 6.180 from Kamke's collection that has been discussed in Example 6.14. Choose the two-parameter subgroups of type $\mathbf{g}_{25}$ and $\mathbf{g}_{26}$ and generate the respective canonical form. What about subgroups comprising three or four parameters?

Chapter 7

Solution Algorithms

It is assumed now that for a given ode of order not higher than three with nontrivial Lie symmetries the transformation functions from actual variables x and $y(x)$ to canonical variables u and $v(u)$ are known. This implies that its infinitesimal symmetry generators are explicitly known in canonical variables from the tabulation in Section 5.1 *without* further calculations. What remains to be done is to solve the canonical equation, and to generate the solution in actual variables from it. The term solution always means *general solution*; i.e., a n-parameter family of solutions if n is the order of the equation. This excludes the so-called *singular solutions* which may or may not be specializations of the general solution. The relation between these two concepts is discussed in detail in the article by Buium and Cassidy [23], Section 1.8, and articles by Ritt quoted there.

Before this proceeding is described in detail for the individual symmetry classes, the underlying general principles will be outlined. The following discussion is based on Engel [44], vol. V, *Anmerkungen* by Engel, pages 643-669 and 682-687 where many more details may be found.

Let the canonical ode of order n have an r-parameter symmetry group. The further proceeding depends crucially on the relative values of r and n. At first it is assumed that $r < n$ and $\Delta \neq 0$; the Lie determinant Δ is defined by (5.15). The fundamental invariants Φ_1 and Φ_2 defined on page 201 are of order $r - 1$ and r respectively. In terms of these invariants the canonical equation has the form

$$\Omega\left(\Phi_{r-1}, \Phi_r, \frac{d\Phi_r}{d\Phi_{r-1}}, \dots, \frac{d^{n-r}\Phi_r}{d\Phi_{r-1}^{n-r}}\right) = 0 \tag{7.1}$$

where Ω is an undetermined function. This is a general ode of order $n - r$. It has to be solved first. If this is not possible the main achievement of the symmetry analysis consists in lowering the order of the originally given ode of order n to an equation of order $n - r$ in the independent variable Φ_{r-1} and dependent variable Φ_r. If the solution of (7.1) may be found, however, an equation of order r allowing an r-parameter symmetry group is obtained.

Consequently, the case $n = r$ has to be considered next. It is discussed in full detail for arbitrary values of n in the above mentioned *Anmerkungen* by Engel. The complete answer for $n = 2$ and $n = 3$ that is relevant here is given in the book by Lie [113] on differential equations, Theorem 45 on page 464 and Kapitel 25, in particular Theorem 49 on page 555. It may be formulated as follows.

311

312

THEOREM 7.1   *(Lie 1891) If a second order ode has a two-parameter symmetry group and its symmetry generators are explicitly known, its solution involves two integrations. If a third order ode has a three-parameter solvable symmetry group with known generators, its solution involves three integrations. If the group is not solvable, the solution of a first order Riccati equation is required.*

The proof is based on the representation of the given ode as an equivalent first order pde, and Theorem 2.23 dealing with its solution. The details may be found in Lie's book.

If $n < r$, i.e., if the number of symmetries is higher than the order of the differential equation, the solution process may be further simplified. In many cases the solution may be given explicitly without any integrations involved. The complete answer for arbitrary values of n and r including all exceptions is due to Engel [44]; see in particular pages 651-665. Engel's comprehensive proof is based to a large extent on a case-by-case discussion. Therefore it is not given here. Rather for order $n = 2$ and $n = 3$ all possible symmetry types with $r \geq 3$ or $r \geq 4$ respectively are treated in detail below.

For first order equations the situation is different because point symmetries cannot be determined algorithmically. Therefore, at least a single nontrivial symmetry generator must be known a priori in order to apply the algorithms described in this chapter. Alternatively, the symmetry problem may be specialized to certain classes of equations for which algorithmic methods are available, e. g. for Abel's equation. This is discussed in Section 7.1. Subsequently, equations of order two and nonlinear equations of order three are treated in Sections 7.2 and 7.3 respectively. The subject of the final Section 7.4 are equations of order three that are equivalent to linear ones.

7.1   First Order Equations

In this section it is assumed that in addition to the first order equation to be solved at least a single symmetry generator is explicitly known; it is considered as part of the input to the solution algorithms.

One-Parameter Symmetry. If for a quasilinear first order equation

$$\omega \equiv y' + r(x, y) = 0 \tag{7.2}$$

a single nontrivial symmetry generator is known, it may be possible to introduce new variables u and $v(u)$ according to Theorem 6.1 such that the symmetry transformation becomes the translation in v. In these variables equation (7.2) takes the form

$$v' + s(u) = 0. \tag{7.3}$$

Integration yields the representation

$$v + \int s(u)du = C \tag{7.4}$$

of its general solution. If the actual variables are resubstituted into (7.4), the general solution of (7.2) is obtained. The function field for its representation is determined by the integral in (7.4) and the transformation functions between the actual variables x, y and the variables u, v.

According to Lemma 3.4 it is not guaranteed that the canonical form (7.3) from the knowledge of a single symmetry generator can be obtained. If this is not possible, the following lemma which is also due to Lie [113], Kapitel 6, §1, may be applied.

LEMMA 7.1   If a first order ode $y' + r(x,y) = 0$ allows the nontrivial symmetry generator $U = \xi(x,y)\partial_x + \eta(x,y)\partial_y$, the expression $\dfrac{1}{\eta + r\xi}$ is an integrating factor. Consequently, there holds

$$\phi = \oint \frac{rdx + dy}{r\xi + \eta}. \tag{7.5}$$

PROOF   Let $\phi(x,y) = C$ be the general solution of $y' + r(x,y) = 0$. By definition of a symmetry generator, it obeys $\xi\phi_x + \eta\phi_y = 1$. Moreover, as a solution, ϕ has to satisfy $\phi_x + y'\phi_y = \phi_x - r\phi_y = 0$. Eliminating the first order derivatives from these two expressions yields $\phi_x = \dfrac{r}{\eta + r\xi}$, $\phi_y = \dfrac{1}{\eta + r\xi}$. Substituting them into the right hand side of $d\phi = \phi_x dx + \phi_y dy$ yields $d\phi = \dfrac{dy + rdx}{\eta + r\xi}$ which is a total differential by construction. Consequently, $\dfrac{1}{\eta + r\xi}$ is an integrating factor and the representation is obtained. ◻

The various steps described above are applied in the subsequent algorithm for solving equation (7.2) with a known symmetry generator.

ALGORITHM 7.1   *LieSolve1.1(ω, U).* Given a quasilinear first order ode ω in the form (7.2) with the known symmetry generator U, its general solution is returned.

$S1$: *Determine transformation functions.* Applying Lemma 3.4 determine the transformation of ω to canonical form (7.3). If it cannot be determined return goto $S4$.

$S2$: *Canonical form.* Applying the result of step $S1$, generate the canonical form (7.3).

$S3$: *Return solution.* Integrate the canonical form. Resubstitute the actual variables and return the result.

$S4$: *Use integrating factor.* Evaluate the integral in (7.5) and return the result.

314

EXAMPLE 7.1   For equation 1.15 from the collection by Kamke the canonical form $v' + \dfrac{1}{u^2 - 1} = 0$ has been obtained in Example 6.1 by the transformation $x = v$ and $y = u + v^2$. It may be integrated to $v + \frac{1}{2}\log\frac{u-1}{u+1} = C$. Resubstituting the actual variables by $u = y - x^2$ and $v = x$ finally yields

$$y = \frac{C(x^2 - 1)e^{-2x} - x^2 - 1}{Ce^{-2x} - 1}.$$ ☐

Two-Parameter Symmetries. If for a first order equation (7.2) the generators of a two-parameter symmetry group are explicitly known, its general solution may often be obtained by algebraic operations as the next result shows.

THEOREM 7.2   *Let an equation (7.2) have two symmetries with generators $U_i = \xi_i(x,y)\partial_x + \eta_i(x,y)\partial_y$ for $i = 1, 2$, and define $F \equiv \dfrac{\eta_2 + r\xi_2}{\eta_1 + r\xi_1}$. If $\dfrac{\partial F}{\partial x} \neq 0$ and $\dfrac{\partial F}{\partial y} \neq 0$, its general solution is $F(x,y) = C$.*

PROOF   The symmetry type $\mathbf{g}_{26}$ is considered first, i.e., $[U_1, U_2] = 0$. Let the canonical generators in variables u and $v \equiv v(u)$ be $V_1 = \partial_u$ and $V_2 = \partial_v$. By Lemma 3.3 the transformation functions $x = \phi(u, v)$ and $y = \psi(u, v)$ obey the relations

$$V_1\phi = \phi_u = \xi_1, \quad V_2\phi = \phi_v = \xi_2, \quad V_1\psi = \psi_u = \eta_1, \quad V_2\psi = \psi_v = \eta_2$$

from which the first derivative $y' = \dfrac{\eta_1 + \eta_2 v'}{\xi_1 + \xi_2 v'}$ is obtained. Substitution into (7.2) and some simplifications yields $v' + \dfrac{\eta_2 + r\xi_2}{\eta_1 + r\xi_1} = 0$. According to Theorem 6.2 the canonical form for this symmetry class is $v' + a = 0$. This is assured if the relation given above holds.

The symmetry type $\mathbf{g}_{25}$ is considered next, i.e., now $[U_1, U_2] = U_1$. From the canonical generators $V_1 = \partial_v$ and $V_2 = u\partial_u + v\partial_v$, the relations

$$V_1\phi = \phi_v = \xi_1, \quad V_1\psi = \psi_v = \eta_1,$$

$$V_2\phi = u\phi_u + v\phi_v = \xi_2, \quad V_2\psi = u\psi_u + v\psi_v = \eta_2$$

follow from which the first order derivatives

$$\phi_u = \frac{\xi_2 - v\xi_1}{u}, \quad \phi_v = \xi_1, \quad \psi_u = \frac{\eta_2 - v\eta_1}{u}, \quad \psi_v = \eta_1$$

of ϕ and ψ are obtained. They lead to the expression $y' = \dfrac{\eta_2 - \eta_1 v + \eta_1 uv'}{\xi_2 - \xi_1 v + \xi_1 uv'}$ for the first derivative. Substitution into (7.2) and rearranging terms yields now $v'u - v + \dfrac{\eta_2 + r\xi_2}{\eta_1 + r\xi_1} = 0$. This represents the correct canonical form $v' + a = 0$ if $\dfrac{\eta_2 + r\xi_2}{\eta_1 + r\xi_1} - v$ is proportional to u; i.e., if there holds $\dfrac{\eta_2 + r\xi_2}{\eta_1 + r\xi_1} = v + au = C$.

☐

Solution Algorithms 315

If F defined in Theorem 7.2 is identically a constant, the above theorem cannot be applied for the solution procedure. In this case the canonical form

$$v' + a = 0 \tag{7.6}$$

has to be determined by Theorem 6.2 from which the solution may always be obtained. Lie [113] discusses the two alternatives in detail on page 124, Satz 1. On pages 126 and 127 he explains the different behavior by geometrical considerations. A solution scheme based on the preceding discussion may be designed as follows.

ALGORITHM 7.2   *LieSolve1.2*(ω, U_1, U_2). Given a quasilinear first order ode ω in the form (7.2) with two known symmetry generators U_1 and U_2, its general solution is returned.

$S1$: *Generate F.* From the coefficients of U_1 and U_2 generate F defined in Theorem 7.2. If F depends explicitly on x and y, return it.

$S2$: *Return result.* Generate the canonical form (7.6). Determine the solution in actual variables from it and return the result.

It should be noted that due to the explicit knowledge of two symmetry generators this algorithm never fails.

EXAMPLE 7.2   Consider the equation $y' + \dfrac{(x-y)^2 + 1}{(x-y)^2 - 1} = 0$ with the two symmetry generators $U_1 = \partial_x + \partial_y$ and $U_2 = y\partial_x + x\partial_y$ of a type $\mathbf{g}_{25}$ group. Substituting the coefficients of these generators into the definition of F in Theorem 7.2 yields after some simplifications the general solution in the form of $(x + y + C)(x - y) = 1$. ◻

EXAMPLE 7.3   The same equation as in the preceding example is considered. If only the generator $U = \partial_x + \partial_y$ is known, the solution scheme of algorithm *LieSolve1* has to be applied. By Lemma 3.4 the canonical variables are obtained as $x = v - u$, $y = v$ with canonical equation $v' - \dfrac{u^2 + 1}{2u^2} = 0$. Integration and resubstitution of the original variables yields the same solution as above. However, knowing only a single symmetry generator requires an additional integration. ◻

Abel's Equation. If one or two symmetry generators of an Abel equation may be found, its solution may be obtained by applying the above results for general first order equations. However, due to the special structure, the more explicit answer given in the subsequent theorem is usually more convenient.

THEOREM 7.3   *The two rational normal forms of Theorem 4.14 are considered separately.*

i) If the rational normal form is $y' + Ay^3 + By = 0$, the general solution is

$$y = \frac{1}{\sqrt{2\bar{B}(\int \frac{A}{B} dx + C)}} \tag{7.7}$$

where $\bar{B} = \exp(2 \int B dx)$ and C is the integration constant.

ii) *If the rational normal form is $y' + Ay^3 + By + 1 = 0$ and allows a one-parameter symmetry with generator*

$$U = \frac{1}{A^{1/3}} \left(\partial_x - \frac{1}{3} \frac{A'}{A} y \partial_y \right),$$

its canonical variables are $u = A^{1/3} y$ and $v = \int A^{1/3} dx$. They lead to the canonical form

$$v' + \frac{1}{u^3 + ku + 1} = 0 \tag{7.8}$$

where $k = \sqrt[3]{K}$, and K is the absolute invariant (4.51). From (7.8) its general solution is obtained by integration as

$$v + \int \frac{du}{u^3 + ku + 1} = C. \tag{7.9}$$

If $k = 0$ the integral may be performed with the result

$$v + \int \frac{du}{u^3 + 1} = v + \frac{1}{3} \log \frac{u+1}{\sqrt{u^2 - u + 1}} + \frac{1}{\sqrt{3}} \arctan \frac{2u-1}{\sqrt{3}} = C. \tag{7.10}$$

PROOF   The transformation functions $u \equiv \sigma(x, y)$ and $v \equiv \rho(x, y)$ obey the system $\sigma_x \xi + \sigma_y \eta = 0$ and $\rho_x \xi + \rho_y \eta = 1$. With the above values for ξ and η the solution $\sigma = A^{1/3} y$ and $\rho = \int A^{1/3} dx$ follows. The first derivative y' is transformed according to $y' = -\dfrac{\rho_x - \sigma_x v'}{\rho_y - \sigma_y v'}$ (compare (B.6) and (B.3)) with the result $y' = \dfrac{1}{v'} - \dfrac{1}{3} \dfrac{A'}{A} y$. Substituting this expression into the rational normal form yields after some simplifications (7.8).   □

The various alternatives that may occur for solving an Abel equation according to this theorem are combined in the following algorithm. The fundamental difference compared to *LieSolve1.1* and *LieSolve1.2* is the fact that it is not required to know any symmetry generator in advance. Rather it is automatically determined whenever this is possible.

ALGORITHM 7.3   *SolveAbel(ω)*. Given an Abel equation ω, its general solution is returned if a symmetry may be found, or *failed* otherwise.

$S1$: *Second kind equation?* If ω is an Abel equation of second kind, apply the transformation (4.40).

$S2$: *Rational normal form.* Transform ω to rational normal form.

$S3$: *Bernoulli case?* If the result of $S2$ is a Bernoulli equation, determine its solution from (7.7) and goto $S5$.

$S4$: *Symmetry?* If $K(A, B) = constant$, determine its solution from (7.9) or (7.10). Otherwise return *failed*.

Solution Algorithms 317

$S5$: *Backtransformation.* Apply the reverse transformations performed in $S1$ and $S2$ and return the result.

Next to Riccati's equation, Abel's equations are the most thoroughly studied differential equations in the history of mathematics. In the course of time several integrable cases have been found by some kind of heuristics, a short discussion of them may be found in Kamke [85], pages 24-28. It turns out that most of them may be subsumed under the symmetry analysis described here.

Scalizzi (1917) describes the integrable case of (4.38) for the constraint

$$a_0 = \frac{1}{3} \left(\frac{a_2}{a_3} \right)' + \frac{1}{3} \frac{a_1 a_2}{a_3} - \frac{2}{27} \frac{a_2^3}{a_3^2}.$$

From the last relation in (4.43) it is obvious that this condition identifies the rational normal form allowing a two-parameter symmetry with the solution (7.7).

Chiellini (1931) found an integrable case of (4.38) for $a_0 = a_1 = 0$, $(\frac{a_3}{a_2})' = const. \cdot a_2$ and $a_2 \neq 0$, $a_3 \neq 0$. Under these constraints the coefficients (4.43) are $b_3 = a_3$, $b_1 = -\frac{a_2^2}{3a_3}$ and $b_0 = const. \cdot \frac{a_2^3}{a_3^2}$; i.e., due to $b_0 \neq 0$ a rational normal form corresponding to case $ii)$ of Theorem 4.14 is obtained. Applying the relation $\frac{a_3'}{a_3} - \frac{a_2'}{a_2} = const. \cdot \frac{a_2^2}{a_3}$ which is a consequence of the above constraints it follows that the absolute invariant $K(A, B)$ is constant. Consequently, there is a one-parameter symmetry group with generator

$$U = \frac{a_3}{a_2^2} \left[\partial_x - \left(2\frac{a_2'}{a_2} - \frac{a_3'}{a_3} \right) y\partial_y \right].$$

Chini (1924) observed that the condition $\Phi_3 = 0$ for the invariant defined in Theorem 4.15 leads to a Bernoulli equation. Evaluating it for the equation with coefficients b_k in the proof of Theorem 4.14 yields $\Phi_3 = 3b_0b_3^2$. Because $b_3 \neq 0$ it follows $b_0 = 0$, i.e., the Bernoulli rational normal form. For $\Phi_3 \neq 0$ Chini showed that under the condition

$$a_3\Phi_3' + (3a_1a_3 - a_2^2 - 3a_3')\Phi_3 = 3\alpha\Phi_3^{5/3}$$

the integral representations described in case $ii)$ of Theorem 7.3 exist. Evaluating the above constraint for the rational normal form of this case shows that it is equivalent to the condition $K(A, B) = const.$; i.e., the corresponding equation allows a symmetry generator.

For the equation of the second kind (4.39) Abel himself described several integrable cases if certain relations between the coefficients are satisfied; e.g., if $f_1 = 2f_2 - g'$ the transformed equation becomes

$$y' + (f_0 - f_2g^2 + gg')y^3 + f_2y = 0$$

318

which is a Bernoulli equation.

Finally Todorov and Kristev [177] consider an Abel equation of the second kind (4.39) and impose the constraints

$$f_0 = (g^3 - 1)f_3 + gg', \qquad f_1 = 3f_3 g^2 - g', \qquad f_2 = 3f_3 g.$$

The signs of the f_k's have been reversed in order to agree with the notation in (4.39). They imply the rational normal form $y' + f_3^2 y^3 + \frac{f_3'}{f_3} y + 1 = 0$ with a vanishing absolute invariant and the integral representation (7.10) (equation (14) in Todorov and Kristev [177]).

EXAMPLE 7.4   For equation 1.46 of Kamke's collection the rational normal form $z' - x^2 z^3 + \frac{2}{x^2} z = 0$ has been obtained in Example 5.7. The solution of this Bernoulli equation is obtained from (7.7) in step $S3$. Resubstituting the original variable $z = y - \frac{1}{x^2}$ yields the final answer

$$y = \frac{e^{2/x}}{\sqrt{-2 \int x^2 e^{4/x} dx + C}} + \frac{1}{x^2}. \qquad \square$$

It is not always advisable to generate the canonical form (7.8) because it may lead to unnecessary algebraic numbers due to the cubic root in $k = \sqrt[3]{K}$. The next example shows how they may be avoided.

EXAMPLE 7.5   Equation 1.255 $\left(y - \frac{3}{x}\right)y' + \frac{1}{x}y^2 - \frac{1}{x^2}y = 0$ from Kamke's collection is an Abel equation of the second kind. Introducing the new dependent variable z by $y = \frac{1}{z} + \frac{3}{x}$ (compare equation (4.39) and the transformation given there) leads to $z' - \frac{6}{x^3}z^3 - \frac{2}{x^2}z^2 - \frac{1}{x}z = 0$. It is transformed into rational normal form $w' - \frac{32}{19683}\frac{w^3}{x^3} - \frac{7}{9}\frac{w}{x} + 1 = 0$ by $z = -\frac{1}{243}(4w + 27x)$. Its absolute invariant is $K = -\frac{729}{4}$; consequently, $k = -\frac{9}{2}\sqrt[3]{2}$. In order to avoid the cubic root, it is better to introduce canonical variables by $u = \frac{2y}{9x}$ and $v = -\frac{2}{9}\log x$ with inverse $x = \exp\left(-\frac{9}{2}v\right)$ and $y = \frac{9}{2}u\exp\left(-\frac{9}{2}v\right)$. It yields

$$v' + \frac{1}{\frac{4}{27}u^3 - u - 1} = v' + \frac{\frac{27}{4}}{(u + \frac{3}{2})^2(u - 3)} = 0$$

with the solution $v = \frac{1}{3}\log\frac{2u + 3}{u - 3} - \frac{3}{2u + 3} + C$. Resubstituting the original variables w, z and finally y yields the solution of the originally given equation in the form $xy^3 = C\exp(xy)$. $\qquad \square$

7.2 Second Order Equations

The subjects of this section are second order equations that are linear in the highest derivative. A cursory glance at the various collections of solved examples shows that they have been investigated more thoroughly than any other family of equations. Lie [113] himself devoted to them almost completely his book on symmetries of differential equations. These results are discussed in detail in the current section.

One-Parameter Symmetry. If a second order quasilinear ode belongs to symmetry class $\mathcal{S}_1^2$ it is equivalent to an equation of the form

$$v'' + r(u, v') = 0 \tag{7.11}$$

with $r(u, v')$ an undetermined function of its arguments. The substitution $w = v'$ in (7.11) yields

$$w' + r(u, w) = 0. \tag{7.12}$$

Because there are no constraints on $r(u, w)$, this is a general first order equation. It is the first example of an equation of the form (7.1) because the number of symmetries is lower than the order of the equation. As explained in the introduction to this chapter, solving it is a new problem that has to be handled as described in the preceding Section 7.1. The further proceeding depends on whether or not this first order equation can be solved. The algorithm below applies these steps. The solution procedure may terminate in step $S2$ if σ cannot be obtained, or in step $S3$ if equation (7.12) cannot be solved.

ALGORITHM 7.4   *LieSolve2.1(ω)*. Given a quasilinear second order ode in symmetry class $\mathcal{S}_1^2$, its general solution or an equivalent first order equation is returned; or *failed* if neither may be found.

$S1$: *Janet basis for symmetries.* Generate a Janet basis for the determining system of its Lie symmetries.

$S2$: *Determine transformation functions.* If case a) of Theorem 6.7 applies choose σ. If case b) applies solve the first order equation for σ. If it cannot be obtained return *failed*. Determine ρ by integration.

$S3$: *Canonical form.* Applying σ and ρ obtained in $S2$ generate the canonical form (7.12) and apply the solution algorithms of Section 7.1 to it. If a solution has not been found, return the canonical form equation together with the transformation functions σ and ρ.

$S4$: *Return solution.* From the solution obtained in step $S3$, construct the solution of ω and return the result.

EXAMPLE 7.6   For equation 6.90 of Kamke's collection the canonical form has been obtained in Example 6.5. Substituting $v' = w$ yields

$$w' + u(u + 5)w^3 - (u + 3)w^2 + \tfrac{1}{4}w = 0.$$

320

This is an Abel equation the solution of which cannot be obtained. $\square$

EXAMPLE 7.7   The canonical form (6.2) for equation 6.98 of Kamke's collection leads to $w' + 2(2u - 1)w^2 + w = 0$. This Bernoulli equation has the general solution $w = \dfrac{1}{C_1 e^x - 4x - 2}$ and finally $v = \displaystyle\int \dfrac{du}{C_1 e^u - 4u - 2} + C_2$. If the actual variables $u = \dfrac{y}{x^2}$ and $v = \log x$ are resubstituted, an integral representation of the general solution of equation 6.98 is obtained. $\square$

Two-Parameter Symmetries. The canonical form of a second order quasilinear ode in the symmetry class $S_{2,1}^2$ has been shown to be

$$v'' + r(v') = 0 \tag{7.13}$$

where $r(v')$ is an undetermined function of v'. In order to solve this equation it is rewritten as

$$v'' + r(v') = \frac{dv'}{du} + r(v') = 0 \ \text{ or } \ \frac{dv'}{r(v')} + du = 0. \tag{7.14}$$

Integration yields

$$R(v') + u = C_1 \quad \text{with} \quad R(v') = \int \frac{dv'}{r(v')}. \tag{7.15}$$

If $r(v')$ is such that the latter integration may be performed, the original equation is reduced to a first order equation. Depending on the form of $R(v')$ it may be possible to continue the solution procedure. In any case the general solution may always be represented in parameter form as

$$u + R(p) = C_1, \quad v + pR(p) = \int R(p)dp + C_2. \tag{7.16}$$

This is a consequence of the fact that (7.15) is a first order differential equation that does not contain the dependent variable explicitly; i.e., it allows the symmetry ∂_v. Details may be found in Kamke [86], vol. I, page 49; see also Exercise 7.4. If $u \equiv \sigma(x, y)$ and $v \equiv \rho(x, y)$ are resubstituted, the solution in actual variables is obtained.

In the subsequent algorithm the various steps for solving an equation in symmetry class $S_{2,1}^2$ are put together. Only step $S2$ may fail if the transformation to canonical form is not Liouvillian.

ALGORITHM 7.5   *LieSolve2.2.1(ω)*. Given a quasilinear second order ode ω in symmetry class $S_{2,1}^2$, its general solution is returned explicitly or in parameter form. Or *failed* if it cannot be determined.

S1 : *Equations for σ and ρ*. Generate a Janet basis for the symmetry generators of ω and set up system (6.3).

Solution Algorithms 321

$S2$: *Determine transformation functions.* Determine an independent pair of Liouvillian solutions of the system obtained in step $S1$. If it cannot be found return *failed*.

$S3$: *Generate canonical form.* By means of the transformation functions obtained in $S2$ generate the canonical form (7.13).

$S4$: *Return solution.* Using $r(v')$ from step $S3$ construct the solution (7.16), resubstitute actual variables and return the result.

EXAMPLE 7.8   In the preceding chapter, a canonical form of Kamke's equation 6.227 has been determined in Example 6.7. Substituting the respective value for $r(v')$ into (7.15) and (7.16) and performing the integrations yields

$$u + \log \frac{p+2}{p+1} + \frac{p}{p+2} = \bar{C}_1, \quad v + \log(p+1) - \frac{2p}{p+2} = \bar{C}_2.$$

If the canonical variables are resubstituted, the parameter representation

$$x = C_1 \frac{1}{p+2} \exp \frac{p}{p+2}, \quad y = C_2 \frac{p+1}{p+2} \exp\left(-\frac{p}{p+2}\right)$$

for x and y is obtained. ⬜

In Exercise 7.6 the same answer will be obtained from the canonical form of Example 6.8.

The other two-parameter symmetry class is $\mathcal{S}_{2,2}^2$. A quasilinear second order ode in this symmetry class has the canonical form

$$uv'' + r(v') = 0 \tag{7.17}$$

where again $r(v')$ is an undetermined function of v'. Rewriting it as

$$v'' + \frac{r(v')}{u} = \frac{dv'}{du} + \frac{r(v')}{u} = 0 \quad \text{or} \quad \frac{dv'}{r(v')} + \frac{du}{u} = 0 \tag{7.18}$$

it may be integrated once with the result

$$R(v') + \log u = C_1 \quad \text{with} \quad R(v') = \int \frac{dv'}{r(v')}. \tag{7.19}$$

The same general remarks apply to this equation as for (7.15). If a further integration is possible, the general solution of the canonical equation follows explicitly. Otherwise the parameter representation

$$u = C_1 e^{-R(p)}, \quad v = C_1 (p e^{-R(p)} - \int e^{-R(p)} dp) + C_2 \tag{7.20}$$

in canonical variables or, after substitution of $\sigma(x, y)$ and $\rho(x, y)$ in actual variables is obtained. As in the previous case, details may be found in Kamke [86], vol. I, page 49.

322

A solution scheme for equations in symmetry class $\mathcal{S}_{2,2}^2$ is realized in terms of the following algorithm.

ALGORITHM 7.6   *LieSolve2.2.2(ω)*. Given a quasilinear second order ode ω in symmetry class $\mathcal{S}_{2,2}^2$, its general solution is returned explicitly or in parameter form.

S1 : Equations for σ and ρ. Generate a Janet basis for the symmetry generators of ω.

S2 : Determine transformation functions. Determine the transformation functions σ from (6.6), and ρ from (6.7) or (6.8).

S3 Generate canonical form. By means of the transformtion functions obtained in *S2* generate the canonical form (7.17).

S4 : Return solution. Construct the solution (7.20), resubstitute the actual variables and return the result.

According to Theorem 6.3, for any equation in this symmetry class a canonical form may be obtained. Consequently, a solution in parameter form is guaranteed and the above algorithm *LieSolve2.2.2(ω)* does not have a *failed* exit. In those cases where (7.19) may be solved for v', the solution may be expressed as an integral as it is the case in the following example.

EXAMPLE 7.9   The canonical form for Kamke's equation 6.159 has been determined in Example 6.9. With the above notation there follows

$$r(v') = -\tfrac{3}{2}v'(v'^2 - 1) \quad \text{and} \quad R(v') = \tfrac{2}{3}\log v' - \tfrac{1}{3}\log(v'^2 - 1).$$

Substituting it into (7.20) leads to $u^3 v'^2 = C_1(v'^2 - 1)$. Solving for v' and integrating leads to

$$v = C_1 \int \frac{du}{\sqrt{u^3 + C_1^2}} + C_2.$$

If the actual variables $u = \dfrac{1}{\sqrt{y}}$ and $v = x$ are resubstituted, and the integration constant C_1 is replaced by its inverse, the solution of the original equation is obtained in the form

$$x + \tfrac{1}{2} \int \frac{dy}{\sqrt{y\sqrt{y}}\sqrt{y\sqrt{y} + C_1}} = C_2. \qquad \qquad □$$

Three-Parameter Symmetries. In this subsection the number of symmetries exceeds the order of the respective ode by one and the remarks of the second paragraph on page 312 apply. In the first symmetry class $\mathcal{S}_{3,1}^2$ to be discussed the canonical form is

$$v''(u - v) + 2v'(v' + a\sqrt{v'} + 1) = 0 \qquad (7.21)$$

where a is a constant, $a \neq 0$, or in rational form

$$v''^2(u - v)^2 + 4v''v'(v' + 1)(u - v) + 4v'^4 + 4(2 - a^3)v'^3 + 4v'^2 = 0. \quad (7.22)$$

Solution Algorithms   323

Two subsequent integrations of the canonical form (7.21) yield

$$\frac{(u-v)\sqrt{v'}}{1+a\sqrt{v'}+v'} = \frac{1}{C_1}$$

and finally a third integration leads to the general solution

$$(C_1v + C_2)(C_1u + C_2 + a) + 1 = 0. \tag{7.23}$$

The various steps for solving an equation in this symmetry class are put together in the following algorithm.

ALGORITHM 7.7   *LieSolve2.3.1(ω).* Given a quasilinear second order ode ω in symmetry class $\mathcal{S}_{3,1}^2$, its general solution is returned; or *failed* if it cannot be determined.

$S1$: *Equations for R and S.* Generate Janet basis for the symmetry generators of u. Set up system (6.13), (6.16) or (6.18) respectively for R, or system (6.20) for R and S.

$S2$: *Determine R and S.* From the equations obtained in $S1$ determine rational solutions for R and S. If none can be found return *failed*.

$S3$: *System for transformation functions.* Applying the result of $S2$ set up the system (6.14), (6.17), (6.19) or (6.21) for σ and ρ.

$S4$: *Determine transformation functions.* Determine two independent Liouvillian solutions for the system obtained in $S2$.

$S5$: *Return solution.* Generate the canonical form and determine the value for a from it. Substitute a and the canonical variables into (7.23) and return the result.

EXAMPLE 7.10   An equation in this symmetry class has been considered before in Examples 5.13 and 6.10. Comparing the coefficients of the cubic power of v' in the canonical form of latter example and of (7.22) yields $a = 1$. Substituting this value and the expressions for the canonical variables $u = xy$ and $v = \frac{2}{x}$ into (7.23) yields the solution, after some simplifications, in the form

$$y = \frac{(C_2^2 + C_2 + 1)x + 2C_1(C_2 + 1)}{(2C_1 + C_2 x)C_1 x}. \qquad \Box$$

A quasilinear second order ode in symmetry class $\mathcal{S}_{3,2}^2$ is equivalent to an equation of the form

$$v''v^3 + a = 0 \tag{7.24}$$

where a is constant, $a \neq 0$. In order to obtain the solution, the canonical equation (7.24) is written as $(v'^2)' - \left(\frac{a}{v^2}\right)' = 0$. In this form the integration is obvious and leads to $v'^2 - \frac{a}{v^2} = C_1$. Solving for v' yields $v' = \frac{\sqrt{C_1 v^2 + a}}{v}$. Separation of variables allows it to perform the second integration with the result

$$\frac{v\,dv}{\sqrt{C_1 v^2 + a}} = du \quad \text{or} \quad \frac{1}{C_1}\sqrt{C_1 v^2 + a} - u = C_2.$$

324

The latter relation may be rewritten as

$$C_1 v^2 - C_1^2 (u + C_2)^2 + a = 0. \tag{7.25}$$

Similar to the preceding case the solution procedure comes down to determining the transformation functions of the actual to canonical variables, determining the constant a from the canonical form and substituting it together with $u \equiv \sigma$ and $v \equiv \rho$ into the solution of the latter. Theorem 6.11 explains how the transformation to canonical variables is achieved. These steps are put together in the subsequent algorithm.

ALGORITHM 7.8   *LieSolve2.3.2(w).* Given a quasilinear second order ode w in symmetry class $\mathcal{S}_{3,2}^2$, its general solution is returned; or *failed* if it cannot be determined.

$S1$: *Equations for R.* Generate Janet basis for the symmetry generators of w. If it is of type $\mathcal{J}_{3,4}^{(2,2)}$ set up system (6.22), of type $\mathcal{J}_{3,7}^{(2,2)}$ system (6.23) and of type $\mathcal{J}_{3,6}^{(2,2)}$ system (6.24) for R.

$S2$: *Determine R.* From the equations obtained in $S1$ determine a rational solution for R. If it cannot be found return *failed*.

$S3$: *System for transformation functions.* Applying the result of $S2$ set up the systems (6.22), (6.23) or (6.24) respectively for σ and ρ.

$S4$: *Determine transformation functions.* Determine two independent Liouvillian solutions for the system obtained in $S3$.

$S5$: *Return solution.* Generate the canonical form and determine the value for a from it. Substitute a and the canonical variables into (7.25) and return the result.

EXAMPLE 7.11   The solution procedure for equation 6.133 of Kamke's collection may now be completed. Substituting $u = x - y$, $v^2 = x + y$ and $a = \frac{1}{4}$ as obtained in Example 6.11 into (7.25) yields the general solution in actual variables

$$C_1 (x + y) + C_1^2 (x - y + C_2)^2 + \tfrac{1}{4} = 0. \qquad \qquad \Box$$

For an equation in symmetry class $\mathcal{S}_{3,3}^2(c)$ it is assumed that the parameter c has already be determined. Its canonical form is

$$v'' + a v'^{(c-2)/(c-1)} = 0 \tag{7.26}$$

with a constant $a \neq 0$. In order to yield an integer power v'^m, the parameter c must be determined by $c = 1 + \dfrac{1}{1-m}$. For $|m| \leq 5$ the resulting values of c are given in the following table.

m	-5	-4	-3	-2	-1	0	2	3	4	5
c	7/6	6/5	5/4	4/3	3/2	2	0	1/2	2/3	3/4

Integrating the canonical form (7.26) once yields

$$(c-1)v'^{1/(c-1)} + au = C_1.$$

The second integration gives the final result

$$v = \frac{(1-c)^{1-c}}{ac}(au - C_1)^c + C_2. \tag{7.27}$$

This form is most suitable for applications. The value of the constant c determines the symmetry type completely. In Theorem 6.12 it is expressed in terms of the coefficients of the Janet basis for the symmetries. Combining these results the following algorithm is obtained.

ALGORITHM 7.9   *LieSolve2.3.3(ω)*. Given a second order quasilinear ordinary differential equation in symmetry class $\mathcal{S}_{3,3}^2(c)$, the general solution is returned.

$S1$: *Janet basis for symmetries*. Generate a Janet basis for the symmetry generators of ω.

$S2$: *System for transformation functions*. If the Janet basis obtained in step $S1$ has type $\mathcal{J}_{3,6}^{(2,2)}$, set up system (6.25), (6.26) or (6.27) respectively. If it has type $\mathcal{J}_{3,7}^{(2,2)}$ set up system (6.28).

$S3$: *Determine transformation functions*. Determine two independent Liouvillian solutions for the system obtained in $S2$.

$S4$: *Return solution*. Generate the canonical form (7.26), substitute a and the canonical variables into (7.27), and return the result.

EXAMPLE 7.12   Two equations with this symmetry group have been considered in Example 6.12. In both cases the canonical form is $v'' + v'^4 = 0$ corresponding to $\gamma = \frac{2}{3}$ and $a = 1$. Substituting these values into (7.27) yields

$$v = \frac{3}{2\sqrt[3]{3}}(u - C_1)^{\frac{2}{3}} + C_2$$

for the solution in canonical variables. Inverting the substitution functions $x = \frac{1}{v}$ and $y = \frac{1}{u}$ for the first alternative leads to the solution

$$\left[(8C_2^3 - 9C_1)x^3 + 24C_2^2 x^2 + 24C_2 x + 8\right]y^2 - 18C_1 x^3 y - 9x^3 = 0.$$

In the other case, inverting $x = v - u$ and $y = 3u - 2v$ yields the solution

$$y^3 - 9\left(x + \tfrac{1}{3}C_1 - \tfrac{1}{8}\right)y^2 + 27\left[x^2 + (\tfrac{2}{3}C_1 - \tfrac{1}{6})x + \tfrac{1}{9}C_1^2 - \tfrac{1}{12}C_2\right]y$$

$$+ 27\left[x^3 + (C_1 - \tfrac{1}{6})x^2 + (\tfrac{1}{3}C_1^2 - \tfrac{1}{6}C_2)x - \tfrac{1}{24}C_2^2\right] = 0$$

for the second equation of Example 6.12. $\square$

An equation in symmetry class $\mathcal{S}_{3,4}^2$ has the canonical form

$$v'' - ae^{-v'} = 0 \tag{7.28}$$

326

with constant $a \neq 0$. Two integrations lead immediately to the general solution

$$v = \frac{au + C_1}{a} \left[\log (au + C_1) - 1 \right] + C_2. \tag{7.29}$$

Similar to $S_{3,1}^2$ and $S_{3,2}^2$ there is a single constant that has to be determined from the canonical form of the given equation. It is substituted together with the transformation functions into (7.29) in order to generate the solution for the given equation. The algorithm below performs these steps, it returns the solution for any equation in this symmetry class.

ALGORITHM 7.10   *LieSolve2.3.4(ω)*. Given a second order quasilinear ordinary differential equation in symmetry class $S_{3,4}^2$, the general solution is returned.

$S1$: *Janet basis for symmetries.* Generate a Janet basis for the symmetry generators of ω.

$S2$: *System for transformation functions.* If the Janet basis obtained in $S1$ has type $\mathcal{J}_{3,6}^{(2,2)}$, set up system (6.29), (6.30) or (6.31) respectively. If it has type $\mathcal{J}_{3,7}^{(2,2)}$ set up system (6.32).

$S3$: *Determine transformation functions.* Determine two independent Liouvillian solutions for the system obtained in $S2$.

$S4$: *Return solution.* Generate the canonical form and determine the value for a from it. Substitute a and the canonical variables into (7.29) and return the result.

EXAMPLE 7.13   The transformation functions for an equation in symmetry class $S_{3,4}^2$ have been obtained in Example 6.13 as $\sigma = x^2$ and $\rho = x + y^2$. Substituting the inverse $x = \sqrt{u}$, $y = \sqrt{v - \sqrt{u}}$ leads to a canonical form (7.28) with $a = 1$. Substituting a and the transformation functions into (7.29) yields finally the solution

$$y^2 = (C_1 + x^2) \left[\log (C_1 + x^2) - 1 \right] - x + C_2$$

for the equation considered in Example 5.16   ⬚

Eight-Parameter Projective Symmetry. The canonical form $v''(u) = 0$ with the general solution $v = C_1 + C_2 u$ of an equation with this symmetry is particularly simple because it does not contain any unspecified elements. Solving such an equation therefore comes down to determining the transformation functions σ and ρ. According to Theorem 6.14 this is a two-step procedure. At first a rational solution of the partial Riccati system (6.37) has to be found. Secondly, an independent set of Liouvillian solutions σ and ρ has to be determined by solving (6.36). The various steps of this proceeding are put together in the subsequent algorithm.

$$\text{Solution Algorithms} \qquad\qquad 327$$

ALGORITHM 7.11   *LieSolve2.8(ω).* Given a quasilinear second order ode ω in symmetry class $\mathcal{S}_8^2$. Its general solution is returned; or *failed* if it cannot be determined.

S1 : Determine a and b. From the coefficients of ω generate the system (6.37) and determine a special rational solution. If none is found return *failed*.

S2 : System for transformation functions. Apply the result of *S1* and construct the Janet basis (6.36).

S3 : Determine transformation functions. Find two independent Liouvillian solutions of the Janet basis constructed in *S2*. If none are found return *failed*.

S4 : Return solution. Substitute $u = \sigma(x, y)$ and $v = \rho(x, y)$ into the canonical form $v = C_1 + C_2 u$ and return the result.

EXAMPLE 7.14   Equation 6.180 has been considered already in Example 6.14. Substituting the values for $u \equiv \sigma(x, y)$ and $v \equiv \rho(x, y)$ obtained there into $v = C_1 + C_2 u$ yields after some simplifications the general solution

$$y = \frac{C_2 x^2 - x}{C_1 + C_2 x^2 - x}. \qquad\qquad \square$$

EXAMPLE 7.15   Equation 6.124 has been considered before in Example 6.15. Substituting the values for $u \equiv \sigma(x, y)$ and $v \equiv \rho(x, y)$ obtained there into $v = C_1 + C_2 u$ yields after some simplifications the general solution

$$y = \frac{1}{\sqrt{C_1 e^{-x} + C_2 e^{-2x}}}. \qquad\qquad \square$$

Lie's Second Integration Method. In his book on differential equations Lie [113], Kapitel 20, §4 and 5, a solution scheme for differential equations with symmetries is described that he called *second method of integration*. It is based on the observation that any second order quasilinear ode $y'' = \omega(x, y, y')$ may be written as the system $y' = \frac{dy}{dx}$, $y'' = \frac{dy'}{dx}$ which in turn is equivalent to the linear pde

$$\frac{\partial f}{\partial x} + y'\frac{\partial f}{\partial y} + \omega(x, y, y')\frac{\partial f}{\partial y'} = 0.$$

The application of this second integration scheme is based on the following lemma.

LEMMA 7.2   *(Lie 1891) The differential equation $y'' + r(x, y, y') = 0$ allows the symmetry with the generator $U = \xi\partial_x + \eta\partial_y$ if and only if the linear pde*

$$Af \equiv \frac{\partial f}{\partial x} + y'\frac{\partial f}{\partial y} - r(x, y, y')\frac{\partial f}{\partial y'} = 0$$

allows the first extension of U, i.e., if $[U, A] = \rho(x, y, y')A$ for some function ρ.

328

The proof may be found in the above quoted book by Lie. Based on this lemma, the problem of solving an ode with symmetries is replaced by a linear pde allowing the once extended vector fields and the theory of Chapter 2, Section 5 applies. This scheme is applied in the following example.

EXAMPLE 7.16   Equation 6.159 has been solved in Example 7.9. For this equation $r = -3y^2 - \frac{3}{4y}y'^2$, its symmetry generators are (see Example 5.23) $U_1 = \partial_x$ and $U_2 = x\partial_x - 2y\partial_y$. Therefore the system

$$\partial_x f + y'\partial_y f + \left(3y^2 - \frac{3}{4y}y'^2\right)\partial_{y'} f = 0, \quad \partial_x f = 0, \quad x\partial_x f - 2y\partial_y f - 3y'\partial_{y'} f = 0$$

is obtained. It leads to $\Delta = \frac{3}{2}y'^2 - 6y^3$ and the integral (2.47) yields

$$\Phi_1 = \int \left(3y^2 + \frac{3}{4y}y'^2\right)\frac{dy}{\Delta} - \int y'\frac{dy'}{\Delta} = \log\frac{y\sqrt{y}}{y'^2 - 4y^3}.$$

Its exponential is the first integral $\frac{u^3 v'^2}{v'^2 - 1}$ obtained in Example 7.9 written in actual variables x and y.   ⬚

7.3   Nonlinear Equations of Third Order

In this section a complete discussion of symmetry methods for solving genuinely nonlinear equations of third order is presented, similar to second order equations above. In Lie's book [113] on differential equations there is only a short chapter on third order equations allowing certain three-parameter symmetry groups.

One-Parameter Symmetry. A quasilinear third order ode in symmetry class $\mathcal{S}_1^3$ has the canonical form

$$v''' + r(v'', v', u) = 0. \tag{7.30}$$

Substituting $v' = w(u)$ yields

$$w'' + r(w', w, u) = 0. \tag{7.31}$$

This is a general second order equation for $w(u)$ because $r(w', w, u)$ is an undetermined function of its arguments as it has been explained in the introduction on page 311. If it has a symmetry, the results of the preceding section may be applied for solving it, if not it is returned by the following algorithm performing these steps.

Solution Algorithms 329

ALGORITHM 7.12 *LieSolve3.1(ω)*. Given a quasilinear third order ode ω in symmetry class $\mathcal{S}_1^3$, its general solution or a lower order equation is returned, or *failed* if neither may be found.

$S1$: *Janet basis for symmetries.* Generate a Janet basis for the determining system of the Lie symmetries of ω.

$S2$: *Determine transformation functions.* If case a) of Theorem 6.7 applies choose σ. If case b) applies solve the first order equation for σ. If it cannot be obtained return *failed*. Determine ρ by integration.

$S3$: *Canonical form.* Applying the solutions obtained in $S2$ generate the canonical form (7.31). If it does not allow any nontrivial symmetry return it.

$S4$: *Return solution.* Proceed with the proper solution procedure for a second order equation. If a solution is found, resubstitute the actual variables and return the result.

EXAMPLE 7.17 The canonical form obtained in Example 5.27 leads to the second order equation (7.30)

$$w''w - 3w'^2 - (u - 4)w'w^2 + (2u - 4)w^4 + w^3 = 0.$$

Because it does not have any nontrivial symmetry, this is the best possible answer in the context of Lie's theory. $\quad\Box$

Two-Parameter Symmetries. According to the discussion in the introduction of this chapter, a third order equation with a two-parameter symmetry in general may be reduced to a first order ode of type (7.1). For the four two-parameter symmetry classes discussed there these are the equations (7.32), (7.35), (7.39) and (7.42) below. The first symmetry class to be discussed is $\mathcal{S}_{2,1}^3$. It has the canonical form

$$v''' + r(v'', v') = 0. \tag{7.32}$$

Substituting $v''' = v'' \dfrac{dv''}{dv'}$ yields

$$\frac{dv''}{dv'} + \frac{r(v'', v')}{v''} = 0. \tag{7.33}$$

This is a general first order equation for $v''(v')$ because $r(v'', v')$ is an undetermined function of its arguments. If $v'' = w(v', C_1)$ is its general solution, an integration yields

$$\int \frac{dv'}{w(v', C_1)} = u + C_2. \tag{7.34}$$

If the integration may be performed, the resulting first order ode allows the one-parameter symmetry ∂_v. Therefore the desired solution of (7.32) may be represented in terms of an integral.

330

The various steps are put together in the algorithm below. The solution procedure may fail in step $S2$ if the transformation functions are not Liouvillian, or in step $S4$ if equation (7.33) cannot be solved.

ALGORITHM 7.13   *LieSolve3.2.1(ω).* Given a quasilinear third order ode ω in symmetry class $\mathcal{S}_{2,1}^3$, its general solution or some lower order equation is returned.

$S1$: *Equations for σ and ρ.* Generate a Janet basis for the symmetry generators of ω and set up system (6.3).

$S2$: *Determine transformation functions.* Determine an independent pair of Liouvillian solutions of the system obtained in step $S1$. If it cannot be found return *failed*.

$S3$: *Second order equation.* By means of the transformation functions obtained in $S2$ generate equation (7.33) and determine its solution. If it cannot be found return (7.33).

$S4$: *Integral representation.* Perform the integration at the left hand side of (7.34). If it cannot be done return (7.34).

$S5$: *Return solution.* Solve the first order equation obtained in step $S4$ and return the result. If a solution cannot be found return the first order equation and the transformation functions.

EXAMPLE 7.18   The canonical equation $v''' + 2v''v' + 1 = 0$ of Example 6.17 has the form (7.32). With $v'' = w$ and $v' = z$ the Abel equation of second kind $ww' + 2wz + 1 = 0$ for $w(z)$ is obtained. According to Chapter 4.2 it may be transformed into an Abel equation of first kind for which a symmetry cannot be found. $\Box$

A quasilinear third order ode in symmetry class $\mathcal{S}_{2,2}^3$ has the form

$$v''' + \frac{1}{u^2}r(uv'', v') = 0. \tag{7.35}$$

Substituting first $v' = z$ yields the equation $z'' + \frac{1}{u^2}r(uz', z) = 0$ allowing the one-parameter group $u\partial_u$. By Theorem 6.7 new variables $\bar{w}$ and t defined by $u = e^{\bar{w}}$, $z = t$ with $\bar{w} \equiv \bar{w}(t)$, and finally $\bar{w}'(t) = w(t)$ yield the first order equation

$$w' + w^2 - w^3 r\left(\frac{1}{w}, t\right) = 0. \tag{7.36}$$

If its general solution is $w = w_0(t, C_1)$, the relation

$$u = C_2 \exp\left(\int w_0(v', C_1)dv'\right) \tag{7.37}$$

between v' and u involving two constants follows. If the integration may be performed, the solution procedure may be continued.

The subsequent algorithm performs these steps. The crucial step is $S3$. If it fails the first order equation (7.36) is the best possible answer.

Solution Algorithms 331

ALGORITHM 7.14   *LieSolve3.2.2(ω).* Given a quasilinear third order ode ω in symmetry class $\mathcal{S}_{2,2}^3$, its general solution or some lower order equation is returned.

$S1$: *Equations for σ and ρ.* Generate a Janet basis for the symmetry generators of ω.

$S2$: *Determine transformation functions.* Determine the transformation functions σ from (6.6), and ρ from (6.7) or (6.8).

$S3$: *First order equation.* By means of the transformation functions obtained in $S2$ generate equation (7.36) and determine its solution. If it cannot be found return (7.36).

$S4$: *Integral representation.* Perform the integration at the left hand side of (7.34). If it cannot be done return (7.37).

$S5$: *Return solution.* Solve the first order equation obtained in step $S4$ and return the result. If a solution cannot be found return the first order equation and the transformation functions.

It may occur that the above second order equation for z allows a larger symmetry group than just the one-parameter generator $u\partial_u$ as it is shown in the example below. In these cases it is advantageous to apply this larger symmetry for the further solution procedure.

EXAMPLE 7.19   The canonical equation obtained in Example 6.18 has the form (7.35). It turns out that after substituting $v' = z$ the resulting equation

$$z'' - \frac{3}{z}z'^2 - \frac{5}{u}z' - \frac{z(z^2+4)}{u^2} = 0 \qquad (7.38)$$

belongs to symmetry class $\mathcal{S}_8^2$. Consequently, algorithm *LieSolve2.8* may be applied. It yields the solution $z = \dfrac{4C_1}{\sqrt{C_2 u^4 + u^2 - 4C_1^2}}$, and finally

$$x = \int \frac{dy}{\sqrt{C_1 y^2 + C_2 - \frac{1}{4}y^4}} + C_3$$

in actual variables. In Exercise 7.5 the same result is obtained from the first order equation (7.36). $\quad\square$

A quasilinear third order ode in symmetry class $\mathcal{S}_{2,3}^3$ has the form

$$v''' + v'r\left(\frac{v''}{v'}, u\right) = 0. \qquad (7.39)$$

Substituting first $v' = z$ yields

$$z'' + zr\left(\frac{z'}{z}, u\right) = 0 \quad \text{or} \quad \frac{z''}{z} + r\left(\frac{z'}{z}, u\right) = 0.$$

332

Finally substituting $\frac{z'}{z} = w$ leads to

$$w' + w^2 + r(w, u) = 0. \tag{7.40}$$

If this general first order equation may be solved, and $w = w_0(u, C_1)$ is its general solution, the solution of (7.39) is

$$v = C_2 \int \exp\left(\int w_0(u, C_1)du\right)du + C_3. \tag{7.41}$$

The subsequent algorithm may terminate in step $S2$ without generating the desired solution if the first order ode for σ cannot be solved, or in step $S3$ if equation (7.40) cannot be solved.

ALGORITHM 7.15   *LieSolve3.2.3(w).* Given a quasilinear third order ode w in symmetry class $\mathcal{S}_{2,3}^3$, its general solution or some lower order equation is returned.

$S1$: *Equations for σ and ρ.* Generate a Janet basis for the symmetry generators of w, and set up equation (6.41) for σ and equations (6.39),...., (6.43) for ρ.

$S2$: *Determine transformation functions.* Choose σ or solve (6.41). If σ is not known at this point return *failed*. Otherwise determine independent solution for ρ.

$S3$: *Second order equation.* By means of the transformation functions obtained in $S2$ generate equation (7.40) and determine its solution. If it cannot be found return (7.40).

$S4$: *Integral representation.* Perform the integration at the left hand side of (7.41). If it cannot be done return (7.41).

$S5$: *Return solution.* Solve the first order equation obtained in step $S4$ and return the result. If a solution cannot be found return the first order equation and the transformation functions.

EXAMPLE 7.20   The canonical equation obtained in Example 6.19 has the form (7.39) from which the Abel equation (7.40) $w' + w^3 + w^2 + u + 1 = 0$ follows. Because a symmetry cannot be found for it, this first order equation is the best possible answer in the realm of symmetry analysis. ☐

A quasilinear third order ode in symmetry class $\mathcal{S}_{2,4}^3$ has the form

$$v''' + r(v'', u) = 0. \tag{7.42}$$

Substituting $v'' = w$ the general first order equation

$$w' + r(w, u) = 0 \tag{7.43}$$

is obtained. If $w = w_0(u, C_1)$ is its general solution, the solution of (7.42) is

$$v = \int \int w_0(u, C_1)du\, du + C_2 u + C_3. \tag{7.44}$$

Solution Algorithms   333

The subsequent algorithm applies these steps. It may fail in step $S2$, or a first order equation may be returned in step $S3$ as the best possible answer of the symmetry analysis.

ALGORITHM 7.16   *LieSolve3.2.4(ω)*. Given a quasilinear third order ode ω in symmetry class $S_{2,4}^3$, its general solution or some lower order equation is returned.

$S1$: *Equations for R or S.* Generate a Janet basis for the symmetry generators of ω and set up equations (6.44), (6.45) or (6.46).

$S2$: *Determine R or S.* From the equations obtained in $S1$ determine rational solution for R or S. If none is found return *failed*.

$S2$: *Determine transformation functions.* Applying the solution obtained in $S2$ determine σ and ρ by integration.

$S4$: *First order equation.* Generate the first order equation (7.43) with the help of σ and ρ. If a solution cannot be found return this equation.

$S5$: *Integral representation.* Using the solution obtained in $S4$ return the integral representation (7.44) and the transformation functions.

EXAMPLE 7.21   The canonical equation obtained in Example 6.20 has the form (7.42) from which the Riccati equation (7.43) $w' + uw^2 + w + 1 = 0$ follows. Because a closed form solution cannot be found it is the best possible result of the symmetry analysis. $\Box$

Three-Parameter Symmetries. The general solution procedure for the equations covered in this subsection has been described in Theorem 7.1 in the introduction to this section. According to the classification of symmetries given in Theorem 5.13 and the listing of groups and Lie algebras on page 135, the first two symmetry classes with nonsolvable Lie algebras of symmetry generators require solving a Riccati equation, i.e., equations (7.46) and (7.49) below. In the remaining cases with a solvable symmetry algebra the solution reduces essentially to integrations.

The first symmetry class $S_{3,1}^3$ has a canonical form

$$v''' + \frac{6v''(v'+1)}{u-v} + \frac{6v'(v'^2 + 4v' + 1)}{(u-v)^2} + \frac{v'^2}{(u-v)^2} r\left(\frac{v''(u-v) + 2v'(v'+1)}{v'\sqrt{v'}}\right) = 0.$$
$$(7.45)$$

Introducing $w = \sqrt{v'}$ and $z = \dfrac{v''(u-v) + 2v'(v'+1)}{v'\sqrt{v'}}$ as new variables, the Riccati equation

$$\frac{dw}{dz} + \frac{w^2 - \frac{1}{2}zw + 1}{r(z) - \frac{3}{2}z^2 - 12} = 0 \qquad (7.46)$$

for $w \equiv w(z)$ is obtained. If its general solution may be determined in the form $\Phi(v', z) = C_1$, two more independent relations between u, v, v' and z may be obtained by differentiation, see Lie [109], page 288.

$$(v-u)v'\Phi_{v'} = C_2 \text{ and } (v-u)v'\big[(v-u)v'\Phi_{v'v'} + 2v\Phi_{v'}\big] = C_3. \qquad (7.47)$$

334

From these relations v may be obtained by elimination. These steps are applied in the subsequent algorithm. It may terminate in step $S2$ if a rational solution of the respective Riccati equations may not be found, or in step $S4$ if equation (7.46) cannot be solved.

ALGORITHM 7.17 *LieSolve3.3.1(ω).* Given a quasilinear third order ode ω in symmetry class $\mathcal{S}^3_{3,1}$, its general solution is returned; or *failed* if it cannot be determined.

$S1$: *Equations for R and S.* Generate Janet basis for the symmetry generators of u and set up system (6.13), (6.16) or (6.18) for R, or system (6.20) for R and S respectively.

$S2$: *Determine R and S.* From the equations obtained in $S1$ determine a rational solution for R and S. If none can be found return *failed.*

$S3$: *Determine transformation functions.* Applying the result of $S2$ set up the system (6.14), (6.17), (6.19) or (6.21) for σ and ρ respectively and determine two independent Liouvillian solutions.

$S4$: *Riccati equation.* Applying the transformation functions obtained in step $S3$ generate equation (7.46). If its solution cannot be determined return this equation.

$S5$: *Return solution.* Set up system (7.47), eliminate v, resubstitute the actual variables and return the result.

EXAMPLE 7.22   The canonical equation obtained in Example 6.21 has the form (7.45) with $r(z) = z^2 + 1$ and z as given above. It leads to the Riccati equation (7.49) $w' - \dfrac{2w^2 - zw + 2}{z^2 + 22} = 0$ for $w(z)$. Because a closed form solution cannot be found it is the best possible result of the symmetry analysis. $\quad\Box$

A quasilinear third order ode in symmetry class $\mathcal{S}^3_{3,2}$ has the canonical form

$$v'''v^5 + 3v''v'v^4 + r(v''v^3) = 0. \tag{7.48}$$

If the two lowest invariants $w \equiv vv'$ and $z \equiv v''v^3$ of the two-parameter subgroup $\{\partial_u, 2u\partial_u + v\partial_v\}$ are introduced as new dependent and independent variables and $\dfrac{dw}{du} = vv'' + v'^2$, $\dfrac{dz}{du} = v'''v^3 + 3v''v'v^2$ are applied, (7.48) assumes the form

$$\frac{dw}{dz} + \frac{1}{r(z)}w^2 + \frac{z}{r(z)} = 0. \tag{7.49}$$

This Riccati equation for $w(z)$ has to be solved. If $w = w_0(z, C)$ is its general solution, in the original variables u and v a second order equation of the form

$$w_0(v''v^3, C) - vv' = 0$$

is obtained allowing the above mentioned two-parameter symmetry group. Because it is not in the canonical form as applied in Theorem 5.8, this second

Solution Algorithms   335

order ode is not in canonical form (7.17). Introducing new variables $\bar{u}$ and $\bar{v}(\bar{u})$ by $u = \bar{v} - \bar{u}$ and $v = \sqrt{\bar{u}}$ yields $\partial_{\bar{v}}$ and $\bar{u}\partial_{\bar{u}} + \bar{v}\partial_{\bar{v}}$, and the second order equation

$$w_0 \left(-\frac{2\bar{u}\bar{v}'' + \bar{v}' - 1}{4(\bar{v}' - 1)^3}, C \right) - \frac{1}{2(\bar{v}' - 1)} = 0. \tag{7.50}$$

Under suitable constraints for the function w_0 it may be treated by the algorithm *LieSolve2.2.2* on page 322. These steps are applied in the algorithm below. It may terminate in step $S2$ if a rational solution for R cannot be found, or in step $S5$ if (7.49) cannot be solved, or if (7.50) is too complicated.

ALGORITHM 7.18   *LieSolve3.3.2(ω)*. Given a quasilinear third order ode ω in symmetry class $\mathcal{S}_{3,2}^3$, its general solution is returned; or *failed* if it cannot be found.

$S1$: *Equations for R.* Generate Janet basis for the symmetry generators of ω. Set up system (6.22), (6.23) or (6.24) for R from it.

$S2$: *Determine R.* From the equations obtained in $S1$ determine a rational solution for R. If it cannot be found return *failed*.

$S3$: *System for transformation functions.* Applying the result of $S2$ set up the systems (6.22), (6.23) or (6.24) for σ and ρ.

$S4$: *Determine transformation functions.* Determine two independent Liouvillian solutions for the system obtained in $S3$.

$S5$: *Riccati equation.* Generate equation (7.49) and determine a rational solution. If it cannot be found return this equation.

$S6$: *Return solution.* Employing the solution obtained in $S5$, generate (7.50), apply algorithm *LieSolve2.2.2* to it and return the result.

EXAMPLE 7.23   The canonical equation obtained in Example 6.22 has the form (7.48) with $r(v''v^3) = 2v''v^3 + 1$ and leads to the Riccati equation (7.49) $\frac{dw}{dz} + \frac{1}{2z+1}w^2 + \frac{z}{2z+1} = 0$. Because a closed form solution cannot be found it is the best possible result of the symmetry analysis.   ⬜

A quasilinear third order ode in symmetry class $\mathcal{S}_{3,3}^3$ has the form

$$v''' + v'^{(c-3)/(c-1)} r \left(v'' v'^{-(c-2)/(c-1)} \right) = 0. \tag{7.51}$$

With $v''' = \dfrac{dv''}{dv'} v''$ it may be rewritten as a first order ode for $v''(v')$

$$\frac{dv''}{dv'} v'' + v'^{(c-3)/(c-1)} r \left(v'' v'^{-(c-2)/(c-1)} \right) = 0$$

with the symmetry $(c - 1)v'\partial_{v'} + (c - 2)v''\partial_{v''}$. Applying Theorem 6.1, new variables $\bar{u}$ and $\bar{v}(\bar{u})$ are introduced by $v' = \bar{u}^{-(c-1)/(c-2)} e^{(c-1)\bar{v}}$ and $v'' = e^{(c-2)\bar{v}}$ with the result

$$\bar{v} = \frac{c-1}{c-2} \int \frac{1}{(c-1)r(\bar{u}) + (c-2)\bar{u}^2} \frac{d\bar{u}}{\bar{u}} + C_1 \equiv R(\bar{u}) + C_1. \tag{7.52}$$

336

If the original variables are resubstituted by $\bar{u} = v''v'^{-(c-2)/(c-1)}$ and $\bar{v} = \frac{1}{c-2}\log v''$, the equation

$$R\left(\frac{v''}{v'^{(c-2)/(c-1)}}\right) - \frac{1}{c-2}\log v'' + C_1 = 0 \qquad (7.53)$$

is obtained. If R does not involve integrals, this is a second order ode allowing the two symmetries ∂_u and ∂_v. Under suitable constraints for R it may be treated by the algorithm $LieSolve2.2.1$ on page 322. The subsequent algorithm is designed according to this scheme. The only bottleneck where it may fail is the integral (7.52).

ALGORITHM 7.19 $LieSolve3.3.3(\omega)$. Given a third order quasilinear ordinary differential equation in symmetry class $\mathcal{S}_{3,3}^3(c)$, its general solution is returned, or equation (7.53) if it cannot be determined.

S1 : *Janet basis for symmetries.* Generate a Janet basis for the symmetry generators of ω and set up system (6.25), (6.26), (6.27) or (6.28) respectively.

S2 : *Determine transformation functions.* Determine two independent Liouvillian solutions for the system obtained in $S2$.

S3 : *Integral representation.* Generate (7.53). If the integration may not be executed in closed form return this integral.

S4 : *Return solution.* If (7.53) is in $\mathcal{S}_{2,2}^2$, apply algorithm $LieSolve2.2.1$ to it and return the result. Otherwise return (7.53).

EXAMPLE 7.24   If into the canonical form $v'''v'^3 + 1 = 0$ of Example 6.23 a new dependent variable is introduced by $v' = w$, the second order equation $w''w^3 + 1 = 0$ in symmetry class $\mathcal{S}_{3,2}^2$ is obtained. Applying (7.25) an additional integration yields its solution

$$v = \tfrac{1}{2}(u + C_2)\sqrt{C_1^2(u + C_2)^2 - 1}$$
$$- \frac{1}{2C_1}\log\left[C_1(u + C_2) + \sqrt{C_1^2(u + C_2)^2 - 1}\right] + C_3.$$

Substituting the canonical variables determined in Example 6.23, the general solution

$$y = \tfrac{1}{2x}(C_2 x + 1)\sqrt{C_1^2(C_2 x + 1)^2 - x^2} + \frac{1}{2C_1}\log x$$
$$- \frac{1}{2C_1}\log C_1(C_2 x + 1) + \sqrt{C_1^2(C_2 x + 1)^2 - x^2} + C_3$$

of the equation in Example 5.34 is finally obtained.   ⬜

A quasilinear third order ode in symmetry class $\mathcal{S}_{3,4}^3$ has the form

$$v''' + e^{-2v'}r(v''e^{v'}) = 0. \qquad (7.54)$$

$$\text{Solution Algorithms} \qquad 337$$

Writing $v''' = \dfrac{dv''}{dv'} v''$ yields the first order equation

$$\frac{dv''}{dv'} + \frac{e^{-2v'}}{v''} r(v''e^{v'}) = 0 \tag{7.55}$$

for $v''(v')$ with the symmetry $\partial_{v'} - v'' \partial_{v''}$. Applying Theorem 6.1, new variables $\bar{u}$ and $\bar{v}(\bar{u})$ are introduced by $v' = \log \bar{u} + \bar{v}$ and $v'' = e^{-\bar{v}}$ with the result

$$\bar{v} = \int \frac{r(\bar{u})}{\bar{u}^2 - r(\bar{u})} \frac{d\bar{u}}{\bar{u}} + C_1 \equiv R(\bar{u}) + C_1. \tag{7.56}$$

Upon resubstitution of the original variables by $\bar{u} = v'' e^{v'}$ and $\bar{v} = -\log v''$, the equation

$$R(v'' e^{v'}) + \log v'' + C_1 = 0 \tag{7.57}$$

is obtained. If R does not involve integrals, this is a second order ode allowing the two symmetries ∂_u and ∂_v. Similar to the preceding case, the only bottleneck where the algorithm below may fail is the integral (7.56).

ALGORITHM 7.20   *LieSolve3.3.4(ω).* Given a third order quasilinear ordinary differential equation ω in symmetry class $\mathcal{S}_{3,4}^3$, the general solution is returned, or equation (7.56) if it cannot be determined.

S1 : Janet basis for symmetries. Generate a Janet basis for the symmetry generators of ω and set up system (6.29), (6.30), (6.31) or (6.32) respectively.

S2 : Determine transformation functions. Determine two independent Liouvillian solutions for the system obtained in $S2$.

S3 : Integral representation. Generate (7.57). If the integration may not be performed return this integral.

S4 : Return solution. Apply algorithm *LieSolve2.2.1* to the equation obtained in $S3$ and return the result.

EXAMPLE 7.25   The canonical equation obtained in Example 6.24 has the form (7.54) with $r(v'' e^{v'}) = v''^2 e^{2v'} + 1$. In step $S2$ the equation $v'' - \sqrt{2}\sqrt{v' - C_1} e^{-v'} = 0$ is obtained. Finally, step $S4$ yields the parameter representation $u = f(p) + C_2$ and $v = \int f(p) dp - pf(p) + C_3$ where

$$f(p) = -\frac{1}{\sqrt{2}} \int \frac{e^p dp}{\sqrt{p - C_1}}. \qquad \qquad \qquad \Box$$

A quasilinear third order ode in symmetry class $\mathcal{S}_{3,5}^3$ has the form

$$v''' + v' r \left(\frac{v''}{v'} \right) = 0. \tag{7.58}$$

Substituting $\dfrac{v''}{v'} = z$ yields the first order equation $\dfrac{dz}{du} + z^2 + r(z) = 0$ from which

$$\int \frac{dz}{z^2 + r(z)} + u = C_1 \tag{7.59}$$

338

follows. If the integration may be performed and the result may be solved for z, i.e., $z = z_0(u, C_1)$, the second order linear equation

$$v'' - z_0(u, C_1)v' = 0 \qquad (7.60)$$

is obtained with the general solution

$$v = C_2 \int \exp \left(\int z_0(u, C_1) du \right) du + C_3. \qquad (7.61)$$

The algorithm below is based on this solution scheme. It may fail in step $S3$ if the integration in (7.59) may not be executed, or if the result of the integration may not be solved for z. In these cases an expression involving v'' and v' is returned. On the other hand, if equation (7.60) may be obtained, the general solution is returned.

ALGORITHM 7.21 $LieSolve3.3.5(\omega)$. Given a quasilinear third order ode ω in symmetry class $S^3_{3,5}$, its general solution is returned.

$S1$: *Janet basis for symmetries.* Generate Janet basis for symmetry generators of ω and set up equations (6.48), (6.49), (6.50), (6.51) or (6.52) respectively.

$S2$: *Determine transformation functions.* From the equations obtained in $S2$ determine two independent Liouvillian solutions.

$S3$: *Integral representation.* Applying the solutions obtained in $S3$ generate the equation (7.59). If $z_0(u, C_1)$ cannot be obtained, return this expression.

$S4$: *Return solution.* Generate equation (7.60) and its solution (7.61), resubstitute actual variables and return the result.

The symmetry class $S^3_{3,5}$ is one of the few cases of third order equations considered by Lie [113], pages 556-558. He uses canonical variables $\bar{u} = v$ and $\bar{v} = u$ such that the group generators have the form $\partial_{\bar{u}}$, $\partial_{\bar{v}}$ and $\bar{u}\partial_{\bar{u}}$ with invariants $\dfrac{\bar{v}''}{\bar{v}'^2}$, $\dfrac{\bar{v}'''}{\bar{v}'^3}$ and the canonical equation $v''' + \bar{v}''r \left(\dfrac{\bar{v}''}{\bar{v}'^3} \right) = 0$. Its integration leads finally to the same answer (7.61). Of particular interest are Lie's remarks on page 556, line 7 to 10 from the bottom, concerning the comparison of this proceeding with his second integration method described on page 327.[1]

EXAMPLE 7.26   The canonical form obtained in Example 6.25 corresponds to $r(z) = z^2 - 2$ and the solution $z = \dfrac{Ce^{4u} + 1}{Ce^{4u} - 1}$. Substituting it into the second order equation $v'' - zv' = 0$ yields

$$v = C_1 + C_2 u^2 + \frac{1}{2} \int \left[\log \left(C_3 e^u - 1 \right) + \log \left(C_3 e^u + 1 \right) + \log \left(C_3^2 e^{2u} + 1 \right) \right] du.$$

---

[1]Translation by the author: *In certain cases this method requires simpler operations than the above... .*

Solution Algorithms

339

The four symmetry classes $S_{3,6}^3$ to $S_{3,8}^3$ and $S_{3,9}^3(c)$ arise from the groups $\mathbf{g}_{17}$ in Chapter 3, page 139. Their common canonical form

$$\frac{dz}{du} + r(z) = 0 \quad \text{where} \quad z(u) \equiv v'' + c_1 v' + c_2 v \tag{7.62}$$

is linear and homogeneous in $v \equiv v(u)$, v' and v'', with constant coefficients c_1 and c_2. This first order equation for z may always be integrated with the result

$$\int \frac{dz}{r(z)} + u + C \equiv R(z) + u + C = 0. \tag{7.63}$$

This relation determines z either explicitly in the form $z = z_0(u, C)$ or implicitly as function of u. In any case v is the solution of the linear equation

$$v'' + c_1 v' + c_2 v = z_0(u, C). \tag{7.64}$$

The left hand side of this equation is always completely reducible (see Theorem 6.16). Therefore its general solution may be written as follows.

$$S_{3,6}^3 : \qquad v = C_1 + C_2 u + \int \int z_0(u, C) du du,$$

$$S_{3,7}^3 : \quad v = C_1 e^u + C_2 u e^u - (u+1) e^{-u} z_0(u, C) + \int (u+1) e^{-u} z_0(u, C) du,$$

$$S_{3,8}^3 : \quad v = C_1 + C_2 e^u + z_0(u, C) + \int z_0(u, C) du - e^u \int e^{-u} z_0'(u, C) du,$$

$$S_{3,9}^3(c) : \quad v = C_1 e^u + C_2 e^{cu} + \frac{e^u}{C} \int e^{-u} z_0(u, C) du - \frac{e^{cu}}{C} \int e^{-cu} z_0(u, C) du. \tag{7.65}$$

Resubstituting the actual variables into these expressions finally yields the solution of the given equation. Due to the similiar course of the solution procedure for all four groups under consideration, the various steps have been put together in the following algorithm. It may fail in step $S3$ if a hyperexponential solution for ρ cannot be found.

ALGORITHM 7.22 *LieSolve3.3.6to9(ω)*. Given a quasilinear third order ode ω in one of the symmetry classes $S_{3,6}^3$ to $S_{3,8}^3$ or $S_{3,9}^3(c)$, its general solution is returned; or *failed* if it cannot be determined.

$S1$: *Equations for transformation functions.* Generate Janet basis for symmetry generators of ω and set up equations for σ and ρ as described in Theorems 6.20, 6.21, 6.22 or 6.23 respectively.

$S2$: *Determine transformation functions.* From the equations obtained in $S1$ determine a pair of independent Liouvillian solutions for σ and hyperexponential solutions for ρ. If the latter may not be found return *failed*.

$S3$: *Canonical form.* Applying the solutions obtained in $S2$ generate the canonical form (7.64).

$S4$: *Return solution.* Substitute the value of z_0 and the actual variables into the appropriate expression (7.65) and return the result.

EXAMPLE 7.27   The canonical form equation $\frac{dz}{dx} + r(z) = 0$ with $r(z) = -\frac{z^2+1}{z}$ that has been obtained in Example 7.27 yields $R(z) = \frac{1}{2}\log z^2 + 1$ and, consequently, $z_0(u, C) = \sqrt{Ce^{-2u} - 1}$. This value for $z_0(u, C)$ has to be substituted into the expressions for v given above. Finally resubstituting the actual variables by $u = \frac{1}{x}$ and $v = y$ and renaming C into C_3 yields the solution

$$y = C_1 + C_2\frac{1}{x} + \int \left(\int \bar{z}_0(x, C_3)\frac{dx}{x^2} \right) \frac{dx}{x^2}$$

of the first ode's in Example 5.37 with $\bar{z}_0(x, C_3) = \sqrt{C_3 \exp\left(-\frac{2}{x}\right) - 1}$. Proceeding along the same lines, the answer for the remaining three ode's is

$$v = C_1 \exp\left(\frac{1}{x}\right) + C_2\frac{1}{x}\exp\left(\frac{1}{x}\right) - \left(1 + \frac{1}{x}\right)\exp\left(-\frac{1}{x}\right)\bar{z}_0(x, C_3)$$

$$+ \int \left(1 + \frac{1}{x}\exp\left(-\frac{1}{x}\right)\bar{z}_0(x, C_3)\right)\frac{dx}{x^2},$$

$$v = C_1 + C_2 \exp\left(\frac{1}{x}\right) + \bar{z}_0(x, C_3) - \int \bar{z}_0(x, C_3)\frac{dx}{x^2}$$

$$+ \exp\left(\frac{1}{x}\right) \int \exp\left(-\frac{1}{x}\right)\bar{z}_0'(x, C_3)dx,$$

$$v = C_1 \exp\left(\frac{1}{x}\right) + C_1 \exp\left(\frac{c}{x}\right) + \frac{1}{c}\left[\exp\left(\frac{1}{x}\right)\int \exp\left(-\frac{1}{x}\right)\bar{z}_0(x, C_3)\frac{dx}{x^2}\right.$$

$$\left. - \exp\left(\frac{c}{x}\right)\int \exp\left(-\frac{c}{x}\right)\bar{z}_0(x, C_3)\frac{dx}{x^2}\right]$$

respectively.   □

A quasilinear third order ode in symmetry class $S_{3,10}^3$ has the canonical form

$$v''' + v''^2 r(v') = 0. \tag{7.66}$$

Substituting $v''' = v''\frac{dv''}{dv'}$ yields the linear first order equation $\frac{dv''}{dv'} + r(v')v'' = 0$ for $v''(v')$. Integration leads to

$$v'' + C_1 \exp\left(-\int r(v')dv'\right) = 0. \tag{7.67}$$

If the integral may be performed, this is a second order ode in symmetry class $S_{2,1}^2$. For special forms of $r(v')$ its solution may be obtained explicitly. In

$$\text{Solution Algorithms} \qquad 341$$

any case, defining $R(p) = \int \exp\left(\int r(p)dp\right)dp$ the expressions (7.16) yield the parameter representation

$$u + C_1 R(p) = C_2, \quad v + C_1 p R(p) = C_1 \int R(p)dp + C_3. \qquad (7.68)$$

The following algorithm is designed according to this scheme. For any equation in this symmetry class at least the parameter representation (7.68) is returned.

ALGORITHM 7.23 $\quad$ *LieSolve3.3.10(ω)*. Given a quasilinear third order ode ω in symmetry class $\mathcal{S}^3_{3,10}$, its general solution is returned either explicitly or as a parameter representation

$\quad$ *S1* : *Equations for transformation functions.* Generate Janet basis for symmetry generators of ω and set up equations (6.56) for σ and ρ.

$\quad$ *S2* : *Determine transformation functions.* From the equations obtained in *S1* determine a pair of independent Liouvillian solutions or *failed* if it may not be found.

$\quad$ *S3* : *Canonical form.* Applying the solutions obtained in *S3* generate the equation (7.67).

$\quad$ *S4* : *Return solution.* Apply algorithm *LieSolve2.2.1* to the equation obtained in *S3* and return the result.

EXAMPLE 7.28 $\quad$ The canonical form of the ode considered in Example 6.27 in symmetry class $\mathcal{S}^3_{3,10}$ corresponds to $r(v') = v' + \frac{1}{v'}$. For this particular choice of r the integrals at the left hand side of (7.67) may be executed in closed form with the result $v'^2 = 2\log(C_1 u + C_2)$. It leads to the general solution

$$v = \sqrt{2}\int \sqrt{\log(C_1 u + C_2)}\,du + C_3.$$

Substituting $u = \frac{1}{x}$ and $v = y$ yields the solution

$$y = -\sqrt{2}\int \sqrt{\log\left(C_1\frac{1}{x} + C_2\right)}\frac{1}{x^2}dx + C_3$$

of the ode in Example 5.38. $\qquad\qquad\qquad\qquad\qquad\qquad\qquad\qquad$ ⬜

Four-Parameter Symmetries. In the remaining part of this section the number of symmetries exceeds the order of the given equation at least by one. Consequently, the solution procedure is simplified. It requires only algebraic operations.

The first symmetry class $\mathcal{S}^3_{4,1}$ has a canonical form

$$v'''v' + av''^2 = 0 \qquad (7.69)$$

with constant $a \neq 0$, $a \neq -\frac{3}{2}$. Dividing by $v''v'$ and performing three consecutive integrations leads to the general solution

$$(a+2)^{a+1}(C_1 v + C_2)^{a+1} = (a+1)^{a+2}(C_1 u + C_3)^{a+2}.$$

342

In this form the invariance under exchange of u and v combined with the replacement of a by $-(a+3)$ is obvious. Usually the explicit form

$$v = (C_1 u + C_2)^{(a+2)/(a+1)} + C_3 \tag{7.70}$$

with new constants C_k is more convenient for solving equations. The constant a together with the transformation functions has to be substituted into (7.70) in order to generate the solution for the given equation.

The subsequent algorithm applies these steps for obtaining the general solution. For any equation in this symmetry class it is guaranteed that the solution is found.

ALGORITHM 7.24 *LieSolve3.4.1(ω).* Given a quasilinear third order ode ω in symmetry class $\mathcal{S}_{4,1}^3$, its general solution is returned.

S1 : *Equations for transformation functions.* Generate Janet basis for symmetry generators of ω and set up equations (6.57), (6.58), (6.59) or (6.60) respectively depending on the type of the Janet basis.

S2 : *Determine transformation functions.* From the equations obtained in S1 generate an independent pair of solutions.

S3 : *Canonical form.* Applying the solutions obtained in S2 generate the canonical form (7.69).

S4 : *Return solution.* Substitute the value of a and the actual variables into (7.70) and return the result.

EXAMPLE 7.29 Substituting the values $a = -\frac{1}{2}$ and $u = xy$, $v = \frac{1}{y}$ into (7.70) ensures the solution $(C_1 xy + C_2)^3 y + C_3 y + 12 C_1 = 0$, i.e., a third order algebraic function. It is easy to see that $a = -\frac{5}{2}$ and $u = \frac{1}{y}$, $v = xy$ lead to the same answer. $\qquad\square$

The canonical form of an ode in symmetry class $\mathcal{S}_{4,2}^3$ is

$$v'''v + 3v''v' + av''\sqrt{v''v} = 0. \tag{7.71}$$

Its symmetry group contains $\{\partial_u, 2u\partial_u + v\partial_v, u^2\partial_u + uv\partial_v\}$ as a subgroup. Writing (7.71) as $v'''v^5 + 3v''v'v^4 + a(v''v^3)^{\frac{3}{2}} = 0$ shows that it is a special case of (7.48) with $r(v''v) = a(v''v^3)^{\frac{3}{2}}$. This special form of r is a consequence of the additional symmetry $v\partial_v$. Applying the same transformation which led to (7.49) yields a special Riccati equation with an additional symmetry. It is more efficient however to introduce the new dependent variable $w = \frac{v'}{v}$ which has to satisfy

$$w'' + 6ww' + 4w^3 + a(w' + w^2)^{3/2} = 0.$$

Its general solution is rational and leads to the general solution

$$v = C_1 \exp \int \frac{a^2(x + C_2)dx}{a^2 x(x + C_3) + 4C_2^2 - (a^2 - 16)C_2(C_2 - C_3)} \tag{7.72}$$

Solution Algorithms 343

of (7.71). Another method for arriving at this result is discussed in Exercise 7.11.

The subsequent algorithm implements the various steps. The desired solution is only returned if rational transformation functions in step $S2$ may be found.

ALGORITHM 7.25 *LieSolve3.4.2(ω).* Given a quasilinear third order ode ω in symmetry class $S_{4,2}^3$, its general solution is returned; or *failed* if it cannot be determined.

$S1$: *Equations for R.* Generate Janet basis for symmetry generators of ω and set up equations for R according to (6.61) or (6.62) depending on the type of the Janet basis.

$S2$: *Determine R.* From the equations obtained in $S1$ determine rational solution for R. If none may be found return *failed*.

$S3$: *Equations for transformation functions.* Applying the solutions obtained in $S2$ set up equations (6.61) or (6.61) for σ and ρ.

$S4$: *Return solution.* Applying the transformation function obtained in $S3$ generate (7.72), resubstitute the actual variables and return the result.

For an ode in symmetry class $S_{4,3}^3(c)$ it is assumed that the value of the parameter c is already known. The canonical form of an equation in this symmetry class is

$$v''' + av''^\alpha = 0 \quad \text{with} \quad \alpha = \frac{c-3}{c-2}. \tag{7.73}$$

Substituting $v'' = w$ leads to $w' + aw^\alpha = 0$. Three successive integrations yield the general solution

$$v = \frac{1}{a^2(2-\alpha)(3-2\alpha)}(\alpha-1)^{(3-2\alpha)/(1-\alpha)}(au + C_1)^{(3-2\alpha)/(1-\alpha)} + C_2 u + C_3. \tag{7.74}$$

The subsequent algorithm implements the various steps. The desired solution is only returned if hyperexponential transformation functions in step $S2$ may be found.

ALGORITHM 7.26 *LieSolve3.4.3(ω).* Given a quasilinear third order ode ω in symmetry class $S_{4,3}^3(c)$, its general solution is returned; or *failed* if it cannot be determined.

$S1$: *Equations for transformation functions.* Generate Janet basis for the symmetry generators of ω and set up equations (6.63), (6.64) or (6.65) respectively depending on the type of Janet basis.

$S2$: *Determine transformation functions.* From the equations obtained in $S1$ generate an independent pair of hyperexponential solutions. If none is found return *failed*.

$S3$: *Canonical form.* Applying the solutions obtained in $S3$ generate the canonical form (7.73).

344

$S4$: *Return solution.* Substitute the value of a and the actual variables into (7.74) and return the result.

EXAMPLE 7.30   Substituting the values $a = 1$, $c = \frac{3}{2}$, i.e., $\alpha = 3$ and $u = \frac{1}{x}$, $v = xy$ into (7.74) ensures the general solution

$$y = \frac{2\sqrt{2}}{3x}\left(C_1 + \frac{1}{x}\right)^{3/2} + \frac{C_2}{x^2} + \frac{C_3}{x}. \qquad \Box$$

Finally the canonical form of an ode in symmetry class $S_{4,4}^3$ is

$$v''' + a\exp\left(-\tfrac{1}{2}v''\right) = 0. \tag{7.75}$$

Substituting $v'' = w$ leads to $w' + a\exp\left(-\frac{1}{2}w\right) = 0$. Similar to the preceding case, three succesive integrations lead to

$$v = (u - C_1)^2\log(u - C_1) + \left(\log\left(-\tfrac{1}{2}a\right) - \tfrac{3}{2}\right)u^2 + C_2 u + C_3. \tag{7.76}$$

The subsequent algorithm implements the various steps. The desired solution is only returned if hyperexponential transformation functions in step $S2$ may be found.

ALGORITHM 7.27   *LieSolve3.4.4(ω).* Given a quasilinear third order ode ω in symmetry class $S_{4,4}^3$, its general solution is returned; or *failed* if it cannot be determined.

$S1$: *Equations for transformation functions.* Generate a Janet basis for the symmetry generators of ω and set up equations (6.66), (6.67), (6.68) or (6.69).

$S2$: *Determine transformation functions.* From the equations obtained in $S1$ generate an independent pair of hyperexponential solutions. If none is found return *failed*.

$S3$: *Canonical form.* Applying the solutions obtained in $S3$ generate the canonical form (7.75).

$S4$: *Return solution.* Substitute the value of a and the actual variables into (7.76) and return the result.

Six-Parameter Symmetry S_6^3. In canonical variables an ode in symmetry class S_6^3 has the form

$$v'''v' - \tfrac{3}{2}v''^2 = 0. \tag{7.77}$$

The subgroup $\{\partial_u, \partial_v, u\partial_u, v\partial_v\}$ may be utilized for its integration. The special value $a = -\frac{3}{2}$ in (7.70) leads to the general solution

$$v = \frac{C_1}{u + C_2} + C_3$$

after a suitable change of the integration constants. The algorithm below applies these steps. It may fail if rational solutions for R and S cannot be found.

$$\text{Solution Algorithms} \qquad 345$$

ALGORITHM 7.28   *LieSolve3.6(ω)*. Given a quasilinear third order ode ω in symmetry class $\mathcal{S}_6^3$, its general solution is returned; or *failed* if it cannot be determined.

$S1$: *Equations for R and S*. If u has the form (5.40) set up the equations from (6.70) and (6.71). If ω has the form (5.41), set up equations from (6.72) and (6.73).

$S2$: *Determine R and S*. From the equations obtained in $S1$ determine a rational solution for R and S. If none may be found return *failed*.

$S3$: *System for transformation functions*. Applying the result of $S2$ set up the system from (6.70), (6.71) or (6.72) and (6.73) for σ and ρ.

$S4$: *Determine transformation functions*. Determine two independent Liouvillian solutions for the system obtained in $S3$.

$S5$: *Return solution*. Substitute the solutions from $S4$ into (7.77) and return the result.

EXAMPLE 7.31   For the equation with symmetry type $\mathcal{S}_6^3$ and structure (5.40) in Example 5.41 the transformation functions $\sigma = xy$ and $\rho = \frac{y}{x}$ have been obtained. Substituting these values into (7.77) leads to the general solution

$$x^3 y^2 + (C_1 x^2 + C_2) y + C_3 x = 0. \qquad \square$$

EXAMPLE 7.32   For the equation in symmetry class $\mathcal{S}_6^3$ and structure (5.41) of Example 5.42 the transformation functions $\sigma = \frac{1}{y}$ and $\rho = \frac{y}{x} e^x$ have been obtained. Substituting these values into (7.77) leads to the general solution

$$C_1 y^2 + (C_2 x e^{-x} + 1) y + C_3 x e^{-x} = 0. \qquad \square$$

7.4   Linearizable Third Order Equations

The equations considered in this section are assumed to be equivalent to a linear ode. Consequently, the methods described in the first section of Chapter 2 may be applied after the canonical form has been obtained.

Four-Parameter Symmetry. The canonical form for any equation in the only symmetry class $\mathcal{S}_{4,5}^3$ of this subsection is the generic linear third order equation in rational normal form

$$v''' + P(u)v' + Q(u)v = 0. \qquad (7.78)$$

The coefficients $P(u)$ and $Q(u)$ may assume quite different shapes by transformations of the structure invariance group $u = F(\bar{u})$ and $v = F'(\bar{u})\bar{v}$. The further proceeding depends on the structure of the differential Galois group of (7.78) which may allow it to obtain solutions explicitly, e. g. if it is reducible with a nontrivial Loewy decomposition. In general, however, it is irreducible

346

and the only information that is guaranteed for equations in this symmetry class is its linearizability.

ALGORITHM 7.29   *LieSolve3.45(ω)*. Given a quasilinear third order ode ω in symmetry class $\mathcal{S}^3_{4,5}$, its general solution is returned; or *failed* if it cannot be determined.

S1 : *Identify case a), b) or c)*. Determine the Janet basis for the Lie symmetries of ω. By means of the criteria given in Theorem 6.31 identify the case that applies for equation ω.

S2 : *Equations for σ and ρ*. From the Janet basis obtained in step S1 construct the system for ρ according to (6.74), (6.76) or (6.78) respectively. In case c) set up the first order ode for σ.

S3 : *Determine σ and ρ*. In case a) and b) choose σ, in case c) determine it from the first order ode. From the equations constructed in step S2 find a hyperexponential solution for ρ, if none exists return *failed*.

S4 : *Canonical form equation*. Applying the transformation functions from step S3 generate a canonical form equation, determine its Loewy decomposition and construct the most general explicit solution from it.

S5 : *Find solution*. Substitute $u = \sigma(x, y)$ and $v = \rho(x, y)$ into the canonical form solution obtained in step S4 and return the result.

EXAMPLE 7.33   The first canonical form (6.80) obtained in Example 6.32 has the Loewy decomposition $\left(v' + \frac{4u}{2u-1}v\right) Lclm\left(v' - 2v, v' - \frac{1}{u}v\right) = 0$ from which the general solution

$$v = C_1 u + C_2 e^{2u} + C_3 \left[e^{2u} \int \frac{u e^{-4u}}{(2u-1)^2} du - u \int \frac{e^{-2u}}{(2u-1)^2} du \right]$$

follows. Substituting the transformation $u = x$ and $v = \frac{x}{y}$ given there yields the solution of the equation in Example 5.43 in the form

$$y = \frac{x}{C_1 x + C_2 e^{2x} + C_3 \left[e^{2x} \int \frac{x e^{-4x}}{(2x-1)^2} dx - x \int \frac{e^{-2x}}{(2x-1)^2} dx \right]}. \qquad (7.79)$$

The second canonical form (6.81) has the Loewy decomposition

$$\left[v' + \left(\frac{1}{u-2} + \frac{1}{u} - \frac{2}{u^2} \right) v \right] Lclm \left(v' - \frac{1}{u}v, v' - \left(\frac{2}{u} - \frac{2}{u^2}\right)v \right) = 0$$

from which the solution

$$v = C_1 u + C_2 u^2 \exp \frac{2}{u} + C_3 \left[u^2 \exp \frac{2}{u} \int \frac{\exp\left(-\frac{4}{u}\right) du}{u(u-2)^2} - u \int \frac{\exp\left(-\frac{2}{u}\right) du}{(u-2)^2} \right]$$

is obtained. Substituting $u = \frac{1}{x}$ and $v = \frac{x}{y}$ yields again (7.79).   ⬜

Solution Algorithms 347

EXAMPLE 7.34   The equation considered in Example 5.44 generates the type $\mathcal{J}_{4,19}^{(2,2)}$ Janet basis, therefore case *b*) applies. If $\sigma = y$ is chosen, the system (6.74) yields the independent solution $\rho = \frac{x}{y}$ and the canonical form (6.80). Choosing $\sigma = \frac{1}{y}$ instead yields $\rho = \frac{1}{xy}$ with the canonical form (6.81). $\square$

EXAMPLE 7.35   The equation considered in Example 5.45 generates the type $\mathcal{J}_{4,17}^{(2,2)}$ Janet basis, therefore case *c*) applies. The first order equation for σ is $y' + 1 = 0$, i.e., $\phi(x,y) = x + y$. Choosing $\sigma = x + y$ yields a system for ρ with the fundamental system $\{0, x, y\}$. Consequently, any linear combination $\rho = c_1 x + c_2 y$ with $c_1 \neq c_2$ yields a possible transformation to a canonical form which turns out to be (6.80). $\square$

Five-Parameter Symmetries. The canonical form for the symmetry classes $\mathcal{S}_{5,1}^3$ and $\mathcal{S}_{5,2}^3(a)$ is $v''' - (a+1)v'' + av' = 0$ with a some number. This canonical form is particularly convenient because it is completely reducible for any value of a without introducing new algebraic numbers. If $a = 0$ it decomposes as

$$v''' - v'' = Lclm\left(v', v' - \frac{1}{u}v, v' - v\right) = 0 \qquad (7.80)$$

with the general solution $v = C_1 + C_2 u + C_3 e^u$, and if $a \neq 0$

$$v''' - (a+1)v'' + av' = Lclm(v', v' - v, v' - av) = 0 \qquad (7.81)$$

with the general solution $v = C_1 + C_2 e^u + C_3 e^{au}$. Substituting ρ and σ for v and u immediately yields the desired solution of the actual equation. These steps are organized in terms of the subsequent algorithm for both $\mathcal{S}_{5,1}^3$ and $\mathcal{S}_{5,2}^3(c)$ symmetries.

ALGORITHM 7.30   *LieSolve3.5(ω).* Given a quasilinear third order ode ω with any five-parameter group of Lie symmetries, its general solution is returned; or *failed* if it cannot be determined.

S1 : Identify case a), b) or c). Determine the Janet basis for the Lie symmetries of ω. By means of the criteria given in Theorem 6.11, identify the case that applies for the equation at issue.

S2 : Determine value of parameter a. Solve equation (5.42) and choose some value of a from it.

S3 : Equations for σ and ρ. From the Janet basis constructed in step $S1$ and the value for a obtained in step $S2$ construct the systems for σ and ρ according to (6.82) and (6.83), (6.84) and (6.32) or (6.85) and (6.86).

S4 : Determine σ and ρ. From the equations constructed in step $S3$ determine σ by integration. Find a hyperexponential solution for ρ, if none exists return *failed*.

S5 : Return solution. Substitute $u = \sigma(x,y)$ and $v = \rho(x,y)$ into the canonical form solution (7.80) or (7.81) and return the result.

348

EXAMPLE 7.36   The transformation functions $\sigma = xy$ and $\rho = \frac{y}{x}$ obtained in Example 6.35 yield the general solution

$$(C_3 x^2 + 1)y + C_2 e^{xy} x + C_1 x = 0$$

for the differential equation considered in Example 5.46.   ▯

EXAMPLE 7.37   The transformation functions $\sigma = y$ and $\rho = \frac{x}{y}$ obtained in Example 6.36 yield the general solution

$$\left(C_3 e^{3x} + C_2 e^y + C_1\right)y + x = 0$$

for the differential equation considered in Example 5.47.   ▯

Seven-Parameter Symmetry. In canonical variables u and $v \equiv v(u)$, a quasilinear third order ode in symmetry class $\mathcal{S}_7^3$ has the form $v''' = 0$ with the obvious solution $v = C_1 + C_2 u + C_3 u^2$. Therefore, the whole solution procedure comes down to generating the canonical form of the given equation, very much like for a second order equation with eight-parameter projective symmetry group. For third order equations the two alternatives of Theorems 5.18 and 6.33 have to be distinguished. The various steps are put together in the subsequent algorithm.

ALGORITHM 7.31   *LieSolve3.7(ω).*   Given a quasilinear third order ode ω with a seven-parameter group of Lie symmetries, its general solution is returned; or *failed* if it cannot be determined.

$S1$: *Identify case a) or case b).* By means of the criteria given in Theorem 5.18 determine which case applies for equation ω.

$S2$: *Equations for σ and ρ.* Determine a solution for S from equations (6.87) or (6.87). If none is found return *failed*; otherwise construct the system for σ from it.

$S3$: *Determine σ and ρ.* From the equations constructed in step $S2$ determine Liouvillian transformation functions σ and ρ.

$S4$: *Return solution.* Substitute $u = \sigma(x, y)$ and $v = \rho(x, y)$ into the canonical form $v = C_1 + C_2 u + C_2 u^2$ and return the result.

EXAMPLE 7.38   For equation 7.8 of Kamke's collection in Example 6.37 the transformation functions $\sigma = x$ and $\rho = \dfrac{2}{\sqrt{y}} - 2$ have been obtained. Substituting them into the above solution for $v''' = 0$ yields the solution $4\dfrac{2}{\sqrt{y}} - 2 = C_1 + C_2 x + C_3 x^2$ or equivalently

$$y = \frac{1}{(C_1 + C_2 x + C_3)^2}$$

after a redefinition of the integration constants.   ▯

EXAMPLE 7.39   For the equation with symmetry type $\mathcal{S}_7^3$ and structure (5.44) in Example 5.49 the transformation functions $\sigma = xy$ and $\rho = \frac{y}{x}$ have

Solution Algorithms 349

been obtained. Substituting these values into the above solution for $v''' = 0$ leads to the general solution $x^3 y^2 + (C_1 x^2 + C_2)y + C_3)x = 0$. ⬜

EXAMPLE 7.40   For the equation with symmetry type $\mathcal{S}_7^3$ and structure (5.44) in Example 5.50 the transformation functions $\sigma = \frac{1}{y}$ and $\rho = \frac{y}{x}e^x$ have been obtained. Substituting these values into the above solution for $v''' = 0$ leads to the general solution $y^3 + (C_1 y^2 + C_2 y + C_3)xe^{-x} = 0$. ⬜

Exercises

EXERCISE 7.1   Apply Lemma 7.1 for solving the equation in Example 6.2.

EXERCISE 7.2   Solve the equation of Example 7.2 by transforming it into the form $v' = C$.

EXERCISE 7.3   Solve the equation of Example 7.2 under the assumption that only the second symmetry generator $U_2 = y\partial_x + x\partial_y$ is known.

EXERCISE 7.4   Derive the expression (7.16) for the solution of (7.13).

EXERCISE 7.5   Generate the solution of the equation considered in Example 7.19 from the first order equation (7.36).

EXERCISE 7.6   Determine the solution of equation Kamke 6.227 from the canonical form obtained in Example 6.8.

EXERCISE 7.7   Solve the equation $y'' + r(y) = 0$ by symmetry methods.

EXERCISE 7.8   Solve the equation $y''y + r(x)y'^2 = 0$ by symmetry methods.

EXERCISE 7.9   Determine the solution of (7.24) from the equivalent pde and compare the result with (7.25).

EXERCISE 7.10   Derive a first order Riccati equation from (7.71) and determine its solution.

EXERCISE 7.11   Derive solution (7.72) from the equivalent pde of (7.71).

Chapter 8

Concluding Remarks

After discussing the various integration methods for ode's developed by Lie it appears appropriate to conclude with a few words of comparison, assess its advantages and limitations, and also put them into proper perspective. Furthermore, several directions of future research are outlined.

Comparison of Lie's Methods. There are basically two alternative methods for solving differential equations using its symmetries. Either an equation with a certain nontrivial symmetry is transformed into a canonical form the solution of which may be determined. Or the infinitesimal symmetry generators are determined explicitly, and are used to determine a set of first integrals from which the solution may be constructed. The relation between these two alternatives is most easily understood from the following drawing. The order n of the given quasilinear ode is two or three.

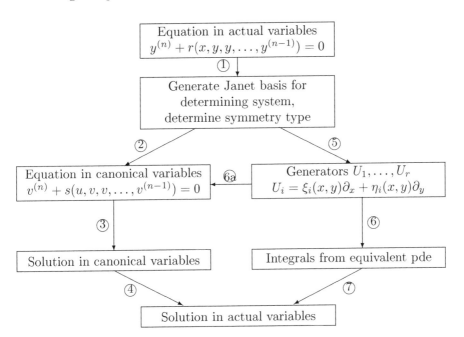

Step 1 takes the given ode in actual variables as input, constructs the determining system from it and transforms it into a Janet basis. To this end

352

only differentiations and arithmetic in the base field are required. From the Janet basis coefficients the symmetry type may be obtained as described in Theorems 5.9 and 5.14. This step is *guaranteed* and it is always performed first.

After that there are two main alternatives. The left branch which is emphasized in this book begins with step 2. At first a system of equations for the transformation functions to canonical form is set up. If an independent pair of solutions for the transformation functions may be obtained, it is applied for generating the canonical form equation for the respective symmetry class. In step 3 its solution, or at least a simplified equation, is determined. In the final step 4 this result is transformed back to the actual variables x and y.

Alternatively the vector fields generating the symmetries may be determined explicitly in step 5. They are applied in step 6 to determine a set of first integrals for the pde which is equivalent to the given ode. In step 7 by elimination the desired solution or some lower order equation may be constructed from the first integrals.

These two main branches are not completely disconnected. One may leave the right branch with step 6a and then proceed along the left branch. There are several reasons why the left branch in the above drawing is preferred.

i) The equations which determine the transformations to canonical form are usually simpler than the equations determining the symmetry generators. For all possible types of equations occurring in the symmetry analysis of second and third order ode's algorithms have been described for decomposing them into components of lowest order.

ii) If the symmetry generators are determined explicitly, they have to be transformed such that they obey the canonical commutators of the respective Lie algebra.

iii) The first integrals for the equivalent pde often lead to unnecessary complicated expressions for the solution of the given ode which are difficult to simplify.

There is an additional, more fundamental reason. In general, solving a differential equation may be considered as an equivalence problem consisting of two parts. At first membership of the given equation to a particular equivalence class has to be decided. Secondly, within this equivalence class the transformation of the given equation to a canonical form has to be determined algorithmically. This scheme is valid independent of whether an equation has symmetries or not. In the former case, the symmetry class of the equation narrows down the possible equivalence classes, and the canonical form for it is already known. These aspects are clearly discussed by Tresse [181] as it is explained in Section 4.3 on page 188.

If there are no symmetries, candidates for equivalence classes are usually specified in terms of differential equations defining certain special functions

Concluding Remarks

like e. g. Painlevé transcendents, Bessel functions, etc. As soon as membership to any of them is established, the transformation to the canonical equation has to be found. These two steps may be combined into a single one. This approach is dicussed in the Dissertation of Berth [10].

Problems for Further Research. It would be highly desirable to admit more general base fields, e. g. elementary extensions by one or more exponentials, angular functions or logarithms, or algebraic functions. Because the base field occurs as coefficient field for the Janet basis of the determining system, and also for the equations determining the transformation to canonical form, the decompositions of these Janet bases over more general coefficient fields must be known first if an algorithmic theory without heuristics is the goal.

For various applications it might be of interest to have a similar scheme available for equations of order four comprising more than one hundred symmetry classes.

Lie himself announced a book on point and contact symmetries of partial differential equations. Due to his early decease he has not been able to realize this project; and so another field of possible research is left. Due to the large number of possible symmetry types that have to be considered, a complete classification is rather voluminous and relies even more on appropriate computer algebra tools. Similar arguments apply to the symmetry analysis of systems of ode's. There have been various attempts to extend Lie's symmetry concept in order to make more equations amenable to a solution procedure. To this end, Barbara Abraham-Shrauner et al. [1] applied nonlocal symmetries. Muriel et al. [134, 49] define the new concept of a potential symmetry.

A completely different approach has been pursued by Zaitsev and Polyanin [191]. As opposed to Lie, they apply *discrete* groups for solving ordinary differential equations. It would be highly interesting to compare their results with those obtained by Lie analysis.

Another comparison concerns the relation between Galois' theory and Lie's symmetry analysis. For second order linear equations there is certainly no chance for any correlation because any such equation has the full eight-parameter projective group as its symmetry group. The situation is different, however, for equations of order three as may be seen from the listings in Appendix D and E. For example, all equations in symmetry class $\mathcal{S}_7^3$ have a Loewy decomposition of type $\mathcal{L}_9^3$. All equations in symmetry class $\mathcal{S}_{5,2}^3(c)$ have Loewy decompositions of type $\mathcal{L}_9^3$ or type $\mathcal{L}_{12}^3$. This suggests that for linear equations of order higher than two there may be some correlations between its symmetry type and the type of its Loewy decomposition, which in turn entails a special Galois group.

Appendix A

Solutions to Selected Problems

In this Appendix the solutions to most exercises are given, although in some cases only a few hints are provided. The reader is strongly advised to try first on his own to find the answers.

2. Linear Differential Equations

2.1 The constraints for the coefficients of the given system of algebraic equations are

$$a_1 = 2b_4a_3b_3 - b_2a_3 + b_3^3 - b_3^2a_4 + b_3a_2 + c_3a_3,$$

$$b_1 = b_2b_3 - a_3b_4^2 - c_4a_3 - b_4b_3^2,$$

$$c_1 = a_2c_4 + a_2b_4^2 - b_2^2 + b_2c_3 - b_3c_2 + a_3b_4^3 + b_4c_4a_3 + b_4^2b_3^2,$$

$$c_2 = a_4c_4 - 2b_2b_4 + a_4b_4^2 + b_4c_3 + b_4^2b_3 - c_4b_3.$$

2.2 If $y_1 = z_1^2$ is differentiated three times, and $z_1'' + \frac{1}{4}pz_1 = 0$ is applied for reduction whenever possible, one obtains $y_1' = 2z_1z_1'$, $y_1'' = -\frac{1}{2}pz_1^2 + 2z_1'^2$ and $y_1''' = -\frac{1}{2}p'z_1^2 - 2pz_1z_1'$. Substitution into the given third order ode shows that y_1 is a solution. Similarly differentiating and substituting $y_2 = z_1z_2$, $y_2' = z_1'z_2 + z_1z_2'$, $y_2'' = -\frac{1}{2}pz_1z_2 + 2z_1'z_2'$ and $y_2''' = -\frac{1}{2}p'z_1^2 - 2pz_1z_1'$ shows that y_2 is a solution as well. The same is true for y_3.

2.3 Let y_1, y_2 and y_3 be a fundamental system for the given equation and W its Wronskian. The general solution may be written as

$$y = C_1y_1 + C_2y_2 + C_3y_3 + \int \frac{r}{W}(y_2y_3' - y_2'y_3)dx$$

$$- \int \frac{r}{W}(y_1y_3' - y_1'y_3)dx + \int \frac{r}{W}(y_1y_2' - y_1'y_2)dx$$

where C_1, C_2 and C_3 are constants. For $y''' = r$ with the fundamental system $y_1 = 1$, $y_2 = x$ and $y_3 = x^2$ there follows

$$y = C_1 + C_2x + C_3x^2 + \frac{1}{2}\int x^2rdx - x\int xrdx + \frac{1}{2}x^2\int rdx.$$

2.4 Reducing $y'' + Py' + Qy = 0$ w.r.t. $y' + py = 0$ and $y' + qy = 0$ yields $Q - pP = p' - p^2$ and $Q - qP = q' - q^2$. Solving for P and Q yields the given expressions. Substituting the fundamental system $y_1 = \exp(-\int pdx)$,

355

356

$y_2 = \exp\left(-\int q\,dx\right)$ and the Wronskian $W = (q - p)\exp\left(-\int(p+q)dx\right)$ into (2.9) yields

$$y = \exp\left(-\int p\,dx\right)\left(\int \exp\left(\int p\,dx\right)\frac{R\,dx}{q-p} + C_1\right)$$
$$+ \exp\left(-\int q\,dx\right)\left(\int \exp\left(\int q\,dx\right)\frac{R\,dx}{q-p} + C_2\right).$$

This representation generalizes the solution of a first order equation in an obvious way.

2.5 For $\mathcal{L}_2^2$, the first solution y_1 is obtained from $(D + a_1)y = 0$, the second from $(D + a_1)y = \bar{y}_2$ with $\bar{y}_2$ a solution of $(D + a_2)y = 0$.

For $\mathcal{L}_3^2$ the two solutions are obtained from $(D + a_i)y = 0$. Linear dependence over the base field would imply a relation $q_1 y_1 + q_2 y_2 = 0$ with q_1, q_2 from the base field. Substituting the solutions this would entail $\frac{q_1}{q_2} = -\exp\int(a_2 - a_1)dx$. Due to the nonequivalence of a_1 and a_2, its difference is not a logarithmic derivative. Consequently the right hand side cannot be rational.

For $\mathcal{L}_4^2$ the equation $(D + a(C))y = 0$ has to be solved; then C is specialized to C_1 and C_2. Substituting $a_i = \frac{r'}{r + C_i} + p$ in the above quotient, the integration may be performed with the result $\frac{q_1}{q_2} = -\frac{r + C_2}{r + C_1}$ which is rational.

2.6 Let $y''' + ay'' + by' + cy = 0$ with fundamental system y_1, y_2, y_3 be the third order linear equation corresponding to the second order Riccati equation $z'' + 3zz' + z^3 + a(z' + z^2) + bz + c = 0$. If $\frac{y_i'}{y_i}$ is rational for a single value of i, or for two or three values of i, but for any pair i, j with $1 \leq i < j \leq 3$, the ratios

$$\frac{C_i y_i' + C_j y_j'}{C_i y_i + C_j y_j} \tag{A.1}$$

are not, the corresponding number of rational solutions of the Riccati equation exists. This is case $i)$ for the second order equation. If there is a pair of indices such that the ratio (2.17) is rational, case $ii)$ or $iii)$ respectively applies depending on whether the logarithmic derivative of the remaining element of the fundamental system is rational or not. Finally, if the ratio

$$\frac{C_1 y_1' + C_2 y_2' + C_3 y_3'}{C_1 y_1 + C_2 y_2 + C_3 y_3}$$

is rational, it may be rewritten as

$$\frac{C y_1'/y_1 + C_2 y_2'/y_1 + y_3'/y_1}{C_1 + C_2 y_2/y_1 + y_3/y_1} = \frac{C_2(y_2/y_1)' + (y_3/y_1)'}{C_1 + C_2(y_2/y_1) + y_3/y_1} + \frac{y_1'}{y_1} \tag{A.2}$$

with the ratios $\frac{C_1}{C_3}$ and $\frac{C_2}{C_3}$ appropriately redefined and $p = \frac{y_1'}{y_1}$, $q = \frac{y_2'}{y_1}$, $r = \frac{y_3'}{y_1}$, $u = \frac{y_2}{y_1}$ and $v = \frac{y_3}{y_1}$. This is case $iv)$.

Solutions to Selected Problems 357

2.7 If the Galois group is Z_1, by Theorem 2.7 the equation has a rational fundamental system $\{y_1, y_2\}$; i.e., it is linearly dependent over the base field. By the result of Exercise 2.4 the Loewy decomposition is of type $\mathcal{L}_4^2$. If the group is Z_2, its second symmetric power has a rational fundamental system y_1^2, $y_1 y_2$ and y_2^2. Consequently, the ratio $\frac{y_1^2}{y_1 y_2} = \frac{y_1}{y_2}$ is rational and the decomposition $\mathcal{L}_4^2$ follows as above.

2.8 Substituting $P(u) \equiv u^2 + B_1 u + b_0 = 0$ into $u' + u^2 + r = 0$ leads to $b_0 = r - \frac{1}{2}(b_1' - b_1^2)$ and $b_1 b_1' - 2b_1' + \frac{1}{2}b_1(b_1^2 - 4b_0)$. If the expression for b_0 is substituted into this latter relation, the constraint

$$b_1'' - 3b_1 b_1' + b_1^3 + 4rb_1 - 2r' = 0$$

follows. With $b_1 = -\frac{z'}{z}$ the third order equation $z''' + 4rz' + 2r'z = 0$ for z is obtained which is the the second symmetric power of $y'' + ry = 0$.

2.9 Substituting z_1 and z_2 into the equations of the given Janet basis leads to a linear system for the coefficients a_1, a_2, b_1 and b_2. Defining

$$w_0 = \begin{vmatrix} z_{1,x} & z_{2,x} \\ z_1 & z_2 \end{vmatrix}, \quad w_1 = \begin{vmatrix} z_{1,y} & z_{2,y} \\ z_1 & z_2 \end{vmatrix}, \quad w_2 = \begin{vmatrix} z_{1,y} & z_{2,y} \\ z_{1,x} & z_{2,x} \end{vmatrix},$$

$$w_3 = \begin{vmatrix} z_{1,xx} & z_{2,xx} \\ z_1 & z_2 \end{vmatrix}, \quad w_4 = \begin{vmatrix} z_{1,yy} & z_{2,yy} \\ z_1 & z_2 \end{vmatrix},$$

$$w_5 = \begin{vmatrix} z_{1,xx} & z_{2,xx} \\ z_{1,x} & z_{2,x} \end{vmatrix}, \quad w_6 = \begin{vmatrix} z_{1,yy} & z_{2,yy} \\ z_{1,y} & z_{2,y} \end{vmatrix}$$

they may be expressed as

$$a_1 = -\frac{w_1}{w_0}, \quad a_2 = \frac{w_2}{w_0}, \quad b_1 = -\frac{w_3}{w_0}, \quad b_2 = \frac{w_5}{w_0}.$$

2.10 With the same notation as in the previous exercise the representations $a_1 = -\frac{z_{1,x}}{z_1} = -\frac{z_{2,x}}{z_2}$, $b_1 = -\frac{w_4}{w_1}$ and $b_2 = \frac{w_6}{w_1}$ are obtained.

2.11 By direct reduction the following Riccati-like system of pde's is obtained.

$$a_{1,x} + A_1 a_1^2 + (B_1 - A_2)a_1 + a_2 - B_2 = 0,$$
$$a_{1,y} + B_1 a_1^2 - a_1 a_2 + (C_1 - B_2)a_1 - C_2 = 0,$$
$$a_{2,x} + A_1 a_1 a_2 + B_1 a_2 - A_3 a_1 - B_3 = 0, \qquad (A.3)$$
$$a_{2,y} - a_2^2 + B_1 a_1 a_2 + C_1 a_2 - B_3 a_1 - C_3 = 0,$$
$$b_1 + A_1 a_1 - A_2 = 0, \quad b_2 + A_1 a_2 - A_3 = 0.$$

358

2.12 Substituting (2.79) into system (2.81) and reordering yields after a few autoreduction steps the following result.

$$w_{0,xx} - A_1 w_{0,y} - \left(\frac{A_{1,x}}{A_1} - A_2 - 2B_1 \right) w_{0,x}$$
$$- \left(\frac{A_{1,x}}{A_1} A_2 - A_{2,x} + 2A_1 B_2 - 2A_2 B_1 \right) w_0 = 0,$$
$$w_{0,xy} + A_2 w_{0,y} - \frac{1}{A_1}(B_{1,x} - A_1 B_2 + A_2 B_1 - A_3 - B_1^2)w_{0,x}$$
$$+ \left[B_{2,x} + A_2 B_2 - B_1 B_2 + B_3 - \frac{A_2}{A_1}(B_{1,x} - B_1^2 + A_2^2 B_1 - A_2 A_3) \right] w_0 = 0,$$
$$w_{0,yy} + (2B_2 + C_1)w_{0,y} - \frac{1}{A_1}(C_{1,x} - A_1 C_2 + 2B_1 B_2 - 2B_3)w_{0,x}$$
$$+ \left[C_{2,x} + 2B_2^2 + 2C_3 - \frac{A_2}{A_1}(C_{1,x} + 2B_1 B_2 - 2B_3) \right] w_0 = 0.$$

2.13 Each term in (2.19) is uniquely determined by an index sequence $k_0, k_1, \ldots, k_{\nu-1}$. If the lexicographic ordering with respect to this sequence is assumed, the leading terms corresponding to these sequences are given in the subsequent table. Furthermore, the sum of the indices is given.

No.	k_0	k_1	k_2	k_3	$\ldots$	$k_{\nu-2}$	$k_{\nu-1}$	$\sum_{j=0}^{\nu-1} k_j(M+j)$
1	ν	0	0	0	$\ldots$	0	0	νM
2	$\nu-2$	1	0	0	$\ldots$	0	0	$(\nu-1)M+1$
3	$\nu-3$	0	1	0	$\ldots$	0	0	$(\nu-2)M+2$
4	$\nu-4$	2	0	0	$\ldots$	0	0	$(\nu-2)M+2$
5	$\nu-4$	0	0	1	$\ldots$	0	0	$(\nu-3)M+3$
$\vdots$	$\vdots$				$\ldots$	$\vdots$		$\vdots$
last	0	0	0	0	$\ldots$	0	1	$M+\nu-1$

The rightmost column contains the order of a pole of order M for the term determined by the values in the preceding columns. Because $\nu \geq 1$, the value νM in the first line is maximal if $M > 1$. For $M = 1$ they are all equal to ν.

2.14 The first member of a fundamental system is $z_1 = \exp\left(- \oint a dx + b dy\right)$. Dividing out $w_1 \equiv z_x + az$ and $w_2 \equiv z_y + bz$ yields the system

$$w_2 + A_1 w_1 = 0, \quad w_{1,x} + (B_1 - a)w_1 = 0, \quad w_{1,y} + (A_{1,x} + A_1 + A_2 - A_1 B_1)w_1 = 0$$

with the solution (see Lemma 2.9)

$$\bar{w}_1 = \exp\left(- \oint (B_1 - a)dx + (A_{1,x} + A_1 + A_2 - A_1 B_1)dy\right), \quad \bar{w}_2 = -A_1 \bar{w}_1.$$

Solutions to Selected Problems 359

The corresponding inhomogeneous system $z_x + az = \bar{w}_1$, $z_y + bz = -A_1\bar{w}_1$ finally yields the second member of a fundamental system (see Lemma 2.10) $z_2 = z_1 \oint \frac{w_1}{z_1}(dx - A_1 dy)$.

2.15 The first member of a fundamental system is $z_{1,1} = \exp(-\oint b\,dx + c\,dy)$, $z_{2,1} = -az_{1,1}$. Dividing out the Loewy factor $w_1 \equiv z_2 + az_1$, $w_2 \equiv z_{1,x} + bz_1$ and $w_3 \equiv z_{1,y} + cz_1$ yields the system

$$w_{1,x} + (aA_1 + C_1)w_1 = 0, \quad w_{1,y} + (aB_1 + D_1)w_1 = 0,$$
$$w_2 + A_1 w_1 = 0, \quad w_3 + B_1 w_1 = 0$$

with the solution (see Lemma 2.9)

$$\bar{w}_1 = \exp\left(-\oint (aA_1 + C_1)dx + (aB_1 + D_1)dy\right), \quad \bar{w}_2 = -A_1\bar{w}_1, \quad \bar{w}_3 = -B_1\bar{w}_1.$$

The corresponding inhomogeneous system

$$z_2 + az_1 = \bar{w}_1, \quad z_{1,x} + bz_1 = -A_1\bar{w}_1, \quad z_{1,y} + cz_1 = -B_1\bar{w}_1$$

finally yields the second member of a fundamental system (see Lemma 2.10)

$$z_{1,2} = -z_{1,1} \oint \frac{\bar{w}_1}{z_{1,1}}(A_1 dx + B_1 dy), \quad z_{2,2} = -az_{1,2}.$$

2.16 Reduction w.r.t. the given two first order systems yields a linear algebraic system for the coeffients of the desired $Lclm$. If $a_1 \neq a_2$, the result is of type $\mathcal{J}_{2,1}^{(1,2)}$,

$$z_y - \frac{b_1 - b_2}{a_1 - a_2}z_x + \frac{a_1 b_2 - a_2 b_1}{a_1 - a_2}z = 0,$$
$$z_{xx} + (a_1 + a_2 - \frac{a_{1,x} - a_{2,x}}{a_1 - a_2})z_x + (a_1 a_2 + \frac{a_1 a_{2,x} - a_2 a_{1,x}}{a_1 - a_2})z = 0.$$

If $a_1 = a_2$ and $b_1 \neq b_2$, the result is of type $\mathcal{J}_{2,1}^{(1,2)}$,

$$z_x + a_1 z = 0, \quad z_{yy} + \left(b_1 + b_2 - \frac{b_{1,y} - b_{2,y}}{b_1 - b_2}\right)z_y + \left(b_1 b_2 - \frac{b_1 b_{2,y} - b_{1,y}b_2}{b_1 - b_2}\right)z = 0.$$

2.17 Introducing a differential indeterminate z, the system of linear pde's

$$z_y - \frac{x}{y}z_x = 0, \quad z_{xx} - \frac{xy-1}{x}z_x - \frac{y}{x}z = 0 \qquad (\text{A.4})$$

is obtained. The right factor leads to $z_x - yz = 0$, $z_y - xz = 0$ with the solution e^{xy}. It is the first basis element for the solution space of (A.4). Introducing coordinates z_1 and z_2 in $\mathcal{D}^2$, the exact quotient generates the pde's $z_2 - \frac{x}{y}z_1 = 0$, $z_{1,x} + \frac{1}{x}z_1 = 0$ and $z_{1,y} = 0$ with the solution $z_1 = \frac{1}{x}$,

$z_2 = \frac{1}{y}$. In order to obtain the second basis element for the solution space of (A.4), the inhomogeneous system $z_x - yz = \frac{1}{x}$, $z_y - xz = \frac{1}{y}$ has to be solved with the result $e^{xy} Ei(-xy)$. The exponential integral is defined by

$$Ei(-x) = -\int_x^\infty e^{-t} \frac{dt}{t}.$$

2.18 A special solution of the inhomogeneous system is searched for in the form $z = C_1 z_1 + C_2 z_2$ where both $C_1 \equiv C_1(x,y)$ and $C_2 \equiv C_2(x,y)$ are undetermined functions of x and y. They will be obtained by an extension of Lagrange's variation of constants. The last equation yields

$$C_{1,x} = -\frac{wz_2}{W}, \quad C_{2,x} = \frac{wz_1}{W} \quad \text{where } W = \begin{vmatrix} z_1, & z_2 \\ z_{1,x}, & z_{2,x} \end{vmatrix}.$$

Substituting the expression for z into the first equation and the equation

$$z_{xy} + (a_{1,x} + a_2 - a_1 b_1)z_x + (a_{2,x} - a_1 b_2)z = v_x - a_1 w$$

that is obtained from it by derivation w.r.t. y yields

$$C_{1,y} z_1 + C_{2,y} z_2 = v, \quad C_{1,y} z_{1,x} + C_{2,y} z_{2,x} = v_x - a_1 w.$$

By elimination the expressions

$$C_{1,y} = \frac{vz_{2,x} - (v_x - a_1 w)z_2}{W}, \quad C_{2,y} = -\frac{vz_{1,x} - (v_x - a_1 w)z_1}{W}$$

for the y-derivatives are obtained. Combining these results, both coefficients C_1 and C_2, and consequently a special solution of the inhomogeneous system, may be obtained as a path integral.

2.19 A special solution of the inhomogeneous system is searched for in the form $z = C_1 z_1 + C_2 z_2$ with

$$C_{1,x} = \frac{vz_{2,y} - v_y z_2}{W}, \quad C_{2,x} = \frac{v_y z_1 - vz_{1,y}}{W} \quad \text{where } W = \begin{vmatrix} z_1, & z_2 \\ z_{1,y}, & z_{2,y} \end{vmatrix},$$

$$C_{1,y} = -\frac{wz_2}{W}, \quad C_{2,y} = \frac{wz_1}{W}.$$

It is most easily obtained from the preceding case by exchange of x and y and substituting $a = 0$.

2.20 A special solution of the inhomogeneous system is searched for in the form $z = C_1 z_1 + C_2 z_2 + C_3 z_3$, $C_k \equiv C_k(x,y)$ for $k = 1, 2, 3$ are undetermined functions of x and y. Lagrange's variation of constants leads to

$$C_{1,x} = \frac{w}{W} \begin{vmatrix} z_2, & z_3 \\ z_{2,x}, & z_{3,x} \end{vmatrix}, \quad C_{1,y} = \frac{1}{W} \begin{vmatrix} v, & z_2, & z_3 \\ v_x, & z_{2,x}, & z_{3,x} \\ v_{xx} - aw, & z_{2,xx}, & z_{3,xx} \end{vmatrix},$$

$$C_{2,x} = -\frac{w}{W}\begin{vmatrix} z_1, & z_3 \\ z_{1,x}, & z_{3,x} \end{vmatrix}, \quad C_{2,y} = \frac{1}{W}\begin{vmatrix} z_1, & v, & z_3 \\ z_{1,x}, & v_x, & z_{3,x} \\ z_{1,xx}, & v_{xx} - aw, & z_{3,xx} \end{vmatrix},$$

$$C_{3,x} = \frac{w}{W}\begin{vmatrix} z_1, & z_2 \\ z_{1,x}, & z_{2,x} \end{vmatrix}, \quad C_{3,y} = \frac{1}{W}\begin{vmatrix} z_1, & z_2, & v \\ z_{1,x}, & z_{2,x}, & v_x \\ z_{1,xx}, & z_{2,xx}, & v_{xx} - aw \end{vmatrix}$$

where

$$W = \begin{vmatrix} z_1, & z_2, & z_3 \\ z_{1,x}, & z_{2,x}, & z_{3,x} \\ z_{1,xx}, & z_{2,xx}, & z_{3,xx} \end{vmatrix}.$$

From these expressions the $C_k(x, y)$ may be expressed in terms of path integrals.

2.21 With the substitutions $x_k \equiv \phi_k(y_1, \ldots, y_n)$, $y_k \equiv \psi(x_1, \ldots, x_n)$ and $g(y_1, \ldots, y_n) \equiv f(\phi_1, \ldots, \phi_n)$, the partial derivatives of f become

$$\frac{\partial f}{\partial x_i} = \frac{\partial g}{\partial y_1}\frac{\partial y_1}{\partial x_i} + \ldots + \frac{\partial g}{\partial y_n}\frac{\partial y_n}{\partial x_i}.$$

Substituting them into (2.45), rearranging terms and defining

$$B_k = a_1\frac{\partial \psi_k}{\partial x_1} + \ldots + a_n\frac{\partial \psi_k}{\partial x_n}$$

for $k = 1, \ldots, n$ yields (2.46).

2.22 With the notation of Theorem 2.20, the equation for $w(x, y, z)$ is $w_y + aw_x + rw_z = 0$. From $dy = \frac{dx}{a}$ there follows $\frac{dy}{dx} = \frac{1}{a}$ and the first integral $\psi_1(x, y)$. Introducing a new variable $\bar{x} = \psi_1(x, y)$ with the inverse $x = \phi(\bar{x}, y)$, the equation for w becomes $w_y + (b(\phi, y)z + c(\phi, y))w_z = 0$. From $dy = \frac{dz}{bz + c}$ there follows $\frac{dz}{dy} - b(\phi, y)z = c(\phi, y)$. It yields a second first integral

$$\psi_2(x, y, z) = \left[z\exp\left(-\int b(\phi, y)dy \right) - \int c(\phi, y)\exp\left(-\int b(\phi, y)dy \right)dy \right]\Big|_{\bar{x} = \psi_1}.$$

The general integral of the originally given equation may be written as $\Phi(\psi_1, \psi_2) = C$ with a constant C. The particular choice of the function Φ determines the actual form of the answer. As an example, consider the equation $z_y + \frac{2x}{3y}z_x = \frac{1}{3y}z$ with the first integrals $\psi_1 = \frac{y^2}{x^3}$ and $\psi_2 = \frac{z}{y^{1/3}}$. Choosing $\Phi = \frac{\psi_2}{\psi_1^{1/3}}$ yields $z = C\frac{y}{x}$.

2.24 Taking the commutator with A of both sides of the assumed relations

362

yields

$$(A, B_k) = (A, \beta^k_{m+1} B_{m+1}) + \ldots + (A, \beta^k_q B_q) + (A, \alpha A)$$
$$= A\beta^k_{m+1} \cdot B_{m+1} + \ldots + A\beta^k_q \cdot B_q$$
$$+ \beta^k_{m+1}(A, B_{m+1}) + \ldots + \beta^k_q(A, B_q) + A\alpha \cdot A.$$

By assumption the commutators involving the B_k in the last line are proportional to A; i.e., the complete expression is a relation of the same type given in the Theorem which is a contradiction. Consequently, $A\beta^k_j = 0$ for all k and j.

2.25 The second equation has the general solution

$$z(x, y, \bar{C}(x)) = \frac{\frac{y}{x}}{\bar{C}(x) + y} - \frac{xy}{x^2 y - 2}.$$

Substitution into (2.55) yields $\bar{C}_x x + \bar{C} = 0$; i.e., $\bar{C} = \frac{C}{x}$ with C a constant from which the general rational solution $z = \frac{y}{xy + C} - \frac{xy}{x^2 y - 2}$ follows.

2.26 Substituting $z = z_1 + \frac{1}{w}$ into (2.50) yields the linear system

$$w_x - (2A_1 z_1 + A_2)w = A_1, \quad w_y - (2B_1 z_1 + B_2)w = B_1$$

for w in complete analogy to a first order ordinary Riccati equation.

2.27 If in the system $z_x + A_1 z^2 + A_2 z + A_3 = 0$, $z_y + B_2 z + B_3 = 0$ the new function w is introduced by $z = \frac{w}{A_1}$, the following system for w is obtained.

$$w_x + w^2 + \left(A_2 - \frac{A_{1,x}}{A_1} \right) w + A_1 A_3 = 0, \quad w_y + \left(B_2 - \frac{A_{1,y}}{A_1} \right) w + A_1 B_3 = 0.$$

The coefficient of w in the latter equation vanishes identically due to the first constraint in (2.51). The second constraint of (2.51) yields for the last term

$$A_1 B_3 = \frac{1}{2} \left(A_2 - \frac{A_{1,x}}{A_1} \right)_y.$$

The y-integration may be performed with the result

$$w = \frac{1}{2} \left(\frac{A_{1,x}}{A_1} - A_2 \right) + C$$

where $C \equiv C(x)$ is a function of x. Substituting this expression for w into the first equation yields the Riccati equation (2.57) for the x-dependence of C. If the terms not containing C are derived w.r.t. y and the constraints (2.51) are applied for eliminating all y-derivatives, zero is obtained. If the solution for C is substituted into the above expression for w and $z = \frac{w}{A_1}$ is used, the representation (2.56) for the solution follows.

$$\text{Solutions to Selected Problems} \qquad 363$$

2.28 Dividing out the exact component $\{z_x \equiv z_1, z_y \equiv z_2\}$ yields $\{z_2 + az_1, z_{1,x} + cz_1, z_{1,y} + (a_x - ac)z_1\}$. The corresponding linear system has the general solution

$$z_1 = \exp\left(-\int cdx + (a_x - ac)dy\right), \quad z_2 = -az_1.$$

It leads to the system $z_x = z_1$, $z_y = -az_1$ with the solution

$$z = \int z_1 dx - az_1 dy = \int \exp\left(-\int cdx + (a_x - ac)dy\right)(dx - ady)$$

which is the non-constant member of a fundamental system.

3. Lie Transformation Groups

3.1 The differential equations

$$\frac{\partial \bar{x}}{\partial a} = 1, \quad \frac{\partial \bar{x}}{\partial b} = 0, \quad \frac{\partial \bar{y}}{\partial a} = 0, \quad \frac{\partial \bar{y}}{\partial b} = 1$$

have the required form; however, there does not exist an identity element.

3.2 Let a_k correspond to the one-parameter group U_k. Then the finite transformations are

$$\bar{x} = \frac{(x + a_1)e^{a_2}}{1 - (x + a_1)e^{a_2}a_4}, \quad \bar{y} = \frac{e^{a_3}y}{1 - (x + a_1)e^{a_2}a_4}$$

3.3 The finite transformations of g_2 are $\bar{x} = a_1 x + a_2 y + a_3$, $\bar{y} = a_4 x + a_5 y + a_6$. Accordingly, three arbitrarily chosen points (x_1, y_1), (x_2, y_2) and (x_3, y_3) are transformed into $\bar{x}_i = a_1 x_i + a_2 y_i + a_3$ and $\bar{y}_i = a_4 x_i + a_5 y_i + a_6$ for $i = 1, 2, 3$. These relations may be solved for the group parameters as follows.

$$a_1 = \frac{1}{D}\begin{vmatrix} \bar{x}_1 & y_1 & 1 \\ \bar{x}_2 & y_2 & 1 \\ \bar{x}_3 & y_3 & 1 \end{vmatrix}, \quad a_2 = \frac{1}{D}\begin{vmatrix} x_1 & \bar{x}_1 & 1 \\ x_2 & \bar{x}_2 & 1 \\ x_3 & \bar{x}_3 & 1 \end{vmatrix}, \quad a_3 = \frac{1}{D}\begin{vmatrix} x_1 & y_1 & 1 \\ x_2 & y_2 & 1 \\ \bar{x}_1 & \bar{x}_2 & \bar{x}_3 \end{vmatrix},$$

$$a_4 = \frac{1}{D}\begin{vmatrix} \bar{y}_1 & y_1 & 1 \\ \bar{y}_2 & y_2 & 1 \\ \bar{y}_3 & y_3 & 1 \end{vmatrix}, \quad a_5 = \frac{1}{D}\begin{vmatrix} x_1 & \bar{y}_1 & 1 \\ x_2 & \bar{y}_2 & 1 \\ x_3 & \bar{y}_3 & 1 \end{vmatrix}, \quad a_6 = \frac{1}{D}\begin{vmatrix} x_1 & y_1 & 1 \\ x_2 & y_2 & 1 \\ \bar{y}_1 & \bar{y}_2 & \bar{y}_3 \end{vmatrix},$$

D is defined by $\begin{vmatrix} x_1 & y_1 & 1 \\ x_2 & y_2 & 1 \\ x_3 & y_3 & 1 \end{vmatrix}$.

3.4 Introducing new variables $u = \dfrac{F_2(x)}{F_1(x)} \equiv \sigma(x)$ and $v = \dfrac{y}{F_1(x)} \equiv \rho(x, y)$, Lemma 3.3 yields vanishing coefficients for all ∂_u and $G_k(u) = \dfrac{F_2(x)}{F_1(x)}$ with x expressed by u for $k = 1, \ldots, r$. Obviously $G_1 = 1$ and $G_2(u) = u$ as required.

3.5 Applying the same notation as in Example 3.1, the following system of equations for the undetermined coefficients is obtained.

$$\alpha_{12}\alpha_{23} - \alpha_{13}\alpha_{22} = \alpha_{11}, \quad \alpha_{11}\alpha_{23} - \alpha_{13}\alpha_{21} = -\alpha_{12},$$
$$\alpha_{12}\alpha_{33} - \alpha_{13}\alpha_{32} = 2\alpha_{21} = 0, \quad \alpha_{11}\alpha_{33} - \alpha_{13}\alpha_{31} = -2\alpha_{22} = 0,$$
$$\alpha_{22}\alpha_{33} - \alpha_{23}\alpha_{32} = \alpha_{31} = 0, \quad \alpha_{21}\alpha_{33} - \alpha_{23}\alpha_{31} = -\alpha_{32} = 0$$

and $\alpha_{13} = \alpha_{23} = \alpha_{33} = 0$. Obviously, it allows only the trivial solution $\alpha_{ij} = 0$ for all i and j.

3.6 For $r = 2$ there holds $\Delta(\omega) = \omega^2 - \left(c_{12}^2 E_1 - c_{12}^1 E_2\right)\omega$, for $r = 3$

$$\Delta(\omega) = \omega^3 - \left[(c_{12}^2 + c_{13}^3)E_1 - (c_{12}^1 - c_{23}^3)E_2 - (c_{13}^1 + c_{23}^2)E_3\right]\omega^2$$
$$+ \left[(c_{12}^2 c_{13}^3 - c_{13}^2 c_{12}^3)E_1^2 - (c_{12}^1 c_{23}^3 - c_{23}^1 c_{12}^3)E_2^2 + (c_{13}^1 c_{23}^2 - c_{23}^1 c_{13}^2)E_3^2 \right.$$
$$- (c_{12}^1 c_{13}^3 - c_{13}^1 c_{12}^3 - c_{12}^2 c_{23}^3 + c_{23}^2 c_{12}^3)E_1 E_2$$
$$+ (c_{12}^1 c_{13}^2 - c_{13}^1 c_{12}^2 + c_{13}^2 c_{23}^3 - c_{23}^2 c_{13}^3)E_1 E_3$$
$$+ \left. (c_{12}^1 c_{23}^2 - c_{13}^1 c_{23}^3 - c_{23}^1 c_{12}^2 + c_{23}^2 c_{13}^3)E_2 E_3\right]\omega.$$

3.7 For $l = 1$, $k = 1$, $\rho_1 \geq 1$. For any value of ρ_1, there are two groups of size $\rho_1 + 2$ corresponding to $\alpha_1 = 0$ or $\alpha_1 = 1$, respectively. Their generators and non-vanishing commutators are

$$\partial_y, x\partial_y, \ldots, x^{\rho_1}\partial_y, \partial_x \quad \text{with} \quad [\partial_x, x^k \partial_y] = kx^{k-1}\partial_y \quad \text{and}$$
$$e^x\partial_y, xe^x\partial_y, \ldots, x^{\rho_1}e^x\partial_y, \partial_x \quad \text{with} \quad [e^x\partial_x, x^k\partial_y] = kx^{k-1}e^x\partial_y + x^k e^x\partial_y.$$

For $l = 2$, $1 \leq k \leq 2$, the constraint $\rho_1 + \rho_2 \geq 0$ becomes redundant. For any pair of values ρ_1 and ρ_2, there are two groups of size $\rho_1 + \rho_2 + 3$ corresponding to $\alpha_1 = 0$ or $\alpha_1 = 1$ respectively. Their generators and nonvanishing commutators are

$$\partial_y, x\partial_y, \ldots, x^{\rho_1}\partial_y, e^{\alpha_2 x}\partial_y, xe^{\alpha_2 x}\partial_y, \ldots, x^{\rho_2}e^{\alpha_2 x}\partial_y, \partial_x \quad \text{with}$$
$$[\partial_x, x^k\partial_y] = kx^{k-1}\partial_y, \quad [\partial_x, x^k e^{\alpha_2 x}\partial_y] = kx^{k-1}e^{\alpha_2 x}\partial_y + \alpha_2 x^k e^{\alpha_2 x}\partial_y$$

and

$$e^x\partial_y, xe^x\partial_y, \ldots, x^{\rho_1}e^x\partial_y, e^{\alpha_2 x}\partial_y, xe^{\alpha_2 x}\partial_y, \ldots, x^{\rho_2}e^{\alpha_2 x}\partial_y, \partial_x \quad \text{with}$$
$$[\partial_x, x^k e^x\partial_y] = kx^{k-1}e^x\partial_y + x^k e^x\partial_y,$$
$$[\partial_x, x^k e^{\alpha_2 x}\partial_y] = kx^{k-1}e^{\alpha_2 x}\partial_y + \alpha_2 x^k e^{\alpha_2 x}\partial_y.$$

3.8 In general, the answer is not unique. For example, the most general valid choice for group $\mathbf{g}_{17}(l = 1, \rho_1 = 1, \alpha_1 = 0)$ is $U_1 = (\beta_1\gamma_3 - \gamma_1\beta_3)\partial_y$, $U_2 = \beta_1\partial_x + \beta_2\partial_y + \beta_3 x\partial_y$, $U_3 = \gamma_1\partial_x + \gamma_2\partial_y + \gamma_3 x\partial_y$ where $\beta_1, \ldots, \gamma_3$ are

Solutions to Selected Problems

arbitrary constants. Choosing $\beta_1 = \gamma_3 = 1$ and 0 for the remaining coefficients leads to special case given below.

$\mathbf{g}_5$: $U_1 = \partial_y$, $U_2 = \partial_x$, $U_3 = y\partial_y$.
$\mathbf{g}_7$: $U_1 = \partial_x$, $U_2 = \partial_y$, $U_3 = x\partial_x + cy\partial_y$.
$\mathbf{g}_8$: $U_1 = \partial_y$, $U_2 = y\partial_y$, $U_3 = y^2\partial_y$.
$\mathbf{g}_{10}$: $U_1 = \partial_x + \partial_y$, $U_2 = x\partial_x + y\partial_y$, $U_3 = x^2\partial_x + y^2\partial_y$.
$\mathbf{g}_{13}$: $U_1 = \partial_x$, $U_2 = x\partial_x + \frac{1}{2}y\partial_y$, $U_3 = x^2\partial_x + xy\partial_y$.
$\mathbf{g}_{15}(r = 3)$: $U_1 = \phi_1\partial_y$, $U_2 = \phi_2\partial_y$, $U_3 = \phi_3\partial_y$.
$\mathbf{g}_{16}(r = 2)$: $U_1 = \phi_1\partial_y$, $U_2 = \phi_2\partial_y$, $U_3 = y\partial_y$.
$\mathbf{g}_{17}(l = 1, \rho_1 = 1, \alpha_1 = 0)$: $U_1 = \partial_y$, $U_2 = \partial_x$, $U_3 = x\partial_y$.
$\mathbf{g}_{17}(l = 1, \rho_1 = 1, \alpha_1 = 1)$: $U_1 = e^x\partial_y$, $U_2 = xe^x\partial_y$, $U_3 = -\partial_x$.
$\mathbf{g}_{17}(l = 2, \rho_1 = \rho_2 = 0, \alpha_1 = 0, \alpha_2 = 1)$: $U_1 = e^x\partial_y$, $U_2 = \partial_y$, $U_3 = -\partial_x$.
$\mathbf{g}_{17}(l = 2, \rho_1 = \rho_2 = 0, \alpha_1 = 1, \alpha_2 = c)$: $U_1 = e^x\partial_y$, $U_2 = e^{cx}\partial_y$, $U_3 = -\partial_x$.
$\mathbf{g}_{20}(r = 1)$: $U_1 = \partial_x - \partial_y$, $U_2 = \partial_x + \partial_y$, $U_3 = x\partial_x + (x + y)\partial_y$.
$\mathbf{g}_{24}$: $U_1 = \partial_x$, $U_2 = \partial_y$, $U_3 = x\partial_x + y\partial_y$.

3.9 The same remark applies as in the preceding exercise.
$\mathbf{g}_6$: $U_1 = -x\partial_x$, $U_2 = \partial_x$, $U_3 = -y\partial_y$, $U_4 = \partial_y$.
$\mathbf{g}_9$: $U_1 = \partial_y$, $U_2 = y\partial_y$, $U_3 = y^2\partial_y$, $U_4 = \partial_x$.
$\mathbf{g}_{14}$: $U_1 = \partial_x$, $U_2 = x\partial_x$, $U_3 = x^2\partial_x + xy\partial_y$, $U_4 = y\partial_y$.
$\mathbf{g}_{15}(r = 2)$: $U_k = \phi_k\partial_y$ for $k = 1, \ldots, 4$.
$\mathbf{g}_{16}(r = 1)$: $U_1 = \partial_y$, $U_2 = x\partial_y$, $U_3 = \phi(x)\partial_y$, $U_4 = y\partial_y$.
$\mathbf{g}_{17}(l = 1, \rho_1 = 2, \alpha_1 = 0)$: $U_1 = -2x\partial_y$, $U_2 = 2\partial_y$, $U_3 = x^2\partial_y$, $U_4 = \partial_x$.
$\mathbf{g}_{17}(l = 1, \rho_1 = 2, \alpha_1 = 1)$:
$U_1 = e^x\partial_y$, $U_2 = xe^x\partial_y$, $U_3 = \frac{1}{2}x^2e^x\partial_y$, $U_4 = -\partial_x$.
$\mathbf{g}_{17}(l = 2, \rho_1 = 0, \rho_2 = 1, \alpha_1 = 0, \alpha_2 = 1)$:
$U_1 = \partial_x$, $U_2 = e^x\partial_y$, $U_3 = xe^x\partial_y$, $U_4 = \partial_y$.
$\mathbf{g}_{17}(l = 2, \rho_1 = 1, \rho_2 = 0, \alpha_1 = 0, \alpha_2 = 1)$:
$U_1 = \partial_x$, $U_2 = x\partial_y$, $U_3 = \partial_y$, $U_4 = e^x\partial_y$.
$\mathbf{g}_{17}(l = 2, \rho_1 = 1, \rho_2 = 0, \alpha_1 = 1, \alpha_2 = \alpha)$:
$U_1 = \partial_x$, $U_2 = e^x\partial_y$, $U_3 = xe^x\partial_y$, $U_4 = e^{\alpha x}\partial_y$.
$\mathbf{g}_{17}(l = 3, \rho_1 = \rho_2 = \rho_3 = 0, \alpha_1 = 0, \alpha_2 = 1, \alpha_3 = c)$:
$U_1 = \partial_x$, $U_2 = e^x\partial_y$, $U_3 = e^{cx}\partial_y$, $U_4 = \partial_y$.
$\mathbf{g}_{17}(l = 3, \rho_1 = \rho_2 = \rho_3 = 0, \alpha_1 = 1, \alpha_2 = \alpha, \alpha_3 = c)$:
$U_1 = \partial_x$, $U_2 = \partial_y$, $U_3 = e^{\alpha x}\partial_y$, $U_4 = e^{cx}\partial_y$.
$\mathbf{g}_{18}(l = 1, \rho_1 = 1, \alpha_1 = 0)$: $U_1 = -\partial_y$, $U_2 = x\partial_y$, $U_3 = \partial_x$, $U_4 = y\partial_y$.
$\mathbf{g}_{18}(l = 2, \rho_1 = \rho_2 = 0)$: $U_1 = \partial_x$, $U_2 = e^x\partial_y$, $U_3 = -y\partial_y$, $U_4 = \partial_y$.
$\mathbf{g}_{19}(r = 2)$: $U_1 = \partial_y$, $U_2 = \partial_x$, $U_3 = x\partial_y$, $U_4 = x\partial_x + cy\partial_y$.
$\mathbf{g}_{20}(r = 2)$: $U_1 = -2\partial_y$, $U_2 = 2x\partial_y$, $U_3 = \partial_x$, $U_4 = x\partial_x + (x^2 + 2y)\partial_y$.
3.10 Let the expanded form of the characteristic equation be

$$\omega(\omega^3 + A\omega^2 + B\omega + C) = 0.$$

For the algebra $\mathfrak{l}_{4,4}$ the form of the roots entails the invariant relations

$$B = \tfrac{5}{16}A^2, \qquad C = \tfrac{1}{32}A^3$$

366

and for $l_{4,5}(c)$

$$B = \frac{c^2 + c - 1}{4c^2} A^2, \quad C = \frac{c-1}{8c^2} A^3.$$

3.11 For gl_2 the answer is $U_1 = -x\partial_y$, $U_2 = \frac{1}{2}(y\partial_y - x\partial_x)$, $U_3 = y\partial_x$ and $U_4 = x\partial_x + y\partial_y$. For sl_2 the generators U_1, U_2 and U_3 are the same as for gl_2.

4. Equivalence and Invariants of Differential Equations

4.1 Let $p_0 y^{(n)} + p_1 y^{(n-1)} + \ldots + p_{n-1} y' + p_n y = 0$ be a linear homogeneous ode with polynomial coefficients, i.e., $p_k \in \mathbb{Q}[x]$. Applying (B.16), the general transformation leading to the desired form is $x = F(u)$, $y = G(u)v$ with

$$G = F'^{n/2} q_0^{-1/(n+1)} \exp\left(-\frac{1}{n+1} \int \frac{q_1}{q_0} F' du\right).$$

$F(u)$ is an undetermined function of u, $q_k(u) = p_k(F)$. According to Hirsch [72], $p_1 = -p_0'$ is a necessary condition such that there is a fundamental system of polynomial integrals which is by no means sufficient. An example is $x^2 y'' - 3xy' + 5y = 0$ with the fundamental system $y_1 = x^2$ and $y_2 = x^2 \log x$. Choosing $F = u$ or $F = \frac{1}{u}$ yields

$$u^{7/3} v'' - \tfrac{7}{3} u^{4/3} v' - \tfrac{11}{9} u^{1/3} v = 0 \quad \text{or} \quad u^{-1/3} v'' + \tfrac{1}{3} u^{-4/3} v' - \tfrac{35}{9} u^{-7/3} v = 0$$

respectively. Although $q_1 = -q_0'$, neither of these equations has a polynomial fundamental system.

4.2 A general point transformation $u = \sigma(x,y)$ and $v = \rho(x,y)$ leads from $v' = 0$ to $y' + \frac{\rho_x}{\rho_y} = 0$. This is the given equation $y' + r(x,y) = 0$ if there holds $\rho_x - r(x,y)\rho_y = 0$; i.e., ρ is the first integral of the given equation. The other transformation function $\sigma(x,y)$ may be chosen arbitrarily. This result makes the degree of arbitrariness for the equation given in the proof of Theorem 4.10 completely explicit.

4.3 If $F(u) = u$, the transformation (4.37) simplifies to

$$x = u, \; y = \frac{1}{a_2}\left(v + \frac{1}{2}\frac{a_2'}{a_2} - \frac{1}{2}a_1\right), \; y' = \frac{v'}{a_2} - \frac{a_2'}{a_2^2}v + \frac{a_2''}{2a_2^2} - \frac{a_1'}{2a_2} + \frac{a_1 a_2'}{2a_2^2} - \frac{a_2'^2}{a_2^3}.$$

Substitution into (4.33) and some simplification yields

$$v' + v^2 + \tfrac{1}{2}z' - \tfrac{1}{4}z^2 + a_0 a_2 \quad \text{where} \quad z = \frac{a_2'}{a_2} - a_1 \quad \text{and} \quad a_k \equiv a_k(u).$$

4.4 By Theorem 4.13 the general form of the desired transformation is $x = F(u)$, $y = G(u)v + H(u)$. Upon substitution into Appell's normal form the constraints $F'G^2 = 1$, $G' = 0$ and $H = 0$ are obtained. They yield the two-parameter group $x = e^a u + b$, $y = e^{-a/2}v$ with Lie algebra $\{\partial_x, \partial_x - \frac{1}{2}\partial_y\}$.

Solutions to Selected Problems 367

4.5 Applying formulas (B.18) to Lie's equation (4.56) yields after some rearrangement

$$v'' + \bar{A}(u,v)v'^3 + \bar{B}(u,v)v'^2 + \bar{C}(u,v)v' + \bar{D}(u,v) = 0;$$

i.e., there holds $\bar{A}(u,v) = -D(x,y)|_{x=v,y=u}$, $\bar{B}(u,v) = -C(x,y)|_{x=v,y=u}$, $\bar{C}(u,v) = -B(x,y)|_{x=v,y=u}$ and $\bar{D}(u,v) = -A(x,y)|_{x=v,y=u}$.

4.6 Applying the transformation $x = \alpha u$, $y = e^{\beta u}v$ to $y''' + cy' + y = 0$ yields

$$v''' + 3\beta v'' + (3\beta^2 + c\alpha^2)v' + (\alpha^3 + \alpha^2\beta c + \beta^3)v = 0.$$

Retaining the structure of the original equations requires $\beta = 0$ and $\alpha^3 = 1$. From the factorization $\alpha^3 - 1 = (\alpha - 1)(\alpha^2 - \alpha - 1)$ the asserted equivalence is obvious.

4.7 If $y''' - (a+1)y'' + ay' = 0$ is transformed by $x = \alpha u$, $y = e^{\beta u}v$, the equation

$$v''' + (3\beta - a\alpha - \alpha)v'' + (3\beta^2 + a\alpha^2 - 2a\alpha\beta - 2\alpha\beta)v' + (\beta^2 + a\alpha^2 - a\alpha\beta - \alpha\beta)\beta v = 0$$

is obtained. In order to retain the structure of the original equation, the coefficient of v must vanish. This yields $\beta(\alpha - \beta)(a\alpha - \beta) = 0$; i.e., $\beta = 0$, $\beta = \alpha$ or $\beta = a\alpha$. Substituting these values into the coefficients of v'' and v' and imposing the conditions for the structural invariance leads to the given five alternatives.

4.8 By Theorem 5.15, case ii), the only nonvanishing coefficients of equation (5.41) are $B_3 = 2p(f)$ and $B_5 = 2q(u)$. Consequently, it belongs to symmetry class $\mathcal{S}_6^3$. The invariant $J_1 = 2p(f)$ in equations (6.72) or (6.73) shows that determining S or R comes down to solving a linear second order equation. If the goal is to solve any given third order equation in $\mathcal{S}_6^3$ this is an important simplification. In the particular application at issue, however, it does not lead toward the desired answer.

4.9 Let the given equation be in rational normal form $y'' + q(x)y = 0$ with $q(x) \in \mathbb{Q}(x)$. In order to decide the desired equivalence, a rational solution of the equation

$$\left(\frac{f''}{f'} \right)' - \frac{1}{2}\left(\frac{f''}{f'} \right)^2 - \frac{1}{2}f'^2 f^2 - (a-1)f'^2 = 2q(x) \tag{A.5}$$

for $f(x)$ has to be found. In a partial fraction expansion of f let the leading term for $x \to \infty$ be $a_N x^N$. The first two terms at the left hand side generate a term $\dfrac{1 - N^2}{x^2}$ from it if $N > 1$, whereas the remaining two terms generate the leading powers $4N - 2$ and $2N - 2$ for $N \geq 1$. If n is the leading power in the partial fraction expansion of $q(x)$, the leading power $4N - 2$ must be matched by it which leads to the bound $N = \dfrac{n+1}{4}$ for the behavior at infinity. For any pole of order M at $x = x_0$, a similar consideration yields the term $\dfrac{1 - M^2}{(x - x_0)^2}$ for $M > 1$ and two more terms of highest power $4M + 2$ and $2M$ for

368

$M \geq 1$. The leading power $4M + 2$ must be matched by a singularity of order m such that $M = \frac{m-2}{4}$. Consequently, all singularities of a rational solution are determined by the function $q(x)$ at the right hand side and its orders are properly bounded. An ansatz with undetermined coefficients within these bounds leads to a linear system any solution of which determines a rational solution for f.

4.10 With the same notation as above, a rational solution has to be found for

$$\left(\frac{f''}{f'} \right)' - \frac{1}{2} \left(\frac{f''}{f'} \right)^2 - 2 \left(\nu^2 - \frac{1}{4} \right) \left(\frac{f'}{f} \right)^2 + 2f'^2 = 2q(x). \qquad (A.6)$$

By a similar argumentation, the bounds for the polynomial part and any finite pole are now $N = \frac{n}{2} + 1$ and $M = \frac{m}{2} - 1$ respectively leading again to a linear system for the undetermined coefficients.

5. Symmetries of Differential Equations

5.1 By Definition 5.3 the left hand side of the relation at issue may be written as

$$U^{(n+1)} D(\phi) = \xi \frac{\partial D(\phi)}{\partial x} + \eta \frac{\partial D(\phi)}{\partial y} + \zeta^{(1)} \frac{\partial D(\phi)}{\partial y'} + \dots$$
$$+ \zeta^{(n)} \frac{\partial D(\phi)}{\partial y^{(n)}} + \zeta^{(n+1)} \frac{\partial D(\phi)}{\partial y^{(n+1)}}. \qquad (A.7)$$

Expanding the right hand side yields

$$D(U^{(n)}) - D(\xi)D(\phi)$$

$$= D \left(\xi \frac{\partial \phi}{\partial x} + \eta \frac{\partial \phi}{\partial y} + \zeta^{(1)} \frac{\partial \phi}{\partial y'} + \dots + \zeta^{(n)} \frac{\partial \phi}{\partial y^{(n)}} \right) - D(\xi)D(\phi)$$

$$= D(\xi) \frac{\partial \phi}{\partial x} + D(\eta) \frac{\partial \phi}{\partial y} + D(\zeta^{(1)}) \frac{\partial \phi}{\partial y'} + \dots + D(\zeta^{(n)}) \frac{\partial \phi}{\partial y^{(n)}}$$

$$+ \xi D \left(\frac{\partial \phi}{\partial x} \right) + \eta D \left(\frac{\partial \phi}{\partial y} \right) + \zeta^{(1)} D \left(\frac{\partial \phi}{\partial y'} \right) + \dots$$

$$+ \zeta^{(n)} D \left(\frac{\partial \phi}{\partial y^{(n)}} \right) - D(\xi)D(\phi).$$

Applying the relations

$$D(\eta) = \zeta^{(1)} + y'D(\xi), \quad D(\zeta^{(k)}) = \zeta^{(k+1)} + y^{(k+1)}D(\xi)$$

$$\text{Solutions to Selected Problems} \qquad 369$$

for $k = 1, \ldots, n$ that follow from (5.10) it may be rewritten as

$$D(\xi)\frac{\partial \phi}{\partial x} + [\zeta^{(1)} + y'D(\xi)]\frac{\partial \phi}{\partial y} + [\zeta^{(2)} + y''D(\xi)]\frac{\partial \phi}{\partial y'} + \ldots$$

$$+ [\zeta^{(n+1)} + y^{(n+1)}D(\xi)]\frac{\partial \phi}{\partial y^{(n)}}$$

$$+ \xi D\left(\frac{\partial \phi}{\partial x}\right) + \eta D\left(\frac{\partial \phi}{\partial y}\right) + \zeta^{(1)}D\left(\frac{\partial \phi}{\partial y'}\right) + \ldots$$

$$+ \zeta^{(n)}D\left(\frac{\partial \phi}{\partial y^{(n)}}\right) - D(\xi)D(\phi).$$

The first term and the second term in the square brackets combined cancel the last term with the result

$$\xi D\left(\frac{\partial \phi}{\partial x}\right) + \eta D\left(\frac{\partial \phi}{\partial y}\right) + \zeta^{(1)}\left[\frac{\partial \phi}{\partial y} + D\left(\frac{\partial \phi}{\partial y'}\right)\right] + \ldots$$

$$+ \zeta^{(n+1)}\left[\frac{\partial \phi}{\partial y^{(n)}} + D\left(\frac{\partial \phi}{\partial y^{(n+1)}}\right)\right].$$

The terms in the square brackets may be simplified using the obvious relations

$$\partial_x D = D\partial_x, \quad \partial_y D = D\partial_y, \quad \partial_{y^{(k)}} D = \partial_{y^{(k-1)}} + D\partial_{y^{(k)}}$$

for $k = 1, \ldots n$ such that they yield the right hand side of (A.7).

5.2 Applying Theorem 5.2 to $y'' + r(x)y = 0$ leads to the determining system

$$\xi_{yy} = 0, \quad \eta_{yy} - 2\xi_{xy} = 0, \quad \eta_{xy} - \tfrac{1}{2}\xi_{xx} + \tfrac{3}{2}ry\xi_y = 0,$$

$$\eta_{xx} - ry(\eta_y - 2\xi_x) + r\eta + r'y\xi = 0.$$

Its general solution contains eight constants and yields the generators

$$U_1 = y_1\partial_y, \quad U_2 = y_2\partial_y, \quad U_3 = y\partial_y,$$

$$U_4 = y_1y\partial_x + y_1'y^2\partial_y, \quad U_5 = y_2y\partial_x + y_2'y^2\partial_y,$$

$$U_6 = y_1^2\partial_x + y_1y_1'y\partial_y, \quad U_7 = y_2^2\partial_x + y_2y_2'\partial_y,$$

$$U_8 = 2y_1y_2\partial_x + (y_1'y_2 + y_1y_2')y\partial_y.$$

5.3 If a symmetry generator is searched for in the form $U = \eta(x)\partial_y$, (5.10) yields $\zeta^{(k)} = \frac{\partial^{(k)}\eta}{\partial x^k}$. Consequently, the determining system for $\eta(x)$ is

$$p_0\eta + p_1\eta_x + p_2\eta_{xx} + \ldots + \frac{\partial^{(n)}\eta}{\partial x^n} = 0;$$

i.e., it is identical to the given equation for y. Now if a symmetry generator is searched for in the form $U = \eta(y)\partial_y$, from equations (5.10) it follows $\zeta^{(1)} =$

370

$y'\eta_y$ and $\zeta^{(k)} = D(\zeta^{(k-1)})$ with $D = y'\partial_y + y''\partial_{y'} + \ldots$, therefore the general structure of $\zeta^{(k)}$ is

$$\zeta^{(k)} = y'\frac{\partial^k \eta}{\partial y^k} + y^{(k)}\eta_y + \ldots$$

where the omitted terms are proportional to derivatives $\dfrac{\partial^i \eta}{\partial y^i}$ with $1 < i < n - 1$. From this the determining system $\eta_{yy} = 0$, $p_n\eta - p_ny\eta_y = 0$ follows with the solution $\eta = y$.

5.4 The most general form that is linear and homogeneous in u, v and w is

$$w_x + u + Pv + Qw = 0, \quad w_y + Ru - v + Sw = 0.$$

The requirement that no new integrability conditions are introduced is assured under the condition $S_x - Q_y - \frac{1}{3}C_y - \frac{1}{3}B_x = 0$.

5.5 The subsequent listing allows one to associate Lie's enumeration of groups 1), $\ldots$, 13) with the notation in this book: 1) $\equiv \mathbf{g}_{10}$, 2) $\equiv \mathbf{g}_{13}$, 3) $\equiv \mathbf{g}_8$, 4) $\equiv \mathbf{g}_7$, 5) $\equiv \mathbf{g}_{17}$ with $l = 2$, $\rho_1 = \rho_2 = 0$, $\alpha_1 = 1$, $\alpha_2 = \alpha \neq 0, 1$, 6) $\equiv \mathbf{g}_{24}$, 7) $\equiv \mathbf{g}_{16}$, 8) $\equiv \mathbf{g}_{20}$, 9) $\equiv \mathbf{g}_{17}$ with $l = 1$, $\rho_1 = 1$, $\alpha_1 = 1$, 10) $\equiv \mathbf{g}_5$, 11) $\equiv \mathbf{g}_{17}$ with $l = 2$, $\rho_1 = \rho_2 = 0$, $\alpha_1 = 0$, $\alpha_2 = 1$, 12) $\equiv \mathbf{g}_{17}$ with $l = 1$, $\rho_1 = 1$, $\alpha_1 = 0$, 13) $\equiv \mathbf{g}_{15}$. For three groups a change of variables has to be applied in order to obtain the generators in Lie's form: 9) is obtained from $\mathbf{g}_{17}$ by the change $x = -\bar{x}$, $y = \bar{y}e^{\bar{x}}$, 11) by the change $x = -\log \bar{x}$, $y = \dfrac{\bar{y}}{\bar{x}}$ and 5) by $x = \dfrac{1}{\alpha - 1}\log \bar{x}$ and $y = \bar{y}\bar{x}^{1/(\alpha-1)}$. Finally, the barred variables have to be replaced by the original ones. In Theorem 5.8 there occur only groups 1), 2), 4) and 8) because there are no second order ode's with a symmetry group different from these. This is related to the definition of a symmetry group as the *maximal* group of transformations leaving the equation invariant; see also the remarks by Lie [113] on page 538.

5.6 In order to avoid an unspecified dependence on y in the transformed function r, $\sigma_y = 0$ is required; i.e., $\sigma(x,y) \equiv f(x)$. The quotient $\dfrac{v''}{v'}$ is reproduced if the two derivatives are proportional to y'' and y' respectively. This is assured if $\rho_x = 0$ and $\rho_{yy} = 0$, i.e., $\rho = a_1y + a_2$. It yields

$$\frac{v''}{v'} = \frac{1}{f'}\frac{y''}{y'} - \frac{f''}{f'^2} \quad and \quad v''' = \frac{a_1}{f'^3}y''' - \frac{3a_1 f''}{f'^4}y'' + \frac{3a_1(f''^2 - f''' f')}{f'^5}y'.$$

If the second term of the expression for v''' is rewritten as $-\dfrac{3a_1 f''}{f'^4}\dfrac{y''}{y'}y'$ its correct structure is obvious.

5.7 In order to avoid an unspecified dependence on y in the transformed function r, $\sigma_y = 0$ is required, i.e., $\sigma = h(x)$. The independence of v'' on y' in the first place requires $\rho_{yy} = 0$ due to the quadratic term in equation (5.5), i.e., it has the form $\rho(x,y) = f(x) + g(x)y$. The coefficients of the linear term in y' and the independence on y vanish if f and g satisfy

$$2\frac{g'}{g} - \frac{h''}{h'} = 0, \quad \frac{g''}{g'} - \frac{h''}{h'} = 0$$

with the general solution $g = \frac{1}{C_1 x + C_2}$, $h = C_3 g + C_4$, and, consequently,

$$\sigma = \frac{(C_1 x + C_2)C_4 + C_3}{C_1 x + C_2}, \quad \rho = \frac{y}{C_1 x + C_2}.$$

The function $f(x)$ remains undetermined. Introducing new constants by

$$C_1 = -\frac{a_2 a_4}{a_3}, \quad C_2 = \frac{1 - a_1 a_2 a_4}{a_3}, \quad C_3 = \frac{1}{a_4}, \quad C_4 = -\frac{a_3}{a_4}$$

yields the required answer.

5.8 Both conditions are subsumed under the constraint $\Psi_3' = 0$ with Ψ_3 defined in Theorem 4.9 as a simple calculation shows. They may also be obtained from the Janet basis for the symmetry generators as it has been shown by Schwarz [166].

5.9 Equation (5.46) is completely reducible into first order factors

$$\mathrm{Lclm}(D - \tfrac{3}{2}\tfrac{3}{x-2} + \tfrac{3}{x}, \; D - \tfrac{1}{2} - \tfrac{1}{x+2} - \tfrac{2}{x-2} + \tfrac{3}{x},$$
$$D + \tfrac{1}{2} - \tfrac{2}{x+2} - \tfrac{1}{x-2} + \tfrac{3}{x}, \; D + \tfrac{3}{2} - \tfrac{3}{x+2} + \tfrac{3}{x})y = 0$$

where $D \equiv \frac{d}{dx}$. From this representation the answer is obvious.

6. Transformation to Canonical Form

6.1 Denoting the symmetry algebra of symmetry class $\mathcal{S}_1^2$ by $\mathcal{L}_1^2$, symmetry class $\mathcal{S}_{2,1}^2$ by $\mathcal{L}_{2,1}^2$, etc., the diagram

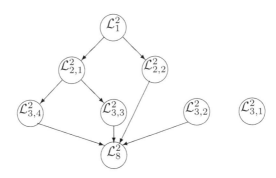

describes the relationship between the various Lie algebras. An arrow from $\mathcal{L}_i$ to $\mathcal{L}_j$ means on the one hand that $\mathcal{L}_i \subset \mathcal{L}_j$. On the other hand it implies

372

that the undetermined elements in the canonical form corresponding to the former symmetry algebra may be specialized such that the symmetry algebra enlarges to the latter. Accordingly, the arrows originating from $\mathcal{L}_1^2$ imply the relations $r_u(u, v') = 0$ or $u r_u(u, v') + r(u, v') = 0$, respectively, for the undetermined function $r(u, v')$ in the canonical form for the symmetry class $\mathcal{S}_1^2$ as given in Theorem 6.4. The arrows originating from $\mathcal{L}_{2,1}^2$ imply that the undetermined function $r(v')$ in the canonical form for symmetry class $\mathcal{S}_{2,1}^2$ has the special form required for symmetry class $\mathcal{S}_{3,3}^2$ or $\mathcal{S}_{3,4}^2$. The remaining arrows imply that the respective undetermined elements vanish such that the canonical form $v'' = 0$ for symmetry class $\mathcal{S}_8^2$ is obtained.

6.2 The desired transformation must have the form $u = a_1 \bar{u} + a_2 \bar{v} + a_3$, $v = a_4 \bar{u} + a_5 \bar{v} + a_6$. Substituting it into the canonical form of Example 6.7 leads to the algebraic system

$$a_1(4a_1^3 + 8a_1^2 a_4 + 5a_1 a_4^2 + a_4^3) = 0, \quad a_2(4a_2^3 - 8a_2^2 a_4 + 5a_2 a_4^2 - a_4^3) = 0,$$

$$(a_1 + a_2)a_4^2 = 1.$$

There are 24 alternatives. A particularly simple one is $a_1 = 0$, $a_2 = a_4 = 1$, $a_5 = -1$, a_3 and a_6 arbitrary.

6.3 For symmetry class $\mathcal{S}_{2,1}^2$ the system (6.4) may be solved for the Janet basis coefficients with the result

$$a_1 = b_2 = I_3, \quad a_2 = I_4, \quad b_1 = I_1, \quad c_1 = d_2 = -I_5, \quad c_2 = -I_6, \quad d_1 = -I_2.$$

These relations have obviously the form given in Theorem 6.6. For symmetry class $\mathcal{S}_{2,2}^2$, system (6.9) yields

$$a_1 = -I_3 - K_2, \quad a_2 = -I_4 - K_1, \quad b_1 = -I_1, \quad b_2 = -I_3,$$

$$c_1 = I_5, \quad c_2 = I_6, \quad d_1 = I_2 - K_2, \quad d_2 = I_5 - K_1.$$

They may be rewritten in the form given in Theorem 6.6 as follows.

$$I_1 = -b_1, \quad I_2 = d_1 - a_1 + b_2, \quad I_3 = -b_2, \quad I_4 = d_2 - a_2 - c_1,$$

$$I_5 = c_1, \quad I_6 = c_2, \quad K_1 = c_1 - d_2, \quad K_2 = b_2 - a_1.$$

6.4 Applying the algorithm *Symmetries* on page 199 to $y'' + r(y, y') = 0$ generates the Janet basis $\eta = \xi_x = \xi_y = 0$; i.e., case b) of Theorem 6.7 with $a = b = c = 0$. Consequently, $\phi = y$ and $G = 0$, and the general transformation may be written in the form $x = Cv + f(u)$, $y = g(u)$ with suitably defined functions f and g. Together with

$$y' = \frac{g'}{Cv' + f'}, \quad y'' = \frac{-Cg'v'' + Cg''v' + f'g'' - f''g'}{(Cv' + f')^2}$$

the desired transformation follows. The special values $C = 1$, $f = 0$ and $g = u$ generate the variables exchange discussed after the proof of Theorem 6.7.

$$\textit{Solutions to Selected Problems} \qquad 373$$

6.5 The vector fields determined by $\xi_x - \frac{2}{x}\xi = 0$, $\xi_y = 0$, $\eta_x = 0$ and $\eta_y - \frac{2}{y}\eta = 0$ generate a two-parameter subgroup of the full symmetry group because it reduces the Janet basis (5.32) to zero. Applying the scheme described in Theorem 6.8, the system $\sigma_{xx} + \frac{1}{x}\sigma_x = 0$, $\sigma_{xy} = 0$, $\sigma_{yy} + \frac{2}{y}\sigma_y = 0$ is obtained with the fundamental system $\{1, \frac{1}{x}, \frac{1}{y}\}$. The transformation $x = \frac{1}{u}$ and $y = \frac{1}{v}$ yields $v''v' - 1 = 0$; i.e., a canonical form $v'' + r(v') = 0$ corresponding to symmetry class $\mathcal{S}_{2,1}^2$. In order to obtain the special value $r(v') = v'^4$, the genuine symmetry class has to be known.

6.6 The vector fields determined by $\xi_x - \frac{\eta + \xi}{x + y} = 0$, $\xi_y = 0$, $\eta_x = 0$ and $\eta_y - \frac{\eta + \xi}{x + y}\eta = 0$ generate a two-parameter subgroup of the full symmetry group because it reduces the Janet basis given in Example 5.14 to zero. Evaluating the integrals given in Theorem 6.9 yields for the transformation functions $\sigma = x + y$ and $\rho = y$. Substituting $x = u - v$ and $y = v$ into the given equation yields the canonical form $uv'' - 2v'^3 - 3v'^2 - v' = 0$. The canonical form given in Example 6.11 cannot be obtained by specialization from it.

6.7 Three generators of a sub-algebra of the full symmetry algebra are $U_1 = \frac{1}{2}x\partial_x$, $U_2 = x^2(y - 1)^2\partial_y$ and $U_3 = y(y - 1)\partial_y$ with commutators $[U_1, U_2] = U_2$, $[U_1, U_3] = 0$ and $[U_2, U_3] = -U_2$. A $\mathbf{g}_{26}$ subgroup is generated by U_1 and U_3. From its Janet basis

$$\xi_x + \frac{1}{x}\xi = 0, \quad \xi_y = 0, \quad \eta_x = 0, \quad \eta_y - \frac{2y - 1}{y(y - 1)}\eta = 0$$

Theorem 6.8 yields $\sigma_{xx} + \frac{1}{x}\sigma_x = 0$, $\sigma_{xy} = 0$, $\sigma_{yy} + \frac{2y - 1}{y(y - 1)}\sigma_y = 0$ with fundamental system $\{1, \log x, \log \frac{y - 1}{y}\}$. It leads to the transformation functions $u = \log x$ and $v = \log \frac{y - 1}{y}$ with the inverse $x = e^u$, $y = \frac{1}{1 - e^v}$ from which the canonical form $v'' - v'^2 - 3v' - 1 = 0$ is obtained.

Choosing the two-parameter group of type $\mathbf{g}_{25}$ generated by U_1 and U_2 with Janet basis $\xi_x - \frac{1}{x}\xi = 0$, $\xi_y = 0$, $\eta_x - \frac{2}{x}\eta = 0$, $\eta_y - \frac{2}{y - 1}\eta = 0$, by Theorem 6.8 there follows $u = \frac{1}{x^2}$, $v = \frac{1}{x^2(y - 1)}$. From its inverse $x = \frac{1}{\sqrt{u}}$ and $y = \frac{u}{v} + 1$ the canonical form $uv'' + \frac{1}{2}v' + \frac{1}{2} = 0$ follows, i.e., an inhomogeneous linear ode.

The three generators U_1, U_2 and U_3 generate a three-parameter group with Lie algebra $\mathfrak{l}_{3,4}$. This symmetry type is *not* included in the classification given in Theorem 5.8 because by definition a symmetry class is defined by the *maximal* group leaving it invariant.

7. Solution Algorithms

7.1 For the ode in Example 6.2 the equation

$$\frac{dy + (y^2 - 2x^2y + x^4 - 2x - 1)dx}{y^2 - 2x^2y + x^4 - 1} = 0.$$

is obtained. Upon integration it yields the same answer as given there. In this case, the integrations for obtaining the canonical form are simpler.

7.2 In Example 6.2 the transformation functions $x \equiv \phi(u,v) = u - v + \frac{1}{2v}$ and $y \equiv \psi(u,v) = u - v - \frac{1}{2v}$ have been shown to generate the canonical form $v' = \frac{1}{3}$. Substituting its inverse $u = \frac{1}{x-y} + \frac{x+y}{2}$, $v = \frac{1}{x-y}$ into the canonical form solution $v = \frac{2}{3}u + C$ yields the solution in actual variables as given in Example 6.2.

7.3 According to Lemma 3.4, canonical coordinates are $u = y^2 - x^2$ and $v = \frac{1}{2}\log\frac{y+x}{y-x}$. They lead to the canonical form $v' + \frac{1}{2u}\frac{u+1}{u-1} = 0$. Integrating and resubstituting x and y yield the solution given in Example 6.2.

7.4 Let $x = g(y')$ be a first order ode and consider y' as a parameter, say p, such that $p = y' = h(x)$ with $h = g^{-1}$. It follows $y = \int h(x)dx + C$ or $y = \int pg'(p)dp + C$. Partial integration yields $y = pg(p) - \int g(p)dp + C$. Substituting u and v for x and y respectively, $-R(p) + C_1$ for g, and C_2 for C leads to (7.16).

7.5 Substituting $u = e^{\bar{w}}$, $z = t$ with $\bar{w} \equiv \bar{w}(t)$, and then $\bar{w}' = \bar{w}$ into (7.38) yields $\bar{w}' + t(t^2 + 4)\bar{w}^3 + 6\bar{w}^2 + \frac{3}{t}\bar{w} = 0$. This is an Abel equation. Its rational normal form $w' + t(t^2 + 4)w^3 + \frac{3t}{t^2 + 4}w = 0$ is obtained by substituting (see the proof of Theorem 4.14) $\bar{w} = w - \frac{2}{t(t^2 + 4)}$ and leaving t unchanged. This Bernoulli equation has the general solution

$$w = \left[C(t^2 + 4)^3 + \tfrac{1}{4}(t^2 + 4)^2 t^2\right]^{-\frac{1}{2}}.$$

Resubstitution of the original variables and redefining the integration constants finally yields the solution given in Example 7.19.

7.6 Now $r(p) = \frac{p(p+1)^2}{p-1}$, $R(p) = \log\frac{p+1}{p} + \frac{2p}{p+1}$ and $\int R(p)dp = (p-1)\log(p+1) - p\log p + 2p$. Substitution into (7.16) yields

$$u + \log\frac{p+1}{p} + \frac{2p}{p+1} = \bar{C}_1, \quad v + \log(p+1) - \frac{2p}{p+1} = \bar{C}_2$$

and finally

$$x = \frac{C_1}{p+1}\exp\frac{2p}{p+1}, \quad y = \frac{C_2 p}{p+1}\exp-\frac{2p}{p+1}.$$

7.7 For generic $r(y)$, a type $\mathcal{J}_{1,2}^{(2,2)}$ Janet basis with $a = b = c = 0$ is obtained for the determining system. By Theorem 6.7, case b), the transformation to canonical form is achieved by $x = v$, $y = u$ with the result

$$\text{Solutions to Selected Problems} \qquad 375$$

$v'' - v'^3 r(u) = 0$. Two integrations and resubstitution of the actual variables leads to the integral representation

$$\int \frac{dy}{\sqrt{C_1 - 2\int r(y)dy}} = x + C_2$$

of the general solution.

7.8 For generic $r(x)$, a type $\mathcal{J}_{1,1}^{(2,2)}$ Janet basis with $a = 0$, $b = -\frac{1}{y}$ is obtained for the determining system. By Theorem 6.7, case a), the transformation to canonical form is achieved by $x = u$, $y = e^v$ with the result $v'' + r(u)v'^2 + v' = 0$. This Riccati equation for v' has the general solution

$$v' = \frac{e^{-u}}{\int r(u)e^{-u}du + C_1}.$$

Integration and resubstitution leads to the general solution in actual variables

$$y = C_2 \exp \int \frac{e^{-x}dx}{\int r(x)e^{-x}dx + C_1}.$$

7.9 The first integrals following from the equivalent pde may be found in Lie [113], page 540. In the notation applied in this book they yield the relations

$$u^2 + \frac{v^2 - 2uvv'}{v'^2 - \dfrac{a}{v^2}} = \bar{C}_1, \quad u - \frac{vv'}{v'^2 - \dfrac{a}{v^2}} = \bar{C}_2.$$

Elimination of v' leads to

$$v^4 - a\frac{u^4 - 4\bar{C}_1 u^3 + (4\bar{C}_1^2 - 2\bar{C}_2)u^2 + 4\bar{C}_1\bar{C}_2 u + \bar{C}_2^2}{\bar{C}_1^2 + \bar{C}_2} = 0.$$

This representation is equivalent to (7.25) if $C_1^2 = \frac{a}{\bar{C}_1}^2 + \bar{C}_2$, $C_2 = -\bar{C}_1$. This result shows that the solutions obtained from the first integrals of the equivalent pde may be unnecessarily complicated.

Determine the solution of (7.24) from the equivalent pde and compare the result with (7.25).

7.10 Write the equation to be solved as $y' + \dfrac{1}{ax\sqrt{x}}y^2 + \dfrac{1}{a\sqrt{x}} = 0$. The symmetry generator $x\partial_x + \frac{1}{2}y\partial_y$ yields the canonical variables $u = \dfrac{y}{\sqrt{x}}$, $v = 2\log y$.

Substituting the inverse $x = \dfrac{1}{u^2}e^v$, $y = e^{v/2}$ yields $v' = \dfrac{4}{u}\dfrac{u^2 + 1}{2u^2 + au + 2}$. An integration leads to $v = 2\log u + \dfrac{4ai}{\sqrt{a^2 - 16}} \arctan\dfrac{4u + a}{i\sqrt{a^2 - 16}} + C$. Applying the relation $\arctan x = -\dfrac{i}{2}\log\dfrac{1 + ix}{1 - ix}$, it assumes the form

$$v = 2\log u + \frac{2a}{\sqrt{a^2 - 16}}\log\frac{\sqrt{a^2 - 16} + a + 4u}{\sqrt{a^2 - 16} - a - 4u} + C.$$

376

Resubstitution of the original variables and some simplifications lead to

$$y = \frac{a\sqrt{x}}{2} \frac{(c - \frac{1}{2})x^c - (c + \frac{1}{2})C}{x^c + C}$$

where $c = \frac{\sqrt{a^2 - 16}}{2a}$.

7.11 The pde for (7.71) is $Af = 0$ with $A = \partial_u + v'\partial_v + v''\partial_{v'} + w\partial_{v''}$ and $w = -\frac{v''}{v}(3v' + a\sqrt{v''v})$. The linear relation between A and the four vector fields generating the symmetry group yields the first integrals

$$\Phi_1 = u - \frac{2vv'}{D}, \quad \Phi_2 = 2u - \frac{4vv' + av\sqrt{v''v}}{D}, \quad \Phi_3 = u^2 + \frac{2v^2 - (4vv' + av\sqrt{v''v})u}{D}$$

with $D = 2vv'' + 2v'^2 + av'\sqrt{v''v}$. Solving the relations $\Phi_i = C_i$, $i = 1, 2, 3$, for $\frac{v'}{v}$ yields the integrand in (7.71).

Appendix B

Collection of Useful Formulas

For ease of reference in this appendix those formulas are put together that occur frequently in the main text. Moreover, they might be of general interest for any person working in this field.

Transformations of the Plane. Let a coordinate transformation from x, y to u, v and its inverse be defined by

$$x = \phi(u,v), \quad y = \psi(u,v) \quad u = \sigma(x,y), \quad v = \rho(x,y). \tag{B.1}$$

They entail the following relations for the first order partial derivatives.

$$\begin{pmatrix} \sigma_x & \sigma_y \\ \rho_x & \rho_y \end{pmatrix} = \begin{pmatrix} \phi_u & \phi_v \\ \psi_u & \psi_v \end{pmatrix}^{-1} = \frac{1}{\phi_u \psi_v - \phi_v \psi_u} \begin{pmatrix} \psi_v & -\phi_v \\ -\psi_u & \phi_u \end{pmatrix} \tag{B.2}$$

and

$$\begin{pmatrix} \phi_u & \phi_v \\ \psi_u & \psi_v \end{pmatrix} = \begin{pmatrix} \sigma_x & \sigma_y \\ \rho_x & \rho_y \end{pmatrix}^{-1} = \frac{1}{\sigma_x \rho_y - \sigma_y \rho_x} \begin{pmatrix} \rho_y & -\sigma_y \\ -\rho_x & \sigma_x \end{pmatrix}. \tag{B.3}$$

It is assumed that $\Delta \equiv \sigma_x \rho_y - \sigma_y \rho_x \neq 0$ and $\phi_u \psi_v - \phi_v \psi_u \neq 0$. The first and second order derivatives of ϕ and ψ are

$$\phi_u = \frac{\rho_y}{\Delta}, \quad \phi_v = \frac{-\sigma_y}{\Delta}, \quad \psi_u = \frac{-\rho_x}{\Delta}, \quad \psi_v = \frac{\sigma_x}{\Delta}, \tag{B.4}$$

$$\begin{aligned}
\phi_{uu} &= \frac{1}{\Delta^3}[\rho_y(\rho_{xy}\Delta - \rho_y\Delta_x) - \rho_x(\rho_{yy}\Delta - \rho_y\Delta_y)], \\
\phi_{uv} &= \frac{1}{\Delta^3}[-\sigma_y(\rho_{xy}\Delta - \rho_y\Delta_x) + \sigma_x(\rho_{yy}\Delta - \rho_y\Delta_y)], \\
\phi_{vv} &= \frac{1}{\Delta^3}[\sigma_y(\sigma_{xy}\Delta - \sigma_y\Delta_x) - \sigma_x(\sigma_{yy}\Delta - \sigma_y\Delta_y)], \\
\psi_{uu} &= \frac{1}{\Delta^3}[-\rho_y(\rho_{xx}\Delta - \rho_x\Delta_x) + \rho_x(\rho_{xy}\Delta - \rho_x\Delta_y)], \\
\psi_{uv} &= \frac{1}{\Delta^3}[\sigma_y(\rho_{xx}\Delta - \rho_x\Delta_x) - \sigma_x(\rho_{xy}\Delta - \rho_x\Delta_y)], \\
\psi_{vv} &= \frac{1}{\Delta^3}[-\sigma_y(\sigma_{xx}\Delta - \sigma_x\Delta_x) + \sigma_x(\sigma_{xy}\Delta - \sigma_x\Delta_y)].
\end{aligned} \tag{B.5}$$

If $y \equiv y(x)$ and $v \equiv v(u)$, the substitutions (B.1) generate the following transformation of the derivatives y', y'' and y'''.

$$y' = \frac{\psi_u + \psi_v v'}{\phi_u + \phi_v v'}, \tag{B.6}$$

377

$$y'' = \frac{1}{(\phi_u + \phi_v v')^3} \{ (\phi_u \psi_v - \phi_v \psi_u)v'' + (\phi_v \psi_{vv} - \phi_{vv}\psi_v)v'^3$$

$$+ \left[\phi_u \psi_{vv} - \phi_{vv}\psi_u + 2(\phi_v \psi_{uv} - \phi_{uv}\psi_v) \right] v'^2$$

$$+ \left[\phi_v \psi_{uu} - \phi_{uu}\psi_v + 2(\phi_u \psi_{uv} - \phi_{uv}\psi_u) \right] v' + \phi_u \psi_{uu} - \phi_{uu}\psi_u \}.$$
(B.7)

Higher order derivatives are obtained from (5.6). In general the structure of the kth derivative is

$$y^{(k)} = \frac{1}{(\phi_u + \phi_v v')^{k+1}} \left[(\phi_u \psi_v - \phi_v \psi_u)v^{(k)} - \frac{m_k \phi_v (\phi_u \psi_v - \phi_v \psi_u)v'' v^{(k-1)}}{\phi_u + \phi_v v'} \right.$$

$$\left. + \frac{P(v')v^{(k-1)} + Q(v', \dots, v^{(k-2)})}{(\phi_u + \phi_v v')^{k-2}} \right].$$
(B.8)

P and Q are polynomial in their arguments and furthermore depend on u and v; m_k is a natural number. If $\phi_v = 0$, this expression simplifies to

$$y^{(k)} = \frac{1}{\phi_u^k} \left[\psi_v y^{(k)} + k v' v^{(k-1)} + \frac{F(u,v)v^{(k-1)} + G(v', \dots, v^{(k-2)})}{\phi_u^{k-1}} \right]. \quad (B.9)$$

The expressions (B.8) and (B.9) are due to Neumer [137].

Various special cases that occur frequently in this book lead to simplifications of these expressions. If

$$x = F(u), \quad y = G(v), \quad F' = \frac{dF}{du} \quad \text{and} \quad G' = \frac{dG}{dv} \quad (B.10)$$

the derivatives are

$$y' = \frac{G'}{F'}v', \quad y'' = \frac{1}{F'^3}(F'G'v'' + F'G''v'^2 - F''G'v'),$$

$$y''' = \frac{1}{F'^5}(F'^2 G'v''' + 3F'^2 G''v'v'' + F'^2 G'''v'^3 - 3F''F'G'v'' \quad (B.11)$$

$$-3F''F'G''v'^2 + 3F''^2 G'v' + F'''F'G'v').$$

If

$$x = F(u), \quad y = G(u)v + H(u), \quad F' = \frac{dF}{du}, \quad G' = \frac{dG}{du} \quad \text{and} \quad H' = \frac{dH}{du} \quad (B.12)$$

the derivatives are

$$y' = \frac{G}{F'}\left(v' + \frac{G'}{G}v + H \right),$$

$$y'' = \frac{G}{F'^2}\left[v'' + \left(2\frac{G'}{G}F' - F'' \right) F'^2 v' \right.$$

$$\left. + \left(\frac{G''}{G}F' - \frac{G'}{G}F'' \right) F'^2 v + (H''F' - H'F'')\frac{F'^2}{G} \right].$$
(B.13)

Collection of Useful Formulas

If
$$x = F(u), \quad y = G(u)v, \quad F' = \frac{dF}{du} \text{ and } G' = \frac{dG}{du} \tag{B.14}$$

the derivatives are

$$y' = \frac{G}{F'}\left(v' + \frac{G'}{G}v \right),$$

$$y'' = \frac{G}{F'^2}\left[v'' + \left(2\frac{G'}{G} - \frac{F''}{F'} \right)v' + \left(\frac{G''}{G} - \frac{F''}{F'}\frac{G'}{G} \right)v \right],$$

$$y''' = \frac{G}{F'^3}\Big\{ v''' + 3\left(\frac{G'}{G} - \frac{F''}{F'} \right)v''$$

$$+ \left[3\left(\frac{F''}{F'} \right)^2 + 3\frac{G''}{G} - 6\frac{F''}{F'}\frac{G'}{G} - \frac{F'''}{F'} \right]v'$$

$$+ \left[3\left(\frac{F''}{F'} \right)^2\frac{G'}{G} + \frac{G'''}{G} - \frac{F'''}{F'}\frac{G'}{G} - 3\frac{F''}{F'}\frac{G''}{G} \right]v \Big\}. \tag{B.15}$$

For some applications the explicit form of the coefficients of the three leading derivatives are required. They may be seen from

$$y^{(n)} = \frac{G}{F'^n}\Big\{ v^{(n)} + n\left[\frac{G'}{G} - \frac{n-1}{2}\frac{F''}{F'} \right]v^{(n-1)}$$

$$+ \frac{n(n-1)}{2}\left[\frac{G''}{G} - (n-1)\frac{F''}{F'}\frac{G'}{G} \right.$$

$$\left. - \frac{n-2}{3}\frac{F'''}{F'} + \frac{(n-2)(n+1)}{4}\left(\frac{F''}{F'} \right)^2 \right]v^{(n-2)} \Big\} + \cdots \tag{B.16}$$

Upon further specialization $G = F'$, it simplifies to

$$y^{(n)} = \frac{1}{F'^{n-1}}\Big\{ v^{(n)} - \frac{n(n-3)}{2}\frac{F''}{F'}v^{(n-1)}$$

$$- \frac{n(n-1)}{2}\left[\frac{n-5}{3}\frac{F'''}{F'} - \frac{n^2-5n+2}{4}(\frac{F''}{F'})^2 \right]v^{(n-2)} \Big\} + \cdots \tag{B.17}$$

The independent variable x and the dependent variable $y \equiv y(x)$ are exchanged by the transformation $x = v$, $y = u$ with $v \equiv v(u)$. The derivatives up to order four are transformed by

$$y' = \frac{1}{v'}, \quad y'' = -\frac{v''}{v'^3}, \quad y''' = -\frac{v'''}{v'^4} + \frac{3v''^2}{v'^5}, \tag{B.18}$$

$$y^{(4)} = -\frac{v^{(4)}}{v'^5} + \frac{10v'''v''}{v'^6} - \frac{15v''^3}{v'^6}.$$

Symmetric Powers of Second Order Equations. The listing below gives the k-th symmetric powers of $y'' + ry = 0$ from $k = 2$ to $k = 6$.

$$k = 2: \quad y''' + 4ry' + 2r'y = 0,$$

$$k = 3: \quad y^{(4)} + 10ry'' + 10r'y' + (3r'' + 9r^2)y = 0,$$

$$k = 4: \quad y^{(5)} + 20ry''' + 30r'y'' + (18r'' + 64r^2)y' + (4r''' + 64r'r)y = 0,$$

$$k = 5: \quad y^{(6)} + 35ry^{(4)} + 70r'y''' + (63r'' + 259r^2)y'' + (28r''' + 518r'r)y'$$
$$+ (5r^{(4)} + 155r''r + 130r'^2 + 225r^3)y = 0,$$

$$k = 6: \quad y^{(7)} + 56ry^{(5)} + 140r'y^{(4)} + (168r'' + 784r^2)y''' + (112r''' + 2352r'r)y''$$
$$+ (40r^{(4)} + 1408r''r + 1180r'^2 + 2304r^3)y'$$
$$+ (6r^{(5)} + 312r'''r + 708r''r' + 3456r'r^2)y = 0,$$

Janet Basis Types. The Janet basis types that are required in earlier parts of this book are listed, beginning with the Janet bases of type $\mathcal{J}^{(1,2)}$ of order 1, 2 and 3. Their coherence conditions may be found on page 59. The dependent variable is z, the independent variables are x and y, the *grlex* term order with $y > x$ is always applied.

$$\mathcal{J}_1^{(1,2)}: \quad z_x + az = 0, \quad z_y + bz = 0.$$

$$\mathcal{J}_{2,1}^{(1,2)}: \quad z_y + a_1 z_x + a_2 z = 0, \quad z_{xx} + b_1 z_x + b_2 z = 0.$$

$$\mathcal{J}_{2,2}^{(1,2)}: \quad z_x + a_1 z, \quad z_{yy} + b_1 z_y + b_2 z = 0.$$

$$\mathcal{J}_{3,1}^{(1,2)}: \quad z_y + a_1 z_x + a_2 z = 0, \quad z_{xxx} + b_1 z_{xx} + b_2 z_x + b_3 z = 0.$$

$$\mathcal{J}_{3,2}^{(1,2)}: \quad \begin{aligned} z_{xx} + a_1 z_y + a_2 z_x + a_3 z &= 0, \quad z_{xy} + b_1 z_y + b_2 z_x + b_3 z = 0, \\ z_{yy} + c_1 z_y + c_2 z_x + c_3 z &= 0. \end{aligned}$$

$$\mathcal{J}_{3,3}^{(1,2)}: \quad z_x + a_1 z = 0, \quad z_{yyy} + b_1 z_{yy} + b_2 z_y + b_3 z = 0.$$

The Janet bases of type $\mathcal{J}^{(2,2)}$ of order 1, 2 and 3 are given next. Their coherence conditions may be found on pages 64 to 67. The dependent variables are z_1 and z_2. The term order is *grlex* with $z_2 > z_1$ and $y > x$.

$$\mathcal{J}_{1,1}^{(2,2)}: \quad z_1 = 0, \quad z_{2,x} + az_2 = 0, \quad z_{2,y} + bz_2 = 0.$$

$$\mathcal{J}_{1,2}^{(2,2)}: \quad z_2 + az_1 = 0, \quad z_{1,x} + bz_1 = 0, \quad z_{1,y} + cz_1 = 0.$$

$$\mathcal{J}_{2,1}^{(2,2)}: \quad z_1 = 0, \quad z_{2,y} + a_1 z_{2,x} + a_2 z_2 = 0, \quad z_{2,xx} + b_1 z_{2,x} + b_2 z_2 = 0.$$

$$\mathcal{J}_{2,2}^{(2,2)}: \quad z_1 = 0, \quad z_{2,x} + a_1 z_2 = 0, \quad z_{2,yy} + b_1 z_{2,y} + b_2 z_2 = 0.$$

$$\mathcal{J}_{2,3}^{(2,2)}: \quad \begin{aligned} z_{1,x} + a_1 z_2 + a_2 z_1 &= 0, \quad z_{1,y} + b_1 z_2 + b_2 z_1 = 0, \\ z_{2,x} + c_1 z_2 + c_2 z_1 &= 0, \quad z_{2,y} + d_1 z_2 + d_2 z_1 = 0. \end{aligned}$$

$$\mathcal{J}_{2,4}^{(2,2)}: \quad z_2 + a_1 z_1 = 0, \ z_{1,y} + b_1 z_{1,x} + b_2 z_1 = 0, \ z_{1,xx} + c_1 z_{1,x} + c_2 z_1 = 0.$$

$$\mathcal{J}_{2,5}^{(2,2)}: \quad z_2 + a_1 z_1 = 0, \ z_{1,x} + b_1 z_1 = 0, \ z_{1,yy} + c_1 z_{1,y} + c_2 z_1 = 0.$$

$$\mathcal{J}_{3,1}^{(2,2)}: \quad z_1 = 0, \ z_{2,y} + a_1 z_{2,x} + a_2 z_2 = 0, \ z_{2,xxx} + b_1 z_{2,xx} + b_2 z_{2,x} + b_3 z_2 = 0.$$

$$\mathcal{J}_{3,2}^{(2,2)}: \begin{array}{l} z_1 = 0, \ z_{2,xx} + a_1 z_{2,y} + a_2 z_{2,x} + a_3 z_2 = 0, \\ z_{2,xy} + b_1 z_{2,y} + b_2 z_{2,x} + b_3 z_2 = 0, \\ z_{2,yy} + c_1 z_{2,y} + c_2 z_{2,x} + c_3 z_2 = 0. \end{array}$$

$$\mathcal{J}_{3,3}^{(2,2)}: \quad z_1 = 0, \ z_{2,x} + a_1 z_2 = 0, \ z_{2,yyy} + b_1 z_{2,yy} + b_2 z_{2,y} + b_3 z_2 = 0.$$

$$\mathcal{J}_{3,4}^{(2,2)}: \begin{array}{l} z_{1,x} + a_1 z_2 + a_2 z_1 = 0, \ z_{1,y} + b_1 z_2 + b_2 z_1 = 0, \\ z_{2,y} + c_1 z_{2,x} + c_2 z_2 + c_3 z_1 = 0, \ z_{2,xx} + d_1 z_{1,x} + d_2 z_2 + d_3 z_1 = 0. \end{array}$$

$$\mathcal{J}_{3,5}^{(2,2)}: \begin{array}{l} z_{1,x} + a_1 z_2 + a_2 z_1 = 0, \ z_{1,y} + b_1 z_2 + b_2 z_1 = 0, \\ z_{2,x} + c_1 z_2 + c_2 z_1 = 0, \ z_{2,yy} + d_1 z_{2,y} + d_2 z_2 + d_3 z_1 = 0. \end{array}$$

$$\mathcal{J}_{3,6}^{(2,2)}: \begin{array}{l} z_{1,y} + a_1 z_{1,x} + a_2 z_2 + a_3 z_1 = 0, \ z_{2,x} + b_1 z_{1,x} + b_2 z_2 + b_3 z_1 = 0, \\ z_{2,y} + c_1 z_{1,x} + c_2 z_2 + c_3 z_1 = 0, \ z_{1,xx} + d_1 z_{1,x} + d_2 z_2 + d_3 z_1 = 0. \end{array}$$

$$\mathcal{J}_{3,7}^{(2,2)}: \begin{array}{l} z_{1,x} + a_1 z_2 + a_2 z_1 = 0, \ z_{2,x} + b_1 z_{1,y} + b_2 z_2 + b_3 z_1 = 0, \\ z_{2,y} + c_1 z_{1,y} + c_2 z_2 + c_3 z_1 = 0, \ z_{1,yy} + d_1 z_{1,y} + d_2 z_2 + d_3 z_1 = 0. \end{array}$$

$$\mathcal{J}_{3,8}^{(2,2)}: \quad z_2 + a_1 z_1 = 0, \ z_{1,x} + b_1 z_1 = 0, \ z_{1,yyy} + c_1 z_{1,yy} + c_2 z_{1,y} + c_3 z_1 = 0.$$

$$\mathcal{J}_{3,9}^{(2,2)}: \begin{array}{l} z_2 + a_1 z_1 = 0, \ z_{1,xx} + b_1 z_{1,y} + b_2 z_{1,x} + b_3 z_1 = 0, \\ z_{1,xy} + c_1 z_{1,y} + c_2 z_{1,x} + c_3 z_1 = 0, \ z_{1,yy} + d_1 z_{1,y} + d_2 z_{1,x} + d_3 z_1 = 0. \end{array}$$

$$\mathcal{J}_{3,10}^{(2,2)}: \quad z_2 + a_1 z_1 = 0, \ z_{1,y} + b_1 z_{1,x} + b_2 z_1 = 0, \ z_{1,xxx} + c_1 z_{1,xx} + c_2 z_{1,x} + c_3 z_1 = 0.$$

A partial listing of Janet basis types of order four is given next; they are required for the symmetry class identification on page 227.

$$\mathcal{J}_{4,2}^{(2,2)}: \begin{array}{l} z_1 = 0, \ z_{2,xy} + a_1 z_{2,xx} + a_2 z_{2,y} + a_3 z_{2,x} + a_4 z_2 = 0, \\ z_{2,yy} + b_1 z_{2,xx} + b_2 z_{2,y} + b_3 z_{2,x} + b_4 z_2 = 0, \\ z_{2,xxx} + c_1 z_{2,xx} + c_2 z_{2,y} + c_3 z_{2,x} + c_4 z_2 = 0. \end{array}$$

$$\mathcal{J}_{4,9}^{(2,2)}: \begin{array}{l} z_{1,y} + a_2 z_{1,x} + a_3 z_2 + a_4 z_1 = 0, \ z_{2,y} + b_1 z_{2,x} + b_2 z_{1,x} + b_3 z_2 + b_4 z_1 = 0, \\ z_{1,xx} + c_1 z_{2,x} + c_2 z_{1,x} + c_3 z_2 + c_4 z_1 = 0, \\ z_{2,xx} + d_1 z_{2,x} + d_2 z_{1,x} + d_3 z_2 + d_4 z_1 = 0. \end{array}$$

$$\mathcal{J}_{4,10}^{(2,2)} : \begin{array}{c} z_{1,y} + a_2 z_{1,x} + a_3 z_2 + a_4 z_1 = 0, \quad z_{2,x} + b_2 z_{1,x} + b_3 z_2 + b_4 z_1 = 0, \\ z_{1,xx} + c_1 z_{2,y} + c_2 z_{1,x} + c_3 z_2 + c_4 z_1 = 0, \\ z_{2,yy} + d_1 z_{2,y} + d_2 z_{1,x} + d_3 z_2 + d_4 z_1 = 0. \end{array}$$

$$\mathcal{J}_{4,12}^{(2,2)} : \begin{array}{c} z_{1,x} + a_3 z_2 + a_4 z_1 = 0, \quad z_{2,x} + b_2 z_{1,y} + b_3 z_2 + b_4 z_1 = 0, \\ z_{1,yy} + c_1 z_{2,y} + c_2 z_{1,y} + c_3 z_2 + c_4 z_1 = 0, \\ z_{2,yy} + d_1 z_{2,y} + d_2 z_{1,y} + d_3 z_2 + d_4 z_1 = 0. \end{array}$$

$$\mathcal{J}_{4,14}^{(2,2)} : \begin{array}{c} z_{2,x} + a_1 z_{1,y} + a_2 z_{1,x} + a_3 z_2 + a_4 z_1 = 0, \\ z_{2,y} + b_1 z_{1,y} + b_2 z_{1,x} + b_3 z_2 + b_4 z_1 = 0, \\ z_{1,xx} + c_1 z_{1,y} + c_2 z_{1,x} + c_3 z_2 + c_4 z_1 = 0, \\ z_{1,yy} + d_1 z_{1,y} + d_2 z_{1,x} + d_3 z_2 + d_4 z_1 = 0, \\ z_{1,yy} + e_1 z_{1,y} + e_2 z_{1,x} + e_3 z_2 + e_4 z_1 = 0. \end{array}$$

$$\mathcal{J}_{4,17}^{(2,2)} : \begin{array}{c} z_2 + a_4 z_1 = 0, \quad z_{1,xy} + b_1 z_{1,xx} + b_2 z_{1,y} + b_3 z_{1,x} + b_4 z_1 = 0, \\ z_{1,yy} + c_1 z_{1,xx} + c_2 z_{1,y} + c_3 z_{1,x} + c_4 z_1 = 0, \\ z_{1,xxx} + d_1 z_{1,xx} + d_2 z_{1,y} + d_3 z_{1,x} + d_4 z_1 = 0. \end{array}$$

$$\mathcal{J}_{4,19}^{(2,2)} : \begin{array}{c} z_2 = 0, \quad z_{1,xx} + a_2 z_{1,y} + a_3 z_{1,x} + a_4 z_1 = 0, \\ z_{1,xy} + b_2 z_{1,y} + b_3 z_{1,x} + b_4 z_1 = 0, \\ z_{1,yyy} + c_1 z_{1,yy} + c_2 z_{1,y} + c_3 z_{1,x} + c_4 z_1 = 0. \end{array}$$

Finally, the following fifth order Janet bases are required for symmetry class identification on page 238.

$$\mathcal{J}_{5,1}^{(2,2)} : \begin{array}{c} z_{1,x} + a_4 z_2 + a_5 z_1 = 0, \quad z_{1,y} + b_4 z_2 + b_5 z_1 = 0, \\ z_{2,xy} + c_1 z_{2,xx} + c_2 z_{2,y} + c_3 z_{2,y} + c_4 z_2 + c_5 z_1 = 0, \\ z_{2,yy} + d_1 z_{2,xx} + d_2 z_{2,y} + d_3 z_{2,y} + d_4 z_2 + d_5 z_1 = 0, \\ z_{2,xxx} + e_1 z_{2,xx} + e_2 z_{2,y} + e_3 z_{2,y} + e_4 z_2 + e_5 z_1 = 0. \end{array}$$

$$\mathcal{J}_{5,2}^{(2,2)} : \begin{array}{c} z_{2,x} + a_2 z_{1,y} + a_3 z_{1,x} + a_4 z_2 + a_5 z_1 = 0, \\ z_{2,y} + b_2 z_{1,y} + b_3 z_{1,x} + b_4 z_2 + b_5 z_1 = 0, \\ z_{1,xx} + c_1 z_{1,yy} + c_2 z_{1,y} + c_3 z_{1,x} + c_4 z_2 + c_5 z_1 = 0, \\ z_{1,xy} + d_1 z_{1,yy} + d_2 z_{1,y} + d_3 z_{1,x} + d_4 z_2 + d_5 z_1 = 0, \\ z_{1,yyy} + e_1 z_{1,yy} + e_2 z_{1,y} + e_3 z_{1,x} + e_4 z_2 + e_5 z_1 = 0. \end{array}$$

$$\mathcal{J}_{5,3}^{(2,2)} : \begin{array}{c} z_{2,x} + a_2 z_{1,y} + a_3 z_{1,x} + a_4 z_2 + a_5 z_1 = 0, \\ z_{2,y} + b_2 z_{1,y} + b_3 z_{1,x} + b_4 z_2 + b_5 z_1 = 0, \\ z_{1,xy} + c_1 z_{1,xx} + c_2 z_{1,y} + c_3 z_{1,x} + c_4 z_2 + c_5 z_1 = 0, \\ z_{1,yy} + d_1 z_{1,xx} + d_2 z_{1,y} + d_3 z_{1,x} + d_4 z_2 + d_5 z_1 = 0, \\ z_{1,xxx} + e_1 z_{1,xx} + e_2 z_{1,y} + e_3 z_{1,x} + e_4 z_2 + e_5 z_1 = 0. \end{array}$$

Appendix C

Algebra of Monomials

In this appendix various properties of semigroups of monomials are described. Details and proofs may be found in the article by Janet [83]; see also Castro and Moreno [27]. The notation applied here is independent from the rest of the book.

Let $x_1, x_2, \ldots, x_n$ be n indeterminates equipped with the *lex* ordering $x_1 < x_2 < \ldots < x_n$. Monomials in the x_i are denoted by $m = x_1^{\alpha_1} \ldots x_n^{\alpha_n}$ with $(\alpha_1, \ldots, \alpha_n) \in \mathbb{N}^n$ or, if several monomials are to be distinguished, by $m_i = x_1^{\alpha_1^i} \ldots x_n^{\alpha_n^i}$. Because the product of two monomials is again a monomial, and the product is commutative and associative, they form a commutative semigroup under multiplication. Because $(1, \ldots, 1)$ is the identity element, they even form a monoid. The connection of the results described in this appendix with partial differential equations is established by the identification of the degree vector $(\alpha_1, \ldots, \alpha_n)$ with the partial derivative $\frac{\partial^{\alpha_1 + \ldots + \alpha_n}}{\partial x_1^{\alpha_1} \ldots \partial x_n^{\alpha_n}}$.

Let $\mathcal{M}$ be a finite set of such monomials. They generate an ideal in this monoid. This ideal should not be mistaken for an ideal in the polynomial ring $k[x_1, \ldots, x_n]$ with k some number field. The totality of all monomials of the monoid decomposes into those which are contained in the ideal generated by $\mathcal{M}$ and those which are not.

For the applications in the main part of this book a unique representation of the members of both subsets is required. To this end for both subsets a decomposition into *classes* will be obtained. The classes are defined as the entirety of all multiples of any monomial of a *complete set* with respect its *multipliers*, i.e., a certain subset of its variables. It is the main subject of this appendix to define a procedure for constructing such a decomposition into classes for any given set of monomials.

The finiteness of various algorithms dealing with monomials is based on the following lemma which is proved in Janet's article [83], pages 69-70.

LEMMA C.1　*A set of monomials such that none of its elements is a multiple of any other monomial is necessarily finite.*

Let $\mathcal{M} = \{m_1, m_2, \ldots\}$ be a finite set of monomials and $a_1 < \ldots < a_k$ the degrees in x_n which do actually occur in $\mathcal{M}$. $\mathcal{M}_\alpha \subset \mathcal{M}$ denotes those monomials in $\mathcal{M}$ which occur for a fixed value of $\alpha \in \{a_1, \ldots, a_k\}$. The first step toward the decomposition mentioned above is the subdivision of variables

384

and monomials as described in the following definitions.

DEFINITION C.1   (*Multipliers and non-multipliers*) *The leading variable x_n is called a multiplier for $m_i \in \mathcal{M}$ if $\alpha_n^i = a_k$; otherwise it is called a non-multiplier. If $n > 1$, the same procedure is applied to the monomials of $\mathcal{M}_{a_j}$, $j = 1, \ldots, k$, w.r.t. $n - 1$ variables $x_1, \ldots, x_{n-1}$.*

DEFINITION C.2   (*Classes*) *The totality of all multiples of a monomial $m_i \in \mathcal{M}$ with respect to its multipliers is called the class defined by this monomial and is denoted by $\overline{m}_i$. The union of the classes of all monomials in $\mathcal{M}$ is denoted by $\overline{\mathcal{M}}$.*

It should be noted that this definition of multipliers is only meaningful with respect to a fixed ordering for the variables. In general there will be monomials in the ideal generated by $\mathcal{M}$ which do not belong to any class at all.

EXAMPLE C.1   Consider the following set $\mathcal{M}$ comprising four monomials in three variables.

$Monomial$	$Multipliers$	$Monomial$	$Multipliers$
$x_1^2 x_2^2 x_3^3$	x_1, x_2, x_3	$x_1^3 x_2 x_3$	x_1, x_2
$x_1^3 x_3^3$	x_1, x_3	$x_2 x_3$	x_2

The monomial $x_1 x_2 x_3$ obviously may not be represented as the product of any monomial in $\mathcal{M}$ and some of its multipliers; and therefore does not belong to any class. ⬜

The special sets of monomials which have the property that any monomial in the ideal generated by it belongs at least to a single class are defined next.

DEFINITION C.3   (*Complete system*) *A set of monomials $\mathcal{M}$ is called complete if any monomial in the ideal generated by $\mathcal{M}$ belongs to at least one of its classes.*

Due to this property of complete sets it is desirable to have a procedure which determines for any given set of monomials an equivalent complete set, i.e., a complete set generating the same ideal as the given one. The following three theorems due to Janet are the prerequisite for such an algorithm.

THEOREM C.1   *A set of monomials $\mathcal{M}$ is complete if and only if*

i) *The sets $\mathcal{M}_{a_1}, \ldots, \mathcal{M}_{a_k}$ are complete with respect to $x_1, \ldots, x_{n-1}$.*

ii) *For any monomial $m \in \mathcal{M}$ there holds $m/x_n^\alpha \in \mathcal{M}_\alpha \to x_n \cdot m \in \overline{\mathbf{M}}_{\alpha+1}$ for $\alpha < a_k$.*

THEOREM C.2   *Let $\mathcal{M}$ be a set of monomials. If the product of any monomial in $\mathcal{M}$ with any of its non-multipliers is contained in $\overline{\mathcal{M}}$, then $\mathcal{M}$ is complete.*

Algebra of Monomials   385

THEOREM C.3   *Any finite set of monomials $\mathcal{M}$ may be extended to a complete set generating the same ideal as $\mathcal{M}$.*

The proof of these theorems may be found in Janet [83], page 74 ff.

An obvious consequence of Theorem C.1 is the constraint $a_{i+1} = a_i + 1$ for $1 \leq i \leq k-1$ for complete sets. Based on these results the following algorithm *Complete* may be designed which is also due to Janet. The notation is the same as above.

ALGORITHM C.1   *Complete($\mathcal{M}$).* Given a set $\mathcal{M} = \{m_1, m_2, \ldots\}$ of monomials, the complete set corresponding to $\mathcal{M}$ is returned.

$S1$: *Find Multipliers & Non-Multipliers.* Determine multipliers and non-multipliers for all monomials in $\mathcal{M}$.

$S2$: *Products with Non-Multipliers.* Generate the set $\mathcal{M}_0$ of monomials which are the product of an m_i w.r.t. any one of its non-multipliers.

$S3$: *In Classes?* Remove all those elements from $\mathcal{M}_0$ which are contained in $\bar{\mathcal{M}}$.

$S4$: *Termination?* If $\mathcal{M}_0$ is empty return $\mathcal{M}$, else set $\mathcal{M} := \mathcal{M} \cup \mathcal{M}_0$ and goto $S1$.

EXAMPLE C.2   If this algorithm is applied to example C.1, the following complete set of monomials is obtained.

Monomial	Multipliers	Monomial	Multipliers
$x_1^2 x_2^2 x_3^3$	x_1, x_2, x_3	$x_1^3 x_2 x_3^2$	x_1, x_2
$x_1 x_2^2 x_3^3$	x_2, x_3	$x_1^2 x_2 x_3^2$	x_2
$x_2^2 x_3^3$	x_2, x_3	$x_1 x_2 x_3^2$	x_2
$x_1^3 x_2 x_3^3$	x_1, x_3	$x_2 x_3^2$	x_2
$x_1^2 x_2 x_3^3$	x_3	$x_1^3 x_2 x_3$	x_2
$x_1 x_2 x_3^3$	x_3	$x_1^2 x_2 x_3$	x_2
$x_2 x_3^3$	x_3	$x_1 x_2 x_3$	x_2
$x_1^3 x_3^3$	x_3, x_1	$x_2 x_3$	x_2

Complete sets of monomials have several properties that make them important tools for analyzing sets of pde's. They are collected in the following theorem.

THEOREM C.4   *A complete set of monomials $\mathcal{M}$ has the following properties.*

i) *Any monomial in the ideal generated by $\mathcal{M}$ belongs exactly to a single of its classes.*

ii) *The product of any monomial in $\mathcal{M}$ by one of its non-multipliers is equal to the product of some other monomial by its multipliers.*

386

Let $\mathcal{M}$ denote a complete set of monomials. To obtain a description of the monomials which are *not* contained in the ideal generated by $\mathcal{M}$, a new set of monomials, the *complementary set* $\mathcal{N}$, will be constructed as follows. If $a_n > 0$ then $x_n^\alpha \in \mathcal{N}$ for $0 \le \alpha < a_1$ with multipliers $x_1, x_2, \ldots x_{n-1}$. For $1 \le i \le k$, consider $\mathcal{M}_{a_i}$. Let $b_1 < b_2 < \ldots b_j$ be the exponents of x_{n-1} which do occur in $\mathcal{M}_{a_i}$. If $b_1 > 0$ then the monomials $x_n^{a_i} x_{n-1}^\beta \in \mathcal{N}$ for $0 \le \beta < b_1$. Its multipliers are $x_1, x_2 \ldots x_{n-2}$ if $i < k$, and in addition x_n if $i = k$. This proceeding is applied recursively all the way down until $n = 1$ is reached.

Appendix D

Loewy Decompositions of Kamke's Collection

In this appendix the types of Loewy decompositions of linear second and third order equations in x and $y(x)$ with coefficients in $\mathbb{Q}$ from Kamke's collection are listed. Excluded are equations containing undetermined functions, more than two parameters, or a parameter in an exponent. For these equations usually a more sophisticated analysis is required in order to determine the constraints leading possibly to different decomposition types. If there are parameters involved, the given decomposition applies to unconstrained parameter values. Any relations between them may change the decomposition type. For inhomogeneous equations only the left hand side is taken into account. From the given nontrivial decompositions fundamental systems may be obtained by integration according to the formulas on page 31.

Equations of Second Order. These are the equations of Chapter 2 of Kamke's collection. The tabulation below is organized by decomposition type as described in Corollary 2.3. For types $\mathcal{L}_k^2$, $k = 2$, 3 or 4 the numbers of the corresponding equations are listed.

For irreducible equations of type $\mathcal{L}_1^2$ a further distinction is made according to the type of Galois group *of its rational normal form* as discussed in Chapter 2 on page 35. If the Galois group is SL_2, the equivalence to an equation defining a special function is given whenever possible. The following equations defining special functions $w(z)$ are taken into account.

$$\text{Bessel}: \quad w'' + \tfrac{1}{z}w' - \left(1 - \frac{\nu^2}{z^2} \right) w = 0,$$

$$\text{Hypergeometrical}: \quad w'' + \frac{(\alpha + \beta + 1)z - \gamma}{z(z - 1)}w' + \frac{\alpha\beta}{z(z - 1)}w = 0.$$

$$\text{Legendre}: \quad w'' + \frac{2z}{z^2 - 1}w' - \frac{\nu(\nu + 1)}{z^2 - 1})w = 0.$$

$$\text{Weber}: \quad w'' - zw' - kw = 0.$$

$$\text{Whittaker}: \quad w'' - \left(\frac{1}{4} - \frac{k}{z} + \frac{4m^2 - 1}{4z^2} \right) w = 0.$$

$$\text{Confluent Hypergeometrical}: \quad w'' - \left(1 - \frac{a}{z} \right) w' - \frac{b}{z}w = 0.$$

387

388

$\mathcal{L}_1^2$: Galois group SL_2, equivalent to Bessel equation. 2.10, 2.14, 2.15, 2.86, 2.94, 2.95, 2.102, 2.103, 2.104, 2.105, 2.106, 2.123, 2.130, 2.149, 2.155, 2.161, 2.162, 2.164, 2.165, 2.167, 2.169, 2.170, 2.172, 2.173, 2.180, 2.185, 2.189, 2.197, 2.200, 2.206, 2.216, 2.272, 2.274, 2.347.

$\mathcal{L}_1^2$: Galois group SL_2, equivalent to Hypergeometrical equation. 2.12, 2.252, 2.258, 2.265, 2.287, 2.291, 2.293, 2.294, 2.406.

$\mathcal{L}_1^2$: Galois group SL_2, equivalent to Legendre equation. 2.226, 2.231, 2.239, 2.240, 2.241, 2.244, 2.245, 2.249, 2.256, 2.269, 2.311, 2.313.

$\mathcal{L}_1^2$: Galois group SL_2, equivalent to Weber equation. 2.42, 2.44, 2.46, 2.87, 2.131, 2.132, 2.139.

$\mathcal{L}_1^2$: Galois group SL_2, equivalent to Whittaker equation. 2.16, 2.37, 2.114, 2.154, 2.195, 2.273.

$\mathcal{L}_1^2$: Galois group SL_2, equivalent to Confluent Hypergeometrical equation. 2.12, 2.92, 2.96, 2.107, 2.110, 2.113, 2.134, 2.138, 2.190.

$\mathcal{L}_1^2$: Imprimitive Galois group D_m with $m > 24$, or D. 2.130, 2.135, 2.222, 2.288, 2.289.

$\mathcal{L}_1^2$: Imprimitive Galois group D_3. 2.290, 2.292.

$\mathcal{L}_2^2$: 2.11, 2.13, 2.39, 2.40, 2.41, 2.43, 2.45, 2.50, 2.56, 2.57, 2.58, 2.59, 2.93, 2.109, 2.121, 2.122, 2.133, 2.136, 2.141, 2.158, 2.166, 2.181, 2.182, 2.192, 2.194, 2.196, 2.198, 2.199, 2.208, 2.209, 2.223, 2.225, 2.230, 2.234, 2.237, 2.238, 2.242, 2.246, 2.251, 2.253, 2.254, 2.257, 2.262, 2.263, 2.264, 2.267, 2.270, 2.271, 2.280, 2.299, 2.300, 2.304, 2.308, 2.310, 2.324, 2.326, 2.319, 2.321, 2.328, 2.332, 2.334, 2.336, 2.338, 2.345, 2.347a, 2.351, 2.355, 2.358, 2.365, 2.378, 2.379, 2.384, 2.386, 2.387, 2.399.

$\mathcal{L}_3^2$: 2.2,... 2.7, 2.35, 2.47, 2.49, 2.51, 2.53, 2.91, 2.98, 2.100, 2.101, 2.111, 2.112, 2.115, 2.119, 2.125, 2.126, 2.129, 2.148, 2.150, 2.151, 2.152, 2.160, 2.165a, 2.176, 2.179, 2.191, 2.193, 2.201, 2.202, 2.203, 2.204, 2.211, 2.243, 2.247, 2.255, 2.276, 2.277, 2.279, 2.282, 2.320, 2.331, 2.337, 2.342, 2.350, 2.353, 2.354, 2.361, 2.370, 2.380, 2.400, 2.401, 2.404, 2.405.

$\mathcal{L}_4^2$, Galois group Z_1: 2.1, 2.146, 2.147, 2.174, 2.183, 2.184, 2.229, 2.250, 2.281, 2.312, 2.322, 2.323, 2.340, 2.346, 2.366.

$\mathcal{L}_4^2$, Galois group Z_2: 2.159, 2.168, 2.175, 2.186, 2.227, 2.229, 2.266, 2.284, 2.286, 2.391.

In the subsequent listing the number of the equation and its decomposition type is given. If the equation is irreducible and it is equivalent to any of the known equations defining a special function, the transformation to this equation is given; as usual $D = \frac{d}{dx}$.

Loewy Decompositions

389

2.1: $y'' = Lclm\left(D - \frac{1}{C+x}\right)y = 0$. Type $\mathcal{L}_4^2$.

2.2,...,2.5: $y'' + y = Lclm(D - \cot x, D + \tan x)y = 0$. Type $\mathcal{L}_3^2$.

2.6: $y'' - y = Lclm(D + 1, D - 1)y = 0$. Type $\mathcal{L}_3^2$.

2.7: $y'' - 2y = Lclm(D + \sqrt{2}, D - \sqrt{2})y = 0$. Type $\mathcal{L}_3^2$.

2.8: $y'' + a^2 y = Lclm(D + ia, D - ia)y = 0$. Type $\mathcal{L}_3^2$.

2.9: $y'' + ay = Lclm(D + i\sqrt{a}, D - i\sqrt{a})y = 0$. Type $\mathcal{L}_3^2$.

2.11: $y'' - (x^2 + 1)y = (D + x)(D - x)y = 0$. Type $\mathcal{L}_2^2$.

2.13: $y'' - (a^2 x^2 + a)y = Lclm(D + ax, D - ax)y = 0$. Type $\mathcal{L}_2^2$.

2.35: $y'' + ay' + by = Lclm\left(D + \frac{1}{2}a + \sqrt{\frac{1}{4}a^2 - b}, D + \frac{1}{2}a - \sqrt{\frac{1}{4}a^2 - b}\right)y = 0$.
Type $\mathcal{L}_3^2$.

2.39: $y'' + xy' + y = D(D + x)y = 0$. Type $\mathcal{L}_2^2$.

2.40: $y'' + xy' - y = \left(D + x + \frac{1}{x}\right)\left(D - \frac{1}{x}\right)y = 0$. Type $\mathcal{L}_2^2$.

2.41: $y'' + xy' + (n+1)y = \left(D + \frac{p_n'}{p_n}\right)\left(D + x - \frac{p_n'}{p_n}\right)y = 0$. Type $\mathcal{L}_2^2$; p_n monic
polynomial of degree n; $p_1 = x$, $p_2 = x^2 - 1$, $p_3 = x^3 - 3x$, $p_4 = x^4 - 6x^2 + 3$.

2.42: $y'' + xy' - ny = 0$. Type $\mathcal{L}_1^2$. Weber equation, $k = -n$, $x = iz$, $y = w$.

2.43: $y'' - xy' + 2y = \left(D - x + \frac{1}{x+1} + \frac{1}{x-1}\right)\left(D - \frac{1}{x+1} - \frac{1}{x-1}\right)y = 0$.
Type $\mathcal{L}_2^2$.

2.44: $y'' - xy' - ay = 0$. Type $\mathcal{L}_1^2$. Weber equation.

2.45: $y'' - xy' + (x - 1)y = (D - x + 1)(D - 1)y = 0$. Type $\mathcal{L}_2^2$.

2.46: $y'' - 2xy' + ay = 0$. Type $\mathcal{L}_1^2$. Weber equation, $k = -\frac{1}{2}a$, $x = \frac{1}{\sqrt{2}}z$,
$y = w$.

2.47: $y'' + 4xy' + (4x^2 + 2)y = Lclm\left(D + 2x - \frac{1}{x}, D + 2x\right)y = 0$. Type $\mathcal{L}_3^2$.

2.49: $y'' - 4xy' + (4x^2 - 1)y = Lclm(D - 2x - \theta) = \exp x^2$, $\theta^2 + 1 = 0$.
Type $\mathcal{L}_3^2$.

2.50: $y'' - 4xy' + (4x^2 - 2)y = (D - 2x)(D - 2x)y = 0$. Type $\mathcal{L}_2^2$.

2.51: $y'' - 4xy' + (4x^2 - 3)y = Lclm(D - 2x + 1, D - 2x - 1)y = 0$.
Type $\mathcal{L}_3^2$.

2.53: $y'' + 2axy' + a^2 x^2 y = Lclm(D + ax - \theta) = 0$, $\theta^2 - a = 0$. Type $\mathcal{L}_3^2$.

2.56: $y'' - x^2 y' + xy = \left(D - x^2 + \frac{1}{x}\right)\left(D - \frac{1}{x}\right)y = 0$. Type $\mathcal{L}_2^2$.

2.57: $y'' - x^2 y' - (x^2 + 2x + 1)y = (D + 1)(D - x^2 - 1)y = 0$. Type $\mathcal{L}_2^2$.

2.58: $y'' - (x^3 + x^2)y' + (x^5 - 2x)y = (D - x^3)(D - x^2)y = 0$. Type $\mathcal{L}_2^2$.

2.59: $y'' + x^4 y' - x^3 y = \left(D + x^4 + \frac{1}{x}\right)\left(D - \frac{1}{x}\right)y = 0$. Type $\mathcal{L}_2^2$.

2.86: $y'' + \frac{9}{4}xy = 0$. Type $\mathcal{L}_1^2$. Bessel equation, $\nu = \frac{1}{3}$, $x = z^{2/3}$, $y = z^{1/3}w$.

2.87: $y'' - \frac{1}{4}(x^2 + a)y = 0$. Type $\mathcal{L}_1^2$. Weber equation, $k = -\frac{1}{4}(a+1)$, $x = z$,
$y = \exp(-\frac{1}{4}z^2)w$.

2.91: $y'' + y = Lclm(D - \theta)y = 0$, $\theta^2 + 1 = 0$. Type $\mathcal{L}_3^2$.

2.93: $y'' + \frac{1}{x}y' = (D + \frac{1}{x})Dy = 0$. Type $\mathcal{L}_2^2$.

2.94: $y'' + \frac{1}{x}y' + \frac{a}{x}y = 0$. Type $\mathcal{L}_1^2$. Bessel equation, $\nu = 0$, $x = \frac{z^2}{4a}$, $y = w$.

2.95: $y'' + \frac{1}{x}y' + \lambda y = 0$. Type $\mathcal{L}_1^2$. Bessel equation, $\nu = 0$, $x = \frac{z}{\sqrt{\lambda}}$, $y = w$.

2.96: $y'' + \frac{1}{x}y' + (1 + \frac{c}{x})y = 0$. Type $\mathcal{L}_1^2$. Confluent Hypergeometric equation, $a = 1$, $b = \frac{1}{2}(1 + ic)$, $x = \frac{1}{2}iz$, $y = \exp(-\frac{1}{2}z)w$.

2.98: $y'' - \frac{1}{x}y' - ax^2y = Lclm(D - \theta x) = 0$, $\theta^2 - a = 0$. Type $\mathcal{L}_3^2$.

2.100: $y'' + \frac{2}{x}y' - y = Lclm(D + 1 + \frac{1}{x}, D - 1 + \frac{1}{x})y = 0$. Type $\mathcal{L}_3^2$.

2.101: $y'' + \frac{2}{x}y' + ay = Lclm(D - \theta + \frac{1}{x}) = 0$, $\theta^2 + a = 0$. Type $\mathcal{L}_3^2$.

2.104: $y'' + \frac{a}{x}y' + \frac{\beta}{x}y = 0$. Type $\mathcal{L}_1^2$. Bessel equation, $\nu = \alpha - 1$, $x = \frac{z^2}{4\beta}$, $y = \left(\frac{z^2}{4\beta}\right)^{(1-\alpha)/2}w$.

2.105: $y'' + \frac{a}{x}y' + by = 0$. Type $\mathcal{L}_1^2$. Bessel equation, $\nu = \frac{1-a}{2}$, $x = -\frac{z}{\sqrt{b}}$, $y = \left(-\frac{z}{\sqrt{b}}\right)^{\nu}w$.

2.106: $y'' + \frac{a}{x}y' + bx^{\alpha-1}y$. Type $\mathcal{L}_1^2$. Bessel equation, $\nu = \frac{1-a}{1+\alpha}$, $x = \left(\frac{(\alpha+1)^2z^2}{4b}\right)^{1/(\alpha+1)}$, $y = \left(\frac{(\alpha+1)^2z^2}{4b}\right)^{(1-a)/(2\alpha+2)}w$.

2.109: $y'' - y' - \frac{1}{x}y = (D + \frac{1}{x})(D - 1 - \frac{1}{x})y = 0$. Type $\mathcal{L}_2^2$.

2.111: $y'' - (1 + \frac{1}{x})y' + \frac{1}{x}y = Lclm(D - 1, D - 1 + \frac{1}{x+1})y = 0$. Type $\mathcal{L}_3^2$.

2.112: $y'' - (1 + \frac{1}{x})y' - (2 - \frac{2}{x})y = Lclm(D + 1 - \frac{1}{x+\frac{1}{3}}, D - 2)y = 0$. Type $\mathcal{L}_3^2$.

2.115: $y'' - (3 - \frac{2}{x})y' + (2 - \frac{3}{x})y = Lclm(D - 1 + \frac{1}{x}, D - 2 + \frac{1}{x})y = 0$. Type $\mathcal{L}_3^2$.

2.119: $y'' - \frac{2(ax+b)}{x}y' + \frac{a(ax+2b)}{x}y = Lclm(D - a + \frac{2b+1}{x}, D - a)y = 0$. Type $\mathcal{L}_3^2$.

2.121: $y'' - (x - 1)y' + (1 - \frac{1}{x})y = (D - x + 1 + \frac{1}{x})(D - \frac{1}{x})y = 0$. Type $\mathcal{L}_2^2$.

2.122: $y'' - (x - 1 - \frac{2}{x})y' - (x + 3)y = (D + 1 + \frac{1}{x})(D - x)y = 0$. Type $\mathcal{L}_2^2$.

2.125: $y'' + (4x - \frac{1}{x})y' - 4x^2y = Lclm(D - 2\alpha x, D - 2\beta x)y = 0$ where $\alpha + \beta + 2 = 0$, $\alpha\beta + 1 = 0$. Type $\mathcal{L}_3^2$.

2.126: $y'' + (2ax^2 - \frac{1}{x})y' + (ax^4 + ax)y = Lclm(D + ax^2 - \frac{2}{x}, D + ax^2)y = 0$. Type $\mathcal{L}_3^2$.

2.129: $y'' - \frac{4x-9}{x-3}y' + \frac{3x-6}{x-3}y =$

$$Lclm\Big(D - 1, D - 3 - \frac{12x^2 - 28x + 50}{4x^3 - 42x^2 + 150x - 183}\Big)y = 0. \text{ Type } \mathcal{L}_3^2.$$

2.131: $y'' - \frac{x-1}{2x}y' + \frac{a}{2x}y = 0.$ Type $\mathcal{L}_1^2.$ Weber equation, $k = -2a$,
$x = z^2$, $y = w$.

2.133: $y'' - \frac{3x-4}{2x-1}y' + \frac{x-3}{2x-1}y = \big(D - \frac{1}{2} + \frac{\frac{5}{2}}{x - \frac{1}{2}}\big)(D - 1)y = 0.$ Type $\mathcal{L}_2^2.$

2.134: $y'' - \frac{x+c}{4x}y = 0.$ Type $\mathcal{L}_1^2.$ Confluent Hypergeometric equation,
$a = 2, b = 1 + \frac{c}{4}, x = z, y = z\exp\big(-\frac{1}{2}z\big)w.$

2.136: $y'' + \frac{1}{x}y' - \frac{x+2}{4x}y = \big(D + \frac{1}{2} + \frac{1}{x}\big)\big(D - \frac{1}{2}\big)y = 0.$ Type $\mathcal{L}_2^2.$

2.139: $y'' + \frac{1}{2x}y' - \frac{x+a}{16x}y = 0.$ Type $\mathcal{L}_1^2.$ Weber equation, $k = -\frac{1}{4}(a + 2)$
$x = z^2, y = \exp\big(-\frac{1}{4}z^2\big)w.$

2.141: $y'' + \frac{3a + bx}{ax}y' + \frac{3b}{ax}y = \big(D + \frac{3}{x}\big)\big(D + \frac{b}{a}\big)y = 0.$ Type $\mathcal{L}_2^2.$

2.146: $y'' - \frac{6}{x^2}y = Lclm\Big(D - \frac{5x^4}{x^5 + C} + \frac{2}{x}\Big)y = 0.$ Type $\mathcal{L}_4^2.$

2.147: $y'' - \frac{12}{x^2}y = Lclm\Big(D - \frac{7x^6}{x^7 + C} + \frac{3}{x}\Big)y = 0.$ Type $\mathcal{L}_4^2.$

2.148: $y'' + \frac{a}{x^2}y = Lclm\big(D - \frac{\theta}{x}\big)y = 0, \theta^2 - \theta + a = 0.$ Type $\mathcal{L}_3^2.$

2.149: $y'' + \frac{ax + b}{x^2}y = 0.$ Type $\mathcal{L}_1^2.$ Bessel equation, $\nu = \sqrt{1 - 4b}$,
$x = \frac{z^2}{4a}, y = \frac{z}{2\sqrt{a}}w.$

2.150: $y'' + \big(1 - \frac{2}{x^2}\big)y = Lclm\big(D + \theta - \frac{1}{x+\theta} + \frac{1}{x}\big)y = 0.$ Type $\mathcal{L}_3^2.$

2.151: $y'' - \big(a + \frac{2}{x^2}\big)y = Lclm\Big(D + a - \frac{\sqrt{a}}{x} + \frac{1}{x^2}, D - a - \frac{\sqrt{a}}{x} - \frac{1}{x^2}\Big)y = 0.$
Type $\mathcal{L}_3^2.$

2.152: $y'' + \big(1 - \frac{6}{x^2}\big)y = Lclm\Big(D - \theta + \frac{2}{x} - \frac{2x + 3\theta}{x^2 + 3\theta x - 3}\Big)y = 0.$ Type $\mathcal{L}_3^2.$

2.153: $y'' + \big(a - \frac{\nu(\nu - 1)}{x^2}\big)y = 0.$ Type $\mathcal{L}_1^2.$ Bessel equation, $\nu + \frac{1}{2}$,
$x = -\frac{z}{\sqrt{a}}, y = \frac{i\sqrt{z}}{a^{1/4}}w.$

2.155: $y'' + \big(ax^{k-2} - \frac{b(b-1)}{x^2}\big)y = 0.$ Type $\mathcal{L}_1^2.$ Bessel equation,
$\nu^2 = \frac{1}{k}\sqrt{1 + 4b(b - 1)}, x = \big(\frac{k^2z^2}{4a}\big)^{1/k}, y = \big(\frac{k^2z^2}{4a}\big)^{1/(2k)}w.$

2.158: $y'' + \frac{a}{x^2}y' - \big(b^2 + \frac{ab}{x^2}\big)y = Lclm\big(D + b + \frac{a}{x^2}, D - b\big)y = 0.$ Type $\mathcal{L}_2^2.$

2.159: $y'' - \frac{1}{x}y' - y = Lclm\big(D - \frac{1}{x^2 + C} + \frac{1}{x}\big)y = 0.$ Type $\mathcal{L}_4^2.$

2.160: $y'' + \frac{1}{x}y' + \frac{a}{x^2}y = Lclm\big(D - \frac{\theta - \frac{1}{2}}{x}\big)y = 0, \theta^2 + a + \frac{1}{4} = 0.$ Type $\mathcal{L}_3^2.$

2.161: $y'' + \frac{1}{x}y' - (\frac{1}{x} + \frac{a}{x^2})y = 0$. Type $\mathcal{L}_1^2$. Bessel equation, $\nu = 2\sqrt{a}$, $x = -\frac{1}{4}z^2$, $y = w$.

2.164: $y'' + \frac{1}{x}y' + (\lambda - \frac{\nu^2}{x^2})y$. Type $\mathcal{L}_1^2$. Bessel equation, $x = \frac{z}{\sqrt{\lambda}}$, $y = w$.

2.165: $y'' + \frac{1}{x}y' + 4(x^2 - \frac{\nu^2}{x^2})y = 0$. Type $\mathcal{L}_1^2$. Bessel equation, $x = \sqrt{z}$, $y = w$.

2.165a: $y'' + (\frac{1}{x} + \frac{a}{x^2})y' - y = Lclm(D - \frac{1}{x} + \frac{a}{x^2}, D - 1 + \frac{1}{x+a})y = 0$. Type $\mathcal{L}_3^2$.

2.166: $y'' - \frac{1}{x}y' + \frac{1}{x^2}y = D(D - \frac{1}{x})y = 0$. Type $\mathcal{L}_2^2$.

2.167: $y'' - \frac{1}{x}y' + (ax^{m-2} + \frac{b}{x^2})y = 0$ Type $\mathcal{L}_1^2$. Bessel equation, $\nu^2 = \frac{4(1-b)}{m^2}$, $x = (\frac{m^2 z^2}{4a})^{1/m}$, $y = (\frac{m^2 z^2}{4a})^{1/m} w$.

2.168: $y'' + \frac{2}{x}y' = Lclm(D - \frac{1}{x+C} + \frac{1}{x}) = 0$. Type $\mathcal{L}_4^2$.

2.169: $y'' + \frac{2}{x}y' + (\frac{a}{x} - \frac{b^2}{x^2})y = 0$. Type $\mathcal{L}_1^2$. Bessel equation, $\nu^2 = 1 + 4b^2$, $x = \frac{z^2}{4a}$, $y = \frac{2\sqrt{a}w}{z}$.

2.170: $y'' + \frac{2}{x}y' + (a + \frac{b}{x^2})y = 0$. Type $\mathcal{L}_1^2$. Bessel equation, $\nu^2 = \frac{1}{4} - b$, $x = \frac{z}{\sqrt{a}}$, $y = \frac{a^{1/4}w}{\sqrt{z}}$.

2.174: $y'' - \frac{2}{x}y' + \frac{2}{x^2}y = Lclm(D + 1 - \frac{1}{x} - \frac{1}{x+C})y = 0$. Type $\mathcal{L}_4^2$.

2.175: $y'' - \frac{2}{x}y' - \frac{4}{x^2}y = Lclm(D + \frac{1}{x} - \frac{5x^4}{x^5 + C})y = 0$. Type $\mathcal{L}_4^2$.

2.176: $y'' - \frac{2}{x}y' + (1 + \frac{2}{x^2})y = Lclm(D - i - \frac{1}{x}, D - i + \frac{1}{x})y = 0$. Type $\mathcal{L}_3^2$.

2.179: $y'' - \frac{2}{x}y' + (a + \frac{2}{x^2})y = Lclm(D - \theta - \frac{1}{x})y = 0$, $\theta^2 + a = 0$. Type $\mathcal{L}_3^2$.

2.180: $y'' + \frac{3}{x}y' + (1 + \frac{1-k^2}{x^2})y = f(x)$. Type $\mathcal{L}_1^2$. Bessel equation, $\nu = 0$, $x = z$, $y = -\frac{w}{z}$.

2.181: $y'' + \frac{3x-1}{x^2}y' + \frac{1}{x^2}y = (D + \frac{2}{x})(D + \frac{1}{x} - \frac{1}{x^2})y = 0$. Type $\mathcal{L}_2^2$.

2.182: $y'' - \frac{3}{x}y' + \frac{4}{x^2}y = (D - \frac{1}{x})(D - \frac{2}{x})y = \frac{5}{x}$. Type $\mathcal{L}_2^2$.

2.183: $y'' - \frac{3}{x}y' - \frac{5}{x^2}y = Lclm(D - \frac{6x^5}{C + x^6} + \frac{1}{x})y = x^2 \log x$. Type $\mathcal{L}_4^2$.

2.184: $y'' - \frac{4}{x}y' + \frac{6}{x^2}y = Lclm(D + \frac{1}{C + x} + \frac{2}{x})y = x^2 - 1$. Type $\mathcal{L}_4^2$.

2.185: $y'' + \frac{5}{x}y' - (2x - \frac{4}{x^2})y = 0$. Type $\mathcal{L}_1^2$. Bessel equation, $\nu = 0$,

$$x = -\tfrac{1}{2}(3z)^{2/3}, \; y = -\frac{4(3z)^{2/3}w}{9z^2}.$$

2.186: $y'' - \frac{5}{x}y' + \frac{8}{x^2}y = Lclm\big(D - \frac{2}{x} - \frac{2x}{x^2+C}\big)y = 0.$ Type $\mathcal{L}_4^2$.

2.189: $y'' + \frac{a}{x}y' + \big(bx^{m-2} + \frac{c}{4x^2}\big)y = 0.$ Type $\mathcal{L}_1^2$. Bessel equation,

$$\nu = \tfrac{1}{m}\sqrt{(1-a)^2 - 4c}, \; x = \Big(\frac{m^2 z^2}{4b}\Big)^{1/m}, \; y = \Big(\frac{m^2 z^2}{4b}\Big)^{(1-a)/(2m)} w.$$

2.191: $y'' - y' - \frac{2}{x^2}y = Lclm\Big(D + 1 - \frac{1}{2(x+2)} + \frac{1}{2x}, D - \frac{1}{x-2} + \frac{1}{x}\Big)y = 0.$

Type $\mathcal{L}_3^2$.

2.192: $y'' + \big(1 - \frac{1}{x^2}\big)y' - \frac{1}{x^2}y = \big(D - \frac{1}{x^2}\big)(D+1)y = 0.$ Type $\mathcal{L}_2^2$.

2.193: $y'' + \big(1 + \frac{1}{x}\big)y' + \big(\frac{1}{x} - \frac{9}{x^2}\big)y = Lclm\Big(D + 1 + \frac{3}{x} - \frac{3x^2 + 18x + 36}{x^3 + 9x^2 + 36x + 60},$

$$D + \frac{3}{x} - \frac{2x - 8}{x^2 - 8x + 20}\Big)y = 0. \text{ Type } \mathcal{L}_3^2.$$

2.194: $y'' + \big(1 + \frac{1}{x}\big)y' + \big(\frac{3}{x} - \frac{1}{x^2}\big)y = \big(D + \frac{1}{x-3} + \frac{2}{x}\big)\big(D + 1 - \frac{1}{x-3} - \frac{1}{x}\big)y = 0.$

Type $\mathcal{L}_2^2$.

2.196: $y'' - \big(1 - \frac{1}{x}\big)y' + \big(\frac{1}{x} - \frac{1}{x^2}\big)y = \big(D - 1 + \frac{2}{x}\big)\big(D - \frac{1}{x}\big)y = 0.$ Type $\mathcal{L}_2^2$.

2.198: $y'' - \big(1 - \frac{2}{x}\big)y' - \big(\frac{3}{x} + \frac{2}{x^2}\big)y = \big(D + \frac{3}{x}\big)\big(D - 1 - \frac{1}{x}\big)y = 0.$ Type $\mathcal{L}_2^2$.

2.199: $y'' - \big(1 + \frac{4}{x}\big)y' - \frac{4}{x^2}y = D\big(D - 1 - \frac{4}{x}\big)y = 0.$ Type $\mathcal{L}_2^2$.

2.201: $y'' + \big(2 + \frac{1}{x}\big)y' - \frac{4}{x^2}y =$

$$Lclm\Big(D - \frac{2(x-1)}{x^2 - 2x + \frac{3}{2}}, D + 2 - \frac{1}{x + \frac{3}{2}} + \frac{2}{x}\Big)y = 0. \text{ Type } \mathcal{L}_3^2.$$

2.202: $y'' - \big(2 + \frac{2}{x}\big)y' + \big(\frac{2}{x} + \frac{2}{x^2}\big)y = Lclm\big(D - \frac{1}{x}, D - 2 - \frac{1}{x}\big)y = 0.$ Type $\mathcal{L}_3^2$.

2.203: $y'' + ay' - \frac{2}{x^2}y = Lclm\big(D - \frac{a}{ax-2} + \frac{1}{x}, D + a + \frac{a}{ax+2} + \frac{1}{x}\big)y = 0.$

Type $\mathcal{L}_3^2$.

2.204: $y'' + \big(a + 2b\big)y' + \big(ab + b^2 - \frac{2}{x^2}\big)y =$

$$Lclm\Big(D + a + b - \frac{1}{ax+2} + \frac{1}{x}, D + ab - \frac{2b}{x} + \frac{1}{x}\Big)y = 0. \text{ Type } \mathcal{L}_3^2.$$

2.208: $y'' + xy' + \big(1 - \frac{2}{x^2}\big)y = \big(D + x - \frac{1}{x}\big)\big(D + \frac{1}{x}\big)y = 0.$ Type $\mathcal{L}_2^2$.

2.209: $y'' + \frac{x^2 + 2}{x}y' + \frac{x^2 - 2}{x^2}y = D\big(D + x + \frac{2}{x}\big)y = 0.$ Type $\mathcal{L}_2^2$.

2.211: $y'' + 4xy' + \big(4x^2 + 2 + \frac{1}{x^2}\big)y = Lclm(D + 2x - \frac{\theta}{x})y = 0, \; \theta^2 - \theta + 1 = 0.$

Type $\mathcal{L}_3^2$.

2.223: $y'' + \frac{x}{x^2 + 1}y' - \frac{9}{x^2 + 1}y =$

$$\Big(D + \frac{2x}{x^2 + \frac{3}{4}} + \frac{x}{x^2 + 1} + \frac{1}{x}\Big)\Big(D - \frac{2x}{x^2 + \frac{3}{4}} + \frac{1}{x}\Big)y = 0. \text{ Type } \mathcal{L}_2^2.$$

2.225: $y'' + \frac{x}{x^2+1}y' + \frac{1}{x^2+1}y = \left(D + \frac{1}{x} - \frac{x}{x^2+1}\right)\left(D - \frac{1}{x}\right)y = 0$. Type $\mathcal{L}_2^2$.

2.227: $y'' - \frac{2x}{x^2+1}y' + \frac{2}{x^2+1}y = Lclm\left(D - \frac{1}{x} - \frac{1 + \frac{1}{x^2}}{C + x - \frac{1}{x}}\right)y = 0$. Type $\mathcal{L}_4^2$.

2.229: $y'' + \frac{4x}{x^2+1}y' + \frac{2}{x^2+1}y = Lclm\left(D - \frac{1}{C+x} + \frac{2x}{x^2+1}\right)y = \frac{2\cos x - 2x}{x^2+1}$. Type $\mathcal{L}_4^2$.

2.230: $y'' + \frac{ax}{x^2+1}y' + \frac{a-1}{x^2+1}y = \left(D + \frac{2x}{x^2+1}\right)\left(D + \frac{(a-2)x}{x^2+1}\right)y = 0$. Type $\mathcal{L}_2^2$.

2.234: $y'' + \left(\frac{\frac{1}{2}}{x+1} + \frac{\frac{1}{2}}{x-1}\right)y' = \left(D + \frac{\frac{1}{2}}{x+1} + \frac{\frac{1}{2}}{x-1}\right)Dy = 0$. Type $\mathcal{L}_2^2$.

2.237: $y'' + \left(\frac{1}{x+1} + \frac{1}{x-1}\right)y' = \left(D + \frac{1}{x+1} + \frac{1}{x-1}\right)Dy = 0$. Type $\mathcal{L}_2^2$.

2.238: $y'' + \frac{2x}{x^2-1}y' = \left(D + \frac{2x}{x^2-1}\right)Dy = 0$. Type $\mathcal{L}_2^2$.

2.242: $y'' - \frac{3x+1}{x^2-1} - \frac{x}{x+1}y = \left(D + \frac{2}{x-1} - \frac{1}{x+1}\right)\left(D + 1 - \frac{2}{x+1}\right)y = 0$. Type $\mathcal{L}_2^2$.

2.243: $y'' + \frac{4x}{x^2-1} + \frac{x^2+1}{x^2-1}y = Lclm\left(D - \theta + \frac{1}{x+1} + \frac{1}{x-1}\right)y = 0,\ \theta^2+1 = 0$. Type $\mathcal{L}_3^2$.

2.247: $y'' + a\left(\frac{1}{x+1} + \frac{1}{x-1}\right)y' - \frac{a(a-1)}{2}\left(\frac{1}{x+1} - \frac{1}{x-1}\right)y = $
$Lclm(D + \frac{a-1}{x+1}, D + \frac{a-1}{x-1})y = 0$. Type $\mathcal{L}_3^2$.

2.250: $y'' + \left(\frac{4}{x+a} + \frac{4}{x-a}\right)y' - \frac{1}{a}\left(\frac{4}{x+a} + \frac{4}{x-a}\right) = $
$Lclm\left(D - \frac{12C(a^2+x^2)x + 9a^4 + 54a^2x^2 + 9x^4}{C(a^4 + 2a^2x^2 - x^4) + 3x(3a^4 - 2a^2x^2 - x^4)}\right)y = 0$. Type $\mathcal{L}_4^2$.

2.251: $y'' - \left(\frac{2}{x+1} - \frac{1}{x}\right)y = \left(D - \frac{2}{x+1} + \frac{1}{x-1} + \frac{1}{x}\right)\left(D - \frac{1}{x-1}\right)y = 0$. Type $\mathcal{L}_2^2$.

2.253: $y'' + \left(\frac{2}{x+1} + \frac{1}{x}\right)y = \left(D + \frac{1}{x+1} + \frac{1}{x}\right)\left(D - \frac{1}{x+1} + \frac{1}{x}\right)y = 0$. Type $\mathcal{L}_2^2$.

2.254: $y'' + \left(1 - \frac{2}{x+2}\right)y' - \left(6 - \frac{3}{x+2} + \frac{3}{x-1}\right)y = $
$\left(D + 3 - \frac{2}{x+2} + \frac{1}{x-1}\right)\left(D - 2 + \frac{1}{x-1}\right)y = 0$. Type $\mathcal{L}_2^2$.

2.255: $y'' + a\left(\frac{1}{x-1} - \frac{1}{x}\right)y' - \left(\frac{2}{x-1} - \frac{2}{x}\right)y = $
$Lclm\left(D + \frac{a-1}{x-1} - \frac{a+1}{x}, D - \frac{2x + 2a - 1}{x^2 + ax - 2 + a(a-1)/2}\right)y = 0$.
Type $\mathcal{L}_3^2$.

2.256: $y'' + \frac{(2x-1)}{x(x-1)}y' - \frac{\nu(\nu+1)}{x(x-1)}y = 0$. Type $\mathcal{L}_1^2$. Legendre equation,
$x = \frac{1-z}{2},\ y = w$.

$$\text{Loewy Decompositions} \qquad\qquad 395$$

2.257: $y'' + \dfrac{(a+1)x + b}{x(x-1)} y' = \left(D + \dfrac{a+b+1}{x-1} - \dfrac{b}{x}\right) Dy = 0.$ Type $\mathcal{L}_2^2$.

2.262: $y'' + \dfrac{x^2 + x - 1}{(x-1)^2} y' - \dfrac{x+2}{(x-1)^2} y = \left(D - \dfrac{1}{x+1} - \dfrac{1}{x + 2x + 1}\right)(D+1)y = 0.$
Type $\mathcal{L}_2^2$.

2.263: $y'' + \dfrac{3x - 1}{(x^2 + 3x)^2} y' + \dfrac{1}{(x^2 + 3x)} y =$

$\left(D + \dfrac{2x + 3}{x^2 + 3x}\right)\left(D + \dfrac{\frac{7}{3}}{x+3} - \dfrac{\frac{4}{3}}{x}\right) y = 0.$ Type $\mathcal{L}_2^2$.

2.264: $y'' + \dfrac{x^2 + x + 1}{x^2 + 3x + 4} y' - \dfrac{2x + 3}{x^2 + 3x + 1} y =$

$\left(D + \dfrac{2x + 1}{x^2 + x + 3} - \dfrac{2x + 3}{x^3 + 3x + 4}\right)\left(D - \dfrac{2x + 1}{x^2 + x + 3}\right) y = 0.$ Type $\mathcal{L}_2^2$.

2.265: $y'' - \dfrac{2x - 3}{(x-1)(x-3)} y' + \dfrac{1}{(x-1)(x-3)} y = 0.$ Type $\mathcal{L}_1^2$.
Hypergeometric equation, $\alpha = -\frac{1}{2}(3 + \sqrt{5})$, $\beta = -\frac{1}{2}(3 - \sqrt{5})$. $\gamma = -1$,
$x = z + 1$, $y = w$.

2.266: $y'' - \dfrac{1}{x-2} y' - \dfrac{3}{(x-2)^2} y = Lclm\left(D - \dfrac{4(x-1)^3)}{C + (x-2)^4} - \dfrac{1}{x-2}\right) y = 0.$
Type $\mathcal{L}_4^2$.

2.267: $y'' - \dfrac{\frac{1}{2}\nu + x^2 - \frac{5}{2}}{x^2} y' + \dfrac{2x - \frac{1}{2}}{x^2} y = \left(D + \dfrac{2}{x}\right)\left(D - 1 + \dfrac{1}{2x} - \dfrac{\nu}{2x^2}\right) y = 0.$
Type $\mathcal{L}_2^2$.

2.269: $y'' + \dfrac{(2\nu + 5)x - 2\nu - 3}{2x(x-1)} y' + \dfrac{\nu + 1}{2x(x-1)} y = 0.$ Type $\mathcal{L}_1^2$. Legendre
equation, inverse of $z = \dfrac{x+1}{2\sqrt{x}}$ and $w = x^{(\nu+1)/2} y$.

2.270: $y'' + \dfrac{5x^2 + 21/2x + 4}{x^2 + 3x + 2} y' + \dfrac{6x^2 + 17/2x + 4}{x^2 + 3x + 2} y =$
$\left(D + \dfrac{4x^2 + 11x + 4}{2x^2 + 6x + 4}\right)\left(D + 3 - \dfrac{4}{x+2}\right) y = 0.$ Type $\mathcal{L}_2^2$.

2.271: $y'' + \dfrac{1}{4x^2} y = \left(D + \dfrac{1}{2x}\right)\left(D - \dfrac{1}{2x}\right) y = 0.$ Type $\mathcal{L}_2^2$.

2.272: $y'' + \left(a^2 + \dfrac{1}{4x^4}\right) y = 0.$ Type $\mathcal{L}_1^2$. Bessel equation, $\nu = 0$,
$x = \frac{z}{a}$, $y = \left(\frac{z}{a}\right)^{1/2} w$.

2.274: $y'' + \dfrac{1}{x} y' + \dfrac{x - \nu^2}{4x^2} y = 0.$ Type $\mathcal{L}_1^2$. Bessel equation, $x = z^2$, $y = w$.

2.276: $y'' + \dfrac{1}{x} y' - \dfrac{4x^2 + 1}{4x^2} y = Lclm\left(D + 1 + \dfrac{1}{2x}, D - 1 + \dfrac{1}{2x}\right) y = \dfrac{\sqrt{x}e^x}{x}.$
Type $\mathcal{L}_3^2$.

2.277: $y'' + \dfrac{1}{x} y' - \dfrac{ax^2 + 1}{4x^2} y = Lclm\left(D + \dfrac{\sqrt{a}}{2} - \dfrac{1}{2x}, D - \dfrac{\sqrt{a}}{2} + \dfrac{1}{2x}\right) y = 0.$
Type $\mathcal{L}_3^2$.

2.279: $y'' + \dfrac{5}{4x} y' - \dfrac{1}{4x^2} y = Lclm\left(D - \dfrac{\theta}{x}\right) y = 0,\ \theta^2 + \frac{1}{4}\theta - \frac{1}{4} = 0.$ Type $\mathcal{L}_3^2$.

2.280: $y'' + \frac{2}{x}y' - \frac{x^2 + 3x + \frac{3}{4}}{x^2}y = \left(D + 1 + \frac{5}{2x}\right)\left(D - 1 - \frac{1}{2x}\right)y = 0$. Type $\mathcal{L}_2^2$.

2.281: $y'' - \frac{2x-1}{x}y' - \frac{4x^2 - 4x - 1}{4x^2}y = Lclm\left(D - \frac{2x-1}{2x} - \frac{1}{C+x}\right)y = 0$.
Type $\mathcal{L}_4^2$.

2.282: $y'' + xy' + \left(\frac{1}{4}x^2 + \frac{1}{2} - 6\right)y = Lclm\left(D + \frac{1}{2}x + \frac{2}{x}, D + \frac{1}{x} - \frac{3}{x}\right)y = 0$.
Type $\mathcal{L}_3^2$.

2.284: $y'' - \frac{2}{2x+1}y' - \frac{12}{(2x+1)^2}y =$
$Lclm\left(D - \frac{1}{x+\frac{1}{2}} + \frac{4x^3 + 6x^2 + 3x + \frac{1}{2}}{C + x^4 + 2x^3 + \frac{3}{2}x^2 + \frac{1}{2}x}\right)y = 0$. Type $\mathcal{L}_4^2$.

2.286: $y'' + \frac{3}{3x-1}y' - \frac{9}{(3x-1)^2}y =$
$Lclm\left(D - \frac{2x - \frac{2}{3}}{C + x^2 - \frac{2}{3}x} + \frac{1}{x - \frac{1}{3}}\right)y = 0$. Type $\mathcal{L}_4^2$.

2.287: $y'' + \frac{3(2x-1)}{9x(x-1)}y' - \frac{20}{9x(x-1)}y = 0$. Type $\mathcal{L}_1^2$. Hypergeometric
equation, $\alpha = \frac{1}{3}$, $\beta = -\frac{20}{3}$.

2.290: $y'' + \frac{27x}{27x^2 + 4}y' - \frac{3}{27x^2 + 4}y = 0$. Type $\mathcal{L}_1^2$. Galois group D_3.

2.299: $y'' + \frac{2a^2 x}{a^2 x^2 - 1}y' = \left(D + \frac{a}{ax+1} + \frac{a}{ax-1}\right)Dy = 0$. Type $\mathcal{L}_2^2$.

2.300: $y'' + \frac{2a^2 x}{a^2 x^2 - 1}y' - \frac{2a^2}{a^2 x^2 - 1}y = \left(D + \frac{a}{ax+1} + \frac{a}{ax-1} + \frac{1}{x}\right)\left(D - \frac{1}{x}\right)y = 0$.
Type $\mathcal{L}_2^2$.

2.304: $y'' + \frac{1}{x^2}y' - \frac{2x+3}{x^3}y = \left(D - \frac{1}{x}, D + \frac{1}{x} + \frac{1}{x^2}\right)y = 0$. Type $\mathcal{L}_2^2$.

2.307: $y'' + \frac{x+1}{x^2}y' - \frac{2}{x^3}y = Lclm\left(D - \frac{1}{x+1}, D + \frac{1}{x} + \frac{1}{x^2} - \frac{1}{x+1}\right)y = 0$.
Type $\mathcal{L}_2^2$.

2.308: $y'' - \frac{1}{x}y' + \frac{1}{x^2}y = D\left(D - \frac{1}{x}\right)y = \frac{\log(x)^3}{x^3}$. Type $\mathcal{L}_2^2$.

2.310: $y'' + \frac{3}{x}y' + \frac{1}{x^2}y = \left(D + \frac{2}{x}\right)\left(D + \frac{1}{x}\right)y = 1$. Type $\mathcal{L}_2^2$.

2.311: $y'' + \frac{2x^2 + 1}{x(x^2 + 1)}y' - \frac{\nu(\nu+1)x}{x^2 + 1}y = 0$. Type $\mathcal{L}_1^2$. Legendre equation,
$x = \sqrt{z^2 - 1}$, $y = w$.

2.312: $y'' + \frac{2x^2 - 2}{x^3 + x}y' - \frac{2}{x^2 + 1}y = Lclm\left(D - \frac{3x^2}{C + x^3} + \frac{2x}{x^2 + 1}\right)y = 0$.
Type $\mathcal{L}_4^2$.

2.319: $y'' - \frac{1}{x^3 + x}y' - \frac{6}{x^2 + 1}y = \left(D + \frac{2x}{x^2 + 1} + \frac{1}{x}\right)\left(D - \frac{x}{x^2 + 1} - \frac{2}{x}\right)y = 0$.
Type $\mathcal{L}_2^2$.

2.320: $y'' - \frac{x^3 + 3x^2 - 2x - 1}{x^3 - 2x}y' + \frac{x^4 + 4x + 2}{x^3 - 2x}y =$

$Lclm\left(D - 1 - \frac{2}{x}, D - \frac{1}{x-1}\right)y = 0$. Type $\mathcal{L}_3^2$.

2.321: $y'' - \frac{2x+1}{x^2+x}y' + \frac{2x+1}{x^3+x^2}y = \left(D - \frac{1}{x+1}\right)\left(D - \frac{1}{x}\right) = 0$. Type $\mathcal{L}_2^2$.

2.322: $y'' + \frac{6x+4}{x^2+x}y' + \frac{6x+2}{x^3+x^2}y = Lclm\left(D - \frac{1}{x+C} + \frac{3x+2}{x(x+1)}\right)y = 0$.
Type $\mathcal{L}_4^2$.

2.323: $y'' + \frac{2x-1}{x^2-x}y' - \frac{2x+2}{x^3-x^2}y = Lclm\left(D - \frac{3x^2-6x+3}{C+x^3-3x^2+3x} + \frac{2}{x}\right)y = 0$.
Type $\mathcal{L}_4^2$.

2.324: $y'' - \frac{5x-4}{x^2-x}y' + \frac{9x-6}{x^3-x^2}y = \left(D - \frac{1}{x-1} - \frac{1}{x}\right)\left(D - \frac{3}{x}\right)y = 0$. Type $\mathcal{L}_2^2$.

2.326: $y'' + \frac{1}{x+1}y' + \frac{1}{x^3+2x^2+x}y = \left(D + \frac{1}{x}\right)\left(D + \frac{1}{x+1} - \frac{1}{x}\right)y = 0$.
Type $\mathcal{L}_2^2$.

2.328: $y'' - \left(\frac{2}{x} - \frac{2}{x-1} + \frac{2}{(x-1)^2}\right)y = \left(D + \frac{1}{x} - \frac{1}{x-1}\right)\left(D - \frac{1}{x} + \frac{1}{x-1}\right)y = 0$.
Type $\mathcal{L}_2^2$.

2.331: $y'' + \left(\frac{\frac{1}{2}}{x-2} - \frac{1}{x}\right)y' - \left(\frac{\frac{1}{8}}{x-2} - \frac{\frac{1}{8}}{x} - \frac{\frac{3}{4}}{x^2}\right)y =$
$Lclm(D - \frac{\frac{1}{2}}{x-2} + \frac{1}{2x}, D - \frac{1}{2x})y = 0$. Type $\mathcal{L}_3^2$.

2.332: $y'' - \frac{1}{x+1}y' + \frac{3x+1}{4x^3+4x^2}y = \left(D - \frac{1}{x+1} + \frac{1}{2x}\right)\left(D - \frac{1}{2x}\right)y = 0$.
Type $\mathcal{L}_2^2$.

2.333: $y' + \frac{3x-1}{2x(x-1)}y' - \frac{\nu(\nu+1)}{4x^2}y = 0$. Type $\mathcal{L}_1^2$. Legendre equation,
$x = 2z^2 - 1 + 2z\sqrt{z^2-1}$, $y = w$.

2.334: $y'' - \frac{5x-4}{x^2-x}y' + \frac{9x-6}{x^3-x^2}y = \left(D + \frac{1}{x-1} - \frac{1}{x}\right)\left(D - \frac{3}{x}\right)y = 0$. Type $\mathcal{L}_2^2$.

2.336: $y'' - \frac{3x-4}{(x-1)(2x-1)^2}y =$
$$\left(D + \frac{1}{x-1} - \frac{\frac{1}{2}}{x-\frac{1}{2}}\right)\left(D - \frac{1}{x-1} + \frac{\frac{1}{2}}{x-\frac{1}{2}}\right)y = 0. \text{ Type } \mathcal{L}_2^2.$$

2.337: $y'' + \frac{3x+a+2b}{2(x+a)(x+b)}y' + \frac{a-b}{4(x+a)^2(x+b)^2}y =$
$Lclm\left(D + \frac{1}{2(x+a)}, D - \frac{a-b}{2(x+a)(x+b)}\right)y = 0$. Type $\mathcal{L}_3^2$.

2.338: $y'' - \frac{18x-3}{9x^2-2x}y' - \frac{3}{9x^3-2x^2}y =$
$Lclm\left(D + \frac{1}{x} + \frac{2x-\frac{1}{3}}{x^2-\frac{1}{3}x+\frac{1}{54}}, D + \frac{3}{2x} + \frac{\frac{3}{2}}{x-\frac{2}{9}}\right)y = 0$.
Type $\mathcal{L}_3^2$.

2.340: $y'' - \frac{2ax+b}{ax^2+bx}y' + \frac{2ax+6b}{ax^3+bx^2}y = Lclm\left(D - \frac{1}{x+C} + \frac{ax+b}{ax^2+bx}\right)y = 0$.
Type $\mathcal{L}_4^2$.

2.342: $y'' + \frac{a}{x^4}y = Lclm\left(D + \frac{1}{x} + \frac{i\sqrt{a}}{x^2}, D + \frac{1}{x} - \frac{i\sqrt{a}}{x^2}\right)y = 0$. Type $\mathcal{L}_3^2$.

2.345: $y'' + \frac{1}{x^3}y' - \frac{2}{x^4}y = \left(D + \frac{1}{x}\right)\left(D - \frac{1}{x} + \frac{1}{x^3}\right)y = 0$. Type $\mathcal{L}_2^2$.

2.346: $y'' - \frac{2}{x^2}y' + \frac{2x+1}{x^4}y = Lclm\left(D - \frac{1}{x^2} - \frac{1}{x+C}\right)y = 0$. Type $\mathcal{L}_4^2$.

2.347: $y'' + \frac{1}{x}y' + \frac{1}{x^4}y = 0$. Type $\mathcal{L}_1^2$. Bessel equation, $\nu = 0$, $x = \frac{1}{z}$, $y = w$.

2.347a: $y'' + \frac{1}{x}y' + \frac{x-1}{x^4}y = \left(D + \frac{1}{x} + \frac{1}{x^2}\right)\left(D - \frac{1}{x^2}\right)y = 0$. Type $\mathcal{L}_2^2$.

2.350: $y'' + \frac{2}{x}y' + \frac{a^2}{x^4}y = Lclm\left(D - \frac{\theta}{x^2}\right)y = 0$, $\theta^2 + a^2 = 0$. Type $\mathcal{L}_3^2$.

2.351: $y'' + \frac{2x^2+1}{x^3}y' - \frac{1}{x^4}y = \left(D + \frac{2}{x}\right)\left(D + \frac{1}{x^3}\right)y = 0$. Type $\mathcal{L}_2^2$.

2.353: $y'' - \left(\frac{2}{x} - \frac{1}{x^3}\right)y' + \frac{2}{x^4}y = \left(D + \frac{1}{x} - \frac{4x^3 + 4x}{x^4 + 2x^2 - 1}\right)$

$\left(D + \frac{\frac{1}{21}x^3 - \frac{23}{7}x}{x^4 + 3x^2 + 3} + \frac{\frac{2}{7}}{x+1} + \frac{\frac{2}{7}}{x-1} + \frac{3}{x}\right)y = 0$. Type $\mathcal{L}_2^2$.

2.354: $y'' - \left(\frac{2}{x} - \frac{1}{x^3}\right)y' + \frac{1}{x^4}y =$

$\left(D - \frac{4}{x} + \frac{1}{x^3} + \frac{2x}{x^2 - \frac{1}{5}}\right)\left(D + \frac{2}{x} - \frac{2}{x^2 - \frac{1}{5}}\right)y = 0$. Type $\mathcal{L}_2^2$.

2.355: $y'' + \left(\frac{\frac{4}{3}x - \frac{2}{3}}{x^2 - x + 1} + \frac{\frac{2}{3}}{x+1} - \frac{1}{x}\right)y' - \frac{x}{x^3+1}y =$

$\left(D + \frac{2x-1}{x^2 - x + 1} + \frac{1}{x+1} - \frac{1}{x}\right)\left(D - \frac{x^2}{x^3+1}\right)y = 0$. Type $\mathcal{L}_2^2$.

2.358: $y'' - \frac{x^2 - 2}{x^3 - x}y' - \frac{x^2 - 2}{x^4 - x}y = \left(D + \frac{\frac{1}{2}}{x+1} + \frac{\frac{1}{2}}{x-1} - \frac{1}{x}\right)\left(D - \frac{1}{x}\right)y = 0$.
Type $\mathcal{L}_2^2$.

2.359: $y'' + \frac{2x^3}{x^2(x^2 - 1)}y' + \frac{\nu(\nu + 1)}{x^2(x^2 - 1)}y = 0$. Type $\mathcal{L}_1^2$. Legendre equation,

$x = \frac{1}{z}$, $y = w$.

2.361: $y'' - \frac{2x}{x^2 - 1}y' - \frac{a^2(x^2 - 1) + a(3x^2 - 1)}{x^4 - x^2}y =$

$Lclm\left(D + \frac{a}{x}, D - \frac{a+1}{x} - \frac{2x}{x^2 - (a + \frac{3}{2})/(a + \frac{1}{2})}\right)y = 0$. Type $\mathcal{L}_3^2$.

2.365: $y'' + \frac{a}{(x^2 + 1)^2}y = D\left(D - \frac{1}{x+1} - \frac{1}{x-1} + \frac{x}{x^2 + 1}\right)y = 0$. Type $\mathcal{L}_2^2$

2.366: $y'' + \frac{2x}{x^2 + 1}y' + \frac{1}{(x^2 + 1)^2}y = Lclm\left(D - \frac{1}{x+C} + \frac{x}{x^2 + 1}\right)y = 0$.
Type $\mathcal{L}_4^2$

2.370: $y'' + \frac{2x}{x^2 - 1}y' - \frac{a^2}{(x^2 - 1)^2}y =$

$Lclm\left(D + \frac{a}{2}\left(\frac{1}{x+1} - \frac{1}{x-1}\right), D - \frac{a}{2}\left(\frac{1}{x+1} - \frac{1}{x-1}\right)\right)y = 0$. Type $\mathcal{L}_3^2$.

$$\text{Loewy Decompositions} \qquad 399$$

2.378: $y'' + \dfrac{2(x+1)}{x(x-1)}y' - \dfrac{2(x^2 - x - 1)}{x^2(x-1)^2}y = \left(D + \dfrac{3}{x-1}\right)\left(D + \left(\dfrac{1}{x-1} - \dfrac{2}{x}\right)\right)y = 0.$

Type $\mathcal{L}_2^2$.

2.379: $y'' - \dfrac{12}{(x+1)^2(x^2 + 2x + 3)}y =$

$\left(D - \dfrac{4}{(x+1)(x^2+2x+3)}\right)\left(D + \dfrac{4}{(x+1)(x^2+2x+3)}\right)y = 0.$ Type $\mathcal{L}_2^2$.

2.380: $y'' + \dfrac{m(m-1)a^2}{x^2(x-a)^2}y = Lclm\left(D + \dfrac{m}{x-a} - \dfrac{m-1}{x}, D + \dfrac{m-1}{x-a} - \dfrac{m}{x}\right)y = 0.$

Type $\mathcal{L}_3^2$.

2.384: $y'' - \dfrac{(ax-b)^2 - 6bx - x^2}{4x^2}y = \left(D - \dfrac{(ax-b)^2 - 2bx - x^2}{(2a+1)x^3 - 2bx^2}\right)$

$\left(D + \dfrac{a+1}{2x} - \dfrac{b}{2x^2} - \dfrac{1}{x - b/(a+1)}\right)y = 0.$ Type $\mathcal{L}_2^2$.

2.386: $y'' - \dfrac{18}{(2x+1)^2(x^2 + x + 1)}y =$

$\left(D - \dfrac{2x+1}{x^2 + x + 1} + \dfrac{2}{x + \frac{1}{2}}\right)\left(D + \dfrac{2x+1}{x^2 + x + 1} - \dfrac{2}{x + \frac{1}{2}}\right)y = 0.$ Type $\mathcal{L}_2^2$.

2.387: $y'' - \dfrac{\frac{3}{4}}{(x^2 + x + 1)^2}y = \left(D + \dfrac{x + \frac{1}{2}}{x^2 + x + 1}\right)\left(D - \dfrac{x + \frac{1}{2}}{x^2 + x + 1}\right)y = 0.$

Type $\mathcal{L}_2^2$.

2.390: $y'' + \dfrac{\frac{3}{16}}{x^2(x-1)^2}y = Lclm\left(D - \dfrac{\frac{3}{4}}{x-1} - \dfrac{\frac{1}{4}}{x}, D - \dfrac{\frac{3}{4}}{x} - \dfrac{\frac{1}{4}}{x-1}\right)y = 0.$

Type $\mathcal{L}_3^2$.

2.391: $y'' - \dfrac{7ax^2 + 5}{ax^3 + x}y' + \dfrac{15ax^2 + 5}{ax^4 + x^2}y =$

$Lclm\left(D - \dfrac{(8ax^3(ax^2 + 1))/(2ax^2 + 1)^2}{C + (2ax^4)/(2ax^2 + 1)} + \dfrac{6ax^2 + 1}{x(2ax^2 + 1)}\right)y = 0.$ Type $\mathcal{L}_4^2$.

2.397: $y'' + \dfrac{1}{x^4}y' - \dfrac{1}{x^5}y = \left(D - \dfrac{1}{x} - \dfrac{1}{x^4}\right)\left(D - \dfrac{1}{x}\right)y = 0.$ Type $\mathcal{L}_2^2$.

2.399: $y'' - \dfrac{3x+1}{x^2 - 1}y' + \dfrac{12(x+1)^2}{(x-1)^2(x + \frac{5}{3})^2}y =$

$\left(D - \dfrac{1}{x+1} - \dfrac{\frac{1}{2}}{x-1} + \dfrac{\frac{1}{2}}{x + \frac{5}{3}}\right)\left(D - \dfrac{\frac{3}{2}}{x-1} - \dfrac{\frac{1}{2}}{x + \frac{5}{3}}\right)y = 0.$ Type $\mathcal{L}_2^2$.

2.400: $y'' - \dfrac{1}{x}y' + \dfrac{a}{x^6}y = Lclm\left(D - \dfrac{2}{x} + \dfrac{\sqrt{a}}{x^3}, D - \dfrac{2}{x} - \dfrac{\sqrt{a}}{x^3}\right)y = 0.$ Type $\mathcal{L}_3^2$.

2.401: $y'' + \dfrac{3x^2 + a}{x^3}y' - \dfrac{a^2 m(m-1)}{x^6}y =$

$Lclm\left(D - \dfrac{a(m-1)}{x^3}, D + \dfrac{am}{x^3}\right)y = 0.$ Type $\mathcal{L}_3^2$.

2.404: $y'' + \dfrac{2x^2 + 1}{x^3}y' - \dfrac{2x^2 - 1}{4x^6}y =$

$Lclm\left(D + \dfrac{1}{x} + \dfrac{1}{2x^3}, D + \dfrac{1}{2x^3}\right)y = 0.$ Type $\mathcal{L}_3^2$.

2.405: $y'' + \dfrac{2x^2 + 1}{x^3}y' - \dfrac{8x^4 + 10x^2 + 1}{4x^6}y =$
$Lclm\left(D - \frac{1}{x} - \frac{1}{2x^3}, D - \frac{2}{x} - \frac{1}{2x^3}\right)y = 0$. Type $\mathcal{L}_3^2$.

2.406: $y'' + \frac{27}{16}\dfrac{x}{(x^3 - 1)^2}y = Lclm\left(D + \frac{1}{x} + \frac{1}{2x^3}, D + \frac{1}{2x^3}\right)y = 0$. Type $\mathcal{L}_1^2$.

Equations of Third Order. These equations are from Chapter 3 of Kamke's collection. The tabulation below is organized by decomposition type $\mathcal{L}_k^3$, $k = 1, \ldots, 12$ for third order equations as described in Corollary 2.4.

$\mathcal{L}_1^3 : 3.6, 3.41, 3.54, 3.57, 3.77,$ $\mathcal{L}_2^3 : 3.54, 3.76,$

$\mathcal{L}_3^3 : 3.73,$ $\mathcal{L}_5^3 : 3.33, 3.58,$

$\mathcal{L}_6^3 : 3.45,$ $\mathcal{L}_7^3 : 3.37, 3.38, 3.39, 3.56, 3.65, 3.68, 3.70,$

$\mathcal{L}_8^3 : 3.27, 3.46, 3.63, 3.74, 3.75,$ $\mathcal{L}_9^3 : 3.1, 3.4, 3.5, 3.16, 3.17, 3.29, 3.47, 3.64,$

$3.66, 3.78,$

$\mathcal{L}_{10}^3 : 3.32, 3.35, 3.42,$ $\mathcal{L}_{12}^3 : 3.18, 3.71.$

Similar to equations of order two, in the listing below the decomposition of equations of Chapter 3 is given if its type is *different* from $\mathcal{L}_1^3$.

3.1: $y''' + \lambda y = Lclm(D - \theta)y = 0$, $\theta^3 + \lambda = 0$. Type $\mathcal{L}_9^3$.

3.4: $y''' + 3y' - 4y = Lclm(D - 1, D - \theta)y = 0$, $\theta^2 + \theta + 4 = 0$. Type $\mathcal{L}_9^3$.

3.5: $y''' - a^2 y' = Lclm(D + a, D - a, D)y = 0$. Type $\mathcal{L}_9^3$.

3.16: $y''' - 2y'' - 3y' + 10y = Lclm(D + 2, D - \theta)y = 0$, $\theta^2 + 4\theta + 5 = 0$. Type $\mathcal{L}_9^3$.

3.17: $y''' - 2y'' - a^2 y' + 2a^2 y = Lclm(D + a, D - a, D - 2)y = 0$. Type $\mathcal{L}_9^3$.

3.18: $y''' - 3ay'' + 3a^2 y' - a^3 y = Lclm\left(D - a - \dfrac{2x + C_1}{x^2 + C_1 x + C_2}\right)y = 0$.
Type $\mathcal{L}_{12}^3$.

3.27: $y''' - 2y'' - \frac{11}{4}y' - \frac{3}{4}y = (D - 3)Lclm\left(D + \frac{1}{2} - \frac{1}{x + C}\right)y = 0$. Type $\mathcal{L}_8^3$.

3.29: $y''' + \frac{3}{x}y'' + y = Lclm\left(D + 1 + \frac{1}{x}, D - \theta + \frac{1}{x}\right)y = 0$, $\theta^2 - \theta + 1 = 0$.
Type $\mathcal{L}_9^3$.

3.32: $y''' - \dfrac{x + 2\nu}{x}y'' - \dfrac{x - 2\nu - 1}{x}y' - \dfrac{x - 1}{x}y =$
$Lclm\left(D^2 - \dfrac{2\nu + 1}{x}D - 1, D - 1\right)y = 0$. Type $\mathcal{L}_{10}^3$.

3.33: $y''' + \left(x - \frac{3}{x}\right)y'' + 4y' + \frac{2}{x}y = Lclm\left(D - \frac{1}{x + C} + \frac{1}{x}\right)(D + x - \frac{5}{x})y = 0$.
Type $\mathcal{L}_5^3$.

3.35: $y''' - \dfrac{2(x + \nu - 2)}{x}y'' + \left(x + \dfrac{x + \frac{3}{\nu} - \frac{5}{2}}{x}\right)y' - \dfrac{\nu - \frac{1}{2}}{x}y =$
$Lclm\left(D^2 - \dfrac{x + 2\nu - 1}{x}D + \dfrac{\nu - \frac{1}{2}}{x}, D - 1\right)y = 0$. Type $\mathcal{L}_{10}^3$.

Loewy Decompositions

3.37: $y''' - y'' - \frac{2}{x^2 - 2x}y' + \frac{2}{x^2 - 2x}y =$

$(D + \frac{1}{x-2} + \frac{1}{x})Lclm(D - 1, D - \frac{2}{x})y = 0.$ Type $\mathcal{L}_7^3$.

3.38: $y''' - \frac{8x}{2x-1}y' + \frac{8}{2x-1}y = (D + 2 + \frac{1}{x-1/2})Lclm(D - 2, D - \frac{1}{x})y = 0.$
Type $\mathcal{L}_7^3$.

3.39: $y''' + \frac{x+4}{2x-1}y'' + \frac{2}{2x-1}y' =$

$(D + \frac{1}{x} + \frac{1}{x - \frac{1}{2}})Lclm(D + \frac{1}{2} + \frac{\frac{1}{4}}{x - \frac{1}{2}}, D)y = 0.$ Type $\mathcal{L}_7^3$.

3.42: $y''' - \frac{1}{x}y'' + (x + \frac{1}{x})y' = Lclm(D^2 - \frac{1}{x}D + 1, D)y = 0.$ Type $\mathcal{L}_{10}^3$.

3.45: $y''' + \frac{4}{x}y'' + (x + \frac{2}{x})y' + \frac{3}{x}y = (D + \frac{3}{x})(D^2 + \frac{1}{x}D + 1)y = 0.$ Type $\mathcal{L}_6^3$.

3.46: $y''' + \frac{5}{x}y'' + \frac{4}{x^2}y' = (D + \frac{3}{x})Lclm(D - \frac{1}{x+C} + \frac{1}{x})y = 0.$ Type $\mathcal{L}_8^3$.

3.47: $y''' + \frac{6}{x}y'' + \frac{6}{x^2})y' = Lclm(D + \frac{2}{x}, D + \frac{1}{x}, D)y = 0.$ Type $\mathcal{L}_9^3$.

3.54: $y''' - \frac{x^3 - 6}{x}y'' - \frac{2x^3 - 6}{x^2}y' + 2y = (D^2 - (x^2 - \frac{4}{x})D + \frac{2}{x^2})(D + \frac{2}{x})y = 0.$
Type $\mathcal{L}_2^3$.

3.55: $y''' + \frac{8x}{x^2 + 1}y'' + \frac{10}{x^2 + 1}y' =$

$Lclm\left(D + \frac{4x}{x^2 + 1} - \frac{4x^3 + 3C_2x^2 + 4x + 3C_2}{x^4 + C_2x^3 + 2x^2 + 3C_2x + C_1}\right)y = 0.$ Type $\mathcal{L}_{12}^3$.

3.56: $y''' + \frac{2x}{x^2 + 2}y'' + y' + \frac{2x}{x^2 + 2}y = Lclm(D - \frac{2}{x}, D + \theta)y = 0, \theta^2 + 1 = 0.$
Type $\mathcal{L}_9^3$.

3.58: $y''' + (\frac{1}{4} + \frac{7}{2x} - \frac{1}{4x^2})y'' + (\frac{1}{x} + \frac{1}{x^2})y' + \frac{1}{2x^2}y =$

$Lclm(D + \frac{1}{C+x} - \frac{2}{x})(D + \frac{1}{4} - \frac{1}{2x} - \frac{1}{4x^2})y = 0.$ Type $\mathcal{L}_5^3$.

3.63: $y''' + \frac{3}{x}y'' - \frac{2}{x^2}y' + \frac{2}{x^3}y = (D + \frac{1}{x})Lclm(D - \frac{3x^2}{C + x^3} + \frac{2}{x})y = 0.$
Type $\mathcal{L}_8^3$.

3.64: $y''' + \frac{3}{x}y'' - \frac{a^2 - 1}{x^2})y = Lclm(D + \frac{a}{x}, D - \frac{1}{x}, D)y = 0.$ Type $\mathcal{L}_9^3$.

3.65: $y''' - \frac{4}{x}y'' + (1 + \frac{8}{x^2})y = Lclm(D - \frac{2}{x}, D + \theta + \frac{1}{x})y = 0, \theta^2 + 1 = 0.$
Type $\mathcal{L}_9^3$.

3.66: $y''' + \frac{6}{x}y'' + (a - \frac{12}{x^3})y = Lclm(D - \frac{\alpha}{x} + \frac{2}{x^2})y = 0, \alpha^3 + a = 0.$ Type $\mathcal{L}_9^3$.

3.68: $y''' + \frac{x+3}{x}y'' + \frac{5x - 30}{x^2}y' + \frac{4x + 30}{x^3}y = (D + \frac{1}{x} + \frac{2x - 60}{x^2 - 60x + 450})$

$Lclm\left(D + \frac{6}{x} - \frac{4x^3 - 252x^2 + 4032x - 20160}{x^4 - 84x^3 + 2016x^2 - 20160x + 75600}, D + 1 + \frac{6}{x} - \frac{p}{q}\right)y = 0.$

$p := 8x^7 + 196x^6 + 2700x^5 + 25500x^4 + 171600x^3 + 801360x^2 + 2358720x$
$+3326400, q := x^8 + 28x^7 + 450x^6 + 5100x^5 + 42900x^4 + 267120x^3$

$+1179360x^2 + 3326400x + 4536000$. Type $\mathcal{L}_7^3$.

3.70: $y''' + \left(\frac{3x}{x^2+1} + \frac{3}{x}\right)y'' - \left(\frac{12x}{x^2+1} - \frac{12}{x}\right)y =$
$\left(D + \frac{2x}{x^2+1} + \frac{3}{x}\right) Lclm\left(D - \frac{2x}{x^2+\frac{1}{2}}, D - \frac{x}{x^2+1} + \frac{1}{x}\right)y = 0$. Type $\mathcal{L}_7^3$.

3.71: $y''' - \left(\frac{1}{x+3} + \frac{2}{x}\right)y'' - \left(\frac{\frac{4}{3}}{x+3} - \frac{\frac{4}{3}}{x} - \frac{2}{x^2}\right)y' - \left(\frac{\frac{2}{3}}{x+3} - \frac{\frac{2}{3}}{x} + \frac{2}{x^2}\right)y =$
$Lclm\left(D - \frac{C_1 + 2C_2x + 3x^2}{C_1(x+1) + C_2x^2 + x^3}\right)y = 0$. Type $\mathcal{L}_{12}^3$.

3.73: $y''' - \left(\frac{2}{x+1} + \frac{2}{x}\right)y'' - \left(\frac{6}{x+1} + \frac{6}{x} + \frac{4}{x^2}\right)y' - \left(\frac{8}{x+1} - \frac{8}{x} + \frac{8}{x^2} + \frac{4}{x^3}\right)y =$
$\left(D - \frac{2}{x+1} + \frac{1}{x}\right)\left(D - \frac{1}{x}\right)(D - \frac{2}{x})y = 0$. Type $\mathcal{L}_3^3$.

3.74: $y''' - \frac{1}{x}y'' + \frac{1}{x^2}y' = DLclm\left(D - \frac{2x}{C+x^2}\right)y = 0$. Type $\mathcal{L}_8^3$.

3.75: $y''' - \left(\frac{2}{x} + \frac{2x}{x^2+1}\right)y'' + \left(\frac{4}{x^2} + \frac{6}{x^2+1}\right)y' - \left(\frac{8}{x} + \frac{4}{x^3} - \frac{8x}{x^2+1}\right)y =$
$\left(D - \frac{2x}{x^2+1} + \frac{1}{x+1} + \frac{1}{x-1}\right)Lclm\left(D - \frac{3Cx^2 + C + 2x}{Cx^3 + Cx + x^2}\right)y = 0$. Type $\mathcal{L}_8^3$.

3.76: $y''' + \frac{1}{x^4}y'' - \frac{2}{x^6}y' = \left(D^2 + (\frac{2}{x} + \frac{1}{x^4})D + \frac{2}{x^5}\right)\left(D - \frac{2}{x}\right)y = 0$. Type $\mathcal{L}_2^3$.

3.78: $y''' - \left(2 - \frac{4}{x} - \frac{1}{x^2} + \frac{4x^3 + 4x + 2}{x^4 + 2x^2 + 2x + 1}\right)y''$
$+ \left(1 - \frac{8}{x} - \frac{2}{x^2} + \frac{8x^3 + 8x + 4}{x^4 + 2x^2 + 2x + 1}\right)y' + \left(\frac{4}{x} + \frac{1}{x^2} - \frac{4x^3 + 4x + 2}{x^4 + 2x^2 + 2x + 1}\right)y =$
$Lclm\left(D - 1, D - 1 - \frac{1}{x}, D + \frac{1}{x^2}\right)y = 0$. Type $\mathcal{L}_9^3$.

Appendix E

Symmetries of Kamke's Collection

In this appendix the symmetry classes of a large number of quasilinear second order, and linear and quasilinear equations of third order from Kamke's collection are listed. Equations that are not rational in all arguments, contain undetermined functions, more than two parameters or parameters in an exponent, are excluded. If there are parameters involved it is assumed that they are unconstrained, any relations between them may change the symmetry class.

Equations of Second Order. These are the second order nonlinear equations of Chapter 6. The following tabulation is organized by symmetry class; for each class the numbers of the equations which it contains are given.

$Trivial$: 6.3, 6.5, 6.6, 6.8, 6.9, 6.27, 6.108, 6.142, 6.144, 6.145, 6.147, 6.171, 6.211, 6.212.

$\mathcal{S}_1^2$: 6.4, 6.10, 6.11, 6.21, 6.26, 6.31, 6.40, 6.45, 6.47, 6.50, 6.56, 6.73, 6.74, 6.79, 6.80, 6.82, 6.86, 6.87, 6.89, 6.90, 6.92, 6.94, 6.96, 6.98, 6.105, 6.106, 6.118, 6.119, 6.120, 6.141, 6.143, 6.153, 6.156, 6.160, 6.172, 6.189, 6.190, 6.219, 6.226, 6.233, 6.240, 6.245.

$\mathcal{S}_{2,1}^2$: 6.227, 6.228, 6.229, 6.232, 6.239, 6.243.

$\mathcal{S}_{2,2}^2$: 6.1, 6.2, 6.7, 6.12, 6.30, 6.32, 6.42, 6.43, 6.57, 6.78, 6.97, 6.104, 6.109, 6.110, 6.111, 6.127, 6.130, 6.137, 6.140, 6.146, 6.154, 6.155, 6.159, 6.174, 6.188, 6.205, 6.237.

$\mathcal{S}_{3,1}^2$: none.

$\mathcal{S}_{3,2}^2$: 6.57, 6.81, 6.133, 6.138, 6.158, 6.162, 6.163, 6.183, 6.209, 6.244

$\mathcal{S}_{3,3}^2$: 6.71.

$\mathcal{S}_{3,4}^2$: none.

$\mathcal{S}_8^2$: 6.93, 6.99, 6.107, 6.117, 6.124, 6.125, 6.126, 6.128, 6.134, 6.135, 6.150, 6.151, 6.157, 6.164, 6.168, 6.169, 6.173, 6.175, 6.176, 6.178, 6.179, 6.180, 6.181, 6.182, 6.184, 6.185, 6.186, 6.191, 6.192, 6.193, 6.194, 6.195, 6.206, 6.208, 6.210, 6.214.

In the subsequent listing the number of the equation, its symmetry class and, if it is different from $\mathcal{S}_8^2$, its symmetry generators are given. Whenever possible, the solution is given explicitly or it is indicated what type of transcendentals for its representation are required.

403

6.1: $y'' - y^2 = 0$. Symmetry class $\mathcal{S}_{2,2}^2$. Generators $\{\partial_x, x\partial_x - 2y\partial_y\}$.

Solution: $x = \displaystyle\int \frac{dy}{\sqrt{\frac{2}{3}y^3 - C_1}} + C_2$.

6.2: $y'' - 6y^2 = 0$. Symmetry class $\mathcal{S}_{2,2}^2$. Generators $\{\partial_x, x\partial_x - 2y\partial_y\}$.
Solution: Weierstraß' $\mathcal{P}$-function.

6.3: $y'' - 6y^2 - x = 0$. Trivial symmetry. Solution: Painlevé's Transcendent.

6.4: $y'' - 6y^2 + 4y = 0$. Symmetry class $\mathcal{S}_1^2$. Generator ∂_x.
Solution: Elliptic integral.

6.5: $y'' + ay^2 + bx + c = 0$. Trivial symmetry. Solution: Painlevé's Transcendent.

6.6: $y'' - 2y^3 - xy + a = 0$. Trivial symmetry. Solution: Painlevé's Transcendent.

6.7: $y'' - ay^3 = 0$. Symmetry class $\mathcal{S}_{2,2}^2$. Generators $\{\partial_x, x\partial_x - y\partial_y\}$.
Solution: Elliptic integral.

6.8: $y'' - 2ay^3 + 2abxy - b = 0$. Trivial symmetry. Reduction to $y' + ay^2 - bx = 0$.

6.9: $y'' + ay^3 + bxy + cy + d = 0$. Trivial symmetry. Solution: Painlevé's Transcendent.

6.10: $y'' + ay^3 + by^2 + cy + d = 0$. Symmetry class $\mathcal{S}_1^2$. Generator ∂_x.

Solution: $x = -\displaystyle\int \frac{dy}{\sqrt{\frac{1}{2}ay^4 + \frac{2}{3}by^3 + cy^2 + 2dy + C_1}} + C_2$.

6.11: $y'' + ax^\nu y^n = 0$. Symmetry class $\mathcal{S}_1^2$. Generator $\partial_x - \dfrac{\nu+1}{n-1}\partial_y$.

6.12: $y'' + (n+1)a^{2n}y^{2n+1} = 0$. Symmetry class $\mathcal{S}_{2,2}^2$.

Generators $\{\partial_x, x\partial_x - \frac{1}{n}y\partial_y\}$. Solution: $x = \displaystyle\int \frac{dy}{y\sqrt{(ay)^{2n} - 1 + C_1}} + C_2$.

6.21: $y'' - 3y' - y^2 - 2y = 0$. Symmetry class $\mathcal{S}_1^2$. Generator ∂_x.

6.23: $y'' + 5ay' + 6ya^2 - 6y^2 = 0$. Symmetry class $\mathcal{S}_{2,2}^2$.
Generators $\{\partial_x, e^{ax}(\partial_x - 2ay\partial_y\}$.
Solution: $y = a^2 C_1^2 e^{-2ax}\mathcal{P}(C_1 e^{-ax} + C_2, 0, -1)$.

6.24: $y'' + 3ay' + 2a^2 y - 2y^3 = 0$. Symmetry class $\mathcal{S}_{2,2}^2$.
Generators $\{\partial_x, e^{ax}(\partial_x - ay\partial_y)\}$.
Solution: $y = -iaC_1 e^{-ax} sn_{k^2=-1}(C_1 e^{-ax} + C_2)$.

6.26: $y'' + ay' + by^n + \dfrac{a^2 - 1}{4}y = 0$. Symmetry class $\mathcal{S}_1^2$. Generator ∂_x.

6.27: $y'' + ay' + bx^\nu y^n = 0$. Trivial symmetry.

6.30: $y'' + yy' - y^3 = 0$. Symmetry class $\mathcal{S}_{2,2}^2$. Generators $\{\partial_x, x\partial_x - y\partial_y)\}$.

Solution: $y = C_1 \dfrac{\mathcal{P}'(u,0,1)}{\mathcal{P}(u,0,1)}$ where $u = C_1 x + C_2$.

6.31: $y'' + yy' - y^3 + ay = 0$. Symmetry class $\mathcal{S}_1^2$. Generator ∂_x.

Symmetries of Kamke's Collection 405

Solution: $\frac{1}{2}\sqrt{\frac{a}{3}}\dfrac{\mathcal{P}'(u,12,C1)}{\mathcal{P}(u,12,C_1)-1}$ with $u=\frac{x}{2}\sqrt{\frac{a}{3}}+C_2$.

6.32: $y''+yy'+2ay'-y^3+ay^2+2a^2y=0$. Symmetry class $\mathcal{S}^2_{2,2}$.
Generators $\{\partial_x,e^{ax}(\partial_x-ay\partial_y)\}$.

Solution: $y=C_1e^{-ax}\dfrac{\mathcal{P}'(u,0,1)}{\mathcal{P}(u,0,1)}$ where $u=\dfrac{C_1}{a}e^{-ax}+C_2$, $a\neq 0$.

6.40: $y''-3yy'-(3ay^2+4a^2y+b)=0$. Symmetry class $\mathcal{S}^2_1$. Generator ∂_x.

6.42: $y''-2ayy'=0$. Symmetry class $\mathcal{S}^2_{2,2}$. Generators $\{\partial_x,x\partial_x-y\partial_y\}$.

6.43: $y''+ayy'+by^3=0$. Symmetry class $\mathcal{S}^2_{2,2}$. Generators $\{\partial_x,x\partial_x-y\partial_y\}$.

Solution: $x=\displaystyle\int\dfrac{dy}{uy^2\log y}+C_2$, u is defined by $\displaystyle\int\dfrac{udu}{2u^2+au+b}+\log y=C_1$.

6.45: $y''+ay'^2+by=0$. Symmetry class $\mathcal{S}^2_1$. Generator ∂_x.

Solution: $x=2a^2\displaystyle\int\dfrac{dy}{2a^2C_1e^{-2ay}+b(1-2ay)}+C_2$.

6.47: $y''+ay'^2+by'+cy=0$. Symmetry class $\mathcal{S}^2_1$. Generator ∂_x.
Reduction to Abel's equation.

6.50: $y''+ayy'^2+by=0$. Symmetry class $\mathcal{S}^2_1$. Generator ∂_x.

Solution: $x=\displaystyle\int\dfrac{dy}{\sqrt{C_1\exp(-ay^2)-b/a}}+C_2$.

6.56: $y''+ay(y'+1)^2=0$. Symmetry class $\mathcal{S}^2_1$. Generator ∂_x.

Solution: $x=\displaystyle\int\sqrt{\dfrac{ay^2+C_1}{1-ay^2-C_2}}dy+C_2$.

6.57 $y''-a(xy'-y)^\nu=0$. $\nu=1$: Symmetry class $\mathcal{S}^2_8$.

Solution: $y=C_1x+C_2x\displaystyle\int\exp\left(\tfrac{1}{2}x^2\right)\dfrac{dx}{x}$.

$\nu=3$: Symmetry class $\mathcal{S}^2_{3,2}$. Generators $\{y\partial_x,x\partial_y,x\partial_x-y\partial_y\}$.
Solution: $(C_1^2C_2^2+a)y^2-2C_1^2C_2xy+C_2^2x^2-C_1=0$.

$\nu=2$ or $\nu\geq 4$: Symmetry class $\mathcal{S}^2_{2,2}$. Generators $\{x\partial_y,x\partial_x-\dfrac{2}{\nu-1}y\partial_y\}$.
Solution: Parameter representation.

6.58: $y''-kx^ay^by'^c=0$. Symmetry class $\mathcal{S}^2_1$. Generator ∂_x.

6.71: $8y''+9y'^4=0$. Symmetry class $\mathcal{S}^2_{3,2}$. Generators $\{\partial_x,\partial_y,x\partial_x+\tfrac{2}{3}y\partial_y\}$.
Solution: $(C_1+y)^3=(C_2+x)^2$.

6.73: $xy''+2y'-xy^n=0$. Symmetry class $\mathcal{S}^2_1$. Generator $x\partial_x-\dfrac{2}{n-1}y\partial_y$.

6.74: $xy''+2y'^2+ax^\nu y^n=0$. Symmetry class $\mathcal{S}^2_8$ for $n=1$; symmetry
class $\mathcal{S}^2_1$ for $n\geq 2$, $\nu\neq n$, generator $x\partial_x-\dfrac{\nu+1}{n-1}y\partial_y$; symmetry class $\mathcal{S}^2_2$
for $\nu=n\geq 2$, additional generator $\partial_x-\dfrac{y}{x}\partial_y$.

6.78: $xy''+yy'-y'=0$. Symmetry class $\mathcal{S}^2_{2,2}$.
Generators $\{x\partial_x,x\log x\partial_x+(y-2)\partial_y\}$. Solution: $y=2-2\tan\left(C_1(\log x-C_2)\right)$.

6.79: $xy''-x^2y'^2+2y'+y^2=0$. Symmetry class $\mathcal{S}^2_1$. Generator $x\partial_x-y\partial_y$.

Solution: $x = C_1 \exp \int \dfrac{d\eta}{C_2 e^{\eta} + 2\eta + 1}$ with $\eta = xy$.

6.80: $xy'' + a(xy' - y)^2 = b$. Symmetry class $\mathcal{S}_1^2$. Generator $x\partial_y$.

6.81: $2xy'' + y'^3 + y' = 0$. Symmetry class $\mathcal{S}_{3,2}^2$.
  Generators $\{\partial_y, x\partial_x + y\partial_y, xy\partial_x + \frac{1}{2}y^2\partial_y\}$.
  Solution: $(y + C_1)^2 = 2C_2 x - C_2^2$.

6.82: $xy'' = a(y^n - y)$. Symmetry class $\mathcal{S}_1^2$. Generator $x\partial_x$.

6.86: $x^2 y'' + a(xy' - y)^2 = bx^2$. Symmetry class $\mathcal{S}_1^2$. Generator $x\partial_y$.

6.87: $x^2 y'' + ayy'^2 + bx = 0$. Symmetry class $\mathcal{S}_1^2$. Generator $x\partial_x + y\partial_y$.

6.89: $y'' + xy'' + y'^2 = 0$. Symmetry class $\mathcal{S}_1^2$. Generator ∂_y.
  Solution: $y = C_1 + C_2 x + (C_2^2) \log x - C_2$.

6.90: $4x^2 y'' - x^4 y'^2 + 4y = 0$. Symmetry class $\mathcal{S}_1^2$. Generator $x\partial_x - 2y\partial_y$.

6.91: $9x^2 y'' + ay^3 + 2y = 0$. Symmetry class $\mathcal{S}_{2,2}^2$.
  Generators $\left\{ x\partial_x, 3x^{2/3}\partial_x - \dfrac{y}{x^{1/3}}\partial_y \right\}$. Solution in terms of elliptic functions.

6.92: $9x^3(y'' + yy' - y^3) + 12xy + 24 = 0$. Symmetry class $\mathcal{S}_1^2$.
  Generator $x\partial_x - y\partial_y$. Solution: Parameter representation.

6.93: $9x^2 y'' - a(xy' - y)^2 = 0$. Symmetry class $\mathcal{S}_8^2$. Solution: $y = \dfrac{x}{a} \log \dfrac{x}{C_1 x + C_2}$.

6.94: $2x^2 y'' + (2x^3 y + 9x^2)y' - 2x^3 y^3 + 3x^2 y^2 + axy + b = 0$.
  Symmetry class $\mathcal{S}_1^2$. Generator $x\partial_x - y\partial_y$.

6.96: $x^4 y'' + a^2 y^\nu = 0$. Symmetry class $\mathcal{S}_1^2$. Generator $x\partial_x + \dfrac{2}{\nu - 1}y\partial_y$.

6.97: $x^4 y'' - 2xyy' - x^3 y' + 4y^2 = 0$. Symmetry class $\mathcal{S}_{2,2}^2$.
  Generators $\{ x\partial_x + 2y\partial_y, x\log x\partial_x + (2y\log x + x^2 - y)\partial_y \}$.
  Solution: $y = [C_2 \tan(C_1 \log x + C_2)]x^2$.

6.98: $x^4 y'' - x^2 y'^2 - x^3 y' + 4y^2 = 0$. Symmetry class $\mathcal{S}_1^2$.
  Generator $x\partial_x + 2y\partial_y$. Solution: $x = \exp \int \dfrac{dy}{x^2(C_1 \exp\left(\frac{y}{x^2}\right) - 4\frac{y}{x^2} - 2} + C_2$.

6.99: $x^4 y'' + (xy' - y)^3 = 0$. Symmetry class $\mathcal{S}_8^2$.
  Solution: $y = C_1 x + x \arcsin \dfrac{C_2}{x}$.

6.104: $yy'' - a = 0$. Symmetry class $\mathcal{S}_{2,2}^2$. Generators $\{\partial_x, x\partial_x + y\partial_y\}$.
  Solution: $x = \int \dfrac{dy}{\sqrt{2a \log y + C_1}}$.

6.105: $yy'' - ax = 0$. Symmetry class $\mathcal{S}_1^2$. Generator $x\partial_x + \frac{3}{2}y\partial_y$.

6.106: $yy'' - ax^2 = 0$. Symmetry class $\mathcal{S}_1^2$. Generator $x\partial_x + 2y\partial_y$.

6.107: $yy'' + y'^2 - a = 0$. Symmetry class $\mathcal{S}_8^2$. Solution $y = \sqrt{ax^2 + C_1 x + C_2}$.

6.108: $yy'' + y'^2 = ax + b$. Trivial symmetry.

6.109: $yy'' + y'^2 - y' = 0$. Symmetry class $\mathcal{S}_{2,2}^2$.
  Generators $\{\partial_x, x\partial_x + y\partial_y\}$. Solution: $x = y + C_1 \log y - C_1 + C_2$.

Symmetries of Kamke's Collection

6.110: $yy'' - y'^2 + 1 = 0$. Symmetry class $\mathcal{S}_{2,2}^2$.
Generators $\{\partial_x, x\partial_x + y\partial_y\}$. Solution: $C_1y = \sin(C_1x) + C_2$.

6.111: $yy'' - y'^2 - 1 = 0$. Symmetry class $\mathcal{S}_{2,2}^2$.
Generators $\{\partial_x, x\partial_x + y\partial_y\}$. Solution: $C_1y = \cosh(C_1x) + C_2$.

6.117: $y''y - y'^2 + ay'y + by^2 = 0$. Symmetry class $\mathcal{S}_8^2$.
Solution $y = \exp(C_1e^{-ax} - \frac{b}{a}x + C_2)$.

6.118: $y''y - y'^2 + ay'y + by^3 - 2ay^2 = 0$. Symmetry class $\mathcal{S}_1^2$. Generator ∂_x.

6.119: $y''y - y'^2 + (ay-1)y' - 2b^2y^3 + 2a^2y^2 + ay = 0$. Symmetry class $\mathcal{S}_1^2$.
Generator ∂_x.

6.120: $yy'' - y'^2 + a(y-1)y' - y(y+1)(b^2y^2 - a^2) = 0$. Symmetry class $\mathcal{S}_1^2$.
Generator ∂_x.

6.124: $y''y - 3y'^2 + 3y'y - y^2 = 0$. Symmetry class $\mathcal{S}_8^2$.
Solution: $(C_1e^x + 1)y^2 + C_2e^{2x} = 0$.

6.125: $y''y - ay'^2 = 0$. Symmetry class $\mathcal{S}_8^2$.
Solution: $y = (C_1x + C_2)^{1/(1-a)}$ for $a \neq 1$, $y = C_1e^{C_2x}$ for $a = 1$.

6.126: $yy'' + ay'^2 + a = 0$. Symmetry class $\mathcal{S}_{2,2}^2$.
Generators $\{\partial_x, x\partial_x + y\partial_y\}$. Solution: $x = \displaystyle\int \frac{dy}{\sqrt{C_1y^{-2a} - 1}} + C_2$.

6.127: $yy'' + ay'^2 + by^3 = 0$. Symmetry class $\mathcal{S}_{2,2}^2$
Generators $\{\partial_x, x\partial_x - 2y\partial_y\}$. Solution: Parameter representation.

6.128: $yy'' + ay'^2 + byy' + cy^2 + dy^{1-a} = 0$. Symmetry class $\mathcal{S}_8^2$.

6.130: $yy'' + ay'^2 + by^2y' + cy^4 = 0$. Symmetry class $\mathcal{S}_{2,2}^2$.
Generators $\{\partial_x, x\partial_x - y\partial_y\}$.

6.133: $y''(y+x) + y'^2 - y' = 0$. Symmetry class $\mathcal{S}_{3,2}^2$.
Generators $\{\partial_x - \partial_y, x\partial_x + y\partial_y, (x^2 - \frac{2}{3}xy - \frac{1}{3}y^2)\partial_x + (\frac{1}{3}x^2 + \frac{2}{3}xy - y^2)\partial_y\}$.
Solution: $y^2 - (3C_2 + \frac{9}{4}C_1 + 2x)y + \frac{9}{4}C_2^2 + 3C_2x - \frac{9}{4}C_1x + \frac{81}{64}C_1 + x^2 = 0$.

6.134: $yy'' - xy'' - 2y'^2 - 2y' = 0$. Symmetry class $\mathcal{S}_8^2$.
Solution $y = C_1 + \dfrac{C_2}{x - C_1}$.

6.135: $yy'' - xy'' + y'^3 + y'^2 + y' + 1 = 0$. Symmetry class $\mathcal{S}_8^2$.
Solution: $(y + C_1)^2 = C_2^2 - (x + C_1)^2$.

6.137: $2yy'' + y'^2 + 1 = 0$. Symmetry class $\mathcal{S}_{2,2}^2$. Generators $\{\partial_x, x\partial_x + y\partial_y\}$.
Solution: $x = C_1(t - \sin t) + C_2, y = C_1(1 - \cos t)$.

6.138: $2yy'' - y'^2 + a = 0$. Symmetry class $\mathcal{S}_{3,2}^2$.
Generators $\{\partial_x, x\partial_x + y\partial_y, x^2\partial_x + 2xy\partial_y\}$.
Solution: $C_1y = (\frac{1}{2}C_1x + C_2)^2 - a$.

6.140: $2y''y - y'^2 - 8y^3 = 0$. Symmetry class $\mathcal{S}_{2,2}^2$.
Generators $\{\partial_x, x\partial_x - 2y\partial_y\}$. Solution: $x + \frac{1}{2}\displaystyle\int \frac{dy}{\sqrt{y(y^2 - C_1)}} = C_2$.

6.141: $2y''y - y'^2 - 8y^3 - 4y^2 = 0$. Symmetry class $\mathcal{S}_1^2$. Generator ∂_x.

408

Solution: Elliptic functions.

6.142: $2y''y - y'^2 - 8y^3 - 4xy^2 = 0$. Trivial symmetry.

6.143: $2y''y - y'^2 + ay^3 + by^2 = 0$. Symmetry class $\mathcal{S}_1^2$. Generator ∂_x.

6.144: $2y''y - y'^2 + ay^3 + 2xy^2 + 1 = 0$. Trivial symmetry.

6.145: $2y''y - y'^2 + ay^3 + bxy^2 = 0$. Trivial symmetry.

6.146: $2yy'' - y'^2 + 3y^4 = 0$. Symmetry class $\mathcal{S}_{2,2}^2$.
Generators $\{\partial_x, x\partial_x - y\partial_y\}$ Solution: Elliptic function.

6.147: $2yy'' - y'^2 - 3y^4 - 8xy^3 - 4(x^2 + a)y^2 + b = 0$. Trivial symmetry.
Solution: Painlevé's Transcendent.

6.150: $2yy'' - 3y'^2 = 0$. Symmetry class $\mathcal{S}_8^2$. Solution: $y = \dfrac{C_1}{x + C_2}$.

6.151: $2yy'' - 3y'^2 - 4y^2 = 0$. Symmetry class $\mathcal{S}_8^2$. Solution: $y = \dfrac{C_1}{\cos{(x + C_2)}^2}$.

6.153: $2yy'' - 6y'^2 + ay^5 + y^2 = 0$. Symmetry class $\mathcal{S}_1^2$. Generator ∂_x
Solution: $x = \displaystyle\int \dfrac{dy}{y\sqrt{C_1 y^4 + ay^3 + \frac{1}{4}}} + C_2$.

6.154: $2yy'' - y'^4 - y'^2 = 0$. Symmetry class $\mathcal{S}_{2,2}^2$. Generators $\{\partial_x, x\partial_x + y\partial_y\}$
Solution: $x = C_1 t + C_2 \sin t + C_2, y = C_1(1 - \cos t)$.

6.155: $4yy'' - 4ay'' + 2y'^2 + 2 = 0$. Symmetry class $\mathcal{S}_{2,2}^2$.
Generators $\{\partial_x, x\partial_x + (y - a)\partial_y\}$.
Solution: $x = \frac{1}{2}\sqrt{(y - a + C_2)(a - y)} - C_2 \arctan\sqrt{\dfrac{a - y}{y - a + C_2}} + C_1$.

6.156: $3yy'' - 2y'^2 = ax^2 + bx + c$. Symmetry class $\mathcal{S}_1^2$.
Generator $(ax^2 + bx + c)\partial_x + \left(3a + \frac{3b}{2a}\right)\partial_y$.

6.157: $3yy'' - 5y'^2 = 0$. Symmetry class $\mathcal{S}_8^2$. Solution: $y = \dfrac{1}{(C_1 x + C_2)^{3/2}}$.

6.158: $4yy'' - 3y'^2 + 4y = 0$. Symmetry class $\mathcal{S}_{3,2}^2$.
Generators $\{\partial_x, x\partial_x + 2y\partial_y, x^2\partial_x + 4xy\partial_y\}$. Solution $x = \frac{1}{2}\displaystyle\int \dfrac{dy}{\sqrt{C_1 y + \sqrt{y}}} + C_2$.

6.159: $4yy'' - 3y'^2 - 12y^3 = 0$. Symmetry class $\mathcal{S}_{2,2}^2$.
Generators $\{\partial_x, x\partial_x - 2y\partial_y\}$. Solution $x = \displaystyle\int \dfrac{\sqrt{C_1 - 4y^3}}{y\sqrt{y}} dy + C_2$.

6.160: $4yy'' - 3y'^2 + ay^3 + by^2 + cy = 0$. Symmetry class $\mathcal{S}_1^2$. Generator ∂_x

6.162: $4yy'' - 5y'^2 + ay^3 = 0$.
Symmetry class $\mathcal{S}_{3,2}^2$. Generators $\{\partial_x, x\partial_x - 2y\partial_y, x^2\partial_x - 4xy\partial_y\}$.
Solution: $2C_1^2 C_2 y^{\frac{1}{4}} + (C_1^2 C_2 - C_1 x^2 - \frac{1}{16}a)\sqrt{y} + C_1^2 = 0$.

6.163: $12yy'' - 15y'^2 + 8y^3 = 0$. Symmetry class $\mathcal{S}_{3,2}^2$.
Generators $\{\partial_x, x\partial_x - 2y\partial_y, x^2\partial_x - 4xy\partial_y\}$.

$$\text{Solution: } y = \frac{6C_2}{[(C_1 + x)^2 + C_2]^2}.$$

6.164: $nyy'' - (n-1)y'^2 = 0$. Symmetry class $\mathcal{S}_8^2$. Solution: $y = (C_1x + C_2)^n$.

6.168: $(ay + b)y'' + cy'^2 = 0$. Symmetry class $\mathcal{S}_8^2$.
Solution: $y = (C_1x + C_2)^{a/(a+c)} - \frac{b}{a}$ if $a+c \neq 0$, $y = C_1 e^{C_2 x} - \frac{b}{a}$ if $a+c = 0$.

6.169: $xyy'' + xy'^2 - yy' = 0$. Symmetry class $\mathcal{S}_8^2$. Solution: $y = \sqrt{\dfrac{1}{C_1 x^2 + C_2}}.$

6.171: $xyy'' - xy'^2 + yy' + axy^4 + by^3 + cy + d = 0$. Trivial symmetry.

6.172: $xyy'' - xy'^2 + ayy' + bxy^3 = 0$. Symmetry class $\mathcal{S}_1^2$.
Generator $x\partial_x - 2y\partial_y$.

6.173: $xyy'' + 2xy'^2 + ayy' = 0$. Symmetry class $\mathcal{S}_8^2$.
$$\text{Solution: } y = \sqrt{\frac{1}{C_1 x^{1-a} + C_2}}.$$

6.174: $xyy'' - 2xy'^2 + yy' + y' = 0$. Symmetry class $\mathcal{S}_{2,2}^2$.
Generators $\{x\partial_x, x\log x\partial_x + y\partial_y\}$.

6.175: $xyy'' - 2xy'^2 + ayy' = 0$. Symmetry class $\mathcal{S}_8^2$.
Solution: $y = \dfrac{1}{C_1 x^{1-a} + C_2}$ for $a \neq 1$, $y = \dfrac{1}{C_1 \log x + C_2}$ for $a = 1$.

6.176: $xyy'' - 4xy'^2 + 5yy' + y' = 0$. Symmetry class $\mathcal{S}_8^2$.
$$\text{Solution: } y = \frac{x}{(C_1 x^3 + C_2)^{1/3}}.$$

6.178: $x(y + x)y'' + xy'^2 - (y - x)y' - y = 0$. Symmetry class $\mathcal{S}_8^2$.
Solution: $y = \sqrt{C_1 x^2 + C_2} - x$.

6.179: $2y''yx - y'^2x + y'y = 0$. Symmetry class $\mathcal{S}_8^2$.
Solution: $y = C_1(\sqrt{x} + C_2)^2$.

6.180: $x^2(y - 1)y'' - 2x^2y'^2 - 2x(y - 1)y' - 2y(y - 1)^2 = 0$.
Symmetry class $\mathcal{S}_8^2$. Solution: $y = \dfrac{C_1 x^2 + C_2 x}{C_1 x^2 + C_2 x - 1}.$

6.181: $x^2(y + x)y'' - (xy' - y)^2 = 0$. Symmetry class $\mathcal{S}_8^2$.
Solution: $y = C_1 x \exp\left(\frac{C_2}{x}\right) - x$.

6.182: $x^2(y - x)y'' - a(xy' - y)^2 = 0$. Symmetry class $\mathcal{S}_8^2$.
Solution: $y = C_1 x\left(\dfrac{x}{C_2 x + a - 1}\right)^{1/(a-1)} + x$.

6.183: $2x^2yy'' - x^2y'^2 + y^2 - x^2 = 0$. Symmetry class $\mathcal{S}_{3,2}^2$. Generators
$\{x\partial_x + y\partial_y, x\log(x)\partial_x + (\log(x) + 1)y\partial_y, x\log(x)^2\partial_x + (\log(x) + 2)\log(x)y\partial_y\}$.
Solution: $y = x(C_1 + \sqrt{4C_1 C_2 - 1}\log x + C_2 \log x^2)$.

6.184: $ax^2yy'' + bx^2y'^2 + cxyy' + dy^2 = 0$ Symmetry class $\mathcal{S}_8^2$.

6.185: $x(x + 1)^2yy'' - x(x + 1)^2y' + 2(x + 1)^2yy' - ax(x + 2)y^2 = 0$.
Symmetry class $\mathcal{S}_8^2$. Solution: $y = C_1(x + 1)^a \exp\left(\frac{C_2}{x}\right)$.

6.186: $8(x^3 - 1)yy'' - 4(x^3 - 1)y'^2 + 12x^2yy' - 3x(x + 2)y^2 = 0$
Symmetry class $\mathcal{S}_8^2$. Solution in terms of hypergeometric functions.

6.188: $y^2y'' - a = 0$. Symmetry class $\mathcal{S}_{2,2}^2$. Generators $\{\partial_x, x\partial_x + \frac{2}{3}y\partial_y\}$.

Solution $x = \displaystyle\int \frac{\sqrt{y}\,dy}{\sqrt{C_1y - 2a}} + C_2$.

6.189: $y^2y''yy'^2 + ax = 0$. Symmetry class $\mathcal{S}_1^2$. Generator $x\partial_x + y\partial_y$.

6.190: $y^2y'' + yyy'^2 = ax + b$. Symmetry class $\mathcal{S}_1^2$. Generator $\left(x + \frac{b}{a}\right)\partial_x + y\partial_y$.

First integral: $y^3y'^3 - 3y^2y'(ax + b) + ay^3 + \frac{1}{a}(ax + b)^3 = C$.

6.191: $y''(y^2 + 1) - y'^2(2y - 1) = 0$. Symmetry class $\mathcal{S}_8^2$.

Solution: $y = \tan\log(C_1x + C_2)$

6.192: $(y^2 + 1)y'' - 3yy'^2 = 0$. Symmetry class $\mathcal{S}_8^2$.

Solution: $y = \dfrac{C_1x + C_2}{\sqrt{C_1x + C_2)^2 - 1}}$

6.193: $(y^2 + x)y'' + 2(y^2 - x)y'^3 + 4yy'^2 + y' = 0$. Symmetry class $\mathcal{S}_8^2$.

Solution: $y = C_1e^{C_2y} - x$.

6.194: $(y^2 + x)y'' - (y'^2 + 1)(xy' - y) = 0$. Symmetry class $\mathcal{S}_8^2$.

Solution: $y = C_1 \exp\arctan\left(C_2\frac{y}{x}\right)$.

6.195: $(y^2 + x^2)y'' - 2(y'^2 + 1)(xy' - y) = 0$. Symmetry class $\mathcal{S}_8^2$.

Solution: $y^2 + C_1y + C_2x + x^2 = 0$.

6.205: $xy^2y'' - a = 0$.

Symmetry class $\mathcal{S}_{2,2}^2$. Generators $\{x\partial_x + \frac{1}{3}y\partial_y, x^2\partial_x + xy\partial_y\}$.

Solution: Parameter representation.

6.206: $(x^2 - a^2)(y^2 - a^2)y'' - (x^2 - a^2)yy'^2 + x(y^2 - a^2)y' = 0$.

Symmetry class $\mathcal{S}_8^2$. Solution $\sqrt{y^2 - a^2} + y = C_1(\sqrt{x^2 - a^2} + x) + C_2$.

6.208: $x^3y^2y'' + (xy' - y)^3(y + x) = 0$. Symmetry class $\mathcal{S}_8^2$.

6.209: $y^3y'' - a = 0$.

Symmetry class $\mathcal{S}_{3,2}^2$. Generators $\{\partial_x, x\partial_x + \frac{1}{2}y\partial_y, x^2\partial_x + xy\partial_y\}$.

Solution: $(C_2x - C_2)^2 + C_1y^2 + a = 0$.

6.210: $(y^3 + y)y'' - (3y^2 - 1)y'^2 = 0$. Symmetry class $\mathcal{S}_8^2$.

Solution: $y = \sqrt{\dfrac{1 - C_1x - C_2}{C_1x + C_2}}$.

6.211: $2y^3y'' + y^4 - a^2xy^2 = 1$. Trivial symmetry.

6.212: $2y^3y'' + y^2y'^2 = ax^2 + bx + c$. Trivial symmetry.

6.214: $(4y^3 - g_2y - g_3)y'' - (6y^2 - \frac{1}{2}g_2)y'^2 = 0$; g_2 and g_3 constant. Symmetry class $\mathcal{S}_8^2$. Solution: $y = \mathcal{P}(C_1x + C_2, g_2, g_3)$; $\mathcal{P}$ Weierstraß' $\mathcal{P}$-function.

6.219: $(y^2 + ax^2 + 2bx + c^2)y'' + dy = 0$. Symmetry class $\mathcal{S}_1^2$.

Generator $(ax^2 + 2bx + c)\partial_x + (xy + \frac{b}{a}y)\partial_y$.

6.226: $y'y'' - x^2yy' - xy^2 = 0$. Symmetry class $\mathcal{S}_1^2$. Generator $y\partial_y$.

6.227: $xy'y'' - yy'' + 4y'^2 = 0$. Symmetry class $\mathcal{S}_{2,1}^2$. Generators $\{x\partial_x, y\partial_y\}$.

Solution: $x = C_1(t - 1)e^{2t}, y = C_2te^{-2t}$.

6.228: $xy'y'' - yy'' - y'^4 - 2y'^2 - 1 = 0$.

Symmetries of Kamke's Collection   411

Symmetry class $\mathcal{S}_{2,1}^2$. Generators $\{x\partial_x + y\partial_y, (x+y)\partial_y\}$.

6.229: $ay''y'x^3 + by^2 = 0$. Symmetry class $\mathcal{S}_{2,1}^2$. Generators $\{x\partial_x, y\partial_y\}$.

6.231: $(2y^2y' + x^2)y'' + 2yy'^3 + 3xy' + y = 0$. Symmetry class $\mathcal{S}_1^2$.
Generator $x\partial_x + y\partial_y$. First integral: $y^2y'^2 + x^2y' + xy = C$.

6.232: $y''y'^2 + y''y^2 + y^3 = 0$. Symmetry class $\mathcal{S}_1^2$. Generators $\{\partial_x, y\partial_y\}$.
Solution: $\log y = \frac{1}{2}\log|\sin(x\sqrt{3} + C_1)| \pm \int \left[(1 + \frac{3}{4}\cot^2(x\sqrt{3} + C_1)\right]^{1/2}dx + C_2$.

6.233: $(y'^2 + a(xy' - y))y'' = b$. Symmetry class $\mathcal{S}_1^2$. Generator $\partial_x - \frac{1}{2}ax\partial_y$.

6.236: $y''^2 + ay + b = 0$.
Symmetry class $\mathcal{S}_{2,2}^2$. Generators $\{\partial_x, x\partial_x + 4(y + \frac{b}{a})\partial_y\}$.

6.237: $a^2y''^2 - 2axy'' + y' = 0$.
Symmetry class $\mathcal{S}_{2,2}^2$. Generators $\{\partial_y, x\partial_x + 3y\partial_y\}$.

6.238: $2(x^2 + 1)y''^2 - x(4y' + y)y'' + 2(y' + x)y' - 2y = 0$. Trivial symmetry.

6.239: $3x^2y''^2 - 2(3xy' + y)y'' + 4y'^2 = 0$. Symmetry class $\mathcal{S}_{2,1}^2$.
Generators $\{x\partial_x, y\partial_y\}$. Solution $y = C_1^2x^2 + C_1C_2x + C_2^2$.

6.240: $(9x^3 - 2x^2)y''^2 - 6x(6x - 1)y'y'' - 6y'y'' + 36xy'^2 = 0$.
Symmetry class $\mathcal{S}_1^2$. Generator $y\partial_y$. Solution $y = C_1^2x^3 + C_1C_2x + C_2^2$.

6.243: $(a^2y^2 - b^2)y''^2 - 2a^2yy'^2y'' + (a^2y'^2 - 1)y'^2 = 0$. Symmetry class $\mathcal{S}_{2,1}^2$.
Generator ∂_x. Solution $y = C_1\exp(C_2x) \pm \frac{1}{a}\sqrt{(b^2 + \frac{1}{C_2^2})}$.

6.244: $(x^2yy'' - x^2y'^2 + y^2)^2 = 4xy(xy' - y)^3$. Symmetry class $\mathcal{S}_{3,2}^2$.
Generators $\{\partial_x + \frac{y}{x}\partial_y, x\partial_x + y\log\frac{x}{y}\partial_y, \partial_y\}$. Solution $y = C_1x\exp\frac{1}{C_2 - x}$.

6.245: $(2yy'' - y'^2)^3 + 32y''(xy'' - y'^3)^3 = 0$. Symmetry class $\mathcal{S}_1^2$.
Generator $x\partial_x + y\partial_y$. Solution $y = \frac{1}{C_1C_2^3}((C_1^2x + 1)^2 + 2C_2^2)$.

Nonlinear Equations of Third Order. These equations are from Chapter 7 of Kamke's collection.

$$\mathcal{S}_1^3 : 7.2, 7.5, 7.6. \quad \mathcal{S}_{2,1}^3 : 7.1. \quad \mathcal{S}_{2,2}^3 : 7.3, 7.4, 7.7.$$

$$\mathcal{S}_{4,4}^3 : 7.12 \quad \mathcal{S}_6^3 : 7.10, 7.11. \quad \mathcal{S}_7^3 : 7.8, 7.9.$$

7.1: $y''' - a^2(y'^5 - 2y'^3 - y') = 0$. Symmetry class $\mathcal{S}_{2,1}^3$. Generators $\{\partial_x, \partial_y\}$.
Solution in parameter representation:

$$x = \frac{1}{a}\int \frac{dp}{\sqrt{C_1 + \frac{1}{3}(p^2 + 1)^3}} + C_2, \quad y = \frac{1}{a}\int \frac{pdp}{\sqrt{C_1 + \frac{1}{3}(p^2 + 1)^3}} + C_3.$$

7.2: $y''' + yy'' - y'^2 = 0$. Symmetry class $\mathcal{S}_1^3$. Generator ∂_x.

7.3: $y''' - yy'' + y'^2 = 0$. Symmetry class $\mathcal{S}_{2,2}^3$. Generators $\{\partial_x, x\partial_x - y\partial_y\}$.
Reduction to Abel equation.

7.4: $y''' + ayy'' = 0$. Symmetry class $\mathcal{S}_{2,2}^3$. Generators $\{\partial_x, x\partial_x - y\partial_y\}$.
Reduction to Abel equation.

7.5: $x^2 y''' + xy'' + (2xy - 1)y' + y^2 = 0$. Symmetry class $\mathcal{S}_1^3$. Generator
$x\partial_x - y\partial_y$.

7.6: $x^2 y''' + xyy'' - xy'' + xy'^2 - yy' + y' = 0$. Symmetry class $\mathcal{S}_1^3 = \{x\partial_x\}$.
First integral: $xy'' + (y - 1)y' = Cx$.

7.7: $yy''' - y'y'' + y^3 y' = 0$. Symmetry class $\mathcal{S}_{2,2}^3$. Generator $\{\partial_x, x\partial_x - y\partial_y\}$.
Solution: $x = \displaystyle\int \frac{dy}{\sqrt{C_1 y^2 + C_2 - \frac{1}{4}y^4}} + C_3$.

7.8: $4y^2 y''' - 18yy'y'' + 15y'^3 = 0$. Symmetry class $\mathcal{S}_7^3$.
Solution: $y = \dfrac{1}{(C_1 x^2 + C_2 x + C_3)^2}$.

7.9: $9y^2 y''' - 45yy'y'' + 40y'^3 = 0$.
Symmetry class $\mathcal{S}_7^3$. Generators $\{\partial_x, x\partial_x, x^2\partial_x - 3xy\partial_y\}$.
Solution: $y^2 = \dfrac{1}{(C_1 x^2 + C_2 x + C_3)^3}$.

7.10: $2y'y''' - 3y''^2 = 0$. Symmetry class $\mathcal{S}_6^3$.
Generators $\{\partial_x, \partial_y, x\partial_x, y\partial_y, x^2\partial_x, y^2\partial_y\}$. Solution: $y = \dfrac{C_1 x + C_2}{C_3 x + C_4}$.

7.11: $y'^2 y''' + y''' - 3y'y''^2 = 0$. Symmetry class $\mathcal{S}_6^3$. Generators
$\{\partial_x, \partial_y, x\partial_x + y\partial_y, y\partial_x - x\partial_y, xy\partial_x - \frac{1}{2}(x^2 - y^2)\partial_y, (x^2 - y^2)\partial_x + 2xy\partial_y\}$.
Solution: $(y - C_1)^2 + (x - C_2)^2 = C_3^2$.

7.12: $y'^2 y''' + y''' - 3y'y''^2 - ay''^2 = 0$. Symmetry class $\mathcal{S}_{4,4}^3$.
Generators $\{\partial_x, \partial_y, x\partial_x + y\partial_y, y\partial_x - x\partial_y\}$.
Solution: $x = C_2 + C_1 e^{-at}(a\cos t - \sin t), y = C_3 + C_1 e^{-at}(a\sin t + \cos t)$.

Linear Equations of Third Order. These equations are from Chapter 3 of
Kamke's collection. There occur the three symmetry classes $\mathcal{S}_{4,5}^3$, $\mathcal{S}_{5,1}^3$ and $\mathcal{S}_7^3$,
and the parametrized class $\mathcal{S}_{5,2}^3(a)$. The symmetry generators are not given
explicitly because they are closely related to the elements of a fundamental
system (see Exercise 5.3). Whenever a closed form for the latter exists it is
also given. It is usually obtained by means of a Loewy decomposition.

$\mathcal{S}_{4,5}^3$: 3.2, 3.7, 3.30, 3.31, 3.32, 3.33, 3.34, 3.35, 3.37, 3.38, 3.39, 3.40,
3.41, 3.42, 3.49, 3.50, 3.51, 3.53, 3.54, 3.56, 3.58, 3.60, 3.65, 3.66,
3.68, 3.70, 3.73, 3.75, 3.76, 3.77.

$\mathcal{S}_{5,1}^3$: 3.27, 3.63, 3.74. $\mathcal{S}_{5,2}^3(a)$:   3.1, 3.4, 3.16, 3.29, 3.71.

$\mathcal{S}_7^3$: 3.6, 3.21, 3.45, 3.47, 3.48, 3.57, 3.61, 3.64.

3.1: $y''' + \lambda y = 0$. Symmetry class $\mathcal{S}_{5,2}^3(\theta)$, $\theta^2 - \theta + 1 = 0$. Fundamental
system: $\{\exp(-\theta_1 x), \exp(\frac{1}{2}\theta_1 x)\cos \frac{1}{2}\theta_1\sqrt{3}x), \exp(\frac{1}{2}\theta_1 x)\sin \frac{1}{2}\theta_1\sqrt{3}x)\}$,
θ_1 real root of $\theta^3 - \lambda = 0$.

3.2: $y''' + ax^3 y + bx = 0$. Symmetry class $\mathcal{S}_{4,5}^3$.

Symmetries of Kamke's Collection

3.3: $y''' = ax^b y$. Symmetry class $\mathcal{S}_{4,5}^3$.

3.4: $y''' + 3y' - 4y = 0$. Symmetry class $\mathcal{S}_{5,2}^3(\theta)$, $\theta^2 - \theta + \frac{2}{5} = 0$. Fundamental system: $\{\exp(x), \exp\left(-\frac{1}{2}x\right)\cos\left(\frac{1}{2}\sqrt{15}x\right), \exp\left(-\frac{1}{2}x\right)\sin\left(\frac{1}{2}\sqrt{15}x\right)\}$.

3.5: $y''' - a^2 y' = 0$. Symmetry class $\mathcal{S}_7^3$. Fundamental system: $\{1, \exp(\pm ax)\}$.

3.6: $y''' + 2axy' + ay = 0$. Symmetry class $\mathcal{S}_7^3$. Fundamental system: $\{u^2, uv, v^2\}$, u and v fundamental system for $w'' + \frac{1}{2}axw = 0$.

3.7: $y''' - x^2 y'' + (a + b - 1)y' - aby = 0$. Symmetry class $\mathcal{S}_{4,5}^3$.

3.16: $y''' - 2y'' - 3y' + 10y = 0$. Symmetry class $\mathcal{S}_{5,2}^3(\theta)$, $\theta^2 - \frac{4}{17}\theta + \frac{4}{17} = 0$. Fundamental system: $\{\exp(-2x), \cos(x)\exp(2x), \sin(x)\exp(2x)\}$.

3.17: $y''' - 2y'' - a^2 y' + 2a^2 y = 0$. Symmetry class: $\mathcal{S}_{5,2}^3(3)$ for $a = 1$, $\mathcal{S}_{5,1}^3$ for $a = 2$, $\mathcal{S}_{5,2}^3(6)$ for $a = 3$, $\mathcal{S}_{5,2}^3(4)$ for $a = 4$. Fundamental system: $\{\exp(2x), \exp(\pm ax)\}$.

3.18: $y''' - 3ay'' + 3a^2 y' - a^3 y = 0$. Symmetry class $\mathcal{S}_{5,2}^3(\theta)$, $\theta^2 - \theta + 1 = 0$. Fundamental system: $\{\exp(ax), x\exp(ax), x^2\exp(ax)\}$.

3.20: $y''' - 6xy'' - (4a + 8x^2 - 2)y' - 8axy = 0$. Symmetry class $\mathcal{S}_7^3$. Fundamental system: $\{u^2, uv, v^2\}$; u and v fundamental system of $w'' - 2xw + aw = 0$.

3.21: $y''' + 3axy'' + 3a^2 x^2 y' + a^3 x^3 y = 0$. Symmetry class $\mathcal{S}_7^3$. Solution: $y = \exp\left(-\frac{1}{2}ax^2\right)\left[C_1 + C_2\cos\left(x\sqrt{3a}\right) + C_3\sin\left(x\sqrt{3a}\right)\right]$.

3.27: $4y''' - 8y'' - 11y' - 3y = 0$. Symmetry class $\mathcal{S}_{5,1}^3$. Fundamental system: $\{\exp(3x), \exp\left(-\frac{1}{2}x\right), x\exp\left(-\frac{1}{2}x\right)\}$.

3.29: $xy''' + 3y'' + xy = 0$. Symmetry class $\mathcal{S}_{5,2}^3(\theta)$, $\theta^2 - \theta + 1 = 0$. Fundamental system: $\{\frac{1}{x}e^{-x}, \frac{1}{x}\exp\left(\frac{1}{2}(1 \pm i\sqrt{3})x\right)\}$.

3.30: $xy''' + 3y'' - ax^2 y = 0$. Symmetry class $\mathcal{S}_{4,5}^3$.

3.31: $xy''' + (a + b)y'' - xy' - ay = 0$. Symmetry class $\mathcal{S}_{4,5}^3$.

3.32: $xy''' + (a + b)y'' - xy' - ay = 0$. Symmetry class $\mathcal{S}_{4,5}^3$. Solution: $y = C_1 e^x + x^{\nu+1}[C_2 J_{\nu+1}(ix) + C_3 Y_{\nu+1}(ix)]$.

3.33: $xy''' - (x + 2\nu)y'' - (x - 2\nu - 1)y' + (x - 1)y = 0$. Symmetry class $\mathcal{S}_{4,5}^3$. Fundamental system: $\{x^5 \exp\left(-\frac{1}{2}x^2\right), x^5 \exp\left(-\frac{1}{2}x^2\right)\int \exp\left(\frac{1}{2}x^2\right)\frac{dx}{x^6}\}$.

3.34: $xy''' + \frac{3}{2}y'' + \frac{1}{2}axy = 0$. Symmetry class $\mathcal{S}_{4,5}^3$.

3.35: $xy''' - 2(x + \nu - 1)y'' + (x + 3\nu - \frac{5}{2})y' - (\nu - \frac{1}{2})y = 0$. Symmetry class $\mathcal{S}_{4,5}^3$. Solution: $y = C_1 e^x + x^\nu \exp\left(\frac{1}{2}x\right)\left[C_2 J_\nu\left(\frac{1}{2}ix\right) + C_3 Y_\nu\left(\frac{1}{2}ix\right)\right]$.

3.37: $(x - 2)xy''' - (x - 2)xy'' - 2y' + 2y = 0$. Symmetry class $\mathcal{S}_{4,5}^3$. Fundamental system: $\{x^2, e^x, e^x \int e^{-x}\left[x(x - 2)\log\frac{x}{x-2} - 2(x - 1)\right]dx\}$.

3.38: $(2x - 1)y''' - 8xy' + 8y = 0$. Symmetry class $\mathcal{S}_{4,5}^3$. Fundamental system: $\{\exp(2x), x\exp(2x), \exp(2x)\int\frac{x\exp(-4x)dx}{4x(x-1)+1} - x\int\frac{\exp(-2x)dx}{4x(x-1)+1}\}$.

3.39: $(2x - 1)y''' + (x + 4)y'' + 2y' = 0$. Symmetry class $\mathcal{S}_{4,5}^3$.

3.40: $x^2y''' - 6y' + ax^2y = 0$. Symmetry class $\mathcal{S}_{4,5}^3$.

3.41: $x^2y''' + (x+1)y'' - y = 0$. Symmetry class $\mathcal{S}_{4,5}^3$.

3.42: $x^2y''' - xy'' + (x^2+1)y' = 0$. Symmetry class $\mathcal{S}_{4,5}^3$.

3.45: $x^2y''' + 4xy'' + (x^2+2)y' + 3xy = 0$. Symmetry class $\mathcal{S}_{4,5}^3$.

3.46: $x^2y''' + 5xy'' + 4y = 0$. Symmetry class: $\mathcal{S}_{5,1}^3$.
Fundamental system: $\{1, \frac{1}{x}, \frac{\log x}{x}\}$.

3.47: $x^2y''' + 6xy'' + 6y' = 0$. Symmetry class $\mathcal{S}_7^3$.
Fundamental system: $\{1, \frac{1}{x}, \frac{1}{x^2}\}$.

3.48: $x^2y''' + 6xy'' + 6y' + ax^2y = 0$. Symmetry class $\mathcal{S}_7^3$.
Fundamental system: $\{\frac{1}{x^2}\exp\theta x\}$, $\theta^3 + a = 0$.

3.49: $x^2y''' - 3(p+q)xy'' + 3p(3q+1)y' - x^2y = 0$. Symmetry class $\mathcal{S}_{4,5}^3$.

3.50: $x^2y''' - 2(n+1)xy'' + (ax^2+6n)y' - 2axy = 0$, $n \in \mathbb{N}$.
Symmetry class $\mathcal{S}_{4,5}^3$.
Fundamental system: $\{ax^2 + 4n - 2, \exp(x\sqrt{-a})P(x), \exp(-x\sqrt{-a})Q(x)\}$,
$P(x)$ and $Q(x)$ polynomials of degree $\leq 2n+2$.

3.51: $x^2y''' - x(x-2)y'' + (x^2 + \nu^2 - \frac{1}{4})y' - (x^2 - 2x + \nu^2 - \frac{1}{4})y = 0$.
Symmetry class $\mathcal{S}_{4,5}^3$.

3.52: $x^2y''' - (x+\nu)xy'' + \nu(2x+1)y' - \nu(x+1)y = 0$.
Symmetry class $\mathcal{S}_{4,5}^3$.

3.53: $x^2y''' - x(x-1)y'' + (x^2 - 2x - \nu^2 + \frac{1}{4})y' - (\nu^2 - \frac{1}{4})y = 0$.
Symmetry class $\mathcal{S}_{4,5}^3$.

3.54: $x^3y''' - (x^4 - 6x)y'' - (2x^3 - 6)y' + 2x^2y = 0$.
Symmetry class $\mathcal{S}_{4,5}^3$.

3.55: $(x^2+1)y''' + 8xy'' + 10y' = 0$. Symmetry class $\mathcal{S}_{4,5}^3$.
Fundamental system: $\left\{ \dfrac{1}{(x^2+1)^2}, \dfrac{x(x^2+3)}{(x^2+1)^2}, \dfrac{x^2(x^2+2)}{(x^2+1)^2} \right\}$.

3.56: $(x^2+2)y''' - 2xy'' + (x^2+2)y' - 2xy = 0$. Symmetry class $\mathcal{S}_{4,5}^3$.
Fundamental system: $\{x^2, \cos x, \sin x\}$.

3.57: $x(x-1)y''' + 3(x - \frac{1}{2})y'' + (ax + \frac{1}{2}b)y' + \frac{1}{2}ay = 0$. Symmetry class $\mathcal{S}_7^3$.
Fundamental system: $\{y_1^2, y_1y_2, y_2^2\}$, y_1 and y_2 hypergeometric functions.

3.58: $4x^2y''' + (x^2 + 14x - 1)y'' + 4(x+1)y' + 2y = 0$.
Symmetry class $\mathcal{S}_{4,5}^3$. Solution:
$$y = \sqrt{x}\exp-\frac{x^2+1}{4x}\left[C_1 + \int\left(C_2x^{-\frac{5}{2}} + C_3x^{-\frac{3}{2}}\right)\exp\frac{x^2+1}{4x}\right].$$

3.60: $x^3y'''(\nu^2 - 1)xy'' + (ax^3 + \nu^2 - 1)y = 0$. Symmetry class $\mathcal{S}_{4,5}^3$.

3.61: $x^2y''' + (4x^3 - 4\nu^2x + x)y' + (4\nu^2 - 1)y = 0$. Symmetry class $\mathcal{S}_7^3$.
Fundamental system: $\{xJ_\nu^2(x), xJ_\nu(x)Y_\nu(x), xY_\nu^2(x)\}$.

3.62: $x^3y''' + (ax^{2\nu} + 1 - \nu^2)xy' + [bx^{3\nu} + a(\nu - 1)x^{2\nu} + \nu^2 - 1]y = 0$.

Symmetries of Kamke's Collection

Symmetry class $\mathcal{S}^3_{4,5}$.

3.63: $x^3 y''' + 3x^2 y'' - 2xy' + 2y = 0$. Symmetry class $\mathcal{S}^3_{5,1}$.
Fundamental system: $\{x, \frac{1}{x^2}, x \log x\}$.

3.64: $x^3 y''' + 3x^2 y'' + (1 - a^2)xy' = 0$. Symmetry class $\mathcal{S}^3_7$.
Fundamental system: $\{1, x^a, \frac{1}{x^a}\}$.

3.65: $x^3 y''' - 4x^2 y'' + (x^2 + 8)xy' - 2(x^2 + 4)y = 0$. Symmetry class $\mathcal{S}^3_{4,5}$.
Fundamental system: $\{x^2, x \cos x, x \sin x\}$.

3.66: $x^3 y''' + 6x^2 y'' + (ax^3 - 12)y = 0$. Symmetry class $\mathcal{S}^3_{4,5}$.
Fundamental system: $\{\frac{\theta x - 2}{x^3} \exp(\theta x)\}$, $\theta^3 + a = 0$.

3.68: $x^3 y''' + (x + 3)x^2 y'' + 5(x - 6)xy' + (4x + 30)y = 0$.
Symmetry class $\mathcal{S}^3_{4,5}$.
Solution: $y = x^5 \frac{d^6}{dx^6} \left\{ x^2 \frac{d^2}{dx^2} \left[x^{-5} e^{-x} \left(\int (C_1 + C_2 \log x) x^4 e^x \, dx + C_3 \right) \right] \right\}$.

3.70: $(x^2 + 1)xy''' + 3(2x^2 + 1)y'' - 12y = 0$. Symmetry class $\mathcal{S}^3_{4,5}$.
Fundamental system: $\{2x^2 + 1, xu, 2x + \frac{2}{3x} - xu \log \frac{u+1}{u-1}\}$, with
$u = \sqrt{x^2 + 1}$.

3.71: $(x+3)x^2 y''' - 3(x+2)xy'' + 6(x+1)y' - 6y = 0$. Symmetry class $\mathcal{S}^3_{5,2}(3)$.
Fundamental system: $\{x + 1, x^2, x^3\}$.

3.73: $(x+1)x^3 y''' - (4x+2)x^2 y'' + (10x+4)xy' - 4(3x+1)y = 0$. Symmetry
class $\mathcal{S}^3_{4,5}$. Fundamental system: $\{x^2, x^2 \log x, x + x^3 + x^2 \log (x)^2)\}$.

3.74: $4x^4 y''' - 4x^3 y'' + 4x^2 y' = 1$. Symmetry class $\mathcal{S}^3_{5,1}$.
Fundamental system: $\{1, x^2, x^2 \log x - \frac{1}{36x}\}$.

3.75: $(x^2 + 1)x^3 y''' - (4x^2 + 2)x^2 y'' + (10x^2 + 4)xy' - 4(3x^2 + 1)y = 0$.
Symmetry class $\mathcal{S}^3_{4,5}$. Fundamental system: $\{x^2, x^3 + x, x^2 \log |x|\}$.

3.76: $x^6 y''' + x^2 y'' - 2y = 0$. Symmetry class $\mathcal{S}^3_{4,5}$.

3.77: $x^6 y''' + 6x^5 y'' + ay = 0$. Symmetry class $\mathcal{S}^3_{4,5}$.

Appendix F

ALLTYPES Userinterface

To a large extent the calculations involved in the analysis of differential equations described in this book can hardly be performed by pencil and paper. Therefore, the userinterface of the computer algebra system ALLTYPES has been developed which is available on the website `www.alltypes.de`. It provides a great number of functions that perform the laborious calculations involved automatically. A complete listing of the functions currently provided may be obtained interactively by submitting

`UserFunctions();`

after the ALLTYPES system has been loaded. A detailed description of an individual function with the name `FunctionName` is obtained by calling

`UserFunction FunctionName;`

In this Appendix the connection to the mathematical background for any particular function is provided. Subsequently, all user interface functions are listed in alphabetical order and the page number where it is explained is given. If there occur several page numbers, they correspond to the various argument types, i.e., the corresponding *Methods* in the ALLTYPES system.

AbsoluteInvariants(LODE|ABEL)
  Returns absolute invariants for linear and Abel's equation.
CharacteristicPolynomial(LIEVEC|LIEALG)
  Returns characteristic polynomial of Lie algebra.
CommutatorTable(LIEVEC)
  Returns commutator table of Lie algebra.
DeterminingSystem(ODE|PDE)
  Returns determining system for Lie symmetries.
ExactQuotient(LODE|LDFMOD)
  Returns exact quotient or failed if it does not exist.
ForsythQuadrinvariants(LODE)
  Returns Forsyth's quadrinvariants for equation of order up to three.
Gcd(LODE|LDFMOD)
  Returns greatest common divisor of the two arguments.
GroebnerBase(DPOLID)
  Returns Gröbner basis in specified term order.
Integral(POLY|PARF|RATF)
  Returns integral of first argument w.r.t. second.

Invariants(LODE)
Returns different kinds of invariants.
JacobianProlongation(LIEVEC)
Returns k-th prolongation of first argument in Jacobian form.
JanetBase(ODE |LDFMOD|LFMOD)
Returns Janet base in specified term order.
JanetResolution(LDFMOD)
Returns Janet resolution.
LaguerreForsythNormalForm(LODE3)
Returns Laguerre-Forsyth normal form.
Lclm(LODE|LDFMOD)
Returns least common left multiple of the two arguments.
LieCanonicalForm(ODE2|ODE3)
Returns canonical form corresponding to Lie symmetry class.
LiouvillianSolutions(LODE|LDFMOD)
Returns Liouvillian solutions following from first order components.
LoewyDecomposition(LODE)
Returns Loewy decomposition into irreducible components.
Prolongation(LDO |LIEVEC)
Returns k-th prolongation of first argument.
RationalNormalForm(LODE| RICCATI|ABEL)
Returns rational normal form.
RationalSolution(UPOLY|LODE|RICCATI)
Returns most general rational solution including constants.
RationalSolutions(UPOLY|LODE|RICCATI)
Returns all rational solutions.
RelativeInvariants(LODE|ABEL)
Returns relative invariants.
SemiSimple?(LIEALG|LIEVEC)
Returns t if Lie algebra is semi-simple.
Socle(LODE|LDFMOD)
Returns largest completely reducible right component.
Solvable?(LIEALG| LIEVEC)
Returns t if Lie algebra is solvable.
Solve(LODE|ODE)
Returns solution or failed *if it cannot be found.*
Symmetries(ODE| PDE)
Returns vector fields of infinitesimal generators.
SymmetryClass(LODE|ODE)
Returns Lie symmetry class $\mathcal{S}_{i,j}^n$.
TresseInvariants(ODE)
Returns Tresse invariants of order not higher than four.

References

[1] Abraham-Shrauner, B., Govinder K. S., Leach, P. G. L., *Integration of Second Order Ordinary Differential Equations Not Possessing Lie Point Symmetries*, Physics Letters **A 203** (1995), 169-174.

[2] Abramowitz, M., Stegun, I., *Handbook of Mathematical Functions*, Dover, New York, 1965.

[3] Adams, W., Loustaunau, P., *An Introduction to Gröbner Bases*, American Mathematical Society, Providence, RI, 1996.

[4] Appell, P., *Sur les équations différentielles algébriques et homogènes par rapport à la fonction inconnus et à ses dérivées*, Comptes Rendus **104** (1887), 1176–1779.

[5] Appell, P., *Sur les invariants des équations différentielles*, Comptes Rendus **105** (1887), 55–58.

[6] Appell, P., *Sur les invariants de quelques équations différentielles*, Journal de Mathématique **5** (1889), 361–423.

[7] Appell, P., *Sur des équations différentielles linéaires transformable en elles-mêmes par un changement de fonction et de variable*, Acta Mathematica **15** (1891), 281–314.

[8] Babich, M. V., Bordag, L. A., *Projective differential geometrical structure of the Painlevé equations*, Preprint NTZ 35/1997.

[9] Beke, E., *Die Irreduzibilität der homogenen linearen Differentialgleichungen*, Mathematische Annalen **45** (1894), 278–294.

[10] Berth, M., *Invariants of Ordinary Differential Equations*, Dissertation, Greifswald, 1999.

[11] Berth, M., Czichowski, G., *Using Invariants to Solve the Equivalence Problem for Ordinary Differential Equations*, Applied Algebra in Engineering, Communication and Computing **11** (1998), 359-376.

[12] Bianci, L., *Lezioni sulla teoria dei gruppi continui finiti di transformazioni*, Nicola Zanichelli, Bologna, 1928.

[13] Bluman, G. W., Kumei, S., *Symmetries of Differential Equations*, Springer, Berlin, 1990.

[14] Blyth, T. S., *Module Theory*, Clarendon Press, Oxford, 1977.

[15] Bordag, L. A., Dryuma, V. S., *Investigation of Dynamical Systems Using Tools of the Theory of Invariants and Projective Geometry*, Z. angew. Math. Phys. **48** (1997), 725-743.

420 *References*

[16] Bourbaki, N., *Eléments d'histoire des mathématiques, Masson Editeur*, Paris, 1984 [English translation, Elements of the History of Mathematics, Springer, Berlin, 1994].

[17] Borel, A., *Essays in the History of Lie Groups and Algebraic Groups*, American Mathematical Society and London Mathematical Society, 2001.

[18] Bouton, C. L., *Invariants of the General Linear Differential Equation and their Relation to the Theory of Continuous Groups*, American Journal of Mathematics **21** (1899), 25–84.

[19] Brioschi, F., *Les invariants des équations différentielles linéaires*, Acta Mathematica **14** (1890), 233–248.

[20] Bronstein, M., *On Solutions of Linear Differential Equations in their Coefficient Field*, Journal of Symbolic Computation **13** (1992), 413-439.

[21] Bronstein, M., *Linear Ordinary Differential Equations, Breaking through the Order 2 Barrier*, Proceedings of the ISSAC 1992, Berkeley, California, ACM Press, 1992; P. S. Wang, Ed., 42-48.

[22] di Bruno, Fáa., *Note sur une nouvelle formule de calcul différentielle*, Quarterly Journal of Pure and Applied Mathematics **1** (1857), 359-360.

[23] Buium A., Cassidy, Ph., *Differential Algebraic Geometry and Differential Algebraic Groups: From Algebraic Differential Equations to Diophantine Geometry*, in: Selected Works of Ellis Kolchin, AMS Press, Providence, Rhode Island, 1999; H. Bass, A. Buium, Ph. Cassidy, Eds.

[24] Campbell, J. E., *Introductory Treatise on Lie's Theory of Finite Continuous Transformation Groups*, 1903 [Reprinted by Chelsea Publishing Company, New York, 1966].

[25] Cartan, É., *Sur la structure des groupes de transformations finis et continus*, Premiere Thèse, Paris, 1894.

[26] Cartan, É., *Sur la réduction à sa forme canonique de la structure d'un groupe de transformations fini et continu*, American Journal of Mathematics **18** (1896), 1-61.

[27] Castro-Jiménez, F. J., Moreno-Frías, M.A., *Janet Bases and Gröbner Bases*, Prepublicaciones del Departamento de Álgebra de la Universidad de Sevilla n° 1, 2000.

[28] Chiellini, A., *Sull' integrazione dell'equazione differentiale* $y' + Py^2 + Qy^3 = 0$, Bolletino Unione Mat. Italiana **10** (1931), 301-307.

[29] Chini, M., *Sull'integrazione di alcune equazioni differenziali del primo ordine*, Rendiconti Istituto Lombardo (2) **57** (1924), 506-511.

[30] Clebsch, A., *Theorie der binären algebraischen Formen*, Teubner, Leipzig, 1872.

[31] Cockle, J., *Correlations of Analysis*, Philosophical Magazine **24**(IV) (1862), 24.

References 421

[32] Cohen, A., *An Introduction to the Lie Theory of One-Parameter Groups*, Hafner Publishing Company, New York, 1911.

[33] Cox, D., Little, J., O'Shea, D., *Ideals, Varieties and Algorithms*, Springer, Berlin, 1991.

[34] Cox, D., Little, J., O'Shea, D., *Using Algebraic Geometry*, Springer, Berlin, 1998.

[35] Curtiss, D. R., *On the Invariants of a Homogeneous Quadratic Differential Equation of the Second Order*, American Journal of Mathematics **25** (1903), 365–382.

[36] Czichowski, G., Fritzsche, B., *Beiträge zur Theorie der Differentialinvarianten*, Teubner Verlagsgesellschaft, 1993.

[37] de Graaf, W. A., *Algorithms for Finite-Dimensional Lie Algebras*, Dissertation, Eindhoven, 1997.

[38] Drach, J., *Essai sur une théorie général de l'intégration et sur la classification des transcendentes*, Gauthier-Villars et Cie, Paris, 1898.

[39] Dresner, L., *Applications of Lie's Theory of Ordinary and Partial Differential Equations*, IOP Publishing, Bristol and Philadelphia, 1999.

[40] Emanuel, G., *Solution of Ordinary Differential Equations by Continuous Groups*, Chapman & Hall/CRC, London, New York, 2001.

[41] Eisenhart, L. P., *Continuous Groups of Transformations*, Princeton University Press, Princeton, 1933 [Reprinted by Dover Publications, New York, 1961].

[42] Emanuel, G., *Solution of Ordinary Differential Equations by Continuous Groups*, Chapman & Hall/CRC, 2000.

[43] *Encyclopaedia of Mathematical Sciences, Lie Groups and Lie Algebras*, vol. I, A. L. Onishchik, Ed., Springer, Berlin, 1993; vol. II, A. L. Onishchik, E. B. Vinberg, Eds., Springer, Berlin, 2000; vol. III, A. L. Onishchik, E. B. Vinberg, Eds.; Springer, Berlin, 1994.

[44] Engel, F., *Gesammelte Abhandlungen Sophus Lie, vol. I to VII*, Teubner, Leipzig, 1922–1960.

[45] Faber, K., *Differentialgleichungen, die eine irreduzible Gruppe von Berührungstransformationen gestatten*, Mitteilungen des Mathematischen Seminars der Universiät Giessen **13** (1924), 1-30.

[46] Faber, K., *Differentialgleichungen, die durch eine Berührungstransformation auf die Form $y''' = 0$ gebracht werden können*, Leipziger Berichte **78** (1926), 9-22.

[47] Forsyth, A. R., *Invariants, Covariants and Quotient-Derivatives Associated with Linear Differential Equations*, Philosophical Transactions **179** (1888), 377–489.

[48] Fritzsche, B., *Sophus Lie, A Sketch of his Life and Work*, Journal of Lie Theory **9** (1999), 1-38.

422 *References*

[49] Gandarias, M. L., Medina, E., Muriel, C., *New Symmetry Reductions for Some Ordinary Differential Equations*, Journal of Nonlinear Mathematical Physics **9** (2002), 47-58.

[50] Gasiorowski, L., *Über die Definitionsgleichungen der endlichen continuierlichen Gruppen von Berührungstransformationen der Ebene*, Prace Matematyczno-Fizyczne **26** (1914), 133-203.

[51] Gat, O., *Symmetry algebras of third-order ordinary differential equations*, Journal of Mathematical Physics, **33** (1992), 2966-2971.

[52] Gerdt, V., Lassner, W., *Isomorphism Verification for Complex and real Lie Algebras by Gröbner Base Technique*, Proceedings of the International Workshop Acireale, Catania, N. H. Ibragimov, N. Torrisi, A. Valenti, Eds.; Kluwer Academic Publishers, Dordrecht, 245-254, 1993.

[53] Gonzáles-López, A., *Symmetry and Integrability by Quadratures of Ordinary Differential Equations*, Physics Letters, **A133** (1988), 190-194.

[54] Gonzáles-López, A., Kamran, N., Olver, P., *Lie Algebras of Vector Fields in the Real Plane*, Proceedings of the London Mathematical Society (3) **64** (1992), 339–368.

[55] Gordan, P., *Invariantentheorie*, Teubner, Leipzig, 1888 [Reprinted by Chelsea Publishing Company, New York, 1987].

[56] Goursat, E., *Sur les invariants des équations différentielles*, Comptes Rendus **107** (1888), 898–900.

[57] Goursat, E., *Leçons sur l'intégration des équations aux dérivées partielles*, Gauthier-Villars et Cie, Paris, 1921.

[58] Grebot, G., *The Characterization of Third Order Ordinary Differential Equations Admitting a Transitive Fiber-Preserving Point Symmetry Group*, Journal of Mathematical Analysis and Applications **206** (1997), 364-388.

[59] Grebot, G., *The Third Order Ordinary Differential Equations Admitting the Painlevé I and II as First Integrals*, Journal of Mathematical Analysis and Applications **220** (1997), 110-124.

[60] Greuel, G. M., Pfister, G., *A Singular Introduction to Commutative Algebra*, Springer, Berlin, 2002.

[61] Grigoriev, D., Schwarz, F. *Factoring and Solving Linear Partial Differential Equations*, Computing **73** (2004), 179-197.

[62] Grigoriev, D., Schwarz, F. *Generalized Loewy-Decomposition of D-Modules*, Proceedings of the ISSAC'05, 163-170, ACM Press, 2005.

[63] Grissom, C., Thomson, G., Wilkens, G., *Linearization of Second Order Ordinary Differential Equations via Cartan's Equivalence Method*, Journal of Differential Equations **77** (1989), 1–15.

[64] Halphen, G. H., *Sur les invariants différentielles*, Thèse présentée à la Faculté de Sciences de Paris, 1878.

References 423

[65] Halphen, G. H., *Mémoire sur la réduction des équations différentielles linéaires aux formes intégrable*, Mémoires Présentés par divers Savants **28** (1878), 1–301.

[66] Halphen, G. H., *Sur la réduction de certaines équations différentielles du premier ordre a la forme linéaire*, Comptes rendus **87** (1882), 741–743.

[67] Halphen, G. H., *Sur les invariants des équations différentielles linéaires du quatrième ordre*, Acta Mathematica **3** (1883), 325–380.

[68] Hawkins, T., *Emergence of the Theory of Lie Groups*, Springer, Berlin, 2000.

[69] Heineck, C., *Invariante Kurvenintegrale bei infinitesimalen Transformationen in drei Veränderlichen x, y und z und deren Verwertung*, Dissertation, Teubner, Leipzig, 1899.

[70] Hereman, W., *Review of Symbolic Software for the Computation of Lie Symmetries of Differential Equations*, Euromath Bulletin **1** (1994), 45–97.

[71] Hill, J. M., *Differential Equations and Group Methods*, CRC Press, Boca Raton, Florida, 1992.

[72] Hirsch, A., *Zur Theorie der linearen Differentialgleichungen mit rationalem Integral*, Dissertation, Königsberg, 1892.

[73] van Hoeij, M., *Factorization of Differential Operators with Rational Function Coefficients*, Journal of Symbolic Computation **24** (1997), 537-561.

[74] Hsu, L., Kamran, N., *Symmetries of Second-Order Ordinary Differential Equations and Élie Cartan's Method of Equivalence*, Letters in Mathematical Physics **15** (1988), 91-99.

[75] Hsu, L., Kamran, N., *Classification of Second-Order Ordinary Differential Equations Admitting Lie Groups of Fibre-Preserving Point Symmetries*, Proceedings of the London Mathematical Society (3) **58** (1989), 387-416.

[76] Hydon, P. E., *Symmetry Methods for Differential Equations*, Cambridge University Press, 2000.

[77] Ibragimov, N. H., *Transformation Groups Applied to Mathematical Physics*, D. Reidel Publishing Company, Dordrecht, 1985.

[78] Ibragimov, N. H., *Elementary Lie Group Analysis of Ordinary Differential Equations*, John Wiley & Sons, New York, 1999.

[79] Imschenetzky, V. G., *Sur l'integration des équations aux dériveées partielles du premier ordre*, Grunerts Archiv für Mathematik **50** (1869), 278.

[80] Ince, E. L., *Ordinary Differential Equations*, Longmans, Green and Co., 1926 [Reprint by Dover Publications Inc., 1956].

[81] Jacobson, N., *The Structure of Rings*, American Mathematical Society, Providence, Rhode Island, 1956.

[82] Jacobson, N., *Lie Algebras*, Interscience Publishers, New York, 1962.

[83] Janet, M., *Les systèmes d'équations aux dérivées partielles*, Journal de mathématiques **83** (1920), 65–123.

424 *References*

[84] Janet, M., *Leçons sur les systèmes d'équations aux derivées partielles*, Gauthier-Villars et Cie, Paris, 1929.

[85] Kamke, E., *Differentialgleichungen, Lösungsmethoden und Lösungen, I. Gewöhnliche Differentialgleichungen*, Akademische Verlagsgesellschaft, Leipzig, 1961.

[86] Kamke, E., *Differentialgleichungen I, Gewöhnliche Differentialgleichungen und II, Partielle Differentialgleichungen*, Akademische Verlagsgesellschaft, Leipzig, 1962.

[87] Kamke, E., *Differentialgleichungen, Lösungsmethoden und Lösungen, II. Partielle Differentialgleichungen*, Akademische Verlagsgesellschaft, Leipzig, 1965.

[88] Kaplansky, I., *An Introduction to Differential Algebra*, Hermann, Paris, 1957.

[89] Killing, W., *Die Zusammensetzung der stetigen endlichen Transformationsgruppen*, Mathematische Annalen **31** (1887), 252-290; **33**, 1-48; **34**, 57-122; **36**, 161-189.

[90] Kolchin, E. R., *Algebraic Matrix Groups and the Picard-Vessiot Theory of Homogeneous Linear Ordinary Differential Equations*, Annals of Mathematics **49** (1948), 1-42.

[91] Kolchin, E. R., *Differential Algebra and Algebraic Groups*, Academic Press, New York and London, 1973.

[92] Kovacic, J., *An Algorithm for Solving Second Order Linear Homogeneous Differential Equations*, Journal of Symbolic Computation **2**(1986), 3-43.

[93] Kowalewski, G., *Einführung in die Theorie der kontinuierlichen Gruppen*, Akademische Verlagsgesellschaft, Leipzig, 1931.

[94] Kowalewski, G., *Integrationsmethoden der Lie'schen Theorie*, Akademische Verlagsgesellschaft, Leipzig, 1933.

[95] Krause, J., Michel, L., *Équations différentielles linéaires d'ordre $n > 2$ ayant une algéébre de Lie de symétrie de dimension $n + 4$*, Comptes Rendus des Academie des Sciences **307** (1988), 905–910.

[96] Krause, J., Michel, L., *Classification of the Symmetries of Ordinary Differential Equations*, Proceedings of the Summer School in Group Theoretical Methods in Physics, Moscow, 1990, Springer LNP 382, 1991.

[97] Kung, J. P. S., Rota, G. C., *Theory of Binary Forms*, Bulletin of the American Mathematical Society **10** (1984), 27-85.

[98] Laguerre, E., *Sur les équations différentielles linéaires du troisième ordre*, Comptes Rendus **88** (1879), 116–119.

[99] Laguerre, E., *Sur quelques invariants des équations différentielles*, Comptes Rendus **88** (1879), 224–227.

[100] Lambek, J., *Lectures on Rings and Modules*, Blaisdell Publishing Company, Waltham, 1966.

References 425

[101] Landau, E., *Über die Faktorisierung linearer Differentialgleichungen*, Journal für die reine und angewandte Mathematik **124** (1902), 115-120.

[102] Leja, F., *Propriétés des équations différentielles ordinaires de 3-ème ordre par rapport aux transformations tangentielles*, Prace Matematyczno-Fizyczne **29** (1918), 179-256.

[103] Leja, F., *Bestimmung der Invarianten der gewöhnlichen Differentialgleichungen 3. Ordnung in bezug auf Punkttransformationen*, Monatshefte für Mathematik und Physik **29** (1918), 203-254.

[104] Li, Z., Schwarz, F., *Rational Solutions of Riccati Like Systems of Partial Differential Equations*, Journal of Symbolic Computation **31** (2001), 691-716.

[105] Li, Z., Schwarz, F., Tsarev, S., *Factoring Zero-dimensional Ideals of Linear Partial Differential Operators*, Proceedings of the ISSAC'02, ACM Press, T. Mora, Ed., 168-175, 2002.

[106] Li, Z., Schwarz, F., Tsarev, S., *Factoring Systems of Linear PDE's with Finite-Dimensional Solution Spaces*, Journal of Symbolic Computation **36** (2002), 443-471.

[107] Lie, S., *Verallgemeinerung und neue Verwertung des Jacobischen Multiplikators*, Christ. Forb. (1874), 255-274 [Gesammelte Abhandlungen, vol. III, 188-206].

[108] Lie, S., *Allgemeine Theorie der partiellen Differentialgleichungen erster Ordnung*, Mathematische Annalen **11** (1877), 464–557 [Gesammelte Abhandlungen, vol. IV, 163-250].

[109] Lie, S., *Klassifikation und Integration von gewöhnlichen Differentialgleichungen zwischen x, y, die eine Gruppe von Transformationen gestatten I, II, III and IV*, Archiv for Mathematik **8** (1883), 187-224, 249-288, 371-458 and **9**, 431-44. [Gesammelte Abhandlungen, vol. V, 240-281, 282-310, 362-427 and 432-446].

[110] Lie, S., *Über unendliche kontinuierliche Gruppen*, Christ. Forh. Aar, Nr. 12 (1883) [Gesammelte Abhandlungen, vol. V, 314-361].

[111] Lie, S., *Über Differentialinvarianten*, Mathematische Annalen **24** (1884), 537–578 [Gesammelte Abhandlungen, vol. VI, 95-138].

[112] Lie, S., *Theorie der Transformationsgruppen I, II and III*, Teubner, Leipzig, 1888 [Reprinted by Chelsea Publishing Company, New York, 1970].

[113] Lie, S., *Vorlesungen über Differentialgleichungen mit bekannten infinitesimalen Transformationen*, Teubner, Leipzig, 1891 [Reprinted by Chelsea Publishing Company, New York, 1967].

[114] Lie, S., *Vorlesungen über continuierliche Gruppen*, Teubner, Leipzig, 1893 [Reprinted by Chelsea Publishing Company, New York, 1971].

[115] Lie, S., *Geometrie der Berührungstransformationen*, Teubner, Leipzig, 1896 [Reprinted by Chelsea Publishing Company, New York, 1977].

426   *References*

[116] Liouville, R., *Sur quelques équations différentielle non linéaires*, Comptes Rendus **103** (1886), 457–460.

[117] Liouville, R., *Sur certaines équations différentielles du premier ordre*, Comptes Rendus **103** (1886), 476-479.

[118] Liouville, R., *Sur une classe d'équations différentielles non linéaires*, Comptes Rendus **103** (1886), 520–523.

[119] Liouville, R., *Sur une classe d'équations différentielles de premier ordre et sur les formations invariantes qui s'y rapportent*, Comptes Rendus **105** (1887), 460–463.

[120] Liouville, R., *Sur une classe d'équations différentielles, parmis lesquelles, en particulier, toutes celles des lignes géodésiques se trouvent comprises*, Comptes Rendus **105** (1887), 1062–1064.

[121] Liouville, R., *Sur quelques équations différentielles non linéaires*, Journal de l'Ecole Polytechnique **57** (1887), 189–250.

[122] Liouville, R., *Sur certaines équations différentielles de premier ordre*, Comptes Rendus **106** (1888), 1648–1651.

[123] Liouville, R., *Sur les lignes géodésiques des surface à courbure constante*, American Journal of Mathematics **10** (1888), 283–292.

[124] Liouville, R., *Sur les invariants de certaines équations différentielles de premier ordre et sur leurs applications*, Comptes Rendus **109** (1889), 560–563.

[125] Liouville, R., *Mémoire sur les invariants de certaines équations différentielles et sur leurs applications*, Journal de l'Ecole Polytechnique **59** (1889), 7–76.

[126] Liouville, R., *Sur une équation différentielles du premier ordre*, Acta Mathematica **27** (1903), 55–78.

[127] Loewy, A., *Über vollständig reduzible lineare homogene Differentialgleichungen*, Mathematische Annalen **56** (1906), 89–117.

[128] Magid, A. R., *Lectures on Differential Galois Theory*, American Mathematical Society, Providence, Rhode Island, 1991.

[129] Mahomed, F. M., Leach, P. G. L., *Symmetry Lie Algebras of n-th Order Ordinary Differential Equations*, Journal of Mathematical Analysis and Applications **151** (1990), 80–107.

[130] Markoff, A., *Sur les équations différentielles linéaires*, Comptes Rendus **113** (1891), 685-688.

[131] Martins, J., *Algebraic Subgroups of $SL(2,C)$*, private communication, 2003.

[132] Maurer, L., Burkhardt, H., *Kontinuierliche Transformationsgruppen*, Encyklopädie der Mathematischen Wissenschaften II.1., Teubner, Leipzig, 401-436, 1916.

[133] Miller, G. A., Blichfeldt, H. F., Dickson, L. E., *Finite Groups*, Dover Publications, New York, 1961.

References 427

[134] Muriel, C., Romero J. L, *New Methods of Reduction for Ordinary Differential Equations*, IMA Journal of Mathematics **66** (2001), 111-125.

[135] Neumer, W., *Über gewöhnliche Differentialgleichungen, die lineare homogene Form erhalten können*, Mitteilungen des Mathematischen Seminars der Universität Giessen **16** (1929), 1-52.

[136] Neumer, W., *Die allgemeinste mit der projektiven G_8 der Ebene ähnliche Gruppe von Berührungstransformationen*, Journal für die reine und angewandte Mathematik **173** (1935), 125-159.

[137] Neumer, W., *Die gewöhnlichen Differentialgleichungen dritter und vierter Ordnung, die lineare homogene Form erhalten können*, Journal für die reine und angewandte Mathematik, I, **176** (1937), 224-249; II **177** (1937), 13-36; III **177** (1937), 65-81.

[138] Noth, G., *Differentialinvarianten und invariante Differentialgleichungen zweier zehngliedriger Gruppen*, Dissertation, Teubner, Leipzig, 1904.

[139] Oaku, T., Shimoyama, T., *A Gröbner Basis Method for Modules over Rings of Differential Operators*, Journal of Symbolic Computation **18** (1994), 223-248.

[140] Olver, P., *Application of Lie Groups to Differential Equations*, Springer, Berlin, 1986.

[141] Olver, P., *Equivalence, Invariants and Symmetries*, Cambridge University Press, Cambridge, UK, 1995.

[142] Olver, P., *Classical Invariant Theory*, Cambridge University Press, Cambridge, UK, 1999.

[143] Ore, Ö., *Formale Theorie der linearen Differentialgleichungen*, Journal für die reine und angewandte Mathematik **167** (1932), 221–234, and **168** (1932), 233-257.

[144] Ovsiannikov, L. V., *Group Analysis of Differential Equations*, Academic Press, New York, 1982.

[145] Peyovitch, T., *Sur les semi-invariants des équations différentielles linéaires*, Bulletin de la Société mathématique de France **53** (1923), 208–225.

[146] Picard, E., *Sur les groupes de transformation des équations différentielles linéaires*, Comptes Rendus **86** (1883), 1131–1134.

[147] Picard, E., *Analogies entre la théorie des équations différentielles linéaires at la théorie des équations algébriques*, Gauthier-Villars, Paris, 1936.

[148] Plesken W., Robertz, D., *Janet's Approach to Presentations and Resolutions for Polynomials and Linear PDEs*, Arch. Math. **84** (2005), 22-37.

[149] Polyanin, A. D., Zaitsev, V. V., *Handbuch der linearen Differentialgleichungen, Exakte Lösungen*, Spektrum Akademischer Verlag, Heidelberg, 1996 [Translated from Russian].

[150] Quadrat, A., *An Introduction to the Algebraic Theory of Linear Systems of Partial Differential Equations*, www-sop.inria.fr/cafe/Alban.Quadrat/Temporaire.html

428 *References*

[151] Rand D., Winternitz, P., Zassenhaus, H., *On the Identification of a Lie Algebra Given by Its Structure Constants. I. Direct Decompositions, Levi Decompositions, and Nilradicals*, Linear Algebra and its Applications **109** (1988), 197-246.

[152] Reid, G., Lisle, I. G., Boulton, A., Wittkopf, A. D., *Algorithmic Determination of Commutation Relations for Lie Symmetry Algebras of PDE's*, Proceedings of the ISSAC'92, ACM Press, G. Gonnet, Ed., 63-68, 1992.

[153] Riquier, C., *Les Systémes d'Équations aux Dérivées Partielles*, Gauthier-Villars, Paris, 1910.

[154] Riverau, P., *Sur les invariants de certaines classes d'équations différentielles homogènes par rapport a la fonction inconnue et a ses dérivées*, Thèses présentées a la Faculté des Sciences de Paris, Gauthiers-Villars, Paris, 1890.

[155] Riverau, P., *Sur les invariants des équations différentielles linéaires*, Annales de la Faculté des Sciences de Toulouse **4** (1890), 1-5.

[156] Riverau, P., *Sur les invariants de quelques équations différentielles*, Journal de Mathématiques pures et appliquées **4**(8) (1892), 233–268.

[157] Sachdev, P. L., *A Compendium on Nonlinear Ordinary Differential Equations*, Wiley-Interscience, New York, 1997.

[158] Scalizzi, P., *Soluzione di alcune equazioni del tipo Abel*, Atti Accad. Lincei **26**(5) (1917), 60-64.

[159] Schlesinger, L., *Ein Beitrag zur Theorie der linearen homogenen Differentialgleichungen dritter Ordnung*, Inauguraldissertation Kiel, Mayer & Müller, Berlin, 1897.

[160] Schlesinger, L., *Handbuch der Theorie der linearen Differentialgleichungen*, vol. I and II, Teubner, Leipzig, 1897.

[161] Schmucker, A., Czichowski, G., *Symmetry Algebras and Normal Forms of Third Order Ordinary Differential Equations*, Journal of Lie Theory **8** (1998), 129-137.

[162] Schwarz, F., *A Factorization Algorithm for Linear Ordinary Differential Equations*, Proceedings of the ISSAC'89, ACM Press, G. Gonnet, Ed., 17–25, 1989

[163] Schwarz, F., *An Algorithm for Determining the Size of Symmetry Groups*, Computing **49** (1992), 95-115.

[164] Schwarz, F., *Symmetries of 2nd and 3rd Order ODE's*, Proceedings of the ISSAC'95, ACM Press, A. Levelt, Ed., 16-25, 1995.

[165] Schwarz, F., *Janet Bases of 2nd Order Ordinary Differential Equations*, Proceedings of the ISSAC'96, ACM Press, Y. N. Lakshman, Ed., 179-187, 1996.

[166] Schwarz, F., *On Third Order Ordinary Differential Equations with Maximal Symmetry Group*, Computing **57** (1996), 273-280.

[167] Schwarz, F., *ALLTYPES, An ALgebraic Language and TYPE System*, Springer, LNAI 1476, 270-283, 1998.

References

429

[168] Schwarz, F., *Solving Third Order Differential Equations with Maximal Symmetry Group*, Computing **65** (2000), 155-167.

[169] Singer, M., *Liouvillian Solutions of n-th Order Homogeneous Linear Differential Equations*, American Journal of Mathematics **103** (1981), 661-682.

[170] Singer, M., *Solving Homogeneous Linear Differential Equations in Terms of Second Order Linear Differential Equations*, American Journal of Mathematics **107** (1985), 663-696.

[171] Singer, M., *Liouvillian Solutions of Linear Differential Equations*, Journal of Symbolic Computation **11** (1991), 251-273.

[172] Singer, M., Ulmer, F., *Galois Groups of Second and Third Order Linear Differential Equations*, Journal of Symbolic Computation **16** (1993), 9-36.

[173] Stäckel, P., *Über Transformationen von Differentialgleichungen*, Journal für reine und angewandte Mathematik **111** (1890), 290–302.

[174] Stephani, H., *Differential equations. Their solution using symmetries*, Cambridge University Press, Cambridge, UK, 1989.

[175] Stubhaug, A., *The Mathematician Sophus Lie*, Springer, Berlin, 2001.

[176] Thomas, J. M., *Matrices of Integers Ordering Derivatives*, Transactions of the American Mathematical Society **33** (1931), 389–410.

[177] Todorov, P. G., Kristev, G. A., *Some Classes of the Reduction and Generalizations of the Abel Equation*, Differensialnye Uravneniya **28** (1992), 2178-2179 (in Russian).

[178] Tresse, A., *Sur les groupes infinis de transformations,* Comptes Rendus **115** (1892), 1003–1006.

[179] Tresse, A., *Sur les invariants différentielles des groupes continus de transformations*, Acta Mathematica **18** (1894), 1–88.

[180] Tresse, A., *Sur les invariants ponctuels de l'équation différentielle ordinaire du second ordre*, Comptes Rendus **120** (1895), 429-431.

[181] Tresse, A., *Détermination des Invariants ponctuels de l'Équation différentielle ordinaire du second ordre*, $y'' = \omega(x, y, y')$), Preisschriften der Fürstlich Jablonowski'schen Gesellschaft, Teubner, Leipzig, 1896.

[182] Ulmer, F., Weil, J. A., *Note on Kovacic's Algorithm*, Journal of Symbolic Computation **22** (1996), 179-200.

[183] Vivanti, G., *Leçons élementaires sur la théorie des groupes de transformations*, Gauthiers-Villars, Paris, 1904.

[184] van der Put, M., *Galois Theory of Differential Equations, Algebraic Groups and Lie Algebras*, Journal of Symbolic Computation **28** (1999), 441-473.

[185] van der Put, M., Singer, M., *Galois Theory of Linear Differential Equations*, Springer, Berlin, 2003.

430   *References*

[186] van der Waerden, B. L., *Gruppen von linearen Transformationen*, Chelsea Publishing Company, New York, 1948.

[187] von Weber, E., *Partielle Differentialgleichungen*, Encyklopädie der Mathematischen Wissenschaften II.1., Teubner, Leipzig, 296-321, 1916.

[188] Wallenberg, G., *Anwendung der Theorie der Differentialinvarianten auf die Untersuchung der algebraischen Integrierbarkeit der linearen homogenen Differentialgleichungen*, Journal für reine und angewandte Mathematik **113** (1894), 1–41.

[189] Wilczynski, E. J., *Projective Differential Geometry of Curves and Ruled Surfaces*, Teubner, Leipzig, 1906 [Reprinted by Chelsea Publishing Company, New York, 1961].

[190] Wussing, H., *Die Genesis des abstrakten Gruppenbegriffes*, VEB Deutscher Verlag der Wissenschaften, Berlin, 1969.

[191] Zaitsev, V. F., Polyanin, A. D., *Discrete-Group Methods for Integrating Equations of Nonlinear Mechanics*, CRC Press, Boca Raton, Florida, 1994.

[192] Zorawski, K., *Über die Integration einer Klasse gewöhnlicher Differentialgleichungen dritter Ordnung*, Krakauer Abhandlungen **34** (1897), 141-205 (Polish).

Index

Abel's equation, 179
absolute invariant, 113, 161
adjoint group, 109
Appell, 179
associated equations, 24
autoreduced system, 49
autoreduction, 47

base field, 8, 11
Bernoulli equation, 180
Bouton, 165

center, 117
centralizer, 117
characteristic equation, 117
characteristic polynomial, 117
characteristic system, 78
coherent, 53
commutation relation, 108
commutator group, 106
complete set of monomials, 384
complete system, 51, 109
completely reducible, 27
component, 70
connected, 109
contact transformation, 159, 195
criticoid, 159

defining equations, 112
dependent variables, 42
derivation operator, 42
derivative, 42
derivative operator, 42
derivative vector, 44
derived Lie algebra, 118
derived series, 118
determining system, 199
Dickson's lemma, 44

differential Galois theory
    for second order equations, 35
differential invariant, 114
differential type, 40
divisor, 70
Drach's equation, 104

elementary extension, 16
Emden-Fowler equation, generalized,
    221
equivalence
    of first order ode's, 176
    of linear second order ode's,
        168
    of linear third order ode's, 172
equivalence class, 160
equivalence problem, 2, 160
equivalence transformation, 160
equivalent rational solutions, 17
essential group parameters, 107
exact quotient, 22
exact quotient module, 70

factorization, 21
full set of solutions, 13
full system of differential invariants,
    115
function, 42
fundamental invariants, 201
fundamental system, 12, 78

gauge, 40
Gcd, 21
Gcrd, 23
general solution, 311
Goursat, 160
graded lexicographic ordering, 45

432   *Index*

graded reverse lexicographic ordering, 45
greatest common divisor, 21
greatest common left divisor, 23
greatest common right divisor, 23
group type, 138

homomorphism of Lie groups, 106
hyperexponential solution, 86

imprimitive Lie group, 108
independent variables, 42
infinitesimal generator, 108
inhomogeneous equation, 14
integrability conditions, 50
integrable Lie algebra, 118
integrable Lie group, 106
integrable pair, 86
integral basis, 78
integrating factor, 313, 349
intersection module, 69
intransitive Lie group, 108
invariant
     absolute, 161
     of a differential equation, 160
     of a Lie group, 113
     relative, 161
     semi-, 161
invariant decomposition, 108
invariants
     absolute of linear ode, 168
     relative of linear ode, 166–168
irreducible, 24
isomorphism of Lie groups, 106

Jacobi normal form, 79
Janet basis, 11
Janet basis
     algorithm for, 53
     classification of, 57, 61
     construction of from solutions, 75
     decomposition of, 99
     definition of, 53
     exact quotient of, 70

for Lie vector fields, 143
type of, 55

k-fold transitive, 108
Krause, 193

Laguerre, 166
Laguerre-Forsyth canonical form, 164
Landau, 26
Lclm, 23
Lcm, 21
leading derivative, 44
leading term, 44
least common left multiple, 23
least common multiple, 21
Levi decomposition, 121
lexicographic ordering, 45
Lie, 10
Lie algebra, 116
Lie algebra
     classical, 120
     classification of, 135
     definition of, 116
     derived, 118
     exceptional, 120
     integrable, 118
     semi-simple, 119
     solvable, 118
Lie group, 106
Lie system, 144
Lie transformation group, 107
Lie's equation, 219
Lie's first theorem, 109
Lie's relations, 143
Lie's second theorem, 110
Lie's third theorem, 110
Lie-determinant, 201
line element, 195
linear ode, 12
Liouville R., 181
Liouville, R., 186
Liouvillian extension, 16
local Lie group, 106
Loewy decomposition, 11
Loewy decomposition

Index 433

type of, 28
Loewy factor, 27
lower equations, 201

matrice de côtes, 44
matrix of weights, 44
Michel, 193

Neumer, 162
normalizer, 117

orbit, 108
order of a derivative, 42
ordinary differential field, 42
ordinary differential ring, 42

parametric derivative, 55
partial differential field, 42
partial differential ring, 42
path curve, 126
Picard-Vessiot extension, 16
point transformation, 159, 194
primitive Lie group, 108
principal derivative, 55
prolongation
    of a symmetry generator, 197
    of a vector field, 114

quotient
    of ordinary differential opera-
        tors, 22

radical, 119
rank
    of a Janet basis, 55
    of a Lie algebra, 117
ranking, 43
rational normal form
    definition of, 162
    of Abel's equation, 180
    of Riccati's equation, 177
reduction, 47
relative invariant, 161
relative syzygies module, 70
Riccati equation, 16
Riccati equation

associated to a linear equation,
    16
Riccati-like partial differential equa-
    tions, 85

semi-invariant, 161
semi-simple Lie algebra, 119
semi-simple Lie group, 106
similar Lie groups, 111
simple Lie algebra, 119
simple Lie group, 106
singular solution, 311
solvable Lie algebra, 118
solvable Lie group, 106
special functions, 169
special rational solutions, 17
Stäckel, 162
stabilizer, 107
stratum, 193
structure constants, 107
structure invariance group
    definition of, 160
    of Abel's equation, 179
    of Lie's equation, 186
    of linear differential equation,
        162
    of Riccati's equation, 177
symmetric power
    and Lie symmetries, 245
    definition of, 33
symmetric product, 33
symmetry, 3, 196
symmetry
    of an ode, 195
symmetry algebra, 193
symmetry class, 4, 7, 193
symmetry group, 193
symmetry type, 4, 193
system of imprimitivity, 108

term, 43
term ordering
    definition of, 43
    grevlex, 45
    grlex, 45

lex, 45
transitive Lie group, 108
Tresse, 188
trivial symmetries, 206
typical differential dimension, 40

unconnected, 109

variation of constants, 14

weight, 161
Wronskian, 12

CPSIA information can be obtained
at www.ICGtesting.com
Printed in the USA
BVHW070044151218
535398BV00015B/211/P